# Binomial Distribution Handbook for Scientists and Engineers

Elart von Collani
Klaus Dräger

# Binomial Distribution Handbook for Scientists and Engineers

Springer Science+Business Media, LLC

Elart von Collani
Klaus Dräger
University of Würzburg
Sanderring 2
D-97070 Würzburg
Germany

**Library of Congress Cataloging-in-Publication Data**
von Collani, Elart, 1944–
Binomial distribution handbook for scientists and engineers / Elart von Collani, Klaus Dräger.
p. cm.
Includes bibliographical references and index.
ISBN 978-1-4612-6666-2 ISBN 978-1-4612-0215-8 (eBook)
DOI 10.1007/978-1-4612-0215-8
1. Binomial distribution. I. Dräger, Klaus.
QA273.6 .V65 2001
512.2'4—dc21 00-046821
CIP

Printed on acid-free paper.



**Additional material to this book can be downloaded from http://extras.springer.com**

ISBN 978-1-4612-6666-2 SPIN 10723723

Production managed by Louise Farkas; manufacturing supervised by Erica Bresler.
Typeset by the author in LaTeX.

9 8 7 6 5 4 3 2 1

**To Claudia and Machiko**

# Contents

# Preface

In 1986, the eminent statistician John Nelder [23] noted in the *Journal of the Royal Statistical Society* (*Series A* ): "If statistics is an applied field and not a minor branch of mathematics, then more than ninety-nine percent of the published papers are useless exercises." Evidently, statistics is an applied field, as its application has pervaded all branches of science and technology, and therefore the question of how to get rid of the huge pile of rubbish that hides the useful things must be discussed.

One possibility is to return to the roots and make an attempt to reformulate the problems aimed at disclosing aberrations, show desirable directions, and develop easy-to-apply methods relevant for application.

This handbook deals with the oldest problem in stochastics and goes back to its deepest roots in order to elaborate on the old problem from a different point of view, thus addressing professional and academic statisticians. New solutions for the old problem are obtained and subsequently presented for the user without assuming mathematical skills and a deep knowledge of statistics. The aim is to offer statistical methods to the potential user in a comparable way as sophisticated technical devices are made available, which can be used by any layman without understanding the underlying physical or chemical principles.

There were two primary motives for writing this handbook. The first one resulted from a consulting project in the course of which proportions had to be estimated. We noticed that the recommended methods are difficult to handle for practitioners and, moreover, they are inaccurate and yield often useless results. This made us think about the "quality" of statistical procedures, and we noticed that the many different criteria used in statistics for assessing the quality of statistical methods have, at most, an auxiliary character. Therefore, we started to look for a way to improve the existing methods, to develop a possibility for evaluating the quality of statistical procedures, and to make their understanding and application easier.

Particularly valuable were Jerzy Neyman's papers on estimation theory, which showed us the direction in which to proceed. By rigorously furthering Neyman's ideas, we finally arrived at something that could be called a geometric approach to statistics or a small sample theory in statistics. This was possible by abandoning the unrealistic assumption of a maximum parameter space and adopting the more realistic aim of developing procedures for appropriately selected bounded parameter ranges. Throughout the handbook, the problems are investigated not asymptotically, but realistically. The approach should be valid for small sample sizes $n$, as are common in many natural sciences and technology. This primary requirement led to problems and questions that are completely different from the investigation of limits and convergence rates. Thus, the handbook represents a first step toward developing statistics as a (natural) science, and further steps must follow.

The handbook illustrates the approach by applying it to the binomial distribution. Scientists and engineers may benefit in several ways:

1. The handbook uses an easy-to-understand terminology, which attempts to reflect the meaning of the relevant concepts in order to avoid ambiguity.

2. The proposed methods have a better quality, particularly for small – and medium – size samples.

3. There are printed tables for a preliminary analysis of data, and there are tables on a CD-ROM for a more detailed analysis with a so far unrealized size.

Statisticians may benefit by the novel approach, which can be applied to any given situation modeled by random variables and aiming to solve a problem related to prediction, estimation, exclusion, and comparison.

According to its aims, the book is organized in the following way. After an introduction into the problem and the state of the art, a unified approach for "estimation and test" theory is developed based on the "measurement & prediction space," which constitutes the basic quantity of the entire theory. The second part of the book is primarily directed at statisticians with a sound background in mathematics. The new approach and the resulting methods are explained in detail and illustrated by examples.

Those readers only interested in applying the methods are referred particularly to the third, fourth, and fifth parts of the book, in addition to the introduction. The third part, *Introduction to the Tables*, contains a brief descriprion of the tables. These include printed tables, which cover samples up to a size of $n = 100$ and tables on a CD-ROM, covering sample sizes up to $n = 1000$. The use of the tables is explained by means of illustrative examples. The fourth part, *Application*, is devoted to examples partly taken from the literature and serving to explain how to apply the tables in various realistic situations. The fifth part, *Tables*, contains the printed tables, which are a small part of the tables on the CD-ROM.

In contrast to traditional methods, almost no transformation or manipulation of data or variables is necessary. The input parameters (measurement range, sample size, confidence level, realization, null hypothesis) are sufficient for immediately obtaining the result. Moreover, the results are "exact," as the selected confidence level or significance level is always met. The measurement and prediction procedures are "most" precise, as they are determined by means of an optimization with respect to precision. The "test" procedures, on the other hand, are determined so as to maximize their success probability or to express it in a more popular way so as to maximize the probability of getting "a significant result."

In this handbook, quantities and their actual values are strictly distinguished. A quantity in the above sense is

- a random variable denoted by capital letters (e.g.,$X$) and its realization denoted by small letters (e.g., $x$);

- a random sample of size $n$ denoted by $\vec{X} = (X_1, \ldots, X_n)$, and its realization denoted by $\vec{x} = (x_1, \cdots, x_n)$; and

- a distribution parameter denoted by $\breve{\theta}$ and its actual value denoted by $\theta$. Here, the distribution parameter of interest is the probability of an event. It is denoted by $\breve{p}$ and its actual value by $p$.

We are aware that the handbook is only one step in the direction of creating a science of statistics, which uses mathematics exclusively as a necessary and most valuable tool. However, despite all our efforts, in many cases we still followed familiar and cherished habits originating from our mathematical past. The same holds for the terminology proposed here, which differs much from the classical one and was developed to make the concepts more understandable. The question of whether our attempt has been successful will be answered by the user. Our aim and hope is that the handbook may arouse the users' interest and demand for better and easier-to-apply statistical methods.

We are deeply indebted to a number of persons who were involved in producing this book, read the first versions of the manuscript, and made many valuable suggestions. Among them we would like to name Monica Dumitrescu from Bucharest University, Christoph Weigand from the University of Applied Science Aachen, Rouven Lepenis, who now works with BMW, Andreas Binder from the University of Applied Sciences Würzburg, and Michael Ratz from University of Würzburg, who professionally programmed and designed the CD-ROM. Finally, we thank Louise Farkas from Birkhäuser Boston for her valuable advice and Margaret Mitchell from Springer-Verlag, New York, for her excellent help in managing the difficulties with LaTeX.

The handbook is an attempt to continue the tradition of Jacob Bernoulli, who founded the science of "stochastics" three hundred years ago, and Jerzy Neyman, who laid the foundation for scientific "statistics." Moreover, it may also be looked upon as a first step in reconciling classical statistics with Bayes statistics.

Würzburg, Germany

Elart von Collani
Klaus Dräger

# Part I

# Introduction

# Chapter 1

# Stochastics

## 1.1 The Science of Stochastics

Since Jakob Bernoulli's[1] masterpiece *Ars Conjectandi* [1] was published in 1713, the science of *stochastics* or, better said, its methods have penetrated not only almost any field of scientific research but are also applied in an ever-increasing intensity in industry, administration, health care, and so forth. Nowadays, there is hardly any science which does not have a particular statistical branch. The reason for this development is very simple: Science investigates different real-world phenomena trying to understand and explain them. Stochastics deals with one of the most interesting and difficult aspect of any real-world phenomenon, namely the uncertainty with respect to its future development. Uncertainty itself is everywhere, and moreover, uncertainty is structured. Sometimes it is large, sometimes it is small, and often it is possible to manipulate it. Hence, uncertainty follows certain laws, which are different from situation to situation. Generally, it is not possible to completely remove uncertainty, even if its laws are known, but knowledge about the laws leads to a better control of uncertainty. In fact, a great part of any human activity aims at controlling the future development by controlling uncertainty.

[1] Jakob (James) Bernoulli (1654-1705), founder of the science of stochastics.

The science of stochastics derives mathematical models for uncertainty, which allow one to take into account and to analyze uncertainty quantitatively and thus enable a better understanding of the phenomenon in question and making controlled predictions about its future developments. Thus, one can state the following

- Uncertainty is a universal aspect of the real world, and the laws of uncertainty are the subject of the science of stochastics.
- The science of stochastics uses mathematical models for describing the laws of uncertainty.
- The science of stochastics develops procedures with stated accuracy (e.g., for determining the actual values of the model parameters).

Thus, stochastics has all the characteristic features of a quantitative or exact natural science.

The most striking point about stochastics is the fact that there is hardly any limitation of its application. It is applied for analyzing games of chance – where it had its origin – within natural sciences, within social sciences, and, particularly, to any decision-making process. By applying stochastical methodology, decisions under uncertainty become rational and transparent as Simon de Laplace[2] noted in his famous booklet *Essai philosophique sur les probabilités* [20, 21]. He concludes his essay with the words: ..., *one will see that there is no science at all more worthy of our consideration, and that it (stochastics) would be a useful part of the system of public education.*

The central problem in stochastics has remained the same since Jakob Bernoulli's times, namely to determine or to *measure* probabilities of events. However measuring something is meaningful only if the reliability of the measurement procedure and its precision are known. Therefore, when Bernoulli discovered a possibility for controlling the accuracy (by means of a

[2]Pierre Simon Marquis de Laplace (1749-1827), French mathematician and natural scientist, made important contributions to stochastics.

first version of the law of large numbers), he was convinced that he had founded a new science that would change the world. Unfortunately, Bernoulli died before he could complete his work. Others, such as Simon Laplace, Ronald Aylmer Fisher,[3] Jerzy Neyman,[4] and Andrej Nikolajewitsch Kolmogorov,[5] continued his research and developed easy-to-apply methods, which today are taught – as Laplace had proposed – in secondary schools. However, in spite of its universally acknowledged usefulness, stochastics is not at all considered an independent science. It is looked upon as a part of other sciences or as a mere methodology. Some (i.e.,those who received an education in mathematics) have the strange belief that it is a branch of mathematics and others (being not so well educated in mathematics) look upon it as being part of other natural or social sciences. Each group of *stochasticians* has its own professional society and even uses its own special terminology. The result of this situation is dramatic:

- The science of *stochastics* is hardly known even among academics.

- *Statistics* being the most widely used part of *stochastics* has a reputation which could not be worse. Frequently, it is looked upon as an ingenious possibility for cheating.

- The development of advances in *stochastics* is dictated by randomly occurring needs in the fields of applications or by pure mathematical research, being more or less unrelated to the objectives in stochastics.

- Doubtful methods are developed and the interpretation of results obtained by stochastical methods remains often vague and ambiguous.

---

[3]Ronald Aylmer Fisher (1890-1962), one of the founders of statistics.

[4]Jerzy Neyman (1894-1981), Polish mathematician who made many important contributions to statistics.

[5]Andrej Nikolajewitsch Kolmogorov (1903-1987), Russian mathematician who made important contributions to probability theory.

Analogously to any other quantitative science, stochastics can be divided into two major branches:

- *Theoretical Stochastics*, which derives the quantitative models and defines the quantities of interest (i.e., the model parameters). This part is usually called *probability theory.*

- *Empirical Stochastics*, which develops scientific procedures connecting the theoretical models with the real world (e.g., by measuring the actual values of model parameters based on empirical experience). This second branch of stochastics is generally known as *statistics.*

One typical situation dealt with in science can be illustrated in the following way. There is a quantity of interest, say $\breve{\theta}$, whose actual value $\theta$ is unknown and cannot be observed directly. Therefore, a method has to be developed for indirectly determining or measuring the actual value $\theta$, for instance by comparing it with a given gauge. Let us denote the thus established method by $\vec{X}$, where $\vec{X}$ represents the outcome of the method. The symbol of a vector is selected in order to indicate that the result may be multidimensional. Clearly, there should be a close relation between the indirect method $\vec{X}$ and the quantity of interest $\breve{\theta}$, or between the outcomes or realization of $\vec{X}$ denoted by $\vec{x}$ and the actual value $\theta$ of the quantity of interest $\breve{\theta}$. However, generally, the relation is not of deterministic nature, but of indeterministic or random nature, which means that any measurement by means of $\vec{X}$ is not exactly reproducible. At this point, the aspect of uncertainty enters the stage, which is dealt with by stochastics. A stochastic model must be developed describing the inherent random variations exhibited by $\vec{X}$. This is done by a probability measure function $\mathbf{P}_{\vec{X}}^{(\theta)}$ given for example by the so-called distribution function of $\vec{X}$

$$F_{\vec{X}}^{(\theta)}(\vec{x}). \tag{1.1}$$

As soon as the relation between the method and the quantity is established by (1.1), the second task of stochastics begins, namely the development of appropriate procedures for example for determining the actual value $\theta$. The method $\vec{X}$ represents the empirical information and the model (1.1) represents the theory. In statistics, $\vec{X}$ is called a random sample and the procedure for determining $\theta$ will be called a measurement procedure.

To repeat, statistics as a branch of stochastics *develops* procedures, for example, for measuring the actual value of some given quantity of interest, based on *random samples.* These procedures are subsequently *applied* to sets of *data* in various areas: academic research, industry, administration, and so forth. The numerical data do not contain any uncertainty and therefore are not the subject of the science of statistics. The data belong to and are used for explaining physical, medical, environmental, or other phenomena. Stochastics does not deal with data, but with the data generation process generally modeled by random variables and having uncertain outcomes. Thus, statistics is not appropriately described by the following:

> "Statistics is the art and science of gathering, analyzing and making inference from data," as given in the *New Encyclopaedia Britannica.*

or

> "Statistics, the science of classification and manipulation of data in order to draw inference," as taken from *Colliers Encyclopaedia.*

These descriptions hold for any quantitative science, except for statistics, which is based on random variables and develops methods for applications to deal with uncertainty.

The methods made available by the statistics may be used for analyzing large or small sets of data, obtained in the context of other sciences. Hence, from the viewpoint of other sciences (i.e., from the viewpoint of application), statistics may be regarded as a methodology (i.e., a set of methods for solving problems), similar to the way a practicing physician may look at medicine, as a set of diagnostic and therapeutic methods, but not as a science aimed at understanding and modeling the complex system of human beings. Therefore, saying that

> Statistics is not a separate discipline like Mathematics, Physics, Chemistry or Biology. It has no problems of its own unlike other disciplines. Its status may be described as the art, logic or technology of solving problems in other disciplines [31]

indicates a fundamental misunderstanding.

This handbook is concerned with one important problem in statistics, namely to develop methods for estimating and testing the unknown value of the probability of a given event. One could argue that this problem has been solved long ago, particularly by Bernoulli himself. Indeed, Bernoulli proposed one solution, which still is recommended in statistical textbooks.

However, applying strictly the principles for developing measurement procedures generally acknowledged in science yields new and, in some sense, better methods for estimating the actual value of a given parameter as shown in [10].

## 1.2 Historical Remarks

In 1930, R.A. Fisher [16] introduced his *fiducial distribution* and about the same time, J. Neyman [25, 26, 27] started to develop the theory of *confidence intervals*.[6] What followed was a heated discussion about the differences of the two concepts (see [28]) and their advantages and disadvantages, which lasted for more than a decade. Whereas Fisher's fiducial theory is hardly applied anymore, Neyman's concept of confidence intervals found general acceptance in the scientific community. The famous controversy between Fisher and Neyman, two of the greatest statisticians of our century, clearly illustrates the main problem within the science of stochastics. It is not so difficult to invent new concepts, models, and methods using mathematical tools.

[6] The term *confidence interval* is a good example of misleading terminology. A confidence interval is not at all an interval, but a system of intervals.

The difficulty starts in trying to find the appropriate meaning of the mathematical model with respect to real-world problems.

Neyman's confidence intervals constitute measurement procedures in the scientific sense, as the procedures' reliability is controlled by the *confidence level* and the achieved precision is accessed by the length of the interval resulting from the application of the measurement procedure.

Already in 1934, the same year of Neyman's pioneering paper, Clopper and Pearson [6] published tables and graphs of confidence intervals for the actual value $p$ of the given probability $\mathbf{P}(E)$ of a specified event $E$. Only 3 years later, a more thorough treatment of the theory of confidence intervals was given by Neyman in [27], which also contained a method for determining "optimal" confidence intervals based on so-called acceptance regions.

In 1952, Hald's statistical tables [18], based on the simple approach used by Clopper and Pearson, were published. Two years later, in 1954, Sterne [34], recalling Neyman's approach, noticed that the confidence intervals of Clopper and Pearson are unnecessarily wide and gave tables of narrower confidence intervals for a few small sample sizes $n$. Sterne's proposal was taken up by Crow [14], who calculated tables for confidence intervals for $1 \leq n \leq 30$ and confidence levels $\beta = 0.90$, $\beta = 0.95$, and $\beta = 0.99$. Some years later, Bunke [4] investigated and computed again new confidence intervals based on acceptance regions for the unknown value $p$ of $\mathbf{P}(E)$. Finally, in 1986, Neumann [24] developed an axiomatic approach. However, these efforts remained more or less unknown to the larger scientific community, particularly in the applied fields.

Clearly, estimation of the actual value of a given probability constitutes the oldest and one of the most frequently occurring problems in statistical applications. Thus, it seems to be surprising that Neyman's proposal (see [2, 3]) for, in some sense, optimal confidence intervals has not come true in the form of a comprehensive table of optimal confidence intervals. One reason, of course, is the computational difficulty; another one is the large size of such tables. However, both the problems have vanished since powerful PCs and CD-ROMs with high storage capacity are available everywhere. But there is another surprising fact. A search for Neyman's concept of acceptance regions in textbooks in statistics reveals that acceptance regions are not treated in the majority of these books, at least not in the context of confidence intervals.

This handbook further develops Neyman's approach for determining confidence intervals for the actual value $p$ identified with the expectation of a given Bernoulli variable $X$. The results given in [10] are applied, yielding procedures which are called *Neyman measurement & prediction procedures.* These procedures are optimal with respect to measurement precision for given reliability $\beta$.

As already mentioned, any statistical measurement procedure for the actual value $p$ of the probability $\mathbf{P}(E)$ of a given event $E$ is necessarily based on a random sample $(X_1, \ldots, X_n)$ for $X$, where $n$ is the sample size. Frequently, the random sample is assumed to consist of $n$ independent copies of $X$, and this approach is also adopted here.

## 1.3 Measurement Procedure and Measurement Range

In any quantitative science, there are quantities of interest with the following features:

- The quantity of interest is clearly defined by the underlying model, thus both its nature and its range of size are known.

- In contrast to the nature and the range of size, the actual value of the quantity is unknown.

The problem then is to determine the unknown value.

Generally, the size of the given quantity essentially determines the measurement procedure. A device for measuring active substances in pharmaceutics, for example, is entirely different than one for measuring the weight of a letter or another one for measuring the number of tons. Each procedure is applicable only in certain situations, which are specified by the procedure's measurement range. If the size of the given parameter falls within the measurement range, the method may be applied, otherwise not.

Hence, development and application are clearly to be distinguished.

- *Development*
  The first step in developing a measurement procedure for a given type of quantity is to divide the overall parameter range into suitable parts. Each part constitutes a possible measurement range. The second step consists of developing a measurement procedure with controlled reliability for each of the measurement ranges.

- *Application*
  The first step for applying a measurement procedure is to select a measurement range according to the size of the given quantity. By selecting an appropriate measurement range, the procedure is specified and may be applied.

It is a surprising fact that the above-outlined principles for developing and applying measurement procedures are ignored for the statistical part of measurements. This handbook will strictly observe the generally acknowledged principles, thus arriving at methods that differ basically from the traditional ones.

This handbook deals with the problem of developing measurement procedures for the actual value $p$ of the probability of a given event $E$. In this case, the overall parameter range is the unit interval $[0, 1]$, and any meaningful measurement range is given by a closed interval $[\underline{p}, \overline{p}] \subset [0, 1]$. Once the measurement procedures are made available for application, the method to be used is fully determined by selecting an appropriate measurement range. As the event is known, the size of its probability is generally known, too. Hence, selection of an appropriate measurement range should not constitute a difficult problem.

# Chapter 2

# Models Related to the Probability of an Event

## 2.1 The Concept of Probability

The concept of *probability* is over two hundred years old and for the whole period of its existence there has been a dispute about its meaning as D.V. Lindley notes in the forword to [15]. There is the "frequentist interpretation" due to R. von Mises, B. de Finetti's, and L. J. Savage's "subjective view" of probability; R. Carnap's "logical probability;" J. M. Kęynes', B. O. Koopman's and B. de Finetti's "comparative probability," and T. L. Fine's "probabilistic reasoning." These approaches seem to shift stochastics into the field of philosophy. Here, a more pragmatic or scientific view is taken, which explains a concept by its aims and effects and handles it by quantified measurement procedures, thus integrating stochastics into science.

In order to describe probability quantitatively, we have to start with an aspect of the real world, which is everywhere and which determines almost everything, namely "uncertainty of events." Everybody has experience with uncertainty, and it is well known that uncertainty is not unstructured, but may adopt very different and clearly distinguishable structures. Sometimes, uncertainty is large, sometimes it is small, and one of the universal aims of mankind is to manipulate uncertainty whenever it is possible. If one knows the structure or rules of uncertainty for given events, then better understanding and better decisions are possible. Therefore, investigating and analyzing uncertainty is a legitimate task, and because of its universal character, it should have high priority. Having recognized "uncertainty" as a real-world aspect, such as, for example, "heat," which is quantified by temperature, uncertainty has to be quantified, too, where quantification means that it can be assessed numerically.

Uncertainty of events may be referred to as the likelihood of their occurrence. Thus, for each event, a quantity must be defined, which reflects the corresponding likelihood, and rules have to be established, which enable one to perform numerical analysis. It was one of the highlights of science when the quantity "probability of an event" was introduced, expressing its degree of occurrence, thus enabling the description and analysis of uncertainty.

Similar to other branches of science, measurement procedures for the unknown values of defined quantities have to be developed, constituting the core of science.

In practical applications, the actual value of the probability of a given event is of interest in two different types of situation. In the first one, interest is to analyze a so-called "random phenomenon" by determining the probability of an event of interest, whereas the second one deals with the determination of an unknown "historical" value of a given quantity by means of an "artificial" random experiment, which is designed in such a way that the probability of a suitably defined event is equal to the historical value of the quantity of interest.

The two types of situation may be described more formally as follows:

- *Random Phenomenon*
  Let $\mathcal{E}$ be a random phenomenon (i.e., a phenomenon with uncertain outcome). The phenomenon results either in the event $E_A$ having the attribute $A$ or in the complementary event $E_{\bar{A}}$ having the attribute $\bar{A}$ (i.e., the complement of $A$). The uncertainty of the future development of $\mathcal{E}$ can be modeled by a so-called Bernoulli experiment, where

the dimension of interest is the *probability* $\mathbf{P}(E_A)$ that an phenomenon results in the event $E_A$ having the attribute $A$.

**Example 2.1**
*Consider a production process with an output of items of a certain product. Let $\mathcal{E}$ denote the random process of producing one item, where each item may be conforming with some specifications. Let $A$ denote the property of being conforming and $E_A$ the event of producing a conforming item; then, the dimension of interest is given by the probability $\mathbf{P}(E_A)$.* •

- *Random Experiment*
  Let $\{e_1, \ldots, e_N\}$ be a finite set of items, frequently called population, where each item has either the attribute $A$ or the complementary attribute $\bar{A}$. Let the dimension of interest be the *proportion* $p$ of elements in the population with attribute $A$. Clearly, there is nothing random with respect to the population $\{e_1, \ldots, e_N\}$. The problem is that the proportion $p$ is unknown. In order to determine $p$, it would be necessary to observe each item $e_i$, $i = 1, \ldots, N$, of the population. If the number of elements $N$ is too large or if observation "destroys" the item, then it is not possible to determine $p$. In such a case, an "artificial" random phenomenon $\mathcal{E}$, called a random experiment, is used to assess the unknown value $p$. The random experiment consists of *drawing randomly* an item out of the population. The random experiment has to be performed in a way that the actual value of the probability $\breve{p}$ of drawing an item having attribute $A$ equals the unknown *proportion* $p$, thus establishing the necessary link between the population and the auxiliary random experiment.

**Example 2.2**
*Let $\{e_1, \ldots, e_N\}$ denote a given lot of products. Each item $e_i$, $i = 1, 2, \ldots, N$, may be conforming or nonconforming*

*with respect to certain specifications. Let p denote the proportion of nonconforming items in the lot. The "artificial" random experiment $\mathcal{E}$ of drawing one item out of the lot defines a Bernoulli experiment with an actual value of the parameter $\breve{p}$ of interest equal to the proportion p.* •

Many real-life situations may be described by one of the above alternatives with the actual value $p$ of the probability of interest $\breve{p}$ being unknown. The problem then is to measure $p$, which in statistical terminology means to estimate it, where estimation is often understood as an approximate measurement procedure. Of course, from a more practical point of view, the second situation is generally simpler in the sense that the experiments can be performed in a controlled way.

The fact that the unknown value $p$ cannot be measured exactly but can only be estimated (approximately) is not at all characteristic of quantities investigated in stochastics, but is a general feature of any *unknown* real-world quantity in any scientific field. Take, for instance, physics, the most *exact* science, and consider one of the simplest cases, namely to measure the unknown length of something. Even this seemingly simple measurement problem can be solved only approximately, and this fact was one of the origins of the science stochastics. The only fundamental difference between measuring a physical quantity and a probability is the fact that human beings are believed to have a conceptual understanding of the length of something, but have definitely no clear understanding of uncertainty. Uncertainty is widely regarded as the result of ignorance, but not as an independent aspect of the real world. Thus, it is induced that removing ignorance would let uncertainty disappear, making a science dealing with uncertainty unnecessary.

**Remark 2.1**
*The belief that "uncertainty" is completely caused by ignorance has led to enormous efforts aiming at removing "ignorance" rather than understanding "uncertainty." These efforts resulted in great advances in many sciences and technologies. However,*

*there are other sciences where the aim of arriving at deterministic models not only causes a lot of additional expenses but probably also prevents advances in many instances.*

•

## 2.2 Random Variables and Data

Often, it is claimed that statistics is the science of "data." However, it should be clear that the properties of statistical procedures do not refer to data, but to random variables. Statistics aims basically at developing measurement procedures for quantities specifying the laws of uncertainty, which are given by probability distributions of random variables. Let $X$ be a random variable and $\breve{\theta}$ the quantity of interest. The measurement procedure for the actual value $\theta$ of $\breve{\theta}$ is based on a random sample $\vec{X} = (X_1, \ldots, X_n)$, where the sample elements should have a close relation to $X$ and $\breve{\theta}$.

Random samples $\vec{X}$ may be classified according to their relation to $\breve{\theta}$. The relation introduced by (1.1) is the joint distribution function of the elements of $\vec{X}$:

$$F^{(\theta)}_{X_1,\ldots,X_n}(x_1, \ldots, x_n). \tag{2.1}$$

The most frequently investigated case is

$$F^{(\theta)}_{X_1,\ldots,X_n}(x_1, \ldots, x_n) = \prod_{j=1}^{n} F^{(\theta)}_X(x_j) \tag{2.2}$$

where $F^{(\theta)}_X(x)$ denotes the distribution function of $X$.

If (2.2) holds for a sample, then $\vec{X}$ consists of independent copies of $X$ and is called an independent random sample for $X$.

Based on samples $\vec{X}$, statistics develops procedures for solving various problems, namely for the following:

- measuring the actual value $\theta$ of $\breve{\theta}$ based on $\vec{X}$,

- predicting the outcome $\vec{x}$ of random variable $\vec{X}$ based on the actual value $\theta$ of $\breve{\theta}$,

- excluding certain probability distributions of $\vec{X}$ by excluding the corresponding values of $\breve{\theta}$,

- deciding for one of a set of alternatives by comparing $\vec{X}$ with the models representing the alternatives.

A procedure is formally a random variable and may be applied in a certain specified situation and results in a numerical value, depending on the realization $(x_1, \ldots, x_n)$ of the sample $\vec{X}$. It should be clear that none of the possible properties of the procedure can be extended to "data," represented by the corresponding realizations. Therefore, it is misleading to call statistics the science of data. The respective arguments would apply much better to physics or any other quantified science where enormous amounts of data are generated and evaluated.

## 2.3 The Model

Following the general habit, we do not distinguish between a "natural" random phenomenon and an "artificial" random experiment defining an event $E$ and its probability $\mathbf{P}(E)$, and we call either of them a *random experiment* denoted by $\mathcal{E}$ in the remainder.

The experiment $\mathcal{E}$ for the simple situation considered here involves the event $E$ of interest, the complementary event, and the probabilities giving the degree or likelihood that the one or the other will occur. $\mathcal{E}$ is appropriately modeled by the "probability space"

$$\left(\Omega, \mathcal{S}, \mathbf{P}\right) \tag{2.3}$$

with

$$\Omega = \left\{e_A, e_{\bar{A}}\right\} \tag{2.4}$$

$$\mathcal{S} = \left\{\emptyset, \{e_A\}, \{e_{\bar{A}}\}, \Omega\right\} \tag{2.5}$$

$$\mathbf{P}^{(p)}(E) = \begin{cases} 0 & \text{for } E = \emptyset \\ p & \text{for } E = \{e_A\} \\ 1-p & \text{for } E = \{e_{\bar{A}}\} \\ 1 & \text{for } E = \Omega. \end{cases} \tag{2.6}$$

$\Omega$ contains the two possible outcomes of the experiment, given by $e_A$ and its complement $e_{\bar{A}}$. $\mathcal{S}$ contains all possible events given by the *impossible* event $\emptyset$, the *certain* event $\Omega$, and the two events of interest, $\{e_A\}$ and $\{e_{\bar{A}}\}$. Finally, the function $\mathbf{P}^{(p)} : \mathcal{S} \to [0,1]$ depends on and defines the actual value $p$ of the probability $\breve{p}$ of interest. Note that in case of an artificial random experiment, the *population* does not explicitly enter the model. Surprisingly, the term *population* is nevertheless widely used in statistics in connection with the model. As the model explicitly does not refer to the population, the use of the term results necessarily in many misunderstandings.

Frequently, the occurrence of the event $\{e_A\}$ is called *success* and the occurrence of the complementary event $\{e_{\bar{A}}\}$ is called *failure*. Accordingly, the probability $\mathbf{P}^{(p)}(\{e_A\})$ is called successprobability, and the problem is to measure the actual value $p$ of the success probability.

Next, we define the random variable $X : \Omega \to I\!R$ by

$$X(\omega) = \begin{cases} 1 & \text{for } \omega = e_A \\ 0 & \text{for } \omega = e_{\bar{A}}. \end{cases} \tag{2.7}$$

By means of $X$, the "real-world model" $\mathcal{E}$ is transferred to a "real-valued model" $\mathcal{E}_{I\!R}$ given by

$$\left(I\!R, \mathcal{B}, \mathbf{P}_X^{(p)}\right) \tag{2.8}$$

with

$$I\!R = \text{set of real numbers} \tag{2.9}$$

$$\begin{aligned} \mathcal{B} &= \sigma\text{-algebra of Borel sets} \\ &= \text{set of real-valued events} \end{aligned} \tag{2.10}$$

$$\mathbf{P}_X^{(p)}(B) = \begin{cases} 0 & \text{for } \{0,1\} \cap B = \emptyset \\ p & \text{for } \{0,1\} \cap B = \{1\} \\ 1-p & \text{for } \{0,1\} \cap B = \{0\} \\ 1 & \text{for } \{0,1\} \cap B = \{0,1\} \end{cases} \tag{2.11}$$

for any $B \in \mathcal{B}$.

Within the real-valued model, everything can be expressed with real numbers and the corresponding methods of real analysis may be directly applied.

It is said that the random variable $X$ is distributed according to the Bernoulli distribution $B(p)$, or the binomial distribution $Bi(1,p)$ defined by

$$\mathbf{P}_X^{(p)}\Big(\{x\}\Big) = \begin{cases} p & \text{for } x = 1 \\ 1-p & \text{for } x = 0 \\ 0 & \text{elsewhere.} \end{cases} \tag{2.12}$$

## 2.4 The Random Sample

The development of a measurement procedure for the actual value $p$ of the probability $\breve{p}$ is based on a random sample, where each sample element $X_i$ refers to a separate random experiment:

$$(X_1, ..., X_n). \tag{2.13}$$

The joint distribution of the random sample $(X_1, ..., X_n)$ depends on the overall experiment consisting of the $n$ experiments described by the sample elements. Generally, there are two cases considered:

- *Independent Experiments:*
  The partial experiments are independent and the experimental conditions are identical. This case is modeled by an independent and identically distributed random sample for $X$, with the meaning that the sample elements $X_1, \ldots, X_n$ are independent random variables each identically distributed like $X$. Thus, the resulting random vector $(X_1, ..., X_n)$ is defined on the $n$-fold product space of $\Big(\Omega, \mathcal{S}, \mathbf{P}\Big)$ denoted by $\Big(\Omega_n, \mathcal{S}_n, \mathbf{P}_n\Big)$ and given by

$$\Big(\Omega \times \cdots \times \Omega, \mathcal{S} \otimes \cdots \otimes \mathcal{S}, \mathbf{P}_X^{(p)} \otimes \cdots \otimes \mathbf{P}_X^{(p)}\Big). \tag{2.14}$$

The joint distribution of $(X_1, ..., X_n)$ is given by

$$\mathbf{P}_{\vec{X}}^{(p)}\left(\{x_1, \cdots, x_n\}\right) = p^{\sum_{i=1}^n x_i}(1-p)^{n-\sum_{i=1}^n x_i}$$
$$\text{with } x_i \in \{0,1\}. \qquad (2.15)$$

From (2.15) it immediately follows that the joint distribution of $(X_1, ..., X_n)$ does not depend explicitly on any $X_i$, but only on the sum $\sum_{i=1}^n X_i$. Thus, $\sum_{i=1}^n X_i$ contains as much information about the joint distribution as the sample itself, and therefore $\sum_{i=1}^n X_i$ is called sufficient for the distribution of $X$ specified by the actual value $p$ of the probability $\breve{p}$.

Therefore, the sum of the sample elements plays an important role in estimating the actual value $p$ of $\breve{p}$. Let

$$X_S = \sum_{i=1}^n X_i. \qquad (2.16)$$

Then $X_S$ adopts values in $\{0, 1, \ldots, n\}$ and is said to be binomial distributed according to $Bi(n, p)$, with

$$\mathbf{P}_{X_S}^{(p)}\left(\{m\}\right) = \binom{n}{m} p^m (1-p)^{n-m}$$
$$\text{for } m = 0, 1, \cdots, n. \qquad (2.17)$$

- *Dependent Experiments:*
Consider the situation of a given set of $N$ elements with $M$ marked elements, and define $n$ partial experiments in the following way: Each experiment consists of drawing randomly one item out of the set without replacing it. A success is defined by drawing a marked element, for which the random variable adopts the value 1. If the drawn element is not marked, the random variable adopts the value 0. Since the drawn elements are not replaced, the single experiments are no longer independent. However, the

sample elements $X_i$ are still copies of $X$ [i.e., distributed according to the Bernoulli distribution $Bi(1,p)$].

The joint distribution of $(X_1, ..., X_n)$ is completely given by

$$\mathbf{P}^{(p)}_{\vec{X}}\left(\{x_1, \ldots, x_n\}\right) = \frac{(N-M)(1-x_1)+Mx_1}{N} \cdots \frac{[N-M-\sum_{i=1}^{n-1}(1-x_i)](1-x_n)+[M-\sum_{i=1}^{n-1}x_i]x_n}{N-n+1} \quad \text{with } x_i \in \{0,1\}. \tag{2.18}$$

Again, it is easily seen that the joint distribution depends only on $\sum_{i=1}^{n} x_i$, as the order of successes only changes the order of the factors in (2.18).

In this case, the sum of the sample elements

$$X_S = \sum_{i=1}^{n} X_i \tag{2.19}$$

is said to be hypergeometrically distributed according to $H(N,M,n)$, with

$$\mathbf{P}^{(p)}_{\vec{X}}\left(\{m\}\right) = \frac{\binom{M}{m}\binom{N-M}{n-m}}{\binom{N}{n}} = \frac{\binom{Np}{m}\binom{N(1-p)}{n-m}}{\binom{N}{n}}. \tag{2.20}$$

It follows that the binomial and the hypergeometric probability distributions are important tools for estimating the actual value $p$ of a given probability $\breve{p}$.

## 2.5 The Binomial Distribution

In the case of an independent random sample, the sum of the sample elements follows a binomial distribution. The following notations are used for the probabilities of the binomial distribution given by (2.17):

$$b_m(n,p) := \binom{n}{m} p^m (1-p)^{n-m} = \mathbf{P}^{(p)}_{X_S}\Big(\{m\}\Big) \quad (2.21)$$

$$B_m(n,p) := \sum_{i=0}^{m} b_i(n,p) = \mathbf{P}^{(p)}_{X_S}\Big(\{0,1,\ldots,m\}\Big) \quad (2.22)$$

$$B_{m_1,m_2}(n,p) := \sum_{i=m_1}^{m_2} b_i(n,p) = \mathbf{P}^{(p)}_{X_S}\Big(\{m_1,\ldots,m_2\}\Big). \quad (2.23)$$

Dealing with a probability distribution, the most frequently arising problems are to compute certain values of the distribution function, or the inverse problem, or to compute the so-called quantiles.

Although it is possible without any difficulties to compute not only the probabilities of any binomial distribution but also to solve the inverse problem by means of a personal computer, different indirect and approximate methods to solve the related problems are proposed in many textbooks.

One important problem consists of determining $p_\alpha$ defined by

$$B_m(n,p_\alpha) = \alpha. \quad (2.24)$$

It is often solved by means of tables of the quantiles of the $F$-distribution utilizing the following relation between the two distributions:

$$B_m(n,p) = 1 - \int_0^{\frac{n-m}{m+1}\frac{p}{1-p}} f_{2(m+1),2(n-m)}(x)dx, \quad (2.25)$$

where $f_{\nu_1,\nu_2}(x)$ denotes the density function of the $F$-distribution with $\nu_1$ degrees of freedom of the nominator and $\nu_2$ degrees of freedom of the denominator. The density function of the

$F$-distribition is given by

$$f_{\nu_1,\nu_2}(x) = \begin{cases} 0 & \text{for } x < 0 \\ \frac{\Gamma(\frac{\nu_1+\nu_2}{2})}{\Gamma(\frac{\nu_1}{2})\Gamma(\frac{\nu_2}{2})}\nu_1^{\frac{\nu_1}{2}}\nu_2^{\frac{\nu_2}{2}}x^{\frac{\nu_1}{2}-1} & \text{for } x \geq 0. \end{cases} \tag{2.26}$$

With (2.25), the problem of determining $p_\alpha$ is solved in the following way:

$$B_m(n, p_\alpha) = \alpha \Leftrightarrow p_\alpha = \frac{F_{1-\alpha;2(m+1),2(n-m)}}{\frac{n-m}{m+1} + F_{1-\alpha;2(m+1),2(n-m)}}, \tag{2.27}$$

where $F_{1-\alpha;2(m+1),2(n-m)}$ denotes the $(1-\alpha)$-quantile of the $F$-distribution with $2(m+1)$ degrees of freedom of the nominator and $2(n-m)$ degrees of freedom of the denominator.

**Example 2.3**
*Consider the problem of solving the equation*

$$B_5(25, p_{0.05}) = 0.05. \tag{2.28}$$

*Relation (2.27) yield:*

$$p_{0.05} = \frac{6F_{0.95;12,40}}{20 + 6F_{0.95;12,40}}. \tag{2.29}$$

*The quantile $F_{0.95;12,40}$ can be taken for instance from [30]:*

$$F_{0.95;12,40} = 2.0035.$$

*Thus,*

$$p_{0.05} = 0.3754. \tag{2.30}$$

•

Besides the above "exact" relation (2.27), several approximations are used to solve (2.24). These approximations are based on limit distributions of the binomial distribution.

- The following limit relation is known as the *Poisson*[1] *Theorem.* Let $a = np$; then,

$$\lim_{\substack{a=\text{const} \\ n\to\infty}} b_m(n,p) = \frac{a^m}{m!}\, e^{-a}. \tag{2.31}$$

The limit defines the so-called Poisson distribution. A random variable $X$ is called Poisson distributed on $\mathbb{N}_0$, denoted by $X \sim Po(a)$ where $\mathbb{N}_0$ is the set of non-negative integers, if

$$\mathbf{P}_X^{(a)}\big(\{m\}\big) = \frac{a^m}{m!}\, e^{-a} \text{ for } m \in \mathbb{N}_0. \tag{2.32}$$

The relation between the Poisson distribution and the $\chi^2$-distribution is similar to that in the case of the binomial and $F$-distributions:

$$\sum_{k=0}^{m} \frac{a^k}{k!}\, e^{-a} = 1 - \int_0^{2a} \frac{1}{2^{m+1}m!}\, x^m e^{-\frac{x}{2}}\, dx. \tag{2.33}$$

The solution of (2.24) is approximated by the solution of

$$\sum_{k=0}^{m} \frac{a_\alpha^k}{k!}\, e^{-a_\alpha} = \alpha \tag{2.34}$$

which is equivalent to

$$a_\alpha = \frac{1}{2}\, \chi^2_{1-\alpha;2(m+1)}, \tag{2.35}$$

where $\chi^2_{1-\alpha;2(m+1)}$ denotes the $(1-\alpha)$-quantile of the $\chi^2$-distribution with $2(m+1)$ degrees of freedom.

**Example 2.4**
*Consider the above given problem* $B_5(25, p_{0.05}) = 0.05$,

[1]Siméon-Denis Poisson, born June 21, 1781, in Pithiviers, Loiret, France, died April 25, 1840, in Paris, France. He made contributions to physics, mechanics and stochastics.

*which shall now be solved approximately using (2.35). We obtain* $\chi^2_{0.95;12} = 12.592$ *(see [30]) and thus*

$$p_{0.05} \approx \frac{a_{0.05}}{25} = \frac{\chi^2_{0.95;12}}{50} = 0.2518. \qquad (2.36)$$

*which evidently is a rather poor approximation.* •

- The next limit relation is known as the *Local Moivre*[2]*-Laplace*[3] *Limit Theorem.*

  Let

$$x = \frac{m - np}{\sqrt{np(1-p)}}. \qquad (2.37)$$

  Then

$$\lim_{\substack{x=\text{const}\\ n\to\infty}} \sqrt{np(1-p)}\, b_m(n,p) = \frac{1}{\sqrt{2\pi}} e^{-\frac{x^2}{2}}. \qquad (2.38)$$

  The Moivre-Laplace Limit Theorem is a special case of the *Central Limit Theorem* and the limit defines the well-known density function of the standardized normal distribution.

  A randon variable $X$ is called standardized normally distributed, denoted by $X \sim N(0,1)$, if

$$\mathbf{P}_X\Big((-\infty, x]\Big) = \frac{1}{\sqrt{2\pi}} \int\limits_{-\infty}^{x} e^{-\frac{y^2}{2}}\, dy. \qquad (2.39)$$

  Using (2.38), the following *Moivre-Laplace Integral Limit Theorem* can be derived (see, for instance, [17]) Let

$$x_1 = \frac{m_1 - np}{\sqrt{np(1-p)}} \qquad (2.40)$$

---

[2] Abraham de Moivre, born May 26, 1667, in Vitry, France, died November 27, 1754, in London. He made ontributions to mathematical analysis, actuarial sciences, and stochastics.

[3] Pierre Simon de Laplace, born March 23, 1749, in Beaumant-en-Auge, France, died March 5, 1827, in Paris, France. He made contributions to mathematics, astronomy, and stochastics.

$$x_2 = \frac{m_2 - np}{\sqrt{np(1-p)}}. \tag{2.41}$$

Define

$$Y = \frac{X_S - np}{\sqrt{np(1-p)}}. \tag{2.42}$$

Then,

$$\begin{aligned} \lim_{n\to\infty} B_{m_1,m_2}(n,p) &= \lim_{\substack{x_1,x_2=\text{const} \\ n\to\infty}} \mathbf{P}_Y\Big([x_1,x_2]\Big) \\ &= \Phi(x_2) - \Phi(x_1), \end{aligned} \tag{2.43}$$

where $\Phi$ denotes the distribution function of a standardized normally distributed random variable.

Equation (2.43) yields the following approximation:

$$B_m(np) \approx \Phi\left(\frac{m-np}{\sqrt{np(1-p)}}\right). \tag{2.44}$$

Similar to that in the case of the Poisson approximation, the solution of (2.24) is approximated by the solution $b_\alpha$ of

$$\Phi\left(\frac{m-nb_\alpha}{\sqrt{nb_\alpha(1-b_\alpha)}}\right) = \alpha. \tag{2.45}$$

From (2.45), the following explicit formula for $b_\alpha$ is obtained:

$$b_\alpha = \begin{cases} \frac{2m+\lambda_\alpha^2}{2(n+\lambda_\alpha^2)} + \sqrt{\left(\frac{2m+\lambda_\alpha^2}{2(n+\lambda_\alpha^2)}\right)^2 - \frac{m^2}{n(n+\lambda_\alpha^2)}} & \text{for } \alpha \le 0.5 \\ \frac{2m+\lambda_\alpha^2}{2(n+\lambda_\alpha^2)} - \sqrt{\left(\frac{2m+\lambda_\alpha^2}{2(n+\lambda_\alpha^2)}\right)^2 - \frac{m^2}{n(n+\lambda_\alpha^2)}} & \text{for } \alpha > 0.5, \end{cases} \tag{2.46}$$

where $\lambda_\alpha$ denotes the $\alpha$-quantile of the standardized normal distribution function, i.e. $\Phi(\lambda_\alpha) = \alpha$.

**Example 2.5**
*For the above example* $B_5(25, p_{0.05}) = 0.05$, *the following approximate solution is obtained by (2.46):*

$$p_{0.05} \approx 0.3577. \tag{2.47}$$

•

A binomially distributed random variable can adopt only integer values in contrast to a normally distributed random variable. This difference may be taken into account by a "continuity correction," which improves the accuracy of the approximation (2.45):

$$\Phi\left(\frac{m - n\tilde{b}_\alpha + 0.5}{\sqrt{n\tilde{b}_\alpha(1 - \tilde{b}_\alpha)}}\right) = \alpha. \tag{2.48}$$

**Remark 2.2** *The Poisson approximation works well for rather small values of* $p$, *whereas the normal approximation works well for values of* $p$ *in the neighborhood of* 0.5. •

## 2.6 The Hypergeometric Distribution

As has been shown, the hypergeometric distribution is an important tool for estimating an unknown proportion $p = \frac{M}{N}$. Drawing the elements randomly without replacement leads to a dependent random sample $(X_1, \ldots, X_n)$, where each sample element is a copy of $X$ [i.e., $X_i \sim B(p)$], but where the sample elements are not independent. In this case, the sum of the random elements $X_S$ is hypergeometrically distributed, denoted by

$$X_S \sim H(N, M, n) = H(N, Np, n) \tag{2.49}$$

with expectation

$$E\Big[X_S\Big] = np \tag{2.50}$$

and variance

$$V\big[X_S\big] = \frac{N-n}{N-1} np(1-p). \tag{2.51}$$

For the hypergeometric distribution, the following notations are introduced:

$$h_m(N,n,p) \;:=\; \frac{\binom{M}{m}\binom{N-n}{n-m}}{\binom{N}{n}} = \mathbf{P}^{(p)}_{X_S}\Big(\{m\}\Big) \tag{2.52}$$

$$\begin{aligned} H_m(N,n,p) \;&:=\; \sum_{i=0}^{m} h_i(N,n,p) \\ &= \mathbf{P}^{(p)}_{X_S}\Big(\{0,1,\ldots,m\}\Big) \end{aligned} \tag{2.53}$$

$$\begin{aligned} H_{m_1,m_2}(N,n,p) \;&:=\; \sum_{i=m_1}^{m_2} h_i(N,n,p) \\ &= \mathbf{P}^{(p)}_{X_S}\Big(\{m_1,\ldots,m_2\}\Big) \end{aligned} \tag{2.54}$$

As an well-known approximation of the probabilities of the hypergeometric distribution, we have

$$\lim_{N\to\infty} h_m(N,n,p) = b_m(n,p). \tag{2.55}$$

By (2.55), the problems related to the hypergeometric distribution can be solved approximately by using the techniques derived for the binomial distribution. The approximation is good for large $N$ compared with $n$.

This handbook is focused on the case of independent samples, and the case of a dependent random sample is only mentioned for completeness. Of course, the methods developed here can easily be extended to the dependent case.

## 2.7 Measuring in the Measurement Range

One of the aims is to develop a procedure for determining the actual value $p$ of the probability $\breve{p}$ for a given measurement range $[\underline{p}, \overline{p}] \subset [0,1]$. As the measurement range is fixed, only the elements of the measurement range are admitted as results of the measurement procedure. Any result that is not an element of $[\underline{p}, \overline{p}]$ makes no sense, because fixing a range is equivalent with defining the numerical values admissible for the procedure.

Any measurement procedure results in a subset of the measurement range. In order to evaluate the results of a procedure and, finally, the procedure itself, the question arises, of how to measure subsets of the measurement range $[\underline{p}, \overline{p}]$. From a more mathematical point of view, the following question has to be answered: Which measure function is appropriate for evaluating sets on the corresponding measurable space

$$\left([\underline{p}, \overline{p}], \mathcal{B}_{[\underline{p}, \overline{p}]}\right), \tag{2.56}$$

where $\mathcal{B}_{[\underline{p}, \overline{p}]}$ denotes the $\sigma$-field of Borel sets over the measurement range $[\underline{p}, \overline{p}]$.

The answer is important, as any serious measurement procedure results in an interval within the measurement range, hopefully including the unknown actual value $p$ of $\breve{p}$. Thus, the problem of evaluating the measurement results is reduced to measure the subsets of the measurement range.

The answer to the question may be explained in two different ways:

- Starting from the result of a measurement procedure leads to the following consideration. The aim of the procedure is to determine reliably and precisely the actual value $p$ of $\breve{p}$. Clearly, the precision of the measurement result is given by the lenghth of the corresponding interval, and hence one arrives necessarily at the Lebesgue measure on $\left([\underline{p}, \overline{p}], \mathcal{B}_{[\underline{p}, \overline{p}]}\right)$.

- Starting not from the result but from the aim of a measurement procedure yields another consideration. The measurement procedure shall be equally valid for any value of the probability $\breve{p}$ within the measurement range. It follows that the procedure should not prefer some of the admitted values over the others; that is each value of the measurement range must be taken into account equally when the procedure is developed. Measuring an element $B$ of $\mathcal{B}_{[\underline{p},\overline{p}]}$ without preferring any value $p \in B$ over the other means necessarily to complete the measurable space to a measure space by adding the Lebesgue measure $\mathcal{L}_{[\underline{p},\overline{p}]}$.

Therefore, the only appropriate measure on $\mathcal{B}_{[\underline{p},\overline{p}]}$ is the Lebesgue measure, or equivalent to $\mathcal{L}_{[\underline{p},\overline{p}]}$, the following normed measure $\Lambda_{[\underline{p},\overline{p}]}$:

$$\Lambda_{[\underline{p},\overline{p}]} = \frac{\mathcal{L}_{[\underline{p},\overline{p}]}}{\overline{p} - \underline{p}}. \tag{2.57}$$

The definition (2.57) implies that for any $p_1$ and $p_2$ with $0 \leq p_1 \leq p_2 \leq 1$

$$\Lambda_{[\underline{p},\overline{p}]}\Big([p_1, p_2]\Big) = \frac{\max\Big[\underline{p}, \min(p_2, \overline{p})\Big] - \min\Big[\overline{p}, \max(p_1, \underline{p})\Big]}{\overline{p} - \underline{p}} \tag{2.58}$$

holds.

**Remark 2.3**
*1. From a purely formal point of view, the measure $\Lambda_{[\underline{p},\overline{p}]}$ may be looked upon as a probability measure of the parameter $\breve{p}$. However, this interpretation assumes $\breve{p}$ to be a random variable defined on a probability space adopting values with probabilities given by the uniform distribution. This idea is completely misleading, as the problem is to develop a precise measurement procedure for the range $[\underline{p}, \overline{p}]$. The measure $\Lambda_{[\underline{p},\overline{p}]}$ enables one to measure in a natural way the precision of the result and guarantees that none of the possible values $p$ of $\breve{p}$ is preferred over the other. Note that the measurement procedure is developed without considering any special situation with respect to $\breve{p}$. The only feature that connects the situation and the measurement procedure in the case of an application is the measurement range, which has to be selected appropriately.*
*2. If, however, $\breve{p}$ represents a second random experiment (besides $\vec{X}$) with probability measure $P_{\breve{p}}^{(a)}$, where $a$ is the actual*

*value of a distribution parameter of* $\breve{p}$ *denoted by* $\breve{a}$*, then a different situation is obtained. For determining the actual value* $a$*, a two-stage random experiment is used with joint distribution* $P^{(a)}_{(\vec{X},\breve{p})}$*. In this case, an appropriate measurement range for* $\breve{a}$ *must be selected, and, again, the measure function on the new range for evaluating any result obtained by applying the procedure is, in essence, the Lebesgue measure.* •

# Chapter 3

# Traditional Estimation Procedures

## 3.1 Theory of Estimation

Having defined the quantity of interest to be the probability $\breve{p}$ of a specified event $E$ leads to the problem of measuring the actual value $p$ of $\breve{p}$. Traditionally, measurement of stochastical quantities is called estimation, which constitutes the core of statistical science. The most successful attempt to develop an estimation theory is due to Jerzy Neyman, who introduced the concept of confidence intervals with guaranteed reliability and thereby identified precision as the ultimate aim of an estimation method. Large parts of the traditional theory of estimation were developed and established by Jerzy Neyman. However, some of his proposals were not taken up by the statistical community. Therefore, a brief outline of Neyman's approach is given here, particularly because the ideas and methodology developed in the next chapters are based essentially on Neyman's results.

### 3.1.1 Neyman's Approach

Neyman introduces in [27] his theory as follows:

> We shall distinguish two aspects of the problems of

estimation: (i) the practical and (ii) the theoretical. The practical aspect may be described as follows:

(ia) The statistician is concerned with a population, $\pi$, which for some reason or other cannot be studied exhaustively. It is only possible to draw a sample from this population which may be studied in detail and used to form an opinion as to the values of certain constants describing the properties of the population $\pi$. For example, it may be desired to calculate approximately the mean of a certain character possessed by the individuals forming the population $\pi$, etc.

(ib) Alternatively, the statistician may be concerned with certain experiments which, if repeated under apparently identical conditions, yield varying results. Such experiments are called random experiments. To explain or describe the machinery of the varying results of random experiments mathematical schemes are drawn to involving one or more parameters, the values of which are not fixed. The statistician is then asked to provide numerical values of these parameters, to be calculated from experimental data and upon the assumption that the mathematical model of the experiment is correct.

In both cases described, the problem with which the statistician is faced is the problem of estimation. This problem consists of determining what arithmetical operations should be performed on the observational data in order to obtain a result, to be called an estimate, which presumably does not differ very much from the true value of the numerical character, either of the population $\pi$, as in (ia), or of the random experiment, as in (ib).

(ii) The theoretical aspect of the problem of statistical estimation consists primarily in putting in a precise form certain vague notions mentioned in (i). It will be noticed that the problem in its practical aspect is not a mathematical problem, and before attempting any mathematical solution we must substitute for (i) another problem, (ii), having a mathematical sense and such that, for practical purposes, it may be considered as equivalent to (i).

In view of the here-developed methods, Neyman's view about the so-called Bayes approach in estimation theory is of interest. Neyman explains:

> The first attempt to solve the problem of estimation is connected with the theorem of Bayes and is applicable when the parameters $\theta_1, \theta_1, \ldots, \theta_\ell$ are themselves random variables.
>
> It is known that, as far as we work with the conception of probability as adopted in this paper, the above theoretically perfect (Bayes) solution may be applied in practice only in quite exceptional cases, and this for two reasons:
>
> (a) It is only very rarely that the parameters $\theta_1, \theta_1, \ldots, \theta_\ell$ are random variables. They are generally unknown constants and therefore their probability law *a priori* has no meaning.
>
> (b) Even if the parameters to be estimated, $\theta_1, \theta_2, \ldots, \theta_\ell$, could be considered as random variables, the elementary probability law *a priori*
>
> $$p(\theta_1, \theta_1, \ldots, \theta_\ell)$$
>
> is usually unknown, and hence the (Bayes) formula cannot be used because of the lack of the necessary data.

Any estimator for $\theta = (\theta_1, \theta_2, \ldots, \theta_\ell)$ is based on a random sample

$$\vec{X} = \left(X_1, X_2, \ldots, X_n\right) \tag{3.1}$$

with image space $\mathcal{X}^n$, and on the probability distribution

$$\mathbf{P}_{\vec{X}}^{(\theta)}. \tag{3.2}$$

After having introduced various principles (e.g., unbiasedness, least squares estimation, and the maximum likelihood principle) Neyman turns to the theory of confidence intervals for $\theta = (\theta_1, \theta_2, \ldots, \theta_\ell)$.

> Denote generally by $\theta$ the parameter to be estimated and by $T$ its estimate, deduced from some principle or another. Whatever the principle, it is obviously impossible to assume that in any particular case $T$ is exactly equal to $\theta$. Therefore, the practical statistician required some measure of the accuracy of the estimate $T$. The generally accepted method of describing this accuracy consists in calculating the estimate, say $S_T^2$, of the variance $V_T$ of $T$ and in writing the result of all the calculations in the form $T \pm S_T$.
>
> Behind this method of presenting the results of estimating $\theta$, there is the idea that the true value of $\theta$ will frequently lie between the value of $T$ minus a certain multiple of $S_T$ and $T$ plus perhaps some other multiple of $S_T$. Therefore, the smaller $S_T$ the more accurate is the estimate $T$ of $\theta$.
>
> If we look through a number of recent statistical publications, we shall find that it is exceedingly rare that the values of unique estimates are given without the $\pm S_T$. We shall find also that the comments on the values of $T$ are largely dependent on those of $S_T$. This shows that what the statisticians have really in mind in problems of estimation is not the idea of

> a unique estimate but that of two estimates having the form, say
>
> $$\underline{\theta} = T - k_1 S_T \quad \text{and} \quad \overline{\theta} = T + k_2 S_T \tag{3.3}$$
>
> where $k_1$ and $k_2$ are certain constants, indicating the limits between which the true value of $\theta$ presumably falls.
>
> In this way the practical work, which is frequently in advance of the theory, brings us to consider the theoretical problem of estimating the parameter $\theta$ by means of the interval $(\underline{\theta}, \overline{\theta})$, extending from $\underline{\theta}$ to $\overline{\theta}$. These limits will be called the lower and upper estimates of $\theta$ respectively. It is obvious that if the values of $k_1$ and $k_2$ in (3.3) are not specified, then the real nature of the two estimates is not determined.

Thus according to Neyman, the true task of statistics is not to provide the practitioner with "point estimators," but with "interval estimators," which include statements on the reliability and the achieved measurement precision.

> Consider the variables (3.1) and assume that the form of their probability law (3.2) is known, that it involves the parameters $\theta_1$, $\theta_1, \ldots, \theta_\ell$, which are constant (not random variables), and that the numerical values of these parameters are unknown. It is desired to estimate one of these parameters say $\theta_1$. By this I shall mean that it is desired to define two functions $\overline{\theta}(\vec{x})$ and $\underline{\theta}(\vec{x}) \leq \overline{\theta}(\vec{x})$, determined and single valued at any point $\vec{x}$ of the sample space, such that if $\vec{x}'$ is the sample point determined by observation, we can (1) calculate the corresponding values of $\underline{\theta}(\vec{x}')$ and $\overline{\theta}(\vec{x}')$, and (2) state that the true value of $\theta_1$, say $\theta_1^0$ is contained within the limits
>
> $$\underline{\theta}(\vec{x}') \leq \theta_1^0 \leq \overline{\theta}(\vec{x}') \tag{3.4}$$

> this statement having some intelligible justification on the ground of probability.
>
> We may also try to select $\underline{\theta}(\vec{x})$ and $\overline{\theta}(\vec{x})$ so that the probability of $\underline{\theta}(\vec{X})$ falling short of $\theta_1^0$ and at the same time of $\overline{\theta}(\vec{X}')$ exceeding $\theta_1^0$, is equal to any number $\alpha$ between zero and unity, fixed in advance. If $\theta_1^0$ denotes the true value of $\theta_1$, then of course this probability must be calculated under the assumption that $\theta_1^0$ *is* the true value of $\theta_1$. Thus we can look for two function $\underline{\theta}(\vec{x}')$ and $\overline{\theta}(\vec{x}')$, such that
>
> $$\mathbf{P}_{\vec{X}}^{(\theta)}\Big(\underline{\theta}(\vec{X}') \leq \theta_1^0 \leq \overline{\theta}(\vec{X}')\Big) = \alpha \qquad (3.5)$$
>
> and require that (3.5) holds good *whatever* the value $\theta_1^0$ of $\theta_1$ and *whatever* the values of the other parameters $\theta_2, \ldots, \theta_\ell$, involved in the probability law of the $X$'s may be.
>
> The functions $\underline{\theta}(\vec{X})$ and $\overline{\theta}(\vec{X})$ satisfying the above conditions will be called the lower and upper confidence limits of $\theta_1$.

Finally, Neyman also offers an ingenious method of how to solve the problem of confidence intervals.

> In order to find the solution of the problem of confidence intervals, let us suppose that it is already solved and that $\underline{\theta}(E)$ and $\overline{\theta}(E)$ are functions determined and single valued in the whole sample space, $W$, such that the equality (3.5) holds good whatever the true values of the parameters $\theta_1, \theta_2, \ldots, \theta_\ell$. It will be convenient to interpret the situation geometrically. For this purpose we shall need to consider the space, $G$, of $n+1$ dimensions which we shall call the general space. The points in this space will be determined by $n+1$ coordinates $x_1, x_2, \ldots, x_n, \theta_1$, the first $n$ of which are the particular values of the

random variables (3.1) and thus determine the position of the sample point, $E$, in the $n$-dimensional space $W$, and the last coordinate $\theta_1$ is one of the possible values of the parameter $\theta_1$ in the probability law $p(E|\theta_1, \ldots, \theta_\ell)$ which we desire to estimate.

Consequently, if we consider any hyperplane, $G(\theta_1)$ in $G$ corresponding to the equation $\theta_1 =$ const., this may be interpreted as an image of the sample space $W$. We notice also that to any point $E$ in the sample space $W$ there will correspond in $G$ a straight line, say $L(E)$, parallel to the axis of $\theta_1$. If $x'_1, x'_2, \ldots, x'_n$ are the coordinates of $E'$, then the line $L(E')$ will correspond to the equations $x_i = x'_i$ for $i = 1, 2, \ldots, n$.

Consider now the functions $\underline{\theta}(E)$ and $\overline{\theta}(E)$. On each line $L(E)$, they will determine two points, say $B(E)$ and $C(E)$ with coordinates

$$x_1, x_2, \ldots, x_n, \underline{\theta}(E) \tag{3.6}$$

and

$$x_1, x_2, \ldots, x_n, \overline{\theta}(E) \tag{3.7}$$

respectively, where $x_1, x_2, \ldots, x_n$ are the coordinates of the sample point $E$. The interval between $B(E)$ and $C(E)$ will be the image of the confidence interval $\delta(E)$ corresponding to the sample point $E$.

In summary, one can state that according to Neyman, the solution of the estimation problem is an interval estimation with prescribed reliability, which can be determined by means of systems of hyperplanes within the general space. The proposal is, in particular, noteworthy, because it solves the problem without using any point estimator and without any asymptotic derivation.

### 3.1.2 Point and Interval Estimation

Before we turn to our problem of measuring the actual value $p$ of a probability of interest $\breve{p}$, let us have a closer look at the traditional estimation theory in general.

Let $\breve{\theta}$ be the quantity or parameter of interest and $\Theta$ the measurement range. The actual value $\theta$ of the parameter $\breve{\theta}$ is unknown and shall be determined ( i.e., measured in a reliable and precise way).

For the purpose of developing a measurement procedure for $\breve{\theta}$, a random experiment $\left(\Omega, \mathcal{S}, \mathbf{P}^{(\theta)}\right)$ is developed and a suitable random variable $X$ with image set $\mathcal{X}$ is selected. The task is to derive a measurement procedure for $\breve{\theta}$ and a given measurement range $\Theta$ based on a random sample $\vec{X} = \left(X_1, \ldots, X_n\right)$ for $X$ with image set $\mathcal{X}^n$, joint distribution $\mathbf{P}_{\vec{X}}^{(\theta)}$, and joint distribution function $F_{\vec{X}}^{(\theta)}(\vec{x})$.

The quantity to be dealt with is the pair $\left(\breve{\theta}, \vec{X}\right)$ which may adopt values in the set $\Theta \times \mathcal{X}^n$. The connection between the parameter $\breve{\theta}$ and the random sample $\vec{X}$ is the fact that the unknown value $\theta$ of $\breve{\theta}$ acts as distribution parameter of $\vec{X}$.

There are two different concepts proposed in estimation theory dealing with the above-outlined problem:

- point estimators

- interval estimators

A point estimator represents an estimation procedure given by an estimation function without prescribed reliability (i.e., without guaranteeing a lower bound for the probability of a correct result). From a scientific point of view, a measurement procedure without stated reliabilty is, strictly speaking, meaningless. Therefore, a point estimator does not represent a scientific measurement procedure. The only secured knowledge about a point estimator is that, generally, it yields a result which is not equal to the unknown value $\theta$. Calling a wrong statement about the value $\theta$ a measurement error implies that a point estimator results in an error with certainty. Thus, the reliability of point estimators is minimum.

The measuring result of a point estimator is a subset of $\Theta$ containing exactly one element. Thus, the precision of a point estimator assessed by means of the measure function $\Lambda_\Theta$ given by (2.57) is maximum. However, maximum precision in conjunction with minimum reliability is meaningless.

Besides reliability and precision, several other properties are investigated with respect to point estimators. Looking at these properties from the viewpoint of the aims of an estimation method leads to the following result:

- *Unbiasedness*: The descriptive meaning of unbiasedness is that the procedure makes no systematic errors. This is, of course, a rather nice looking property. Unfortunately, it does not say anything about the size of the random errors and therefore it is not suitable to evaluate the precision of an estimator.

- *Consistency*: The descriptive meaning of consistency is that large errors become more and more seldom with increasing sample size. Again, this is a very nice property, and a method not having this property should probably be excluded, but being a limiting property, it does not allow any direct statement about the precision of a given procedure.

- *Minimum variance*: The minimum variance property refers to unbiased estimators, neglecting all those that are biased, but nevertheless yield perhaps better results. Therefore, the property of minimum variance also does not characterize a best estimator or even a good estimator from the viewpoint of precision.

In contrast to point estimators, interval estimators state generally the procedure's reliability as well as the achieved precision and therefore can be looked upon as scientific measurement procedures. Unfortunately, Neyman's proposal for constructing confidence intervals, which are called measurement intervals here, was not accepted by the statisticians, who, until now, stick to the familiar approach given by (3.3) (i.e., starting with a point estimator and evaluating its variance generally approximately by means of asymptotic derivations). The problem with such asymptotic properties is that they hold true for $n \to \infty$ and are approximately true for large sample sizes, but they might be not true in the case of small sample sizes.

An interval estimator is traditionally called a confidence interval, which is misleading because an interval estimator consists of a whole family of intervals. Therefore, not the estimator, but only its realizations are intervals. An interval estimator itself constitutes a measurement procedure in a strict scientific sense, and therefore is called *measurement procedure* in the remainder.

The formal definition of a measurement procedure (interval estimator) is as follows:

**Definition 3.1**
*Let $\theta \in \Theta$ be the unknown value of a one-dimensional parameter $\breve{\theta}$ and $\vec{X}$ be a random sample of size $n$ with probability distribution $\mathbf{P}_{\vec{X}}^{(\theta)}$. A* measurement procedure *with* confidence level *$\beta \in (0,1)$ for the unknown value $\theta$ is a pair of random variables $(L_{\breve{\theta}}^{(\beta,n)}, U_{\breve{\theta}}^{(\beta,n)})$ defined as functions of the random sample $\vec{X} = (X_1, \ldots, X_n)$ with the following properties:*

$$\mathbf{P}_{\vec{X}}^{(\theta)}\Big(\{\vec{x} \mid L_{\breve{\theta}}^{(\beta,n)}(\vec{x}) \leq U_{\breve{\theta}}^{(\beta,n)}(\vec{x})\}\Big) = 1 \tag{3.8}$$

$$\mathbf{P}_{\vec{X}}^{(\theta)}\Big(\{\vec{x} \mid L_{\breve{\theta}}^{(\beta,n)}(\vec{x}) \leq \theta \leq U_{\breve{\theta}}^{(\beta,n)}(\vec{x})\}\Big) \geq \beta \tag{3.9}$$

*for any $\theta \in \Theta$ with $\vec{x} \in I\!R^n$.* •

The reliability of a measurement procedure defined by (3.8) and (3.9) is given by the confidence level $\beta$, which means that the procedure will lead to a measurement error at most with probability $1 - \beta$. The procedure's precision is given by the difference $U_{\breve{\theta}}^{(\beta,n)}(X_1, \ldots, X_n) - L_{\breve{\theta}}^{(\beta,n)}(X_1, \ldots, X_n)$.

**Remark 3.1**
*For the interpretation of a confidence interval note that for any realization* $(x_1, \ldots, x_n)$ *of the random sample* $(X_1, \ldots, X_n)$ *yielding the confidence interval* $[\ell_{\breve{\theta}}^{(\beta,n)}, u_{\breve{\theta}}^{(\beta,n)}]$ *of the interval estimator* $(L_{\breve{\theta}}^{(\beta,n)}, U_{\breve{\theta}}^{(\beta,n)})$, *the actual value* $\theta$ *falls with certainty either into the realized interval* $[\ell_{\breve{\theta}}^{(\beta,n)}, u_{\breve{\theta}}^{(\beta,n)}]$ *or not. Thus, the frequently made statement* $\theta \in [\ell_{\breve{\theta}}^{(\beta,n)}, u_{\breve{\theta}}^{(\beta,n)}]$ *with probability* $\beta$ *makes no sense at all. However, there is founded hope (based on the confidence level* $\beta$*) that* $\theta \in [\ell_{\breve{\theta}}^{(\beta,n)}, u_{\breve{\theta}}^{(\beta,n)}]$ *actually holds, or, in other words, that the very probable (with probability at least* $\beta$*) event has happened.* •

## 3.2 Interval Estimator for a Probability

This handbook is concerned with measuring the probability of an event based on a sequence of independent Bernoulli experiments. Thus, the parameter of interest is a probability $\breve{p}$; the actual value $p$ refers to a given event, the random experiment is a Benoulli experiment with a Bernoulli variable $X \sim B(p)$ and the random sample is the vector $\vec{X} = (X_1, \ldots, X_n)$ consisting of $n$ independent and identically like $X$ distributed random variables.

Let $\vec{x} = (x_1, \ldots, x_n)$ be a realization of $\vec{X}$. Because of (2.7), there are exactly $2^n$ different realizations $\vec{x}_0, \vec{x}_1, \ldots, \vec{x}_{2^n-1}$ of $\vec{X}$ and at most $2^n$ different realizations for $L_{\breve{p}}^{(\beta,n)}(\vec{X})$ and $U_{\breve{p}}^{(\beta,n)}(\vec{X})$. Thus, $L_{\breve{p}}^{(\beta,n)}$ and $U_{\breve{p}}^{(\beta,n)}$ may adopt at most $2^n$ different values denoted by

$$\ell_{\breve{p}}^{(\beta,n)}(\vec{x}_0), \ell_{\breve{p}}^{(\beta,n)}(\vec{x}_1), \ldots, \ell_{\breve{p}}^{(\beta,n)}(\vec{x}_{2^n-1})$$

and

$$u_{\breve{p}}^{(\beta,n)}(\vec{x}_0), u_{\breve{p}}^{(\beta,n)}(\vec{x}_1), \ldots, u_{\breve{p}}^{(\beta,n)}(\vec{x}_{2^n-1})$$

respectively.

The following lemma describes a set of possible measurement procedures with confidence level $\beta$ based on a random sample $\vec{X}$ of size $n$.

**Lemma 3.1**
*Let* $\left\{\vec{x}_0, \ldots, \vec{x}_{2^n-1}\right\}$ *be the set of realizations of a random sample* $\vec{X} = \left(X_1, \ldots, X_n\right)$ *of size* $n$ *for* $X_i \sim B(p)$, $i = 1, \ldots, n$, *with* $p \in [\underline{p}, \overline{p}]$ *and*

$$\vec{x}_j = \left(x_j^{(1)}, \ldots, x_j^{(n)}\right) \quad \textit{for} \quad j = 0, \ldots, 2^n - 1$$

$$\sum_{i=1}^{n} x_j^{(i)} \leq \sum_{i=1}^{n} x_{j+1}^{(i)} \quad \textit{for} \quad j = 0, \ldots, 2^n - 2. \tag{3.10}$$

*Then, any sequence*

$$\left\{\left(\ell(\vec{x}_j), \alpha_{\ell,\vec{x}_j}; u(\vec{x}_j), \alpha_{u,\vec{x}_j}\right)\right\}_{j=0,1,\ldots,2^n-1} \tag{3.11}$$

*with*

$$\underline{p} = \ell(\vec{x}_0) \leq \ell(\vec{x}_1) \leq \ldots \leq \ell(\vec{x}_{2^n-1}) \leq \overline{p} \tag{3.12}$$

$$\underline{p} \leq u(\vec{x}_0) \leq u(\vec{x}_1) \leq \ldots \leq u(\vec{x}_{2^n-1}) = \overline{p} \tag{3.13}$$

*and*

$$\alpha_{\ell,\vec{x}_0} = 0 \tag{3.14}$$

$$\alpha_{\ell,\vec{x}_j} = 1 - \mathbf{P}_{\vec{X}}^{(\ell(\vec{x}_j))}\left(\{\vec{x}_0, \ldots, \vec{x}_{j-1}\}\right) \quad \textit{for } j = 1, \ldots, 2^n - 1 \tag{3.15}$$

$$\alpha_{u,\vec{x}_j} = \mathbf{P}_{\vec{X}}^{(u(\vec{x}_j))}\left(\vec{X}_n \in \{\vec{x}_0, \ldots, \vec{x}_j\}\right) \quad \textit{for } j = 0, \ldots, 2^n - 2 \tag{3.16}$$

$$\alpha_{u,\vec{x}_{2^n-1}} = 0 \tag{3.17}$$

$$\alpha_{\ell,\vec{x}_j} + \alpha_{u,\vec{x}_j} \leq \alpha := 1 - \beta \in (0,1) \quad \textit{for } j = 0, \ldots, 2^n - 1 \tag{3.18}$$

*determines a measurement procedure*

$$\left(L_{\check{p}}^{(\beta,n)}(\vec{X}), U_{\check{p}}^{(\beta,n)}(\vec{X})\right)$$

*for the actual value $p$ of a probability $\breve{p}$ with confidence level $\beta$ by setting*

$$L_{\breve{p}}^{(\beta,n)}\left(\vec{X}\right) = \ell(\vec{x}_j) \quad \textit{for } \vec{X} = \vec{x}_j \tag{3.19}$$

$$U_{\breve{p}}^{(\beta,n)}\left(\vec{X}\right) = u(\vec{x}_j) \quad \textit{for } \vec{X} = \vec{x}_j. \tag{3.20}$$

•

**Proof:**

Obviously, there are sequences (3.11) that are a solution of the $(2^{n+1} - 2)$ equations given by (3.15) and (3.16) under the $2^n$ sideconditions given by (3.18). Consider, for instance, the sequence given by $\ell(\vec{x}_j) = \underline{p}$, $u(\vec{x}_j) = \overline{p}$, and $\alpha_{\ell,\vec{x}_j} = \alpha_{u,\vec{x}_j} = 0$ for any $j$; then, the equations and the side conditions are fulfilled.

We have to show that $\left(L_{\breve{p}}^{(\beta,n)}, U_{\breve{p}}^{(\beta,n)}\right)$ fulfills the two conditions given by (3.8) and (3.9).

As

$$\begin{aligned} \ell(\vec{x}_0) &= \underline{p} \leq u(\vec{x}_0) \\ \ell(\vec{x}_{2^n-1}) &\leq \overline{p} = u(\vec{x}_{2^n-1}) \end{aligned}$$

(3.8) is fulfilled for $j = 0$ and $j = 2^n - 1$.

Assume that for $j \in \{1, \ldots, 2^n - 2\}$, we have $\ell(\vec{x}_j) > u(\vec{x}_j)$; then,

$$\begin{aligned} 1 > \alpha &= 1 - \mathbf{P}_{\vec{X}}^{(\ell(\vec{x}_j))}\left(\{\vec{x}_0, \ldots, \vec{x}_{j-1}\}\right) + \mathbf{P}_{\vec{X}}^{(u(\vec{x}_j))}\left(\{\vec{x}_0, \ldots, \vec{x}_j\}\right) \\ &\geq 1 - \mathbf{P}_{\vec{X}}^{(\ell(\vec{x}_j))}\left(\{\vec{x}_0, \ldots, \vec{x}_{j-1}\}\right) \\ &\quad + \mathbf{P}_{\vec{X}}^{(\ell(\vec{x}_j))}\left(\{\vec{x}_0, \ldots, \vec{x}_j\}\right) \\ &\geq 1. \end{aligned}$$

Therefore, the assumption $\ell(\vec{x}_j) > u(\vec{x}_j)$ is wrong, implying that $\left(L_{\breve{p}}^{(\beta,n)}, U_{\breve{p}}^{(\beta,n)}\right)$ fulfills condition (3.8).

Condition (3.9) is fulfilled if for any $p \in [\underline{p}, \overline{p}]$,

$$\mathbf{P}_{\vec{X}}^{(p)}\left(\{\vec{x} | \ell(\vec{x}) > p\} \cup \{\vec{x} | u(\vec{x}) < p\}\right) \leq \alpha \tag{3.21}$$

holds. The events $\{\vec{x} | \ell(\vec{x}) > p\}$ and $\{\vec{x} | u(\vec{x}) < p\}$ are exclusive events; therefore, we can treat them separately.

Define $\ell(\vec{x}_{2^n}) = \overline{p}$; then, for any $p \in [\underline{p}, \overline{p}]$, there is an index $j \in \{0, 1, \ldots, 2^n\}$ with

$$\ell(\vec{x}_j) \leq p \geq \ell(\vec{x}_{j-1})$$

and

$$\begin{aligned}
\mathbf{P}_{\vec{X}}^{(p)}\Big(\{\vec{x}|\ell(\vec{x}) > p\}\Big) &= \mathbf{P}_{\vec{X}}^{(p)}\Big(\{\vec{x}_j, \ldots, \vec{x}_{2^n}\}\Big) \\
&= 1 - \mathbf{P}_{\vec{X}}^{(p)}\Big(\{\vec{x}_0, \ldots, \vec{x}_{j-1}\}\Big) \\
&\leq 1 - \mathbf{P}_{\vec{X}}^{(\ell(\vec{x}_j))}\Big(\{\vec{x}_0, \ldots, \vec{x}_{j-1}\}\Big) \\
&= \alpha_{\ell,\vec{x}_j}\,. \qquad (3.22)
\end{aligned}$$

Next, define $u(\vec{x}_{-1}) = \underline{p}$; then, for any $p \in [\underline{p}, \overline{p}]$, there is an index $j \in \{-1, 0, \ldots, 2^n - 1\}$ with

$$u(\vec{x}_j) \leq p \leq u(\vec{x}_{j+1})$$

and

$$\begin{aligned}
\mathbf{P}_{\vec{X}}^{(p)}\Big(\{\vec{x}|u(\vec{x}) < p\}\Big) &= \mathbf{P}_{\vec{X}}^{(p)}\Big(\{\vec{x}_{-1}, \vec{x}_0, \ldots, \vec{x}_j\}\Big) \\
&\leq \mathbf{P}_{\vec{X}}^{(u(\vec{x}_j))}\Big(\{\vec{x}_{-1}, \vec{x}_0, \ldots, \vec{x}_j\}\Big) \\
&= \alpha_{u,\vec{x}_j}\,. \qquad (3.23)
\end{aligned}$$

From (3.22), (3.23), and (3.18), we immediately obtain

$$\begin{aligned}
&\mathbf{P}_{\vec{X}}^{(p)}\Big(\{\vec{x}|\ell(\vec{x}) > p\} \cup \{\vec{x}|u(\vec{x}) < p\}\Big) \\
&\qquad \leq \alpha_{\ell,\vec{x}_j} + \alpha_{u,\vec{x}_j} = \alpha \qquad (3.24)
\end{aligned}$$

which completes the proof. •

It is easy to see that there are no difficulties in determining measurement procedures $\left(L_{\breve{p}}^{(\beta,n)}, U_{\breve{p}}^{(\beta,n)}\right)$ for any given confidence level $\beta$. The problem is to find a procedure which serves its aim of determining the actual value $p$ of $\breve{p}$ not only reliable, but also most precisely.

The reliability of a measurement procedure is specified by the upper bound for the probability of committing an estimation error, which occurs, if the true value $p$ is smaller than $L$, or larger than $U$. According to the defining relation (3.9), the error probability is controlled by $\beta$, implying that the reliability of a measurement procedure is given by its confidence level $\beta$.

According to the measure function $\Lambda_\Theta$, the precision of a measurement interval is given by its length. Clearly, the precision of the measurement procedure decreases with increasing lengths of the intervals $\left(\ell_{\vec{x}}, u_{\vec{x}}\right)$. Therefore, $\left(L_{\breve{p}}^{(\beta,n)}, U_{\breve{p}}^{(\beta,n)}\right)$ has to be derived so as to guarantee the confidence level and to have the length $U_{\breve{p}}^{(\beta,n)} - L_{\breve{p}}^{(\beta,n)}$ of $\left(L_{\breve{p}}^{(\beta,n)}, U_{\breve{p}}^{(\beta,n)}\right)$ as small as possible. This demand encounters another serious problem, as the difference $U_{\breve{p}}^{(\beta,n)} - L_{\breve{p}}^{(\beta,n)}$ is a random variable, and therefore the meaning of "as small as possible" for two different measurement procedures is unclear. Thus, the task is to develop a possibility to measure the precision $U_{\breve{p}}^{(\beta,n)} - L_{\breve{p}}^{(\beta,n)}$ of a measurement procedure.

## 3.3 The Relative Frequency $\overline{X}$

As noted by Neyman (compare (3.3)), measurement procedures in statistics are often given in the form $(T - k_1 S_T, T + k_2 S_T)$, where $T = T(\vec{X})$ is a point estimator and $S_T$ is its standard deviation. The first step for arriving at such a measurement procedure is to derive a point estimator for the parameter in question. The second step is to determine its variance, which can be achieved exactly only in rare cases. Thus, often an approximation is used for $S_T$. Finally, the factors $k_1$ and $k_2$ are calculated so as to guarantee a given confidence level. Again, $k_1$ and $k_2$ are generally calculated based on approximations of the actual distribution function.

In the case of estimating the actual value $p$ of a probability $\breve{p}$, the unanimoulsy recommended point estimator $TF(\vec{X};\breve{p})$ is the relative frequency:

$$TF\Big(X_1, \ldots, X_n; \breve{p}\Big) = \frac{1}{n}\sum_{i=1}^{n} X_i. \tag{3.25}$$

The relative frequency $TF(X_1, \ldots, X_n; \check{p})$ is a uniformly minimum variance unbiased estimator (UMVUE) and therefore often regarded as an "optimal" estimator. However, as mentioned before, the properties of the relative frequency do not allow one to conclude that with respect to precision, it is a good estimator.

The essential part of the relative frequency is given by the sum

$$X_S = \sum_{i=1}^{n} X_i \tag{3.26}$$

that is, the relative frequency is based on $X_S$ and not on the sample $\vec{X}$ itself. Therefore, any measurement procedure based on the relative frequency is built up on $X_S$, and not on the sample $\vec{X}$, which leads to some important simplifications with respect to the problem of deriving a measurement procedure.

**Remark 3.2** *The relative frequency is one of the central quantities in classical statistics, which sometimes is even called relative frequency theory. The importance of the relative frequency is attributed to the attempt to define the notion of probability of an event as the limit of the relative frequency of occurrences of the event. However, the relative frequency constitutes nothing more than one out of many methods for determining the value of the probability of a given event. Moreover, it will become obvious later that the relative frequency is not at all a very good method, as it often yields useless results. Unfortunately, the unique position of the relative frequency in classical statistics made new developments not based on the relative frequency almost impossible.*

*The approach developed in this handbook for determining the actual value of a probability is independent of the relative frequency and resembles therefore to some extent the Bayesian approach. However, unlike the Bayesian approach, the proposed procedures are more like (objective) technical measurement devices which may be used for determining the actual value of a quantity of interest in different but specified situations. In contrast, Bayesian procedures assume a given situation characterized not by one single value but by a (subjective) prior distribution of the quantity of interest.* •

## 3.4 Measurement Procedures Based on the Relative Frequency

Following the approach indicated by (3.3), the lower and upper bounds $L_{\breve{p}}^{(\beta,n)}$ and $U_{\breve{p}}^{(\beta,n)}$, respectively, of a measurement procedure for the actual value $p$ are traditionally not explicit functions of the random sample $\left(X_1,\ldots,X_n\right)$ as assumed in the lemma, but are functions of the random variable $X_S=\sum_{i=1}^{n} X_i$ or, equivalently, of the relative frequency $\overline{X}=\frac{1}{n}\sum_{i=1}^{n} X_i$. As justification for using $X_S$ or $\overline{X}$, respectively, instead of $\vec{X}$ the fact, is stated that both functions are *sufficient* for $p$.

By using $X_S$ or $\overline{X}$, respectively, instead of $\vec{X}$, the number of realizations is decreased considerably from $2^n$ different realizations to only $n+1$ different realizations, implying that there are, at most, $n+1$ different realizations for the lower and upper bounds $L_{\breve{p}}^{(\beta,n)}(X_S)$ and $U_{\breve{p}}^{(\beta,n)}(X_S)$, respectively, specifying the measurement procedure.

The sum of elements $X_S$ has $n+1$ realizations given by $\{0,1,\ldots,n\}$, implying that there are at most $n+1$ realizations of the bounds $\left(L_{\breve{p}}^{(\beta,n)}(X_S),U_{\breve{p}}^{(\beta,n)}(X_S)\right)$. These $n+1$ realizations are denoted by $[\ell_0,u_0],[\ell_1,u_1],\ldots,[\ell_n,u_n]$, where the index is equal to the realization of $X_S$ [i.e., the number of successes within the random sample $(X_1,\ldots,X_n)$].

The following corollary is an immediate consequence of the lemma.

**Corollary 3.1**

*Let $m\in\{0,1,\ldots,n\}$ denote the realization of $X_S$ being distributed according to $Bi(n,p)$ with $p\in[\underline{p},\overline{p}]$. Then, any sequence*

$$\left\{(\ell_m,\alpha_{\ell,m};u_m,\alpha_{u,m})\right\}_{m=0,1,\ldots,n}$$

*with*

$$\underline{p}=\ell_0\le\ell_1\le\ \ldots\ \le\ell_n\le\overline{p} \tag{3.27}$$

$$\underline{p}\le u_0\le u_1\le\ \ldots\ \le u_n=\overline{p} \tag{3.28}$$

*and*

$$\alpha_{\ell,0} = 0 \tag{3.29}$$

$$\alpha_{\ell,m} = 1 - B_{m-1}(n, \ell_m) \quad \textit{for} \quad m = 1, .., n \tag{3.30}$$

$$\alpha_{u,m} = B_m(n, u_m) \quad \textit{for} \quad m = 0, ..., n-1 \tag{3.31}$$

$$\alpha_{u,n} = 0 \tag{3.32}$$

$$\alpha_{\ell,m} + \alpha_{u,m} = \alpha := 1 - \beta \quad \textit{for} \quad m = 0, ..., n \tag{3.33}$$

*defines a measurement procedure for* $\breve{p}$ *with the measurement range* $[\underline{p}, \overline{p}]$ *and* confidence level $\beta$ *by setting*

$$L_{\breve{p}}^{(\beta,n)}(X_S) = \ell_m \quad \textit{for } X_S = m$$

$$U_{\breve{p}}^{(\beta,n)}(X_S) = u_m \quad \textit{for } X_S = m$$

•

Replacing the random sample $\vec{X}$ by the sum of sample elements $X_S$ or the relative frequency $\overline{X}$ simplifies the problem of determining a measurement procedure, as the number of equations to be solved is reduced from $(2^{n+1} - 2)$ to $2n$, and the number of side conditions from $2^n$ to $n + 1$. However, of course, the question of whether or not the precision decreases, when switching from $\vec{X}$ to $X_S$ or $\overline{X}$ has to be answered.

### 3.4.1 Traditional Measurement Procedures

The traditional measurement procedures for the actual value $p$ of a probability $\breve{p}$ with confidence level $\beta$ have some features in common:

- They assume $\underline{p} = 0$ and $\overline{p} = 1$.
- They are based on the sum of sample elements $X_S$.

- They use specially fixed bounds for the noninclusion probabilities.

The bounds of the noninclusion probabilities are fixed in the following way:

$$\begin{aligned} \alpha_{\ell,0} &= 0 \\ \alpha_{\ell,m} &= \frac{\alpha}{2} \quad \text{for} \quad m = 1, \ldots, n && (3.34) \\ \alpha_{u,m} &= \frac{\alpha}{2} \quad \text{for} \quad m = 0, \ldots, n-1 && (3.35) \\ \alpha_{u,n} &= 0 \end{aligned}$$

yielding the following equations [see (8) and (9)] for the realizations $\ell_m$ of the lower interval bound $L_{\breve{p}}^{(\beta,n)}(X_S)$ and $u_m$ for the upper interval bound $U_{\breve{p}}^{(\beta,n)}(X_S)$:

$$\begin{aligned} \ell_0 &= 0 \\ 1 - B_{m-1}(n, \ell_m) &= \frac{\alpha}{2} \quad \text{for } m = 1, \ldots, n && (3.36) \\ B_m(n, u_m) &= \frac{\alpha}{2} \quad \text{for } m = 0, ..., n-1 && (3.37) \\ u_n &= 1. \end{aligned}$$

The first assumption means that the measurement range is maximum. In many instances, however, it would be of advantage to have a measurement range adapted to the special probability $\breve{p}$ of interest. The maximum measurement range may have some unpleasant consequences in such situations.

- With positive probability, the measurement procedure yields useless results.

- It is likely that the precision of an interval estimator developed assuming the maximum measurement range is smaller than that of an estimator developed for a smaller measurement range.

It is to be expected that either of the above consequences will particularly be valid in the case of small sample sizes.

To base the interval estimator on $X_S$ rather than on the sample $\vec{X}_n = \big(X_1, \ldots, X_n\big)$ could become necessary if only $X_S$ is available. However, if $\vec{X}_n$ is known, then neglecting the knowledge might lead to inferior interval estimators, particularly for small sample sizes $n$.

Fixing the bounds for the noninclusion probabilities as given by (3.34) and (3.35) has the advantage that $2(n-1)$ of the unknown quantities in the system of equations - given in the corollary - do not need to be determined by an algorithm. Moreover, by fixing $\alpha_{\ell,m}$ and $\alpha_{u,m}$, the system of equations (3.36) and (3.37) can be solved by solving each equation independently from the others, in contrast to the general case. But, the binomial distribution is symmetric only in the case of $p = \frac{1}{2}$; therefore, fixing the bounds for the two possible errors $p < L_{\breve{p}}^{(\beta,n)}(X_S)$ and $p > U_{\breve{p}}^{(\beta,n)}(X_S)$ to be equal makes intuitively not much sense, and again it is to be expected that it leads to less precision, especially in the case of small sample sizes $n$.

Often, the measurement procedures derived by solving the equations (3.36) and (3.37) are called *exact* so as to distinguish them from so-called approximate measurement procedures, which do not necessarily meet the requirement with respect to the confidence level $\beta$. The confidence level is a selected upper bound for the probability of committing a measurement error. Thus, it may be looked upon as a classification label for measurement procedures. An evaluation or comparison of different measurement procedures is only meaningful for a given and guaranteed confidence level. For instance, a comparison of a measurement procedure with level $\beta_1 = 0.90$ with another one with level $\beta_2 = 0.95$ is, strictly speaking, meaningless. Thus, the quality of an "approximate" measurement procedure cannot be evaluated in the case of a small sample size, as, then, the approximation might be inferior. Therefore, "approximate" interval estimators cannot be considered as valid ones in a strict sense.

The many difficulties of deriving a good measurement procedure for the actual value $p$ of a probability $\breve{p}$ are reduced within the traditional approach to the problem of solving the equations (3.36) and (3.37).

Many textbooks recommend utilizing the relation between the binomial and the $F$-distribution for finding the solutions by means of tables of the *critical values of the F-distribution.*[1] Evidently, this is neither a convenient nor a very promising method, as the tables of the $F$-distribution are not very comprehensive and, moreover, the relation to be exploited is rather complicated, at least for many practitioners.

It is easier and more convenient to use one of the existing tables, which contain explicit confidence intervals for given sets of $n$ and $\sum_{i=1}^{n} X_i$. The most comprehensive ones are as follows:

- Hald, A. (1962). *Statistical Tables and Formulas.* John Wiley & Sons, London
- Owen, D.B. (1962). *Handbook of Statistical Tables.* Addison-Wesley, Reading, MA
- Müller, P.H., Neumann, P., and Storm, R. (1977). *Tafeln der mathematischen Statistik.* Carl Hansen, München
- Odeh, R.E., and Owen, D.B. (1983). *Attribute Sampling Plans, Tables of Tests and Confidence Limits for Proportions.* Marcel Dekker, New York

The above-listed tables are not comparable in size and meaning to those given in this handbook. Only [8] contains tables which come closer to those given here. The tables of this handbook refer to given measurement ranges, are computed withou any unnecessary side condition, and are complete in the sense that the desired numerical values can be taken directly from the tables without any further computations.

### 3.4.2 Approximate Interval Estimators

Although approximate measurement procedures will not be treated here in detail, they shall be briefly mentioned because of

[1] For example [30], pp. 64 - 87.

their popularity among practitioners and because of their merits in the case of large sample sizes. The cumbersome way for determining "exact" measurement procedures constitutes the main reason for developing approximate interval estimators. Various approximate procedures were proposed, all of them using approximations of the cumulative binomial probabilities $B_x(n,p)$ based on the Poisson distribution or the normal distribution. Each of these approximations uses the same simplified approach as taken for the traditional interval estimator. The measurement range is assumed to be the entire unit interval, the lower and upper bounds are based on $X_S$ or $\overline{X}$, and the bounds for the non-inclusion probabilities are set to be equal (i.e., $\alpha_{\ell,m} = \alpha_{u,m} = \frac{\alpha}{2}$).

For obtaining the approximate measurement procedures, the respective approximations for $B_m(n,p)$ are inserted into (3.36) and (3.37) and the resulting equations are solved for $\ell_m$ and $u_m$, respectively. The most common approximations are solutions of the following equations:

- Approximation 1:

$$1-\Phi\left(\frac{\frac{m}{n}-\ell_m}{\sqrt{\frac{\frac{m}{n}(1-\frac{m}{n})}{n}}}\right)=\frac{\alpha}{2} \text{ and } \Phi\left(\frac{\frac{m}{n}-u_m}{\sqrt{\frac{\frac{m}{n}(1-\frac{m}{n})}{n}}}\right)=\frac{\alpha}{2}$$

- Approximation 2:

$$1-\Phi\left(\frac{m-n\ell_m}{\sqrt{n\ell_m(1-\ell_m)}}\right)=\frac{\alpha}{2}$$

and

$$\Phi\left(\frac{m-nu_m}{\sqrt{nu_m(1-u_m)}}\right)=\frac{\alpha}{2}$$

- Approximation 3:

$$1-\Phi\left(\frac{m-n\ell_m+0.5}{\sqrt{n\ell_m(1-\ell_m)}}\right)=\frac{\alpha}{2}$$

and

$$\Phi\left(\frac{m - nu_m + 0.5}{\sqrt{nu_m(1-u_m)}}\right) = \frac{\alpha}{2}$$

- Approximation 4:

$$1 - \Phi\left(2\sqrt{m+1}\sqrt{1-\ell_m} - 2\sqrt{n-m}\sqrt{\ell_m}\right) = \frac{\alpha}{2}$$

and

$$\Phi\left(2\sqrt{m+1}\sqrt{1-u_m} - 2\sqrt{n-m}\sqrt{u_m}\right) = \frac{\alpha}{2}$$

- Approximation 5:

$$\sum_{k=m}^{\infty} \frac{(n\ell_m)^k}{k!} e^{-n\ell_m} = \frac{\alpha}{2} \text{ and } \sum_{k=0}^{m} \frac{(nu_m)^k}{k!} e^{-nu_m} = \frac{\alpha}{2}$$

The reader must be warned to use Approximation 1, as the probability of a measurement error may exceed the stated value of $1-\beta$ by far, with $\beta$ the confidence level. Approximation 5 exhibits the opposite property: In general, the reliability is much better than required and therefore the length of the resulting intervals is too large.

The three remaining approximations come rather close to values obtained for the *exact* traditional measurement (confidence) intervals.

Approximation 2 yields the following explicit solutions:

$$\ell_m = \frac{2m + \lambda^2_{\frac{\alpha}{2}} - \lambda_{1-\frac{\alpha}{2}}\sqrt{\lambda^2_{\frac{\alpha}{2}} + 4m\left(1-\frac{m}{n}\right)}}{2\left(n + \lambda^2_{\frac{\alpha}{2}}\right)} \quad \text{for} \quad m = 1, \ldots, n \tag{3.38}$$

$$u_m = \frac{2m + \lambda^2_{\frac{\alpha}{2}} + \lambda_{1-\frac{\alpha}{2}}\sqrt{\lambda^2_{1-\frac{\alpha}{2}} + 4m\left(1-\frac{m}{n}\right)}}{2\left(n + \lambda^2_{\frac{\alpha}{2}}\right)} \quad \text{for} \quad m = 0, \ldots, n-1 \tag{3.39}$$

where $\lambda_{\frac{\alpha}{2}}$ and $\lambda_{1-\frac{\alpha}{2}}$ are the $\frac{\alpha}{2}$-quantile and the $(1-\frac{\alpha}{2})$-quantile, respectively, of the standardized normal distribution function.

Approximate measurement procedures have great merits in the case of large sample sizes. In such a case, the deviations due to the approximation become negligible and, at the same time, the simplification compared with the determination of an exact solution is enormous. Hence, the procedures developed in the remainder refer to small and medium sample sizes and the theory proposed can be regarded as part of a *small sample theory*. For large sample sizes, the following features become more and more irrelevant:

- Basing the measurement procedures on the sample $(X_1, \ldots, X_n)$ and not on the sum $X_1 + \ldots + X_n$
- Taking into account the asymmetry of the binomial distribution and not neglecting it
- Developing the measurement procedures for a limited measurement range instead of the maximum measurement range
- Using approximations for the binomial distribution instead of taking the exact expressions

Taking into account these features results in considerable benefits only in the case of small sample sizes. However, the question of what is a small and what is a large sample size cannot be answered generally, but depends on the situation.

The situations in which application of statistical procedures is beneficial cover any field of human society and, therefore, may vary greatly. In the case of a Bernoulli experiment any situation is exclusively characterized by the actual value $p$ of the success probability.

Therefore, it makes sense to develop procedures for situations with small actual value $p$, with medium sized $p$ and with large large actual value $p$. If a procedure shall be applied in a special situation, then the range of possible values of $p$ is generally known to the user. The science of statistics is called upon to provide users with a procedure appropriately adapted to the given situation.

Unfortunately it is not possible in classical statistics to utilize any familiarity with the situation of interest. This fact may lead on the one hand to inferior procedures and on the other hand may weaken the users' confidence in statistical procedures.

# Part II

# Theory

# Chapter 4

# Measurement and Prediction Procedures

## 4.1 The Problem Revisited

There are essentially three characteristic features of any "exact" science.

- There is a "mathematical" model defining certain quantities and the relations between them.
- There are measurement procedures available for assessing the actual values of the model quantities, where a procedure should be called a "measurement procedure" only if the reliability and precision are known.
- Finally, there are procedures, by means of which the future development can be predicted again with stated accuracy.

One main problem in deriving a model and measurement and prediction procedures in any exact science is the aspect of uncertainty, as there are only very few phenomena with negligible uncertainty. Generally, neglecting uncertainty leads to inappropriate models and wrong measurements and predictions.

The principle step in modeling under uncertainty consists of modeling uncertainty itself. Uncertainty of an event is quantified by its probability, which reflects the likelihood of its occurrence. Having modeled uncertainty, measurement procedures are needed for assessing the unknown actual value $p$ of a probability $\breve{p}$, and, undoubtedly, this task constitutes the oldest and most important one in statistics.

Following general practice in any exact science, the measurement procedures to be developed in statistics should not be designed to be applicable in any possible situation with respect to the actual value of the parameter of interest. The theoretically maximum set of parameter values should be divided into appropriate ranges. It is evident that it is sufficient to admit only convex and bounded sets as possible ranges. In the special case given by the probability $\breve{p}$, it makes sense to require that a measurement range is a closed interval (i.e., $[\underline{p}, \overline{p}]$) with $0 \leq \underline{p} < \overline{p} \leq 1$. Of course, for given range, the procedure shall be designed in a way that it works well for any value $p \in [\underline{p}, \overline{p}]$. A range selected for $\breve{p}$ is called the *measurement space* in the remainder.

Once the measurement spaces have been selected, the task of statistics is to develop reliable and precise measurement procedures for each of the given measurement spaces. In the case of an actual application, the user must choose the appropriate measurement space from the set of available ones, in accordance with the known nature of $\breve{p}$. The appropriate measurement space yields immediately the desired measurement procedure.

The actual value $p$ of $\breve{p}$ is modeled by the parameter of a Bernoulli variable $X$, which determines the physical experiments to be performed. The experiments themselves are modeled by a random sample for $X$:

$$\vec{X} = (X_1, \ldots, X_n) \tag{4.1}$$

with realizations $\vec{x} \in \mathcal{X}^n = \{0,1\}^n \subset I\!R^n$. The image set $\mathcal{X}^n$ of $\vec{X}$ contains the possible outcomes of the random experiment. Besides measuring the actual value of a parameter of interest, the second basic task of statistics is to develop prediction procedures for the future outcome of random experiments. Therefore, the image space of $\mathcal{X}^n$ is called the *prediction space*.

Thus, we have arrived at a problem referring to the pair $(\breve{p}, \vec{X})$ with values in the Cartesian product $\left[\underline{p}, \overline{p}\right] \times \mathcal{X}^n$, where the two quantities $\breve{p}$ and $\vec{X}$ are connected to each other, to the measurement space, and to the prediction space by a family of probability distributions on $\mathcal{X}^n$:

$$\left\{\mathbf{P}_{\vec{X}}^{(p)}\right\}_{p\in[\underline{p},\overline{p}]}. \tag{4.2}$$

The Cartesian product of the measurement space and the prediction space $\left[\underline{p}, \overline{p}\right] \times \mathcal{X}^n$ is called the *measurement & prediction space.*[1] Dealing with measurement problems referring to the measurement & prediction space leads to the introduction of the corresponding measurable space:

$$\left([\underline{p}, \overline{p}] \times \{0,1\}^n, \mathcal{B}_{[\underline{p},\overline{p}]} \otimes \mathcal{P}_{\{0,1\}^n}\right), \tag{4.3}$$

where $\mathcal{B}_E$ denotes the $\sigma$-field of Borel sets over $E$ and $\mathcal{P}_E$ the power set over $E$.

As the actual value $p$ of $\breve{p}$ can be measured only indirectly by means of a random experiment and the random vector $\vec{X}$, it is clear that $p$ cannot be determined exactly, but only approximately, and, moreover, the measurement results are not strictly reproducible. Hence, the only way for measuring the value $p$ meaningfully is to determine an upper and a lower bound (i.e., an interval for $p$ within the parameter range $[\underline{p}, \overline{p}]$).

The problem of defining measurement precision is difficult because of the inherent uncertainty. However, in a majority of cases in natural sciences, the random character of the measurement procedures is not appropriately taken into account. If, for instance, in the one-dimensional case, the bounds are given in a nonstochastic way (without stating the reliability) by means of positive real numbers, which have to be added and subtracted to the measured value, then the possibility of measurement errors

[1]The difference between Neyman's "general space" and the here-defined measurement & prediction space is the fact that the latter is restricted to those pairs $(p, \vec{x})$, which may actually occur, whereas Neyman's general space is founded on a maximum parameter range.

is simply neglected. Proceeding like this may be misleading, as it does not take into account the possible event that the true value is smaller than the lower bound or larger than the upper bound, respectively.

Taking into account the random character of a measurement procedure means to model the region of inclusion for the unknown value $p$ in a way guaranteeing that no measurement error is committed with probability of at least $\beta$.

When applying a measurement procedure, two undesirable things may happen:

- A measurement error occurs.
- The measurement procedure results in an interval which is too wide.

The measurement error is controlled by the fixed confidence level $\beta$. Thus, for given $\beta$ and $n$, the problem remains to determine measurement procedures which are not only reliable but also precise.

## 4.2 Measurement & Prediction Space

The problem to be solved refers essentially to the pair $(\breve{p}, \vec{X})$. However, often a function of the sample is taken for deriving measurement and prediction procedures instead of the sample $\vec{X}$. As already mentioned, in the case of Bernoulli experiments, this is the sufficient statistics given by

$$X_S = \sum_{i=1}^{n} X_i \tag{4.4}$$

resulting in the pair $(\breve{p}, X_S)$. In this case, the pair $(\breve{p}, X_S)$ has to be regarded. Each of the two approaches leads to a different problem and a different solution. Therefore, it is worthwhile to investigate both of them.

### 4.2.1 Measurement & Prediction Space for $(\breve{p}, \vec{X})$

The pair $(\breve{p}, \vec{X})$ may adopt values in the measurement & prediction space

$$\left[\underline{p}, \overline{p}\right] \times \left\{0, 1\right\}^n. \tag{4.5}$$

Any procedure and any result concerning the pair $(\breve{p}, \vec{X})$ or one of its elements has a geometric representation within the measurement & prediction space. Therefore, statistics may be regarded as a geometric theory of measurement & prediction spaces or, equivalently, on the corresponding measurable space

$$\mathcal{B}_{[\underline{p},\overline{p}]} \otimes \mathcal{P}_{\{0,1\}^n}. \tag{4.6}$$

Any measurement & prediction space consists of two factors, which are given here by the measurement space $\left[\underline{p}, \overline{p}\right]$ with the $\sigma$-algebra of Borel sets and the prediction space $\{0, 1\}^n$ with the $\sigma$-algebra given by the power set.

The results of any statistical procedure consists of elements of one of the above given $\sigma$-fields. Therefore, an evaluation of a statistical procedure assumes necessarily that subsets of the different spaces can be evaluated by appropriate measure functions. Hence, the availability of suitable measure functions is a necessary condition for assessing the quality of statistical procedures.

The appropriate measure function with respect to the parameter $\breve{p}$ or on the measurable space $\mathcal{B}_{[\underline{p},\overline{p}]}$ has already been identified by the normed Lebesgue measure $\Lambda_{[\underline{p},\overline{p}]}$:

$$\Lambda_{[\underline{p},\overline{p}]}(C) = \int_C \lambda_{[\underline{p},\overline{p}]}(p)dp \quad \text{for any } C \in \mathcal{B}_{[\underline{p},\overline{p}]} \tag{4.7}$$

with density

$$\lambda_{[\underline{p},\overline{p}]}(p) = \begin{cases} \frac{1}{\overline{p}-\underline{p}} & \text{for } p \in [\underline{p}, \overline{p}] \\ 0 & \text{elsewhere.} \end{cases} \tag{4.8}$$

In view of the pair of quantities $\left(\breve{p}, \vec{X}\right)$, the density $\lambda_{[\underline{p},\overline{p}]}(p)$ may be looked upon as the marginal density of $\breve{p}$.

The connection between the two measurable spaces $\left([\underline{p},\overline{p}],\mathcal{B}_{[\underline{p},\overline{p}]}\right)$ and $\left(\{0,1\}^n,\mathcal{P}_{\{0,1\}^n}\right)$ is given by the family of probability distributions $\mathbf{P}_{\vec{X}}^{(p)}$.

Assuming that $\mathbf{P}_{\vec{X}}^{(p)}(A)$ as function of $p \in [\underline{p},\overline{p}]$ is measurable for any fixed $A \in \mathcal{P}_{\{0,1\}^n}$, it follows that $\mathbf{P}_{\vec{X}}^{(p)}$ is the conditional distribution of $\vec{X}$ under the condition $\breve{p} = p$. Another interpretation of $\mathbf{P}_{\vec{X}}^{(p)}$ would be to look upon it as the measure function for subsets of the hyperplane

$$\mathcal{X}_p^n = \left\{(p,\vec{x}) \,\middle|\, \vec{x} \in \mathcal{X}^n\right\}. \tag{4.9}$$

The appropriate joint distribution of $\left(\breve{p},\vec{X}\right)$ or measure function on the measurement & prediction space is obtained by the product measure

$$\Lambda_{[\underline{p},\overline{p}]} \otimes \mathbf{P}_{\vec{X}}^{(p)}. \tag{4.10}$$

For any $M \in \mathcal{B}_{[\underline{p},\overline{p}]} \otimes \mathcal{P}_{\{0,1\}^n}$, consider the projections

$$M_p = \left\{\vec{x}\,\middle|\,(p,\vec{x}) \in M\right\}. \tag{4.11}$$

Then,

$$\begin{aligned}
&\left(\Lambda_{[\underline{p},\overline{p}]} \otimes \mathbf{P}_{\vec{X}}^{(p)}\right)(M) \\
&= \int\limits_{[\underline{p},\overline{p}]} \left(\sum_{\vec{x}\in M_p} \mathbf{P}_{\vec{X}}^{(p)}(\vec{x})\right) \lambda_{[\underline{p},\overline{p}]}(p)dp \\
&= \int\limits_{[\underline{p},\overline{p}]} \left(\sum_{\vec{x}\in M_p} p^{\sum\limits_{i=1}^{n} x_i} (1-p)^{n-\sum\limits_{i=1}^{n} x_i}\right) \frac{dp}{\overline{p}-\underline{p}}.
\end{aligned} \tag{4.12}$$

The above derivation of the joint distribution of $\left(\breve{p},\vec{X}\right)$ starts with the marginal distribution of $\breve{p}$. Another representation of the joint distribution is obtained by starting from the marginal distribution of $\vec{X}$.

Let $Q_{\{0,1\}^n}(\vec{x})$ denote the marginal probability of the event $\{\vec{x}\} \in \mathcal{P}_{\{0,1\}^n}$; then,

$$\begin{aligned} Q_{\{0,1\}^n}(\vec{x}) &= \int\limits_{[\underline{p},\overline{p}]} \mathbf{P}^{(p)}_{\vec{X}}(\vec{x})\,\lambda_{[\underline{p},\overline{p}]}(p)dp \\ &= \frac{\int\limits_{[\underline{p},\overline{p}]} p^{\sum_{i=1}^{n} x_i}(1-p)^{n-\sum_{i=1}^{n} x_i}dp}{\overline{p}-\underline{p}} \\ &\quad \text{for any } \vec{x} \in \{0,1\}^n. \end{aligned} \tag{4.13}$$

The marginal measure $Q_{\{0,1\}^n}$ on $\left(\{0,1\}^n, \mathcal{B}_{\{0,1\}^n}\right)$ is given by

$$Q_{\{0,1\}^n}(A) = \sum_{\vec{x}\in A} Q_{\{0,1\}^n}(\vec{x}) \quad \text{for any } A \in \mathcal{P}_{\{0,1\}^n}. \tag{4.14}$$

Introducing the weight function $w(\vec{x})$ of $\vec{x} \in \{0,1\}^n$,

$$w(\vec{x}) = \int\limits_{[\underline{p},\overline{p}]} \mathbf{P}^{(p)}_{\vec{X}}(\vec{x})\,dp = \int\limits_{[\underline{p},\overline{p}]} p^{\sum_{i=1}^{n} x_i}(1-p)^{n-\sum_{i=1}^{n} x_i}dp \tag{4.15}$$

yields the following representation of the marginal distribution $Q_{\{0,1\}^n}$:

$$Q_{\{0,1\}^n}(\vec{x}) = \frac{w(\vec{x})}{\overline{p}-\underline{p}}. \tag{4.16}$$

By means of the marginal distribution (4.16) of $\vec{X}$, the conditional measure of $\breve{\theta}$ under the condition of $\{\vec{X} = \vec{x}\}$ is obtained:

For any fixed $\vec{x} \in \{0,1\}^n$, the conditional density of $\breve{p}$ is given by

$$\begin{aligned} g^{(\vec{x})}_{\breve{\theta}}(p) &= \frac{\mathbf{P}^{(p)}_{\vec{X}}(\vec{x})\,\lambda_{[\underline{p},\overline{p}]}(p)}{Q_{\{0,1\}^n}(\vec{x})} = \frac{\mathbf{P}^{(p)}_{\vec{X}}(\vec{x})}{w(\vec{x})} \\ &= \frac{p^{\sum_{i=1}^{n} x_i}(1-p)^{n-\sum_{i=1}^{n} x_i}}{w(\vec{x})} \\ &\quad \text{for any } p \in [\underline{p},\overline{p}]. \end{aligned} \tag{4.17}$$

The conditional density of $\breve{p}$ yields for any fixed $\vec{x}$ the conditional measure

$$G_{\breve{p}}^{(\vec{x})}(C) = \int_C g_{\breve{p}}^{(\vec{x})}(p)\, dp \quad \text{for any } C \in \mathcal{B}_{[\underline{p},\overline{p}]}. \tag{4.18}$$

The conditional measure function $G_{\breve{p}}^{(\vec{x})}$ may be looked upon as appropriate measure function on the hyperplanes through $\vec{x}$:

$$\left\{ (p, \vec{x}) \,\middle|\, p \in [\underline{p}, \overline{p}] \right\}. \tag{4.19}$$

Hence, the product measure $Q_{\{0,1\}^n} \otimes G_{\breve{p}}^{(\vec{x})}$ constitutes, besides $\Lambda_{[\underline{p},\overline{p}]} \otimes \mathbf{P}_{\vec{X}}^{(p)}$, another representation of the joint measure of $(\breve{p}, \vec{X})$.

For any $M \in \mathcal{B}_{[\underline{p},\overline{p}]} \otimes \mathcal{P}_{\{0,1\}^n}$, consider the projections

$$M_{\vec{x}} = \left\{ p \middle| \, (p, \vec{x}) \in M \right\}. \tag{4.20}$$

Then,

$$\begin{aligned} &\left( Q_{\{0,1\}^n} \otimes G_{\breve{p}}^{(\vec{x})} \right)(M) \\ &= \sum_{\vec{x} \in \{0,1\}^n} \left( \int_{M_{\vec{x}}} g_{\breve{p}}^{(\vec{x})}(p) dp \right) Q_{\{0,1\}^n}(\vec{x}) \\ &= \sum_{\vec{x} \in \{0,1\}^n} \frac{1}{\overline{p} - \underline{p}} \left( \int_{M_{\vec{x}}} p^{\sum_{i=1}^n x_i} (1-p)^{n - \sum_{i=1}^n x_i} dp \right) \end{aligned} \tag{4.21}$$

The situation with respect to the measure functions on

$$\left( \left[\underline{p}, \overline{p}\right] \times \{0,1\}^n, \mathcal{B}_{[\underline{p},\overline{p}]} \otimes \mathcal{P}_{\{0,1\}^n} \right) \tag{4.22}$$

is briefly summarized in Table 4.1.

**Table 4.1** *The measure functions defined on the measurement & prediction space.*

| quantity | $\breve{p}$ |
|---|---|
| measurable space | $\left([\underline{p},\overline{p}],\mathcal{B}_{[\underline{p},\overline{p}]}\right)$ |
| marginal measure | $\Lambda_\Theta$ |
| conditional measure | $G_{\breve{\theta}}^{(\vec{x})}$ |
| quantity | $\left(\breve{p},\vec{X}\right)$ |
| measurable space | $\left([\underline{p},\overline{p}]\times\{0,1\}^n,\mathcal{B}_{[\underline{p},\overline{p}]}\otimes\mathcal{P}_{\{0,1\}^n}\right)$ |
| joint measure | $\left(\Lambda_\Theta\otimes\mathbf{P}_{\vec{X}}^{(\theta)}\right)$ |

| quantity | $\vec{X}$ |
|---|---|
| measurable space | $\left(\{0,1\}^n,\mathcal{P}_{\{0,1\}^n}\right)$ |
| marginal measure | $Q_{\mathcal{X}^n}$ |
| conditional measure | $\mathbf{P}_{\vec{X}}^{(\theta)}$ |
| quantity | $\left(\vec{X},\breve{p}\right)$ |
| measurable space | $\left(\{0,1\}^n\times[\underline{p},\overline{p}],\mathcal{P}_{\{0,1\}^n}\otimes\mathcal{B}_{[\underline{p},\overline{p}]}\right)$ |
| joint measure | $\left(Q_{\mathcal{X}^n}\otimes G_{\breve{\theta}}^{(\vec{x})}\right)$ |

### 4.2.2 Measurement & Prediction Space for $(\breve{p},X_S)$

Using the sum of the sample elements $X_S=\sum_{i=1}^n X_i$ instead of the sample $\vec{X}=(X_1,\ldots,X_n)$ yields the pair of quantities $(\breve{p},X_S)$ and the measurement & prediction space

$$[\underline{p},\overline{p}]\times\{0,1,\ldots,n\}. \tag{4.23}$$

The situation with respect to the measures is obtained by replacing the measure function $\mathbf{P}_{\vec{X}}^{(p)}$ induced by $\vec{X}$ by the measure function $\mathbf{P}_{X_S}^{(p)}$ induced by $X_S$. The following measure functions are obtained, where $x_S=\sum_{i=1}^n x_i$ denotes a realization of $X_S$:

- Marginal measure on $\mathcal{B}_{[\underline{p},\overline{p}]}$:

$$\Lambda_{[\underline{p},\overline{p}]}(C) = \int_C \frac{dp}{\overline{p} - \underline{p}} \quad \text{for } C \in \mathcal{B}_{[\underline{p},\overline{p}]}. \tag{4.24}$$

- Conditional measure on $\mathcal{P}_{\{0,1,\ldots,n\}}$ under the condition $\breve{p} = p$:

$$\mathbf{P}^{(p)}_{X_S}(x_S) = \binom{n}{x_S} p^{x_S}(1-p)^{n-x_S} \tag{4.25}$$

for $x_S \in \{0, 1, \ldots, n\}$.

- Joint measure on $\mathcal{B}_{[\underline{p},\overline{p}]} \otimes \mathcal{P}_{\{0,1,\ldots,n\}}$ :

$$\begin{aligned} &\left(\Lambda_{[\underline{p},\overline{p}]} \otimes \mathbf{P}^{(p)}_{X_S}\right)(M) \\ &= \int_{[\underline{p},\overline{p}]} \left( \sum_{x_S \in M_p} \binom{n}{x_S} p^{\sum_{i=1}^{n} x_i} (1-p)^{n - \sum_{i=1}^{n} x_i} \right) \frac{dp}{\overline{p} - \underline{p}} \end{aligned} \tag{4.26}$$

for $M \in \mathcal{B}_{[\underline{p},\overline{p}]} \otimes \mathcal{P}_{\{0,1,\ldots,n\}}$ and $M_p = \left\{ x_S \middle| (p, x_S) \in M \right\}$.

- Marginal measure on $\mathcal{P}_{\{0,1,\ldots,n\}}$:

$$Q_{\{0,1,\ldots,n\}}(x_S) = \frac{\int_{[\underline{p},\overline{p}]} \binom{n}{x_S} p^{x_S}(1-p)^{n-x_S} dp}{\overline{p} - \underline{p}} \tag{4.27}$$

for $x_S \in \{0, 1, \ldots, n\}$

- Conditional measure on $\mathcal{B}_{[\underline{p},\overline{p}]}$ under condition $X_S = x_S$:

$$G^{(x_S)}_{\breve{p}}(C) = \int_C \frac{\binom{n}{x_S} p^{x_S}(1-p)^{n-x_S}}{w(x_S)} dp \quad \text{for } C \in \mathcal{B}_{[\underline{p},\overline{p}]} \tag{4.28}$$

with the weight function

$$w(x_S) = \int_{[\underline{p},\overline{p}]} \binom{n}{x_S} p^{x_S}(1-p)^{n-x_S} dp. \tag{4.29}$$

- Joint measure on $\mathcal{P}_{\{0,1,\ldots,n\}} \otimes \mathcal{B}_{[\underline{p},\overline{p}]}$:

$$\left(Q_{\{0,1,\ldots,n\}} \otimes G_{\breve{p}}^{(x_S)}\right)(M) = \sum_{x_S=0}^{n} \frac{1}{\overline{p}-\underline{p}} \left( \int_{M_{x_S}} \binom{n}{x_S} p^{x_S}(1-p)^{n-x_S}\,dp \right) \quad (4.30)$$

for $M \in \mathcal{B}_{[\underline{p},\overline{p}]} \otimes \mathcal{P}_{\{0,1,\ldots,n\}}$ and $M_{x_S} = \left\{p \middle| (p, x_S) \in M\right\}$.

The relation between the two measurement & prediction spaces based on $\vec{X}$ and $X_S$, respectively, is very simple. The random variable $X_S$ defines a partition of the prediction space $\{0,1\}^n$ in disjoint classes, where each vector $\vec{x} = (x_1, \ldots, x_n)$ of one class differs from the others only by permutation of the elements $x_i$. These classes are contracted to one point given by the sum of the elements $x_S$. Thus, the cardinality of the measurement & prediction space for $(\breve{p}, \vec{X})$ is reduced considerably.

## 4.3 The β-Measurement & Prediction Space

The problem dealt with may also be posed in the following way:

1. There is a quantity of interest $\breve{p}$ with unknown actual value $p \in [\underline{p}, \overline{p}]$.

2. There is a random variable of interest $\vec{X}$ with uncertain outcome $\vec{x} \in \{0,1\}^n$.

The aim in the first situation is to determine the actual value $p$ of the probability $\breve{p}$ by means of a set $C \subset [\underline{p}, \overline{p}]$. In the second situation the aim is to make a prediction about the uncertain outcome $\vec{x}$ of the random variable $\vec{X}$ by means of a set $A \subset \{0,1\}^n$. Thus, a measurement procedure and a prediction procedure are needed.

In [10] it is shown that when appropriately defined, the two problems are equivalent in the sense that any solution of the one problem implies a solution of the second problem. Moreover, a quality indicator is derived reflecting the degree of "fitness for use," i.e., quality, of a given procedure.

The whole problem is imbedded into the measurement & prediction space. Any solution of the prediction and measurement problem may be represented by a subset of the measurement & prediction space, called $\beta$-measurement & prediction space. Just as for the measurement & prediction space, the two possibilities to base the procedures on the sample $\vec{X}$ or the sufficient statistics $X_S$ are distinguished.

### 4.3.1 The $\beta$-Measurement & Prediction Space for $(\breve{p}, \vec{X})$

A prediction procedure based on the sample $\vec{X}$ consists of subsets of $\{0,1\}^n$, which are called prediction regions here.[2] Of course, not any subset is appropriate as a prediction region. In order to admit only meaningful subsets, a connected subset of $\{0,1\}^n$ is defined.

**Definition 4.1**

*Let*

$$I(m) = \left\{\vec{x} \,\middle|\, \sum_{i=1}^{n} x_i = m\right\} \tag{4.31}$$

*and* $I'(m)$ *be a subset of* $I(m)$. *A set* $D \subset \{0,1\}^n$ *is called* connected, *if there are two integers* $m_1$ *and* $m_2$ *with* $0 \leq m_1 \leq m_2 \leq n$ *and*

$$D = I'(m_1) \cup I(m_1+1) \cup \cdots \cup I(m_2-1) \cup I'(m_2). \tag{4.32}$$

•

The set $I(m)$ contains all those binary vectors $\vec{x}$, which have exactly $m$ "ones" and $n-m$ "zeros" as elements.

Both the measurement procedure and the prediction procedure are completely determined by the so-called $\beta$-measurement space, which is defined as follows.

[2] In Neyman's terminology, the prediction regions are called acceptance regions.

**Definition 4.2**
*For $n$ and $\beta$ fixed, a set $M^{(\beta,n)}_{\vec{X},\breve{p}} \in \mathcal{B}_{[\underline{p},\overline{p}]} \otimes \mathcal{P}_{\{0,1\}^n}$ is called a* $\beta$-measurement and prediction space *if its projections*

$$A^{(\beta,n)}_{\vec{X}}(p) = \left\{\vec{x}\,\middle|\,(p,\vec{x}) \in M^{(\beta,n)}_{\vec{X},\breve{p}}\right\} \tag{4.33}$$

$$C^{(\beta,n)}_{\breve{p}}(\vec{x}) = \left\{p\,\middle|\,(p,\vec{x}) \in M^{(\beta,n)}_{\vec{X},\breve{p}}\right\} \tag{4.34}$$

*have the following properties:*

1. *$A^{(\beta,n)}_{\vec{X}}(p)$ is for any $p \in [\underline{p},\overline{p}]$ a connected subset of $\{0,1\}^n$ with*
$$\mathbf{P}^{(p)}_{\vec{X}}\left(\vec{X} \in A^{(\beta,n)}_{\vec{X}}(p)\right) \geq \beta \tag{4.35}$$

2. *$C^{(\beta,n)}_{\breve{p}}(\vec{x})$ is for any $\vec{x} \in \{0,1\}^n$ an interval within the measurement space $[\underline{p},\overline{p}]$.* •

**Remark 4.1** *The case of a degenerate interval $C^{(\beta,n)}_{\breve{p}}(\vec{x})$ consisting of only one element is admitted here.* •

For any given $\beta$-measurement & prediction space, the sets $A^{(\beta,n)}_{\vec{X}}(p)$ are called *$\beta$-prediction regions* for $\vec{X}$. Because of (4.35), a $\beta$-prediction region $A^{(\beta,n)}_{\vec{X}}(p)$ is an event with respect to the random variable $\vec{X}$ which occurs with a probability larger than or equal to the confidence level $\beta$. Therefore, we call

$$A^{(\beta,n)}_{\vec{X}} = \left\{A^{(\beta,n)}_{\vec{X}}(p) \mid p \in [\underline{p},\overline{p}]\right\} \tag{4.36}$$

a *$\beta$-prediction procedure* for $\vec{X}$.

For any given $\beta$-measurement & prediction space, the closed intervals

$$\overline{C}^{(\beta,n)}_{\breve{p}}(\vec{x}) = \left[\ell^{(\beta,n)}(\vec{x}), u^{(\beta,n)}(\vec{x})\right]$$

are called *$\beta$-measurement intervals for $\breve{p}$*. From (4.35), the following relation is obtained for any $p \in [\underline{p},\overline{p}]$:

$$\mathbf{P}^{(p)}_{\vec{X}}\left(\{\vec{x}|\ell^{(\beta,n)}(\vec{x}) \leq p \leq u^{(\beta,n)}(\vec{x})\}\right) \geq \beta. \tag{4.37}$$

Therefore, a prediction or a measurement error occurs at most with probability $1-\beta$ meeting the reliability requirement given by the confidence level $\beta$. Therefore,

$$\overline{C}_{\breve{p}}^{(\beta,n)} = \left\{ [\ell^{(\beta,n)}(\vec{x}), u^{(\beta,n)}(\vec{x})] \mid \vec{x} \in \{0,1\}^n \right\} \tag{4.38}$$

is called a *β-measurement procedure* for $p \in [\underline{p}, \overline{p}]$. We may say that the procedure $\overline{C}_{\breve{p}}^{(\beta,n)}$ covers with at least probability $\beta$ the actual value $p \in [\underline{p}, \overline{p}]$.

### 4.3.2 The β-Measurement & Prediction Space for $(\breve{p}, X_S)$

Analogously to the connected subsets of $\{0,1\}^n$, a connected subset of $\{0,1,\ldots,n\}$ is introduced:

**Definition 4.3**
*$D \subset \{0,1,\ldots,n\}$ is called connected subset if there are two integers $m_1$ and $m_2$ with $0 \leq m_1 \leq m_2 \leq n$ and*

$$D = \{m_1, m_1+1, \ldots, m_2\} \tag{4.39}$$

•

Replacing $\vec{X}$ by $X_S$ and $\mathcal{P}_{\{0,1\}^n}$ by $\mathcal{P}_{\{0,1,\ldots,n\}}$ yields the corresponding quantities for $(\breve{p}, X_S)$.

**Definition 4.4**
*For $n$ and $\beta$ fixed, a set $M_{X_S,\breve{p}}^{(\beta,n)} \in \mathcal{B}_{[\underline{p},\overline{p}]} \otimes \mathcal{P}_{\{0,1,\ldots,n\}}$ is called a* β-measurement & prediction space for $X_S$ *if its projections*

$$A_{X_S}^{(\beta,n)}(p) = \left\{ x_S \,\middle|\, (p, x_S) \in M_{X_S,\breve{p}}^{(\beta,n)} \right\} \tag{4.40}$$

$$C_{\breve{p}}^{(\beta,n)}(x_S) = \left\{ p \,\middle|\, (p, x_S) \in M_{X_S,\breve{p}}^{(\beta,n)} \right\} \tag{4.41}$$

*have the following properties:*

1. *$A_{X_S}^{(\beta,n)}(p)$ is for any $p \in [\underline{p}, \overline{p}]$, a connected subset of $\{0,1,\ldots,n\}$ with*

$$\mathbf{P}_{X_S}^{(p)}\left(A_{X_S}^{(\beta,n)}(p)\right) \geq \beta \tag{4.42}$$

2. *$C_{\breve{p}}^{(\beta,n)}(x_S)$ is for any $x_S \in \{0,1,\ldots,n\}$, an interval within the measurement space $[\underline{p},\overline{p}]$.* •

Just as in the case of a $\beta$-measurement & prediction space based on the sample $\vec{X}$, the projections $A_{X_S}^{(\beta,n)}(p)$ are called $\beta$-prediction regions for $X_S$ and the family of $\beta$-prediction regions

$$\left\{A_{X_S}^{(\beta,n)}(p)\right\}_{p\in[\underline{p},\overline{p}]} \tag{4.43}$$

defines a $\beta$-prediction procedure for $X_S$.

The closed projections $\overline{C}_{\breve{p}}^{(\beta,n)}(x_S)$ are $\beta$-measurement intervals, and the family of $\beta$-measurement intervals

$$\left\{\overline{C}_{\breve{p}}^{(\beta,n)}(x_S)\right\}_{x_S\in\{0,1,\ldots,n\}} \tag{4.44}$$

defines a $\beta$-measurement procedure for the parameter $\breve{p}$.

Clearly, the reliability of the above-defined prediction procedure and measurement procedure is given by the confidence level $\beta$.

### 4.3.3 The Relation Between $M_{\vec{X},\breve{p}}^{(\beta,n)}$ and $M_{X_S,\breve{p}}^{(\beta,n)}$

Of special interest is the relation between the $\beta$-measurement & prediction spaces based on the sample $\vec{X}$ and on the sum of its elements $X_S$, respectively.

As a first step, we start with a $\beta$-measurement & prediction space based on $X_S$ and determine the corresponding $\beta$-measurement & prediction space based on $\vec{X}$.

Let $M_{X_S,\breve{p}}^{(\beta,n)}$ be a $\beta$-measurement & prediction space based on $X_S$ with $\beta$-measurement intervals given by $C_{\breve{p}}^{(\beta,n)}(x_S)$ for $x_S \in \{0,1,\ldots,n\}$. Next, consider the $\beta$-measurement & prediction space $M_{\vec{X},\breve{p}}^{(\beta,n)}$ based on the sample $\vec{X}$ and defined by the intervals $C_{\breve{p}}^{(\beta,n)}(x_S)$ in the following way:

$$\begin{aligned} C_{\breve{p}}^{(\beta,n)}(\vec{x}) &= C_{\breve{p}}^{(\beta,n)}\big((x_1,\ldots,x_n)\big) = C_{\breve{p}}^{(\beta,n)}(x_S) \\ &\quad \text{for } x_1+x_2+\ldots+x_n = x_S \end{aligned} \tag{4.45}$$

From (4.45), the prediction regions of $M_{\vec{X},\breve{p}}^{(\beta,n)}$ are obtained for $p \in [\underline{p}, \overline{p}]$:

$$\begin{aligned} A_{\vec{X}}^{(\beta,n)}(p) &= \left\{ (x_1, \ldots, x_n) \,\middle|\, \sum_{i=1}^{n} x_i \in A_{X_s,\breve{p}}(p) \right\} \\ &= A_{X_S}^{(\beta,n)}(p) \end{aligned} \tag{4.46}$$

Clearly, the two $\beta$-measurement & prediction spaces $M_{X_S,\breve{p}}^{(\beta,n)}$ and $M_{\vec{X},\breve{p}}^{(\beta,n)}$ are equivalent in the sense that they yield the same $\beta$-measurement intervals for any realization of the sample and the same $\beta$-prediction region for any $p \in [\underline{p}, \overline{p}]$.

Next, let $M_{\vec{X},\breve{p}}^{(\beta,n)}$ be a $\beta$-measurement & prediction space based on the sample $\vec{X}$. Consider the partition of the prediction space $\{0,1\}^n$ into the sets $I(x_S) = \{\vec{x} \mid \sum_{i=1}^{n} x_i = x_S\}$, with $x_S \in \{0, 1, \ldots, n\}$. Define

$$\tilde{C}_{\breve{p}}^{(\beta,n)}(x_S) = \bigcup_{\vec{x} \in I(x_S)} C_{\breve{p}}^{(\beta,n)}(\vec{x}) \tag{4.47}$$

and for $p \in [\underline{p}, \overline{p}]$,

$$\tilde{A}_{X_s}^{(\beta,n)}(p) = \left\{ x_S \,\middle|\, x_S = \sum_{i=1}^{n} x_i, \vec{x} \in A_{\vec{X}}^{(\beta,n)}(p). \right\} \tag{4.48}$$

From (4.47), it is clear that generally the $\beta$-measurement intervals based on the sum of sample elements $X_S$ are larger than those based on the sample $\vec{X}$, because each $\beta$-measurement interval with respect to $x_S$ is the union of several $\beta$-measuement intervals with respect to $\vec{x}$. Further, (4.48) implies that the $\beta$-prediction regions $\tilde{A}_{\vec{X}}^{(\beta,n)}(p)$ given by

$$\left\{ \vec{x} \,\middle|\, \sum_{i=1}^{n} x_i = x_S \text{ and } x_S \in A_{X_S}^{(\beta,n)}(p) \right\} \tag{4.49}$$

are generally larger than a $\beta$-prediction region $A_{\vec{X}}^{(\beta,n)}(p)$ based directly on $vecX$, because $\tilde{A}_{\vec{X}}^{(\beta,n)}(p)$ has an unnecessarily large probability with respect to $\mathrm{P}_{\vec{X}}^{(p)}$, exceeding, by far, the required reliability $\beta$.

Therefore, it follows that in the Bernoulli case using the sum of sample elements rather than the sample itself for developing measurement and prediction procedures leads generally to a decrease in precision. However, this effect is considerable only for rather small sample sizes.

## 4.4 Quality of a Measurement Procedure

The main problem when developing a statistical procedure is to evaluate its quality with respect to its aim. As the reliability of the procedures is fixed by the given confidence level $\beta$ and the expense by the sample size $n$, only precision has to be achieved. Therefore, the only quality characteristic is the size of the $\beta$-measurement intervals or the size of the $\beta$-prediction regions, respectively. Unfortunately, the size of the measurement intervals depends on the unknown realization of the sample, and the size of the prediction regions depends on the actual value $p$ of the probability $\breve{p}$ of interest. However, any prediction or measurement procedure is completely determined by the corresponding $\beta$-measurement & prediction space and vice versa. Therefore, the quality evaluation may be based on the $\beta$-measurement & prediction space (i.e., on a geometric quantity), and not on the procedure itself.

### 4.4.1 Quality of the $\beta$-Measurement & Prediction Space for $(\vec{X}, \breve{p})$

Any $\beta$-measurement & prediction space $M_{\vec{X},\breve{p}}^{(\beta,n)}$ defines a $\beta$-prediction procedure for $\vec{X}$ and, at the same time, a $\beta$-measurement procedure for $\breve{p}$. The $\beta$-measurement & prediction space is a measurable subset of $[\underline{p}, \overline{p}] \times \{0,1\}^n$. Thus, for evaluating the quality of the corresponding prediction procedure or measurement procedure with respect to precision, the volume of a $\beta$-measurement & prediction space has to be assessed appropriately.

The quality of a $\beta$-measurement procedure depends essentially on the geometrical length of each of the intervals obtained as projections for fixed $\vec{x}$. The normed length is given by

$$\Lambda_{[\underline{p},\overline{p}]}\left(C_{\breve{p}}^{(\beta,n)}(\vec{x})\right) = \frac{u_{\breve{p}}^{(\beta,n)}(\vec{x}) - \ell_{\breve{p}}^{(\beta,n)}(\vec{x})}{\overline{p} - \underline{p}}. \tag{4.50}$$

The normed length (4.50), which formally is the marginal measure of the intervals $C_{\breve{p}}^{(\beta,n)}(\vec{x}) \in \mathcal{B}_{[\underline{p},\overline{p}]}$ with respect to $\breve{p}$, defines a random variable on the prediction space. Which of the intervals is actually realized is determined by the realization $\vec{x}$ of $\vec{X}$ with marginal probability $Q_{\{0,1\}^n}(\vec{x})$.

Therefore, the overall quality of a $\beta$-measurement procedure is given by the expectation of the random variable *normed length of the intervals* $C_{\breve{p}}^{(\beta,n)}(\vec{x})$. The expectation has to be taken with respect to the marginal measure $Q_{\{0,1\}^n}$ of $\vec{X}$. This expectation, called the weighted volume of the $\beta$-measurement & prediction space, is denoted by $V_w\left(M_{\vec{X},\breve{p}}^{(\beta,n)}\right)$ and given by

$$\begin{aligned} V_w\left(M_{\vec{X},\breve{p}}^{(\beta,n)}\right) &= \sum_{\vec{x}\in\{0,1\}^n} \left( \int\limits_{C_{\breve{p}}^{(\beta,n)}(\vec{x})} \lambda_{[\underline{p},\overline{p}]}(p)dp \right) Q_{\{0,1\}^n}(\vec{x}) \\ &= \sum_{\vec{x}\in\{0,1\}^n} \frac{u_{\breve{p}}^{(\beta,n)}(\vec{x}) - \ell_{\breve{p}}^{(\beta,n)}(\vec{x})}{\overline{p} - \underline{p}} \, \frac{\int_{\underline{p}}^{\overline{p}} p^{x_S}(1-p)^{n-x_S}dp}{\overline{p} - \underline{p}}, \end{aligned} \tag{4.51}$$

where $x_S = \sum_{i=1}^{n} x_i$.

Obviously, (4.51) may also be written

$$
\begin{aligned}
&V_w\Big(M^{(\beta,n)}_{\vec{X},\breve{p}}\Big)\\
&= \int\limits_{[\underline{p},\overline{p}]} \left( \sum_{\vec{x}\in A^{(\beta,n)}_{\vec{X}}(p)} Q_{\{0,1\}^n}(\vec{x}) \right) \lambda_{[\underline{p},\overline{p}]}(p)dp\\
&= \int\limits_{[\underline{p},\overline{p}]} \left( \sum_{\vec{x}\in A^{(\beta,n)}_{\vec{X}}(p)} \frac{\int\limits_{[\overline{p},\underline{p}]} \mathbf{P}^{(p)}_{\vec{X}}(\vec{x})dp}{\overline{p}-\underline{p}} \right) \frac{dp}{\overline{p}-\underline{p}}.
\end{aligned}
\tag{4.52}
$$

The relation (4.52) yields another noteworthy interpretation of $V_w$ as a quality indicator of the $\beta$-prediction procedure $A^{(\beta,n)}_{\vec{X}}$. The size of each single $\beta$-prediction region $A^{(\beta,n)}_{\vec{X}}(p)$ is obtained by means of the marginal measure $Q_{\{0,1\}}$:

$$
Q_{\{0,1\}^n}\Big(A^{(\beta,n)}_{\vec{X}}(p)\Big) = \sum_{\vec{x}\in A^{(\beta,n)}_{\vec{X}}(p)} Q_{\{0,1\}^n}(\vec{x})\,. \tag{4.53}
$$

The weighted volume of a $\beta$-measurement & prediction space equals the mean size of the corresponding $\beta$-prediction regions, where the mean is taken with respect to $\Lambda_{[\underline{p},\overline{p}]}$.

There is another quantity referring to the $\beta$-measurement & prediction space, which is of interest in the discrete case, namely the size of $M^{(\beta,n)}_{\vec{X},\breve{p}}$ measured by the joint measure on $[\underline{p},\overline{p}] \times \{0,1\}^n$. This size of the $\beta$-measurement & prediction space is denoted by $V_{prob}\Big(M^{(\beta,n)}_{\vec{X},\breve{p}}\Big)$ and called the *probability volume* of the $\beta$-measurement & prediction space. The probability volume is given by

$$
\begin{aligned}
&V_{prob}\Big(M^{(\beta,n)}_{\vec{X},\breve{p}}\Big)\\
&= \sum_{\vec{x}\in\{0,1\}^n} \left( \int\limits_{C^{(\beta,n)}_{\breve{p}}(\vec{x})} g^{(\vec{x})}_{\breve{p}}(p)dp \right) Q_{\{0,1\}^n}(\vec{x})\\
&= \frac{1}{\overline{p}-\underline{p}} \sum_{\vec{x}\in\{0,1\}^n} \left( \int\limits_{\ell^{(\beta,n)}_{\breve{p}}(\vec{x})}^{u^{(\beta,n)}_{\breve{p}}(\vec{x})} p^{xs}(1-p)^{n-xs}dp \right)
\end{aligned}
\tag{4.54}
$$

$$= \frac{1}{\overline{p} - \underline{p}} \int_{[\underline{p},\overline{p}]} \left( \sum_{\vec{x} \in A_{\vec{X}}^{(\beta,n)}(p)} \mathbf{P}_{\vec{X}}^{(p)}(\vec{x}) \right) dp \tag{4.55}$$

In the continuous case, the $\beta$-measurement & prediction space is derived so that its probability volume is equal to $\beta$. In the discrete case, however, this is generally not possible; that is,the probability volume is generally larger than $\beta$. The more it exceeds the required reliability $\beta$, the larger gets the weighted volume and hence the precision gets worse.

The difference $V_{prob}(M_{\vec{X},\breve{p}}^{(\beta,n)}) - \beta$ (i.e., the difference between the required and the actually achieved reliability) is called the *reliability excess.* In the continuous case, it is always possible to select a $\beta$-measurement & prediction space with zero reliability excess. In the discrete case, the reliability excess is generally positive, and one of the problems is to develop an algorithm for determining a $\beta$-measurement & prediction space with a small *reliability excess.*

### 4.4.2 Quality of the $\beta$-Measurement & Prediction Space for $(X_S, \breve{p})$

For the case in which the $\beta$-measurement & prediction space is based on the sum of sample elements $X_S$, the following expression for the weighted volume is obtained:

$$V_w\left(M_{X_S,\breve{p}}^{(\beta,n)}\right)$$

$$= \sum_{x_S=0}^{n} \frac{u_{\breve{p}}^{(\beta,n)}(x_S) - \ell_{\breve{p}}^{(\beta,n)}(x_S)}{\overline{p} - \underline{p}} \frac{\int_{[\overline{p},\underline{p}]} \mathbf{P}_{X_S}^{(p)}(x_S)dp}{\overline{p} - \underline{p}} \tag{4.56}$$

$$= \sum_{x_S=0}^{n} \frac{u_{\breve{p}}^{(\beta,n)}(x_S) - \ell_{\breve{p}}^{(\beta,n)}(x_S)}{\overline{p} - \underline{p}} \frac{\int_{\underline{p}}^{\overline{p}} \binom{n}{x_S} p^{x_S}(1-p)^{n-x_S} dp}{\overline{p} - \underline{p}}.$$

If the measurement space is maximum (i.e., the unit interval), then (4.56) simplifies to the average length of the $\beta$-

measurement intervals:

$$V_w\left(M_{X_S,\breve{p}}^{(\beta,n)}\right) = \frac{1}{n+1}\sum_{x_S=0}^{n}\left[u_{\breve{p}}^{(\beta,n)}(x_S) - \ell_{\breve{p}}^{(\beta,n)}(x_S)\right]. \qquad (4.57)$$

The probability volume of $M_{X_S,\breve{p}}^{(\beta,n)}$ is given by

$$V_{prob}\left(M_{X_S,\breve{p}}^{(\beta,n)}\right) = \sum_{x_S=0}^{n}\int_{\ell_{\breve{p}}^{(\beta,n)}(x_S)}^{u_{\breve{p}}^{(\beta,n)}(x_S)}\frac{\binom{n}{x_S}p^{x_S}(1-p)^{n-x_S}}{\overline{p}-\underline{p}}dp. \qquad (4.58)$$

Clearly, for any $\beta$-measurement & prediction space $M_{X_S,\breve{p}}^{(\beta,n)}$, there is, in general, a $\beta$-measurement & prediction space $M_{\vec{X},\breve{p}}^{(\beta,n)}$ with a smaller reliability excess and smaller weighted volume (i.e., better quality). However, it is to be expected that the improvement is considerable only in the case of a small sample size.

## 4.5 Neyman $\beta$-Measurement Procedure

Any $\beta$-measurement procedure determines uniquely a $\beta$-measurement space $M_{\vec{X},\breve{p}}^{(\beta,n)}$ or $M_{X_S,\breve{p}}^{(\beta,n)}$, whose quality with respect to precision is given by its weighted volume $V_w$. The smaller the weighted volume of the $\beta$-measurement & prediction space, the more precise is the measurement procedure.

**Definition 4.5**
*A $\beta$-measurement procedure given by the defining intervals*

$$\left\{{}^{*}C_{\breve{p}}^{(\beta,n)}(\vec{x})\right\}_{\vec{x}\in\{0,1\}^n}$$

*is called* Neyman $\beta$-measurement procedure *if the weighted volume of the corresponding $\beta$-measurement & prediction space is minimum within the set of all $\beta$-measurement & prediction spaces.* •

Neyman $\beta$-measurement procedures are optimal with respect to precision for fixed reliability given by the confidence level $\beta$ and fixed expenditure given by the sample size $n$.

Taking into account only those measurement procedures based on the sum of the sample elements $X_S$, we have the following definition.

**Definition 4.6**
*A $\beta$-measurement procedure based on $X_S$ and given by the defining intervals*

$$\left\{ {}^{*}C_{\breve{p}}^{(\beta,n)}(x_S) \right\}_{x_S \in \{0,1,\dots,n\}}$$

*is called* Neyman $X_S$-$\beta$-measurement procedure *if the weighted volume of the corresponding $\beta$-measurement & prediction space is minimum within the set of all $\beta$-measurement & prediction spaces based on $X_S$.* •

Of course, analogously as for $\beta$-measurement procedures, one can define Neyman $\beta$-prediction procedures as the most precise ones for fixed reliability and expenditure.

One important difference to the usual way of constructing statistical measurement procedures is the fact that the above definition is independent of any so-called point estimator and in particular independent of the relative frequency. Another important difference is the defined duality with the corresponding prediction procedure. The prediction procedure assumes complete knowledge about the actual values of the distribution parameter. Thus, determination of the appropriate prediction regions is in the first place a numerical problem. As soon as the prediction regions $A_X^{(\beta)}(p)$ are computed, the measurement intervals are obtained using the relation

$$C_{\breve{p}}^{(\beta)}(\vec{x}) = \left\{ p \,\middle|\, \vec{x} \in A_X^{(\beta)}(p) \right\} \tag{4.59}$$

This has already been noticed by Jerzy Neyman [27], who proposed to calculate $\beta$-confidence intervals by determining so-called $\beta$-acceptance regions $A_{\vec{X}}^{(\beta,n)}(p)$.

## 4.6 Determination of Neyman Measurement Procedures

For determining a Neyman $\beta$-measurement or $\beta$-prediction procedure, the following optimization problem has to be solved.

**Problem 4.1** *For each $p \in [\underline{p}, \overline{p}]$, determine a $\beta$-prediction region such that for the resulting $\beta$-measurement & prediction space, the following is true:*

$$V_w\left(M_{\vec{X},\breve{p}}^{(\beta,n)}\right) \quad = \quad \min \tag{4.60}$$

$$V_{prob}\left(M_{\vec{X},\breve{p}}^{(\beta,n)}\right) \quad \geq \quad \beta \tag{4.61}$$

*when based on the sample $\vec{X}$, and*

$$V_w\left(M_{X_S,\breve{p}}^{(\beta,n)}\right) \quad = \quad \min \tag{4.62}$$

$$V_{prob}\left(M_{X_S,\breve{p}}^{(\beta,n)}\right) \quad \geq \quad \beta \tag{4.63}$$

*when based on the sum of sample elements $X_S$.* •

According to (4.52), the first condition is equivalent to selecting the $\beta$-prediction regions with minimum size:

$$Q_{\{0,1\}^n}\left(A_{\vec{X}}^{(\beta,n)}(p)\right) = \min \tag{4.64}$$

and according to (4.55), the latter condition is equivalent to selecting the $\beta$-prediction regions so as to fulfill the reliability requirement with respect to $\mathbf{P}_{\vec{X}}^{(p)}$. Therefore, the problem may be reformulated in the following way:

**Problem 4.2** *For each $p \in [\underline{p}, \overline{p}]$, determine a $\beta$-prediction region such that the following is true:*

$$Q_{\{0,1\}^n}\left(A_{\vec{X}}^{(\beta,n)}(p)\right) = \min \tag{4.65}$$

$$\mathbf{P}_{\vec{X}}^{(p)}\left(A_{\vec{X}}^{(\beta,n)}(p)\right) \geq \beta \tag{4.66}$$

*when based on the sample $\vec{X}$, and*

$$Q_{\{0,\dots,n\}}\left(A_{X_S}^{(\beta,n)}(p)\right) = \min \tag{4.67}$$

$$\mathbf{P}_{X_S}^{(p)}\left(A_{X_S}^{(\beta,n)}(p)\right) \geq \beta \tag{4.68}$$

*when based on the sum of sample elements $X_S$.* •

The problem may also be formulated in the following more illustrative way. One wishes to buy a quantity of at least $\beta$ of the product $A_{\vec{X}}^{(\beta,n)}$ measured in units of $\mathbf{P}_{\vec{X}}^{(p)}$ at minimum cost measured in units of $Q_{\{0,1\}^n}$. Of course, anybody knows intuitively the solution: One selects the available pieces $A_{\vec{X}}^{(\beta,n)}(\vec{x})$ of the product $A_{\vec{X}}^{(\beta,n)}$ according to their price/quantity relation.

This illustrates a special case of a general selection problem, which is solved by selecting the pairs $(\vec{x}, p)$ or $(x_S, p)$, respectively, to form the $\beta$-measurement & prediction space, according to the ratio of the two measure functions involved.

In the continuous case, the measure function is represented by its density function, and in the discrete case, by its probability function. In statistics, these functions are traditionally called likelihood functions and the quotient of two likelihood functions is called the likelihood ratio. Hence, the above-defined selection problem is solved by selecting the elements of the $\beta$-prediction regions according to the likelihood ratio of the two measure functions involved, starting with the element with the largest value of its likelihood ratio. This procedure is stopped as soon as the reliability requirement with respect to $\mathbf{P}^{(p)}$ is met.

From (4.12) and (4.52), it is easy to see that in the continuous case, the optimal prediction regions may be obtained by selecting the elements of the $\beta$-measurement and prediction space for each value of the parameter of interest separately according to the order given by the likelihood ratio of the two measures $Q$ and $\mathbf{P}^{(\theta)}$.

Proceeding analogously in our discrete case, however, does not generally yield a Neyman $\beta$-prediction procedure, because of the reliability excess. If the selection according to the likelihood ratio yields a $\beta$-measurement & prediction space with a large reliability excess, then a violation of the likelihood ratio order may result in a smaller reliability excess and, thus, in a smaller weighted volume. Again, everybody knows this effect from daily experience. If the available "cheap" units are too large, purchasing them would lead to a larger quantity than the necessary one given by $\beta$. In such a case, it is advisable to buy smaller pieces, even if they are not the cheapest ones.

For determining a Neyman procedure, one should, in fact, compare the likelihood ratios for all elements of the measurement & prediction space simultaneously. As such a complete comparison is too time consuming, even for medium sample sizes, the values tabulated here are obtained by an algorithm similar to that in the continuous case; that is, the selection of the elements for given value of $p$ according to the likelihood ratio

$$\frac{\mathbf{P}_{\vec{X}}^{(p)}\left(\{\vec{x}\}\right)}{Q_{\{0,1\}^n}\left(\{\vec{x}\}\right)} = \frac{p^{x_S}(1-p)^{n-x_S}}{\int\limits_{\underline{p}}^{\overline{p}} p^{x_S}(1-p)^{n-x_S}\,dp} \tag{4.69}$$

where $x_S = \sum\limits_{i=1}^{n} x_i$.

The thus obtained prediction regions are modified by additional rules in a second step in order to have the reliability excess adopt its minimum or at least come close to its minimum. It is to be expected that in a majority of cases, the figures in the tables represent Neyman procedures. If not, then the difference between the weighted volumes of the tabulated procedure and the corresponding Neyman procedure should be negligible.

## 4.7 Limiting Quality of Neyman Procedures

The quality of a measurement procedure is given by the weighted volume of the corresponding $\beta$-measurement & prediction space. The smaller the weighted volume, the better is the precision and, therefore, the quality of the procedure.

Let

$$\left\{ {}^*M^{(\beta,n)}_{\vec{X},\breve{p}} \right\}_{n=1,2,\ldots} \tag{4.70}$$

be the sequence of $\beta$-measurement & prediction spaces belonging to Neyman $\beta$-measurement procedures, and

$$\left\{ {}^*M^{(\beta,n)}_{X_S,\breve{p}} \right\}_{n=1,2,\ldots} \tag{4.71}$$

be the sequence of $\beta$-measurement & prediction spaces belonging to Neyman $X_S$-$\beta$-measurement procedures, and, finally, let

$$\left( M^{(\beta,n)}_{X_S,\breve{p}} \right)_{n=1,2,\ldots} \tag{4.72}$$

be the sequence of $\beta$-measurement & prediction spaces of arbitrary $\beta$-measurement procedure based on $X_S$.

Then,

$$0 \le V_w\left( {}^*M^{(\beta,n)}_{\vec{X},\breve{p}} \right) \le V_w\left( {}^*M^{(\beta,n)}_{X_S,\breve{p}} \right) \le V_w\left( M^{(\beta,n)}_{X_S,\breve{p}} \right) \tag{4.73}$$

is true for any $n \in I\!N$.

Next, consider the $\beta$-measurement & prediction spaces $\left\{ M^{(\beta,n)}_{X_S,\breve{p}} \right\}_{n=1,2,\ldots}$ given by the following $\beta$-measurement intervals:

$$\ell^{(\beta,n)}_{\breve{p}}\left(x_S\right) = \max\left\{ \underline{p}\,; \frac{1}{n}x_S - \sqrt{\frac{k_\beta}{n}} \right\} \tag{4.74}$$

$$u^{(\beta,n)}_{\breve{p}}\left(x_S\right) = \min\left\{ \frac{1}{n}x_S + \sqrt{\frac{k_\beta}{n}}\,; \overline{p} \right\}, \tag{4.75}$$

where $k_\beta \in I\!R^+$ is a constant depending on $\beta$ and assuring that the requirement with respect to the reliability is met by the measurement procedure. The existence of $k_\beta$ is easily proved by means of the Tchebychev inequality. Of course, the bounds (4.74) and (4.75) are essentially the same as those obtained for measurement procedures obtained using the relative frequency in classical statistics.

Then,

$$\begin{aligned} V_w\left(M^{(\beta,n)}_{X_S,\breve{p}}\right) &= \sum_{x_S=0}^{n} \frac{u^{(\beta,n)}_{\breve{p}} - \ell^{(\beta,n)}_{\breve{p}}}{\overline{p}-\underline{p}} \int\limits_{\underline{p}}^{\overline{p}} \frac{\binom{n}{x_S} p^{x_S}(1-p)^{n-x_S}\,dp}{\overline{p}-\underline{p}} \\ &\leq \frac{2\sqrt{\frac{k_\beta}{n}}}{\overline{p}-\underline{p}} \sum_{x_S=1}^{n} \frac{\int\limits_{\underline{p}}^{\overline{p}} \binom{n}{x_S} p^{x_S} 1-p)^{n-x_S}\,dp}{\overline{p}-\underline{p}} \\ &= \frac{2\sqrt{\frac{k_\beta}{n}}}{\overline{p}-\underline{p}} \end{aligned} \qquad (4.76)$$

and hence

$$\lim_{n\to\infty} V_w\left(M^{(\beta,n)}_{X_S,\breve{p}}\right) = 0. \qquad (4.77)$$

The limit (4.77) together with (4.73) implies for a Neyman measurement procedure that

$$\lim_{n\to\infty} V_w\left({}^*M^{(\beta,n)}_{\vec{X},\breve{p}}\right) = 0. \qquad (4.78)$$

The meaning of (4.78) is, of course, the well-known fact that the actual value of a probability can be measured with arbitrarily high reliability and arbitrarily high precision, if the sample size is sufficiently large. But (4.78) also shows that this not only holds for procedures based on the relative frequency and that the latter are not at all the best ones.

Moreover, (4.77) shows that the traditional approach for estimating the actual value $p$ of a probability $\breve{p}$ is asymptotically equivalent with respect to precision to the Neyman approach proposed here. With other words, the traditional approach is justified in a strict sense only asymptotically.

## 4.8 $\beta$-Measurement & Prediction Space for a Large Sample Size

From (4.78) and (4.77), it also becomes clear that the differences between the weighted volumes of $\beta$-measurement & prediction spaces based on the sample $\vec{X}$ and on the sum of sample elements $X_S$, respectively, become negligible for large sample sizes, whereas the numerical difficulties for determining Neyman $\beta$-measurement & prediction spaces based on $\vec{X}$ increase rapidly with the sample size. Therefore, if the sample size is sufficiently large, then the $\beta$-measurement & prediction space should always be based on $X_S$.

Moreover, the differences between an exact $\beta$-measurement procedure and the approximations get smaller and smaller with increasing sample size, and, therefore, simple approximations become more and more attractive from the viewpoint of the numerical determinination of a $\beta$-measurement & prediction space.

For sufficiently large sample size, we suggest taking Approximation 2 as basis for the determination of $\beta$-prediction regions and $\beta$-measurement intervals. In Approximation 2, the binomial distribution $Bi(n,p)$ is approximated by the normal distribution $N\Big(np, np(1-p)\Big)$. Thus, the following $\beta$-prediction regions and $\beta$-measurement intervals are obtained.

- $\beta$-Prediction regions:
  Let the $\beta$-prediction regions based on Approximation 2 for a given value $p \in [\underline{p}, \overline{p}]$ be denoted by

$$\tilde{A}_{X_S}^{(\beta,n)}(p) = \Big\{x_S \,\Big|\, \tilde{\ell}_{X_S}^{(\beta,n)} \leq x_S \leq \tilde{u}_{X_S}^{(\beta,n)}\Big\}. \tag{4.79}$$

  Then,

$$\begin{aligned}
\tilde{\ell}_{X_S}^{[\beta,n)} &= \max\Big\{0\,; np - \lambda_{\frac{1+\beta}{2}}\sqrt{np(1-p)}\Big\} \\
\tilde{u}_{X_S}^{[\beta,n)} &= \min\Big\{np + \lambda_{\frac{1+\beta}{2}}\sqrt{np(1-p)};\ n\Big\}.
\end{aligned}$$

- $\beta$-Measurement intervals:
  Let the $\beta$-measurement intervals based on Approximation 2 for given realization $x_S \in \{0, 1, \ldots, n\}$ be denoted by

$$\tilde{C}_{\breve{p}}^{(\beta,n)}(x_S) = \Big\{p \,\Big|\, \tilde{\ell}_{\breve{p}}^{(\beta,n)}(x_S) \leq p \leq \tilde{u}_{\breve{p}}^{(\beta,n)}(x_S)\Big\}. \tag{4.80}$$

Then,

$$
\begin{aligned}
&\tilde{\ell}_{\breve{p}}^{(\beta,n)}(x_S) \\
&= \max\left\{\underline{p}\,;\ \frac{2x_S + \lambda^2_{\frac{1+\beta}{2}} - \lambda_{\frac{1+\beta}{2}}\sqrt{\lambda^2_{\frac{1+\beta}{2}} + 4x_S\left(1 - \frac{x_S}{n}\right)}}{2\left(n + \lambda^2_{\frac{1+\beta}{2}}\right)}\right\}
\end{aligned}
$$

$$
= \min\left\{\frac{2x_S + \lambda^2_{\frac{1+\beta}{2}} + \lambda_{\frac{1+\beta}{2}}\sqrt{\lambda^2_{\frac{1+\beta}{2}} + 4x_S\left(1 - \frac{x_S}{n}\right)}}{2\left(n + \lambda^2_{\frac{1+\beta}{2}}\right)}\,;\ \overline{p}\right\},
$$

where $\lambda_{\frac{1+\beta}{2}}$ is the $\frac{1+\beta}{2}$-quantile of the standardized normal distribution.

The $\beta$-measurement & prediction space based on Approximation 2 denoted by $\tilde{M}_{X_S,\breve{p}}^{(\beta,n)}$ is either obtained by the union of the direct products $\{p\} \times \tilde{A}_{X_S}^{(\beta,n)}(p)$ or the union of the direct products $\tilde{C}_{\breve{p}}^{(\beta,n)}(x_S) \times \{x_S\}$:

$$
\begin{aligned}
\tilde{M}_{X_S,\breve{p}}^{(\beta,n)} &= \bigcup_{x_S=1}^{n} \left(\tilde{C}_{\breve{p}}^{(\beta,n)}(x_S) \times \{x_S\}\right) && (4.81)\\
&= \bigcup_{p\in[\underline{p},\overline{p}]} \left(\{p\} \times \tilde{A}_{X_S}^{(\beta,n)}(p)\right). && (4.82)
\end{aligned}
$$

The decision of whether the sample size is sufficiently large depends heavily on the situation. For the situations covered in the tables of this handbook, many numerical calculations showed that there is hardly any difference between the exact and the approximate solution for $n > 400$. Therefore, the tables contain the exact solutions for sample sizes up to $n = 400$ and the approximate solution for sample sizes with $n > 400$.

## 4.9 Illustrative Example

For illustrating the concept of the "$\beta$-measurement & prediction space" and also the possibility of taking into account the order of the sample elements $X_i$, $i = 1, \ldots, n$, for constructing a measurement or a prediction procedure, the simplest nontrivial situation with respect to the sample size (i.e., $n = 2$), is selected. In this case, the random sample $\vec{X} = \big(X_1, X_2\big)$ has exactly four different realizations; in other words, the prediction space has four elements:

$$\vec{x}_0 = (0,0), \quad \vec{x}_1 = (0,1), \quad \vec{x}_2 = (1,0), \quad \vec{x}_3 = (1,1).$$

The measurement space for the parameter $\breve{p}$ is assumed to be the whole unit interval $[0,1]$. Fixing the confidence level to be $\beta = 0.90$, the aim is to determine in a first step a system of $\beta$-prediction regions

$$\left\{A_{\vec{X}}^{(0.90,2)}(p)\right\}_{p\in[0,1]}$$

(i.e., a $\beta$-prediction procedure), and in a second step, a system of $\beta$-measurement intervals

$$\left\{C_{\breve{p}}^{(0.90,2)}(\vec{x})\right\}_{\vec{x}\in\{0,1\}^2}$$

(i.e., a $\beta$-measurement procedure).

### 4.9.1 Prediction Procedure Based on $\vec{X}$

The $\beta$-prediction regions $A_{\vec{X}}^{(0.90,2)}(p)$ may be determined successively starting with $p = 0$.

- Clearly, we have

  $$A_{\vec{X}}^{(0.90,2)}(0) = \big\{(0,0)\big\}$$

  as

  $$\mathbf{P}_{\vec{X}}^{(p)}\Big(\{\vec{x}_0\}\,|\,p = 0\Big) = 1.0 \geq \beta$$

- Proceeding to larger values $p$ of the probability $\breve{p}$, we have

  $$\mathbf{P}_{\vec{X}}^{(p)}\Big(\{\vec{x}_0\}\,|\,p\Big) = (1-p)^2 \geq 0.9 \quad \Leftrightarrow$$
  $$p \leq 1 - \sqrt{0.9} = 0.0513.$$

Thus,

$$A_{\vec{X}}^{(0.90,2)}(p) = \{(0,0)\} \quad \text{for } p \in [0, 0.0513].$$

- For values $p$ of the probability $\breve{p}$ exceeding 0.0513, the next realization (i.e., $\vec{x}_1$), is included into $A_{\vec{X}}^{(0.90,2)}(p)$. Hence,

$$\mathbf{P}_{\vec{X}}^{(p)}\Big(\{\vec{x}_0, \vec{x}_1\} \,|\, p\Big) = 1 - p \geq 0.9 \quad \Leftrightarrow \quad p \leq 0.1$$

and

$$A_{\vec{X}}^{(0.90,2)}(p) = \{(0,0),(0,1)\}$$
$$\text{for } p \in (1 - \sqrt{0.9}, 0.1] = (0.0513, 0.1].$$

- For values $p$ exceeding 0.1, the third realization $\vec{x}_2$ is added to $A_{\vec{X}}^{(0.90,2)}(p)$ in order to meet the reliability requirement given by the confidence level $\beta = 0.90$:

$$\mathbf{P}_{\vec{X}}^{(p)}\Big(\{\vec{x}_0, \vec{x}_1, \vec{x}_2\} \,|\, p\Big) = 1 - p^2 \geq 0.9 \quad \Leftrightarrow$$
$$p \leq \sqrt{0.1} = 0.3162.$$

Thus,

$$A_{\vec{X}}^{(0.90,2)}(p) = \{(0,0),(0,1),(1,0)\}$$
$$\text{for } p \in (0.1, \sqrt{0.1}] = (0.1, 0.3162].$$

- A further increase of the value $p$ of the probability $\breve{p}$ means that also the last realization has to be added to $A_{\vec{X}}^{(0.90,2)}(p)$, yielding the certain event. Thus, the question of which values of $p$ the first realization $\vec{x}_0$ may be taken off without violating the reliability requirement has to be solved. This question is immediately answered by utilizing the symmetry relation of binomial coefficients, resulting in

$$A_{\vec{X}}^{(0.90,2)}(p) = \{(0,0),(0,1),(1,0),(1,1)\}$$
$$\text{for } p \in (\sqrt{0.1}, 1 - \sqrt{0.1}) = (0.3162, 0.6838).$$

- Again utilizing the symmetry relation of the binomial coefficients leads to the prediction regions for larger values of $p$:

$$A_{\vec{X}}^{(0.90,2)}(p) = \{(0,1),(1,0),(1,1)\} \quad \text{for } p \in [1-\sqrt{0.1}, 0.9) = [0.6838, 0.9),$$

and

$$A_{\vec{X}}^{(0.90,2)}(p) = \{(1,0),(1,1)\} \quad \text{for } p \in [0.9, \sqrt{0.9}) = [0.9, 0.9487),$$

and, finally,

$$A_{\vec{X}}^{(0.90,2)}(p) = \{(1,1)\} \quad \text{for } p \in [\sqrt{0.9}, 1.0] = [0.9487, 1.0].$$

Table 4.2 displays the prediction regions $\{A_{\vec{X}}^{(0.90,2)}(p)\}$ for confidence level $\beta = 0.90$, sample size $n = 2$ and measurement space $[0,1]$:

**Table 4.2** *Prediction procedure based on $\vec{X}$ for $n = 2$ and $\beta = 0.90$.*

| $p$ | $A_{\vec{X}}^{(0.90,2)}(p)$ |
|---|---|
| $[0.0000, 0.0513]$ | $\{(0,0)\}$ |
| $(0.0513, 0.1000]$ | $\{(0,0),(0,1)\}$ |
| $(0.1000, 0.3162]$ | $\{(0,0),(0,1),(1,0)\}$ |
| $(0.3162, 0.6838)$ | $\{(0,0),(0,1),(1,0),1,1)\}$ |
| $[0.6838, 0.9000)$ | $\{(0,1),(1,0),(1,1)\}$ |
| $[0.9000, 0.9487)$ | $\{(1,0),(1,1)\}$ |
| $[0.9487, 1.0000]$ | $\{(1,1)\}$ |

## 4.9.2 Measurement Procedure Based on $\vec{X}$

By means of the system of prediction regions $\left\{A_{\vec{X}}^{(0.90,2)}(p)\right\}$ given in Table 4.2 the corresponding measurement intervals are immediately obtained (see Table 4.3).

**Table 4.3** *Measurement procedure for* $\breve{p}$ *based on* $\vec{X}$ *for* $n = 2$ *and* $\beta = 0.90$.

| $\vec{x}$ | $\overline{C}_{\breve{p}}^{(0.90,2)}(\vec{x})$ |
|---|---|
| $(0,0)$ | $[0.0000, 0.6838]$ |
| $(0,1)$ | $[0.0513, 0.9000]$ |
| $(1,0)$ | $[0.1000, 0.9487]$ |
| $(1,1)$ | $[0.3162, 1.0000]$ |

## 4.9.3 $\beta$-Measurement & Prediction Space Based on $(\breve{p}, \vec{X})$

The geometric representation of either the prediction procedure and the measurement procedure is given by the $\beta$-measurement & prediction space. For $(\breve{p}, \vec{X})$, it consists of horizontal lines, each representing one possible $\beta$-measurement interval (see Figure 4.1.)

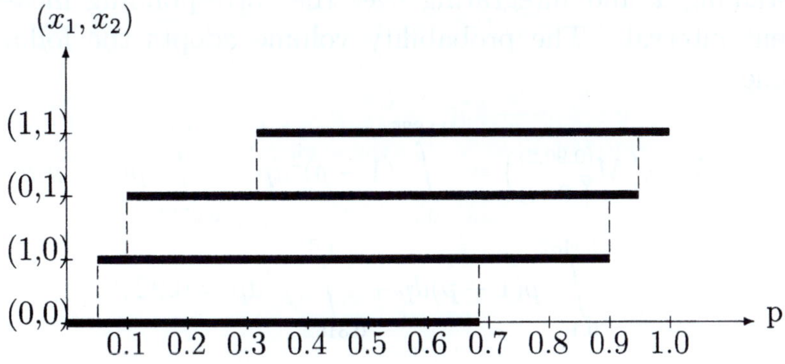


**Figure 4.1** *$\beta$-Measurement & prediction space for* $(\breve{p}, \vec{X})$ *with* $\beta = 0.90$ *and* $n = 2$.

The quality of the measurement or the prediction procedure is specified by the weighted volume of the corresponding $\beta$-measurement and prediction space.

According to (4.51) and because $\overline{p} - \underline{p} = 1$, the weighted volume is obtained as the weighted length of the measurement intervals, where the weights are given by

$$\int_0^1 p^{x_1+x_2}(1-p)^{2-x_1-x_2}dp = \frac{1}{3\binom{3}{x_1+x_2}}. \tag{4.83}$$

Thus, the numerical value of the weighted volume $V_w$ of the $\beta$-measurement & prediction space is easily computed:

$$\begin{aligned} V_w\left(M_{\vec{X},\breve{p}}^{(0.90,2)}\right) &= \frac{1}{3}\left(0.6838 + \frac{1}{2}[0.8487 + 0.8487] + 0.6838\right) \\ &= 0.7388. \end{aligned} \tag{4.84}$$

**Remark 4.2** *The above-determined $\beta$-measurement procedure is a Neyman measurement procedure; that is, it is optimal with respect to the weighted volume $V_w$, which can be interpreted as the expected length of the $\beta$-measurement intervals, and thus as the measurement precision.* •

By means of (4.54), the probability volume of the $\beta$-measurement procedure is computed by inserting the probabilities for obtaining $\vec{x}$ and integrating over the corresponding measurement interval. The probability volume adopts the following value:

$$\begin{aligned} V_{prob}\left(M_{\vec{X},\breve{p}}^{(0.90,2)}\right) &= \int_0^{0.6838} (1-p)^2dp + \int_{0.0513}^{0.9} p(1-p)dp \\ &+ \int_{0.1}^{0.9487} p(1-p)dp + \int_{0.3162}^{1.0} p^2dp = 0.9249. \end{aligned} \tag{4.85}$$

The reliability requirement is given by $\beta = 0.90$, the probability volume of the $\beta$-measurement & prediction space is given by $V_{prob} = 0.9249$, and, therefore, the reliability excess has the value of 0.0249.

### 4.9.4 Prediction & Measurement Procedure Based on $X_S$

The above-determined $\beta$-measurement intervals are based on the sample $\vec{X}$. If not the sample, but the relative frequency $\overline{X}$ or, equivalently, the sum of the sample elements $X_S$ is taken as the basis for the determination, the relations (4.47) and (4.48) can be used to obtain the prediction regions and measurement intervals based on $X_S$.

The $\beta$-prediction regions are obtained by replacing the vectors $\vec{x}$ by the sum of their element $x_S$, and the $\beta$-measurement intervals are obtained by taking for each $x_S$, the union of those $\beta$-measurement intervals belonging to vectors $\vec{x}$ with the sum of their elements equal to $x_S$. The results are displayed in Table 4.4 and Table 4.5.

**Table 4.4** *Prediction procedure based on $X_S$ for $n = 2$ and $\beta = 0.90$.*

| $p$ | $A_{X_S}^{(0.90,2)}(p)$ |
|---|---|
| $[0.0000, 0.0513]$ | $\{0\}$ |
| $(0.0513, 0.3162]$ | $\{0, 1\}$ |
| $(0.3162, 0.6838)$ | $\{0, 1, 2\}$ |
| $[0.6838, 0.9487)$ | $\{1, 2\}$ |
| $[0.9487, 1.0000]$ | $\{2\}$ |

**Table 4.5** *Measurement procedure based on $X_S$ for $n = 2$ and $\beta = 0.90$.*

| $x_S$ of $X_S$ | $\overline{C}_{\breve{p}}^{(0.90,2)}(x_S)$ |
|---|---|
| 0 | $[0.0000, 0.6838]$ |
| 1 | $[0.0513, 0.9487]$ |
| 2 | $[0.3162, 1.0000]$ |

The difference between the two methods is the fact that the probability portions to be added in order to meet the reliability requirement are larger in the case of $X_S$. Therefore, the reliability excess and, simultaneously, the expected length of the measurement intervals are generally larger if computations are based on $X_S$ instead of $\vec{X}$.

The $\beta$-measurement & prediction space based on $X_S$ is displayed in Figure 4.2.

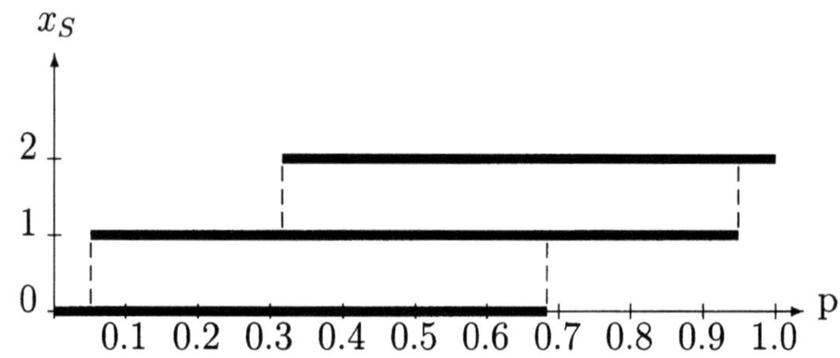


**Figure 4.2** *$\beta$-Measurement & prediction space for $(\breve{p}, X_S)$ with $\beta = 0.90$ and $n = 2$.*

The weighted volume $V_w$ of the $\beta$-measurement and prediction space based on $X_S$, which in this example is the average length of the measurement intervals, is given by

$$V_w\left(M_{X_S,\breve{p}}^{(0.90,2)}\right) = 0.7545 \tag{4.86}$$

which means an increase of about 2% compared to the weighted volume $V_w\left(M_{\vec{X},\breve{p}}^{(0.90,2)}\right)$ obtained based on $\vec{X}$.

The probability volume of the $\beta$-measurement & prediction space adopts the value

$$V_{prob}\left(M_{X_S,\breve{p}}^{(0.90,2)}\right) = 0.9317 \tag{4.87}$$

which means an increase of 27.3% compared to the reliability excess of the procedure based on the sample $\vec{X}$.

## 4.9.5 Traditional Measurement Procedure

Finally, the traditional method, which is based on $X_S$ and assumes symmetric bounds for the error probabilities, yields a third $\beta$-prediction & measurement procedure with the following $\beta$-prediction regions and $\beta$-measurement intervals. (See Tables 4.6 and 4.7.)

**Table 4.6** *Traditional prediction procedure based on $X_S$ for $n = 2$ and $\beta = 0.90$.*

| $p$ | ${}^{trad}A_{X_S}^{(0.90,2)}(p)$ |
|---|---|
| $[0.0000, 0.0253]$ | $\{0\}$ |
| $(0.0253, 0.2236]$ | $\{0, 1\}$ |
| $(0.2236, 0.7764)$ | $\{0, 1, 2\}$ |
| $[0.7764, 0.9747)$ | $\{1, 2\}$ |
| $[0.9747, 1.0000]$ | $\{2\}$ |

**Table 4.7** *Traditional measurement procedure based on $X_S$ for $\beta = 0.90$ and $n = 2$.*

| $x_S$ | ${}^{trad}\overline{C}_{\check{p}}^{(0.90,2)}$ |
|---|---|
| 0 | $[0.0000, 0.7764]$ |
| 1 | $[0.0253, 0.9747]$ |
| 2 | $[0.2236, 1.0000]$ |

The $\beta$-measurement space & prediction for the traditional procedure is displayed in Figure 4.3.

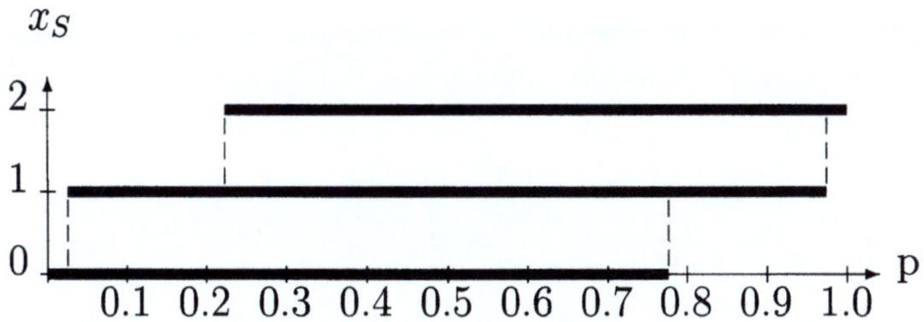


**Figure 4.3** *$\beta$-Measurement & prediction space for $(\check{p}, X_S)$ with $\beta = 0.90$ and $n = 2$.*

The traditional $\beta$-measurement intervals are often called Clopper-Pearson-confidence intervals (see e.g. [19]). For this example, they yield the following weighted volume of the corresponding $\beta$-measurement & prediction space:

$$V_w\left({}^{trad}M_{X_S,\check{p}}^{(0.90,2)}\right) = 0.842, \tag{4.88}$$

which means an increase of 14% compared to that of the Neyman measurement procedure.

The probability volume of the $\beta$-measurement & prediction space defined by the traditional $\beta$-measurement intervals adopts the value

$$V_{prob}\left({}^{trad}M_{X_S,\check{p}}^{(0.90,2)}\right) = 0.9764, \tag{4.89}$$

which constitutes an increase of the reliability excess of about 207% compared to the reliability excess of the Neyman procedure.

# Chapter 5

# Complete Measurement Procedures

## 5.1 Point Estimators

In any quantitative science, the actual values of quantities of interest have to be measured. The measurement devices have uncertain outcomes and, thus, their application establishes random experiments. In order to assess the measurement precision, the inherent uncertainty has to be taken into account by a stochastic model, which incorporates the quantity of interest as part of the distribution parameter.

To make a measurement procedure meaningful, it must necessarily result in bounds for the actual value of the parameter. The closer the bounds are to the actual value, the more precise is the measurement procedure. Thus, as shown in the preceding sections, measurement *precision* is determined by the expected volume of the $\beta$-measurement regions for the model parameter.

A $\beta$-measurement procedure yields a wrong result if the actual value of the parameter falls outside the bounds. The occurrence of such an event is called a measurement error. The measurement *reliability* is represented by an upper bound for the probability of committing a measurement error.

Hence, we may summarize that the lower and upper bounds for the actual value of the parameter of interest determine measurement precision, and the upper bound for committing a measurement error represents measurement reliability. Note that measurement reliability is assumed to be fixed and known in this handbook.

As already mentioned, a measurement procedure for a probability $\breve{p}$ must necessarily be a procedure yielding an interval (*measurement interval*) which covers the actual value $p$ with a probability of at least $\beta$. However, generally, there is also the need for fixing one point $\hat{p}$, which is called *point estimate* as the substitute of $p$. Following the traditional terminology, the procedure (i.e., the sample function for calculating a point estimate) is called *point estimator*, or simply, *estimator* for the parameter of interest.

As any point estimator with unknown, unspecified, or too small measurement reliability is, strictly speaking, meaningless, a point estimator should be used only in conjunction with a $\beta$-measurement procedure and it should, of course, adopt only values within the associated $\beta$-measurement intervals.

A pair consisting of a $\beta$-measurement procedure and a point estimator is called a *complete measurement procedure.* As the theory for determining $\beta$-measurement procedures is already provided in the preceding sections, the only remaining problem is to develop appropriate point estimators. Principally, each point of a $\beta$-measurement interval can be selected as an estimate. However, the aim should be to specify estimates (points) in each $\beta$-measurement interval with some good properties with respect to the unavoidable deviation from the actual value $p$.

A point estimator for the probability $\breve{p}$ and given $\beta$-measurement procedure based on $\vec{X}$ is defined as a function

$$T_{\breve{p}}^{(\beta,n)} : \{0,1\}^n \to [\underline{p}, \overline{p}] \tag{5.1}$$

with

$$T_{\breve{p}}^{(\beta,n)}(\vec{x}) \in C_{\breve{p}}^{(\beta,n)}(\vec{x}). \tag{5.2}$$

As mentioned earlier, a point estimator should complete a $\beta$-measurement procedure, and, therefore, the estimator will depend generally on the confidence level, which is expressed by the notation used for an estimator:

$$T_{\breve{p}}^{(\beta,n)}(\vec{X}) \tag{5.3}$$

in the case that the measurement procedure is based on $\vec{X}$, and

$$T_{\breve{p}}^{(\beta,n)}(X_S) \tag{5.4}$$

if it is based on $X_S$.

## 5.2 Traditional Point Estimator

Long before Jakob Bernoulli wrote his *Ars Conjectandi*, the arithmetic mean $\overline{X}$ of a random sample (2.13) had been used as an estimator of the actual value of a probability. The arithmetic mean of the sample elements is equal to the relative frequency of the occurrence of successes within the sequence of experiments. The relative frequency does not depend on any measurement procedure with a given confidence level, and, therefore, the upper index $\beta$ is omitted in its notation.

$$TF_{\breve{p}}^{(n)}(X_S) = \frac{1}{n}X_S = \frac{1}{n}\sum_{i=1}^{n} X = \overline{X}. \tag{5.5}$$

The relative frequency seems to be not only a very natural estimator for $\breve{p}$, but it is, in some sense, also a very good estimator for the actual value $p$ of the probability of interest.

As already seen, the random variable

$$X_S = X_1 + \ldots + X_n = n\overline{X}$$

being the essential part of the relative frequency, is distributed according to the binomial distribution $Bi(n,p)$, implying

$$nTF_{\breve{p}}^{(n)}(X_S) \sim Bi(n,p).$$

Therefore, the distribution of the traditional point estimator $TF_{\breve{p}}^{(n)}(X_S)$ is given by

$$\mathbf{P}^{(p)}_{TF_{\breve{p}}^{n}(X_S)}\left(\left\{\frac{x_S}{n}\right\}\right) = \binom{n}{x_S} p^{x_S}(1-p)^{n-x_S} \tag{5.6}$$

for $x_S \in \{0, 1, \ldots, n\}$.

As the point estimator of the actual value $p$, the relative frequency of successes $\overline{X}$ has some remarkable properties. It is unbiased (i.e., there is no systematic measurement error), it is consistent (i.e., with increasing sample size large measurement errors are less and less frequent), and it has minimum variance among all unbiased estimators.

Hence, $\overline{X}$ seems to be a very capable point estimator for the actual value of a probability. However, unbiasedness does not say much about the quality of an estimator. It is easy to construct unbiased estimators, which have very poor quality with respect to the aim of determining precisely the actual value of a parameter of interest. Consistency, on the other hand, is a limit property, implying that its significance for small samples remains vague and cannot be used for quantifying the quality of a point estimator. Finally, the property of minimum variance refers to unbiased estimators only and does not admit a comparison with biased estimators.

Thus, the stated "good" properties of the relative frequency as the point estimator for a probability $\breve{p}$ may indicate, but do not prove at all, a good or even best quality.

## 5.3 Measurement Space Conformity

Assume that a letter balance with a measurement range from 0 to 20 grams shows, under certain circumstances, values of pounds and kilograms. Of course, such a letter balance would be ridiculous, and nobody would try to sell it. On the other hand, if somebody would develop a balance for measuring the weight of letters as well as the weight of heavy trucks, the result would either be an extremely expensive device or a very inaccurate device. Therefore,

- a measurement device is generally developed for a fixed range of values,
- the device may only be applied if it is known that the

actual value of the quantity of interest falls within the range, and

- the device will yield only results that are elements of the fixed range.

The development of measurement procedures under uncertainty constitutes one of the main tasks in statistics. However, in contrast to the established principle of developing measurement devices for appropriately fixed ranges, the measurement (estimation) procedures provided in traditional statistics, like the relative frequency, generally assume maximum parameter space, where the maximum is determined by the nature of the quantity of interest with respect to being a distribution parameter. In the case of the relative frequency, it is assumed that any value in the unit interval may be a possible candidate for the actual value $p$ of the probability of interest $\breve{p}$ or, in our terminology, the relative frequency is an estimator for the special case that the measurement space is given by $[\underline{p}, \overline{p}] = [0, 1]$.

However, the event of interest is necessarily known, often implying that one knows, for example, whether it is a rare event or a very frequent event. Therefore, the rough nature of the probability of interest $\breve{p}$ is known, which means that, generally, it is possible to specify a certain part of the unit interval which is sufficient for measuring the actual value $p$ of $\breve{p}$. Hence, it would be desirable – in analogy to any other field of science – to have measurement procedures for a given range of values different from the maximum one. If, however, the measurement space $[\underline{p}, \overline{p}]$ for a probability $\breve{p}$ is not maximum, then the relative frequency adopts, with positive probability, meaningless values outside the given range, and measurement precision is expected not to be at all good.

**Remark 5.1**

*One unpleasant consequence of assuming the maximum parameter space is the fact that totally unacceptable results are obtained with positive probability. To avoid such results, "robust" estimators are developed, instead of omitting the unrealitic assumption of an maximum parameter space.* •

Before the resulting problems are illustrated, a property introduced in [9] and called *measurement space conformity* for point estimators is defined.

**Definition 5.1**
*Let $\breve{\theta}$ be a parameter of interest with actual value $\theta$, and $\vec{X}$ denote a random sample with probability distribution $\mathbf{P}_{\vec{X}}^{(\theta)}$ and prediction space $\mathcal{X}$. A point estimator $T_{\breve{\theta}}^{(\beta,n)}(\vec{X})$ for $\breve{\theta}$ with measurement space $\Theta$ is called* measurement space conform *if*

$$\mathbf{P}^{(\theta)}_{T_{\breve{\theta}}^{(\beta,n)}\left(\vec{X}\right)}\left(\Theta\right) = 1. \tag{5.7}$$

•

If (5.7) is true, then the point estimator adopts values only within the specified space, which means that only estimates are obtained which are reasonable in the given situation.

If the measurement space is not equal to the maximum parameter space, where the maximum is taken over all parameter values leading to a probability distribution, then classical estimators may yield meaningless values. Of course, for a sufficiently large sample size, such an event occurs only with negligible probability, but if the sample size is small or medium, then the probability of a meaningless estimate may become large and even adopt the value 1. Moreover, the precision of an estimator must be reconsidered, as, for instance, the minimum variance property holds only for the case of maximum measurement space, but clearly not for a restricted space.

The above-outlined difficulties occur particularly when using the relative frequency as the estimator for the actual value $p$ of a probability $\breve{p}$. This fact shall be illustrated by a numerical example.

**Example 5.1**
*Let the quantity of interest be the nonconformance probability of a manufacturing process, for which it is known that the nonconformance probability is at most 10%. For estimating the actual*

*value p, the relative frequency is taken as recommended in statistical textbooks. In accordance with the situation, the measurement space is selected to be* $[\underline{p}, \overline{p}] = [0.0, 0.10]$. *Any value larger than 0.10 is looked upon as unreasonable.*

***Table 5.1*** *Illustration of measurement space nonconformity of the relative frequency.*

| *Sample size* | *Estimate* | *Measurement space conformity* |
|---|---|---|
| $n = 1$ | *0* | $0 \in [0.00, 0.10]$ |
| | *1* | $1 \notin [0.00, 0.10]$ |
| $n = 2$ | *0* | $0 \in [0.00, 0.10]$ |
| | 0.5 | $0.5 \notin [0.00, 0.10]$ |
| | *1* | $1 \notin [0.00, 0.10]$ |
| $n = 10$ | *0* | $0 \in [0.00, 0.10]$ |
| | 0.1 | $0.1 \in [0.01, 0.10]$ |
| | 0.2 | $0.2 \notin [0.01, 0.10]$ |
| | ... | ... |
| | *1* | $1 \notin [0.01, 0.10]$ |

*As can be seen from Table 5.1 and as is well known, the estimator for the actual value of a probability given by the relative frequency yields in the case of a small sample size with high probability, results that are not consistent with the given situation and, therefore, have to be discarded.* •

**Example 5.2**
*Next, consider the often occurring case that a probability shall be estimated which is close to 0.5. Let us assume that the measurement space is specified to be* $[0.4, 0.6]$. *Then, for a sample of size* $n = 10$, *only the values* $0.4, 0.5$ *and* $0.6$ *of the relative frequency are consistent with the situation. Any other value has to be discarded or leads to wrong conclusions.* •

The other problem to be considered when constructing an estimator is closely connected to measurement space conformity and the traditional way to construct measurement intervals.

The traditional method starts with a point estimator, which eventually is supplemented by $\beta$-measurement regions with a given confidence level. These are determined by adding to and subtracting appropriate expressions from the estimator. However, this approach works only in some special cases and, even more severe, it is essentially based on the assumption of a maximum parameter space, which implicitly means that any value of the maximum parameter space is reasonable.

In the case of a restricted measurement space, there are always situations in which the classical approach leads to meaningless results, characterized by the fact that the estimate is not an element of the given measurement space and/or the $\beta$-measurement interval has an empty intersection with the measurement space. The problem with the $\beta$-measurement intervals has been solved by means of the $\beta$-measurement & prediction space. Taking a $\beta$-measurement procedure based on a $\beta$-measurement & prediction space yields necessarily meaningful measurement intervals, but combining such a measurement procedure with a traditional point estimator to a complete measurement procedure would lead to the unsatisfactory result that with, positive probability, the estimate would not be an element of the measurement interval. Therefore, in the case of a small sample size, classical estimators cannot be used in conjunction with measurement procedures obtained by the approach proposed in this handbook.

Thus, the task is to develop new point estimators which are measurement space conforming and for which the estimates are, with certainty, elements of the corresponding $\beta$-measurement intervals. This last requirement suggests not starting with a point estimator when developing a complete $\beta$-measurement procedure, but starting with a $\beta$-measurement & prediction space and determining the point estimator correspondingly.

The concept of measurement conformity emphasizes the significance of a bounded measurement space which often is the only assured knowledge about a phenomenon of interest. As a matter of fact most statistical procedures developed in classical statistics do not allow the user to include this knowledge when designing the necessary procedure. Including the measurement space moves statistics nearer to reality and could increase its acceptance among natural scientists.

## 5.4 The $\beta$-Estimator

We have defined a quality indicator of a $\beta$-measurement procedures for given reliability $\beta$ and expense $n$. The quality indicator reflects the precision, and together with the reliability, it specifies the measurement quality. For constructing a (point) estimator and thus completing the procedure, it is natural to start with the $\beta$-measurement procedure and, based on it, define another procedure to determine one single value as a substitute for the actual value $p$ of the probability of interest $\breve{p}$. Of course, only the elements of the given measurement intervals are admissible candidates for being selected as the estimate.

**Definition 5.2**
*Let $C_{\breve{p}}^{(\beta,n)}$ be the $\beta$-measurement procedure for $\breve{p}$ associated with the $\beta$-measurement & prediction space $M_{\vec{X},\breve{p}}^{(\beta,n)}$. A* complete $\beta$-measurement procedure *for $\breve{p}$ consists of the pair*

$$\left(C_{\breve{p}}^{(\beta,n)}, T_{\breve{p}}^{(\beta,n)}\right), \tag{5.8}$$

*where $T_{\breve{p}}^{(\beta,n)} : \{0,1\}^n \to [\underline{p}, \overline{p}]$ is a measurable function which satisfies the following condition:*

$$T_{\breve{p}}^{(\beta,n)}(\vec{x}) \in C_{\breve{p}}^{(\beta,n)}(\vec{x}) \quad \text{for any } \vec{x} \in \{0,1\}^n. \tag{5.9}$$

•

The random variable $T_{\breve{p}}^{(\beta,n)}(\vec{X})$ is called a $\beta$-(point) *estimator for $\breve{p}$* and measurement space $[\underline{p}, \overline{p}]$. Each $\beta$-estimator is associated to a $\beta$-measurement & prediction space. Note that the point estimator $T_{\breve{p}}^{(\beta,n)}$ depends not only on the sample size $n$ but also on the confidence level $\beta$ (i.e., on the reliability of the measurement procedure).

### 5.4.1 The Minimum MSE-$\beta$-Estimator

Generally, point estimators are constructed so as to minimize the mean squared error. Since we deal with $\beta$-measurement & prediction spaces given by $M_{\vec{X},\breve{p}}^{(\beta,n)}$ and defining the $\beta$-measurement intervals $\overline{C}_{\breve{p}}^{(\beta,n)}(\vec{x})$, our goal is to find a $\beta$-point estimator $T_{\breve{p}}^{(\beta,n)}$ specifying one point in each $\beta$-measurement interval as a substitute of $p$. Clearly, any measurement interval and, hence, any point therein as an estimate make sense only in the case that no measurement error has occurred. Therefore, a $\beta$-estimator is looked for with minimum deviation from the actual value of the parameter of interest under the condition that the measurement procedure is applied successfully.

In a first step, the conditional mean squared deviation or *conditional mean squared error* of a $\beta$-estimator is introduced. Let $T_{\breve{p}}^{(\beta,n)}(\vec{X})$ be a $\beta$-estimator and $p$ be the actual value of $\breve{p}$. Then, the conditional mean squared error is defined as the expectation of $\left(T_{\breve{p}}^{(\beta,n)}(\vec{X})-p\right)^2$ with respect to the joint measure of $(\vec{X},\breve{p})$ under the condition of the $\beta$-measurement & prediction space.

Thus, the conditional mean squared error of $T_{\breve{p}}^{(\beta,n)}$ is given by

$$E\left[\left(T_{\breve{p}}^{(\beta,n)}(\vec{X})-p\right)^2 \Big| M_{\vec{X},\breve{p}}^{(\beta,n)}\right]$$

$$= \sum_{\vec{x}\in\{0,1\}^n} \int_{\overline{C}_{\breve{p}}^{(\beta,n)}(\vec{x})} \left(T_{\breve{p}}^{(\beta,n)}(\vec{x})-p\right)^2 \frac{\mathbf{P}_{\vec{X}}^{(p)}(\vec{x})\,\frac{dp}{\overline{p}-\underline{p}}}{V_{prob}\left(M_{\vec{X},\breve{\theta}}^{(\beta,n)}\right)} \tag{5.10}$$

with

$$\overline{C}_{\breve{p}}^{(\beta,n)}(\vec{x}) = \left\{p \,\Big|\, \ell_{\breve{p}}^{(\beta,n)}(\vec{x}) \le p \le u_{\breve{p}}^{(\beta,n)}(\vec{x})\right\}, \tag{5.11}$$

where $V_{prob}\left(M_{\vec{X},\breve{p}}^{(\beta,n)}\right)$ is the probability volume of the $\beta$-measurement and prediction space [i.e., its volume with respect to the joint measure of $(\breve{p},\vec{X})$]. According to (4.54) the probability volume is obtained as

$$V_{prob}\left(M_{\vec{X},\check{\theta}}^{(\beta,n)}\right)$$

$$= \frac{1}{\overline{p}-\underline{p}} \sum_{\vec{x}\in\{0,1\}^n} \int\limits_{\ell_{\check{p}}^{(\beta,n)}(\vec{x})}^{u_{\check{p}}^{(\beta,n)}(\vec{x})} \mathbf{P}_{\vec{X}}^{(p)}\left(\vec{x}\right)dp$$

$$= \frac{1}{\overline{p}-\underline{p}} \sum_{\vec{x}\in\{0,1\}^n} \int\limits_{\ell_{\check{p}}^{(\beta,n)}(\vec{x})}^{u_{\check{p}}^{(\beta,n)}(\vec{x})} p^{\sum_{i=1}^{n} x_i}(1-p)^{n-\sum_{i=1}^{n} x_i} dp. \tag{5.12}$$

Having defined by (5.10) the conditional mean squared error and hence a comparison criterion for $\beta$-estimators, an optimal $\beta$-estimator can be defined.

**Definition 5.3**
*Let $M_{\vec{X},\check{p}}^{(\beta,n)}$ be a $\beta$-measurement & prediction space for determining the actual value $p$ of a probability of interest $\check{p}$. A $\beta$-estimator minimizing the conditional mean squared error*

$$E\left[\left(T_{\check{p}}^{(\beta,n)}(\vec{X})-p\right)^2 \middle| M_{\vec{X},\check{p}}^{(\beta,n)}\right]$$

*is called a* minimum MSE-$\beta$-estimator *and is denoted by $TMS_{\check{p}}^{(\beta,n)}(\vec{X})$.* •

For a given $\beta$-measurement procedure, the minimum MSE-$\beta$-estimator is obtained by means of the following proposition.

**Theorem 5.1** *Let $C_{\check{p}}^{(\beta,n)}$ be a $\beta$-measurement procedure associated with the $\beta$-measurement & prediction space $M_{\vec{X},\check{p}}^{(\beta,n)}$. Then, the minimum MSE-$\beta$-estimator $TMS_{\check{p}}^{(\beta,n)}(\vec{X})$ is given by*

$$TMS_{\check{p}}^{(\beta,n)}(\vec{x}) = \frac{\int\limits_{\ell_{\check{p}}^{(\beta,n)}(\vec{x})}^{u_{\check{p}}^{(\beta,n)}(\vec{x})} p\mathbf{P}_{\vec{X}}^{(p)}(\vec{x})dp}{\int\limits_{\ell_{\check{p}}^{(\beta,n)}(\vec{x})}^{u_{\check{p}}^{(\beta,n)}(\vec{x})} \mathbf{P}_{\vec{X}}^{(p)}(\vec{x})dp} \tag{5.13}$$

*for any $\vec{x} = \{0,1\}^n$* •

**Proof:**
Let

$$V_{prob}\left(M_{\vec{X},\breve{p}}^{(\beta,n)}\right) = \tilde{\beta} \tag{5.14}$$

where, because of the definition of $\beta$-measurement & prediction space, the following relations hold true:

$$\tilde{\beta} \approx \beta, \text{ and } \tilde{\beta} > \beta \tag{5.15}$$

For the conditional expectation of the squared deviation from the actual value $p$ with respect to the joint measure, the following expression is obtained:

$$\begin{aligned}
&\tilde{\beta}(\overline{p} - \underline{p})E\left[\left(T_{\breve{p}}^{(\beta,n)}\left(\vec{X}\right) - p\right)^2 \middle| M_{\vec{X},\breve{p}}^{(\beta,n)}\right] \\
&= \int\limits_{M_{\vec{X},\breve{p}}^{(\beta,n)}} \left(T_{\breve{p}}^{(\beta,n)}(\vec{x}) - p\right)^2 d\mathbf{P}_{\vec{X}}^{(p)}(\vec{x})dp \\
&= \sum_{\vec{x}\in\{0,1\}^n} \left[\int\limits_{\ell_{\breve{p}}^{(\beta,n)}(\vec{x})}^{u_{\breve{p}}^{(\beta,n)}(\vec{x})} \left(T_{\breve{p}}^{(\beta,n)}(\vec{x}) - p\right)^2 \mathbf{P}_{\vec{X}}^{(p)}(\vec{x})dp\right] \\
&= \sum_{\vec{x}\in\{0,1\}^n} \left[\left(T_{\breve{p}}^{(\beta,n)}(\vec{x})\right)^2 \int\limits_{\ell_{\breve{p}}^{(\beta,n)}(\vec{x})}^{u_{\breve{p}}^{(\beta,n)}(\vec{x})} \mathbf{P}_{\vec{X}}^{(p)}(\vec{x})dp \right. \\
&\qquad \left. -2T_{\breve{p}}^{(\beta,n)}(\vec{x}) \int\limits_{\ell_{\breve{p}}^{(\beta,n)}(\vec{x})}^{u_{\breve{p}}^{(\beta,n)}(\vec{x})} p\mathbf{P}_{\vec{X}}^{(p)}(\vec{x})dp + \int\limits_{\ell_{\breve{p}}^{(\beta,n)}(\vec{x})}^{u_{\breve{p}}^{(\beta,n)}(\vec{x})} p^2\mathbf{P}_{\vec{X}}^{(p)}(\vec{x})dp\right]
\end{aligned} \tag{5.16}$$

The very last expression shows that for given $\beta$-measurement intervals, the expectation of the conditional mean squared error depends on the $\beta$-estimator only through the first two terms within the square brackets. Thus, the minimum of (5.16) is adopted for a $\beta$-estimator $T_{\breve{p}}^{(\beta,n)}\left(\vec{X}\right)$ which minimizes for each

$\vec{x} \in \{0,1\}^n$ the difference of these two terms:

$$\left(T_{\breve{p}}^{(\beta,n)}(\vec{x})\right)^2 \int\limits_{\ell_{\breve{p}}^{(\beta,n)}(\vec{x})}^{u_{\breve{p}}^{(\beta,n)}(\vec{x})} \mathbf{P}_{\vec{X}}^{(p)}(\vec{x})dp$$

$$-2(T_{\breve{p}}^{(\beta,n)}\vec{x}) \int\limits_{\ell_{\breve{p}}^{(\beta,n)}(\vec{x})}^{u_{\breve{p}}^{(\beta,n)}(\vec{x})} p\mathbf{P}_{\vec{X}}^{(p)}(\vec{x})dp \qquad (5.17)$$

A $\beta$-estimator $T_{\breve{p}}^{(\beta,n)}\left(\vec{X}\right)$ minimizing (5.17) is optimal with respect to (5.10) for each $\vec{x} \in \{0,1\}^n$ and, therefore, is a minimum MSE-$\beta$-estimator for $\breve{p}$. Hence, a minimum MSE-$\beta$-estimator $TMS_{\breve{p}}^{(\beta,n)}$ for $\breve{p}$ is given by

$$TMS_{\breve{p}}^{(\beta,n)}(\vec{x}) = \frac{\int\limits_{\ell_{\breve{p}}^{(\beta,n)}(\vec{x})}^{u_{\breve{p}}^{(\beta,n)}(\vec{x})} p\mathbf{P}_{\vec{X}}^{(p)}(\vec{x})dp}{\int\limits_{\ell_{\breve{p}}^{(\beta,n)}(\vec{x})}^{u_{\breve{p}}^{(\beta,n)}(\vec{x})} \mathbf{P}_{\vec{X}}^{(p)}(\vec{x})dp} \qquad \text{for } \vec{x} \in \{0,1\}^n \qquad (5.18)$$

•

The minimum MSE-$\beta$-estimator $TMS_{\breve{p}}^{(\beta,n)}(\vec{X})$ has an interesting property with respect to the measurement space $[\underline{p}, \overline{p}]$.

**Property 5.1**
*The minimum MSE-$\beta$-estimator $TMS_{\breve{p}}^{(\beta,n)}\left(\vec{X}\right)$ is an (approximately) unbiased estimator of the center $\frac{\underline{p}+\overline{p}}{2}$ of the measurement space $[\underline{p}, \overline{p}]$.* •

**Remark 5.2**
*The above-stated property holds exactly in the continuous case, as shown in [11].* •

**Proof:**
The estimator is defined on the prediction space. For any actual

value $p \in [\underline{p}, \overline{p}]$, the expectation has to be taken with respect to the marginal measure $Q_{\{0,1\}^n}$:

$$\begin{aligned}
E\left[TMS_{\breve{p}}^{(\beta,n)}\left(\vec{X}\right)\right] &= \sum_{\vec{x}\in\{0,1\}} TMS_{\breve{p}}^{(\beta,n)}(\vec{x})Q_{\{0,1\}^n}(\vec{x}) \\
&= \sum_{\vec{x}\in\{0,1\}^n} \frac{\int\limits_{\ell_{\breve{p}}^{(\beta,n)}(\vec{x})}^{u_{\breve{p}}^{(\beta,n)}(\vec{x})} p\frac{\mathbf{P}_{\vec{X}}^{(p)}(\vec{x})}{\overline{p}-\underline{p}}dp}{\int\limits_{\ell_{\breve{p}}^{(\beta,n)}(\vec{x})}^{u_{\breve{p}}^{(\beta,n)}(\vec{x})} \frac{\mathbf{P}_{\vec{X}}^{(p)}(\vec{x})}{\int_{\underline{p}}^{\overline{p}}\mathbf{P}_{\vec{X}}^{(p)}(\vec{x})dp}dp} \\
&= \sum_{\vec{x}\in\{0,1\}^n} \frac{\int\limits_{\ell_{\breve{p}}^{(\beta,n)}(\vec{x})}^{u_{\breve{p}}^{(\beta,n)}(\vec{x})} p\frac{\mathbf{P}_{\vec{X}}^{(p)}(\vec{x})}{\overline{p}-\underline{p}}dp}{\int\limits_{\ell_{\breve{p}}^{(\beta,n)}(\vec{x})}^{u_{\breve{p}}^{(\beta,n)}(\vec{x})} g_{\breve{p}}^{(\vec{x})}(p)dp.} \qquad (5.19)
\end{aligned}$$

Because

$$\int\limits_{\ell_{\breve{p}}^{(\beta,n)}(\vec{x})}^{u_{\breve{p}}^{(\beta,n)}(\vec{x})} g_{\breve{p}}^{(\vec{x})}(p)dp \approx \beta, \qquad (5.20)$$

we obtain

$$\begin{aligned}
E\left[TMS_{\breve{p}}^{(\beta,n)}\left(\vec{X}\right)\right] &\approx \frac{1}{\beta}\sum_{\vec{x}\in\{0,1\}} \int\limits_{\ell_{\breve{p}}^{(\beta,n)}(\vec{x})}^{u_{\breve{p}}^{(\beta,n)}(\vec{x})} p\frac{\mathbf{P}_{\vec{X}}^{(p)}(\vec{x})}{\overline{p}-\underline{p}}dp \\
&= \frac{1}{\beta}\int\limits_{\underline{p}}^{\overline{p}} \frac{p}{\overline{p}-\underline{p}}\left[\sum_{\vec{x}\in A_{\vec{X}}^{(\beta,n)}} \mathbf{P}_{\vec{X}}^{(p)}(\vec{x})\right]dp. \qquad (5.21)
\end{aligned}$$

Because of

$$\sum_{\vec{x}\in A_{\vec{X}}^{(\beta,n)}} \mathbf{P}_{\vec{X}}^{(p)}(\vec{x}) \approx \beta, \qquad (5.22)$$

we finally get

$$E\left[TMS_{\breve{p}}^{(\beta,n)}\left(\vec{X}\right)\right] \approx \int\limits_{\underline{p}}^{\overline{p}} \frac{p}{\overline{p}-\underline{p}} dp = \frac{\overline{p}+\underline{p}}{2}. \tag{5.23}$$

•

Thus, the minimum MSE-$\beta$-estimator has not only a minimum conditional mean squared error, but is also unbiased in a neighborhood of the center of the measurement space. This is a good feature in view of the habit of selecting a measurement space so as to have the actual but unknown value $p$ of the probability $\breve{p}$ in question be located somewhere in the center.

## 5.5 The Midpoint $\beta$-Estimator

The minimum MSE-$\beta$-estimator has the disadvantage that it is not possible to calculate it without a good mathematical software package. Particularly, in a experimental stage of an investigation, one would like to have an easy-to-compute estimator. A simple but nevertheless rather good estimator for the actual value $p$ of a probability of interest $\breve{p}$ is given by the *midpoint $\beta$-estimator.*

**Definition 5.4**
*Let $M_{\vec{X},\breve{\theta}}^{(\beta,n)}$ be a $\beta$-measurement & prediction space for the actual value $p$ of a probability of interest $\breve{p}$ defining the set of $\beta$-measurement intervals*

$$\left\{[\ell_{\breve{p}}^{(\beta,n)}(\vec{x}), u_{\breve{p}}^{(\beta,n)}(\vec{x})] \,\middle|\, \vec{x} \in \{0,1\}^n\right\}$$

*or, equivalently, the pair of random variables*

$$\left(L_{\breve{p}}^{(\beta,n)}(\vec{X}), U_{\breve{p}}^{(\beta,n)}(\vec{X}).\right) \tag{5.24}$$

*The $\beta$-estimator $TMP_{\breve{p}}^{(\beta,n)}(\vec{X})$ defined by*

$$TMP_{\breve{p}}^{(\beta,n)}(\vec{x}) = \frac{\ell_{\breve{p}}^{(\beta,n)}(\vec{x}) + u_{\breve{p}}^{(\beta,n)}(\vec{x})}{2} \qquad \text{for } \vec{x} \in \{0,1\}^n \tag{5.25}$$

*or, equivalently, by*

$$TMP_{\breve{p}}^{(\beta,n)}(\vec{X}) = \frac{L_{\breve{p}}^{(\beta,n)}(\vec{X}) + U_{\breve{p}}^{(\beta,n)}(\vec{X})}{2} \tag{5.26}$$

*is called* midpoint $\beta$-estimator *for* $\breve{p}$ *and measurement space* $[\underline{p}, \overline{p}]$. •

The midpoint $\beta$-estimator $TMP_{\breve{p}}^{(\beta,n)}(\vec{X})$ is a simple alternative to the minimum MSE-$\beta$-estimator with the convenient property that it is the center of the measurement intervals. Therefore, any measurement result of the complete measurement procedure can be given in the way commonly used in quantitative sciences [i.e., $TMP_{\breve{p}}^{(\beta,n)}(\vec{x}) \pm \epsilon^{(\beta,n)}$, where $\epsilon^{(\beta,n)}$ specifies the achieved precision].

If a Neyman $\beta$-measurement procedure given by its $\beta$-measurement intervals $\left\{{}^*L_{\breve{p}}^{(\beta,n)}(\vec{X}),\ {}^*U_{\breve{p}}^{(\beta,n)}(\vec{X})\right\}$ is available, it is easily seen that the corresponding midpoint $\beta$-estimator ${}^*TMP_{\breve{p}}^{(\beta,n)}(\vec{X})$ as well as the minimum MSE-$\beta$-estimator ${}^*TMS_{\breve{p}}^{(\beta,n)}(\vec{X})$ are consistent and asymptotically unbiased. Again, this illustrates the fact that the unique position of the relative frequency in classical statistics is by no means justified. The relative frequency has benefits in the case of large sample sizes. In this case any of the above proposed estimators are close to the actual value of the probability to be determined, but the relative frequency is easiest to be computed.

**Remark 5.3** *1. Again we would like to draw attention to the fact that the here-derived procedures make no use of the traditional criteria like unbiasedness, minimum variance or consistency, but are based solely on the average precision defined by means of the weighted volume of the $\beta$-measurement & prediction space, and result in complete measurement procedures with estimators, which are better than the traditional ones with respect to the conditional mean squared error. As a trivial consequence the estimators are consistent.*

2. *Moreover, the here developed estimators are in a certain sense robust against outliers, as the measurement space is bounded to those values which are meaningful.*

•

## 5.6 Illustrative Example for $\beta$-Estimators

In the two preceding sections, the minimum MSE-$\beta$-estimator and the midpoint $\beta$-estimator were introduced based on the sample $\vec{X}$. The analogous derivations and statements hold if the measurement procedure is based on the sum of sample elements $X_S$ and not on the sample $\vec{X}$. As usual, the two types of possible estimators are distinguished by their arguments.

Again, the simple example for $\beta = 0.90$, $n = 2$ with measurement space $[0, 1]$ is used to illustrate the proposed $\beta$-estimators.

### Traditional Complete Measurement Procedure

Because of the maximum measurement space $[0, 1]$, the relative frequency $TF_{\breve{p}}^{(n)}(X_S) = \overline{X}$ is measurement space conforming. Therefore, a meaningful complete measurement procedure is obtained by regarding the traditional (Clopper-Pearson) $\beta$-measurement intervals in conjunction with the relative frequency.

In Table 5.2, the point estimates $\overline{x}$, the measurement intervals and finally the conditional mean squared error for the traditional procedure is displayed.

**Table 5.2** *Complete traditional $\beta$-measurement procedure.*

| $\vec{x}$ | $TF_{\breve{p}}^{(2)}(x_S) = \overline{x}$ | Measurement interval |
|---|---|---|
| (0,0) | 0.0000 | $[0.0000, 0.7764]$ |
| (1,0) | 0.5000 | $[0.0253, 0.9747]$ |
| (0,1) | 0.5000 | $[0.0253, 0.9747]$ |
| (1,1) | 1.0000 | $[0.2236, 1.0000]$ |
| average mean squared error: | | 0.0797 |

The value of the conditional mean squared error is obtained in the following way:

$$
\begin{aligned}
&E\left[\left(\overline{X}-p\right)^2 \,\middle|\, {}^{trad}M_{X_S,\breve{p}}^{(0.90,2)}\right] \\
&= \frac{1}{0.9764}\left(\int_{0.0000}^{0.7764} p^2(1-p)^2 dp \right.\\
&\quad + \int_{0.0253}^{0.9747} 2\left(\frac{1}{2}-p\right)^2 p(1-p)dp \\
&\quad \left. + \int_{0.2236}^{1.0000} (1-p)^2 p^2 dp \right) \\
&= 0.0797
\end{aligned}
\tag{5.27}
$$

The values in Table 5.2 illustrate one unpleasant property of the relative frequency as point estimator of a probability. The estimates are often located rather close to the bounds of the measurement interval and far away from their centers. In the case of a measurement space smaller than the unit interval, the estimates will generally fall outside the bounds of the measurement intervals.

## Minimum MSE-$\beta$-Estimator Based on $\vec{X}$

Using the $\beta$-measurement intervals of the Neyman $\beta$-measurement procedure given in Table 4.3, the realizations of the minimum MSE-$\beta$-estimator are obtained:

$$
TMS_{\breve{p}}^{(0.90,2)}\left((0,0)\right) = \frac{\int_{0.0000}^{0.6838} p(1-p)^2 dp}{\int_{0.0000}^{0.6838} (1-p)^2 dp} = 0.2333
$$

$$
TMS_{\breve{p}}^{(0.90,2)}\left((0,1)\right) = \frac{\int_{0.0513}^{0.9000} p^2(1-p) dp}{\int_{0.0513}^{0.9000} p(1-p) dp} = 0.4911
$$

$$TMS_{\breve{p}}^{(0.90,2)}\big((1,0)\big) = \frac{\int\limits_{0.0513}^{0.9000} p^2(1-p)dp}{\int\limits_{0.0513}^{0.9000} p(1-p)dp} = 0.5089$$

$$TMS_{\breve{p}}^{(0.90,2)}\big((1,1)\big) = \frac{\int\limits_{0.3162}^{1.000} p^3 dp}{\int\limits_{0.3162}^{1.000} p^2 dp} = 0.7667$$

With these values and the $\beta$-measurement intervals given in Table 4.3, the average conditional mean squared error is easily obtained.

The corresponding point estimates, measurement intervals, and the conditional mean squared error are shown in Table 5.3.

**Table 5.3** *Complete Neyman measurement procedure with minimum MSE-$\beta$-estimator based on $\vec{X}$.*

| $\vec{x}$ | $TMS_{\breve{p}}^{(0.90,2)}(\vec{x})$ | Measurement interval |
|---|---|---|
| (0,0) | 0.2333 | $[0.0000, 0.6838]$ |
| (1,0) | 0.4911 | $[0.0513, 0.9000]$ |
| (0,1) | 0.5089 | $[0.1000, 0.9487]$ |
| (1,1) | 0.7667 | $[0.3162, 1.0000]$ |
| average mean squared error: | | 0.03624 |

There is a rather large difference between the estimates of the (asymptotic) estimator given by the relative frequency and the small sample estimator given by the minimum MSE-$\beta$-estimator. The quality of the two estimators is reflected by the conditional mean squared error, and the difference is indeed surprisingly high, indicating that for a small sample size, the traditional way of estimating the actual value $p$ of a given probability $\breve{p}$ is not at all good, despite the good asymptotic properties of the relative frequency.

Moreover, the estimates obtained when using the minimum MSE-$\beta$-estimator are located close to the center of the $\beta$-measurement intervals in contrast to a part of the estimates obtained by means of the relative frequency.

### Midpoint $\beta$-Estimator Based on $\vec{X}$

Based on the Neyman $X_S$-measurement procedure given in Table 4.3 and using the simple midpoint $\beta$-estimator, the results obtained are listed in Table 5.4.

**Table 5.4** *Complete Neyman measurement procedure with midpoint $\beta$-estimator based on $\vec{X}$.*

| $\vec{x}$ | $TMP_{\breve{p}}^{(0.90,2)}(\vec{x})$ | Measurement interval |
|---|---|---|
| (0,0) | 0.3419 | $[0.0000, 0.6838]$ |
| (1,0) | 0.4752 | $[0.0513, 0.9000]$ |
| (0,1) | 0.5243 | $[0.1000, 0.9487]$ |
| (1,1) | 0.6581 | $[0.3162, 1.0000]$ |
| average mean squared error: | | 0.04457 |

The difference in the conditional mean squared error compared to the one obtained for the more sophisticated minimum MSE-$\beta$-estimator is small but not negligible. Of course, the estimates of the midpoint $\beta$-estimator are, by definition, located in the center of the corresponding measurement intervals.

### Minimum MSE-$\beta$-Estimator Based on $X_S$

Based on the $\beta$-measurement & prediction space obtained by means of the sum $X_S$, and using the minimum MSE-$\beta$-estimator, the results in Table 5.5 are obtained.

**Table 5.5** *Complete Neyman measurement procedure with minimum MSE-$\beta$-estimator based on $X_S$.*

| $\vec{x}$ | $TMS_{\breve{p}}^{(0.90,2)}(x_S)$ | Measurement interval |
|---|---|---|
| (0,0) | 0.2333 | $[0.0000, 0.6838]$ |
| (1,0) | 0.5000 | $[0.0513, 0.9487]$ |
| (0,1) | 0.5000 | $[0.0513, 0.9487]$ |
| (1,1) | 0.7667 | $[0.3162, 1.0000]$ |
| average mean squared error: | | 0.03731 |

The difference in the conditional mean squared error of the minimum MSE-$\beta$-estimator based on the sample $\vec{X}$ compared to that based on the sum $X_S$ is small, but this was to be expected, because the difference between $\vec{X}$ and $X_S$ in the case of $n = 2$ is rather small, too.

### Midpoint $\beta$-Estimator Based on $X_S$

Based on the Neyman $\beta$-measurement & prediction space for $X_S$ and using the simpler midpoint $\beta$-estimator yields the results given in Table 5.6.

Again, the difference between the results based on the sample $\vec{X}$ and the sum $X_S$ are small due to the extremely small sample size. Nevertheless, a comparison of the average mean squared error of 0.04457 for the procedure based on $\vec{X}$ given in Table 5.4 and the mean squared error of 0.04549 for the procedure based on $X_S$ given in Table 5.6 shows that the quality of the point estimator gets slightly worse when using $X_S$ instead of $\vec{X}$.

**Table 5.6** *Complete Neyman measurement procedure with midpoint $\beta$-estimator based on $X_S$.*

| $\vec{x}$ | $TMP_{\breve{p}}^{(0.90,2)}(x_S)$ | Measurement interval |
|---|---|---|
| (0,0) | 0.3419 | $[0.0000, 0.6838]$ |
| (1,0) | 0.5000 | $[0.0513, 0.9487]$ |
| (0,1) | 0.5000 | $[0.0513, 0.9487]$ |
| (1,1) | 0.6581 | $[0.3162, 1.0000]$ |
| average mean squared error: | | 0.04549 |

### Minimum MSE-$\beta$-Estimator and Traditional Measurement Procedure

Based on the $\beta$-measurement & prediction space corresponding to the traditional $\beta$-measureament procedure and using the minimum MSE-$\beta$-estimator yields the estimates in Table 5.7.

**Table 5.7** *Traditional measurement procedure with minimum MSE-$\beta$-estimator.*

| $\vec{x}$ | $TMS_{\breve{p}}^{(0.90,2)}(x_S)$ | Measurement interval |
|---|---|---|
| (0,0) | 0.2434 | $[0.0000, 0.7764]$ |
| (1,0) | 0.5000 | $[0.0253, 0.9747]$ |
| (0,1) | 0.5000 | $[0.0253, 0.9747]$ |
| (1,1) | 0.7577 | $[0.2236, 1.0000]$ |
| average mean squared error: | | 0.03974 |

Comparing the average mean squared error in Table 5.7 with the one in Table 5.5 indicates that the conditional mean squared error is rather affected by using the minimum MSE-$\beta$-estimator than by the special kind of $\beta$-measurement procedure. In other words, even for the traditional measurement procedure, the minimum MSE-$\beta$-estimator shows very good behavior with respect to the average mean squared error.

## Midpoint $\beta$-Estimator and Traditional Measurement Procedure

Based on the $\beta$-measurement & prediction space corresponding to the traditional measurement procedure and using the midpoint $\beta$-estimator, the results in Table 5.8 are obtained.

**Table 5.8** *Traditional measurement procedure with midpoint $\beta$-estimator.*

| $\vec{x}$ | $TMP_{\breve{p}}^{(0.90,2)}(x_S)$ | Measurement interval |
|---|---|---|
| (0,0) | 0.3882 | $[0.0000, 0.7764]$ |
| (1,0) | 0.5000 | $[0.0253, 0.9747]$ |
| (0,1) | 0.5000 | $[0.0253, 0.9747]$ |
| (1,1) | 0.6118 | $[0.2236, 1.0000]$ |
| average mean squared error: | | 0.05389 |

Also, the midpoint $\beta$-estimator shows, at least in this example, some robustness against the different measurement procedures, although the effect of the nonoptimal measurement procedure is larger than in the case of the minimum MSE-$\beta$-estimator.

### 5.6.1 Conditional Mean Squared Error

The above example shows that in the case of a small sample size, the traditional criteria for point estimators fail to serve the aim of arriving at a precise point estimator sufficiently well. One of the main reasons is the fact that the traditional criteria such as unbiasedness, consistency, and the minimum variance criterion

- do not measure the overall performance of an estimator with respect to precision,
- do not allow a comparison of competing alternatives, and
- refer to many replications and large sample sizes.

Each of the traditional criteria is neither a necessary nor a sufficient condition for high precision.

The conditional mean squared error is defined as the expected squared deviation of the estimator from the actual value to be measured under the condition that the procedure is applied successfully. This criterion does not take into account measurement failure, which is controlled by the confidence level. It measures directly the expected precision of the procedure and is universally applicable. As the example elucidates, the results obtained by taking the conditional mean squared error as the quality criterion for point estimators differ greatly from those obtained by applying the traditional criteria. Moreover, the latter ones lead to more or less useless results in the case of a restricted measurement space.

As a first conclusion, it appears to be necessary that the traditional theory of point estimators has to be reconsidered with respect to properties that actually hold in the case of small samples being typical for many fields of application.

Moreover, it should be noted that an isolated point estimator promotes deterministic thinking, but not *statistical thinking* which has been identified for instance by the late W. Edwards Deming as a prerequisite for solving the problems of our time.

# Chapter 6

# Exclusion Procedures

## 6.1 Rejecting a Null Hypothesis

Roughly speaking, statistics may be divided into estimation theory and test theory. Estimation theory deals with the determination of actual values of probabilities. Test theory deals with developing procedures by means of which certain values of the measurement space $\Theta$ of a probability $\breve{\theta}$ may be excluded. Calling the set of values to be excluded "null hypothesis $H_0$," a test serves for rejecting $H_0$.

Generally, the probability distribution is assumed to be known, except for the actual value of a one-dimensional or multidimensional parameter. In such a situation, the null hypothesis is expressed as a subset of the set of admissible parameter values. In the case investigated here, the set of distributions is given by the values of the measurement space $[\underline{p}, \overline{p}]$ with $0 \leq \underline{p} < \overline{p} \leq 1$. The null hypothesis $H_0$ takes the form

$$H_0 : \quad [\underline{p}^{(0)}, \overline{p}^{(0)}] \quad \text{with } \underline{p} \leq \underline{p}^{(0)} \leq \overline{p}^{(0)} \leq \overline{p}. \tag{6.1}$$

As the word "test" has a rather vague and general meaning and, therefore, may lead (and actually often leads) to misunderstandings, a more precise term, namely *exclusion procedure* instead of *significance tests*, is preferred in this handbook.

Similar to the preceding measurement procedures, the exclusion procedures aim at being a complement to traditional significance tests. They have a special impact on small sample theory and, therefore, may be looked upon as an "exact" alternative to "inexact" and, in some sense, unscientific asymptotic test procedures, where "unscientific" means that the latter procedures adopt the required properties only asymptotically. Nevertheless, they are used also in the case of small samples, often without a sufficient scientific justification.

As mentioned earlier, one of the main differences between the traditional methods and the procedures developed here is the fact that the former are based on the generally unrealistic assumption of a maximum parameter space limited only by formal "mathematical" conditions, which follow from the fact that the quantity of interest serves as a distribution parameter. For any real problem, however, the quantity of interest is known except for its actual value. From the available knowledge about the quantity of interest, it is always possible to specify a convex and bounded set, which contains, with certainty, the actual value of the parameter of interest. Moreover, one should always have in mind that the parameter of interest, although looked upon as a distribution parameter, has generally, at least partially, a "non-distribution" meaning, and any pure distribution parameter often has the character of a "nuisance" parameter.

The primary aim of a measurement procedure is to reduce the measurement space $[\underline{p}, \overline{p}]$ for arriving at a smaller range, which includes the actual value of the probability. However, often the actual value $p$ and thus a range for $p$ is of no or only minor interest. The aim then is to exclude certain probabilities [e.g., for showing that an unknown probability is different (larger/smaller) than another one]. As shown later, it is easy to derive an exclusion procedure for a given null hypothesis if a $\beta$-measurement and prediction space is available. Before we turn to the exclusion procedure, the null hypothesis must be considered.

In traditional test theory, simple null hypotheses, which contain only one single element, play an important role. A reason for this fact might be that a simple hypothesis is very convenient from a mathematical point of view. From a more realistic and scientific point of view, however, it often does not make much sense to exclude only one element, or more generally said, a set of Lebesgue measure zero. Excluding one specific value $p^{(0)}$ of $[\underline{p}, \overline{p}]$ and at the same time admitting that any value arbitrarily close to $p^{(0)}$ is still a possible candidate lacks, strictly speaking, any realistic rationale.

Therefore, from a theoretical point of view, any reasonable null hypothesis with respect to the actual value $p$ of the probability $\breve{p}$ should necessarily be an interval with positive length $[\underline{p}^{(0)}, \overline{p}^{(0)}] \subset [\underline{p}, \overline{p}]$.

However, to exclude one single value makes sense, of course, in the discrete case. Because, in practical applications, the set of discrete portions are approximated by an interval of probabilities, a simple null hypothesis may be meaningful from a practical point of view, in contrast to the above considerations. Therefore, simple hypotheses are not excluded from further considerations, but emphasis is laid on null hypotheses represented by intervals of positive length.

The situation and the problem to be solved may be described in the following way:

- The probability of interest $\breve{p}$ refers to a given event. The actual value $p$ of $\breve{p}$ is unknown. The nature of the event and, thus, of $\breve{p}$ enables one to select a suitable measurement space $[\underline{p}, \overline{p}] \subset [0, 1]$.

- The problem is to exclude a set $[\underline{p}^{(0)}, \overline{p}^{(0)}]$ called null hypothesis and denoted by $H_0$ from the set of all probabilities given by the selected measurement space $[\underline{p}, \overline{p}]$.

- In order to solve the problem, an independent and identically distributed random sample $\vec{X} = (X_1, \ldots, X_n)$ for $X \sim B(p)$ is available.

- An erroneous exclusion of $H_0$ shall only happen with probability not exceeding a given bound $\alpha$, which is called the significance level.

Hence, the aim is to reject the null hypothesis $H_0$ under the condition that an error is committed only with a probability bounded by $\alpha$. An exclusion of $[\underline{p}^{(0)}, \overline{p}^{(0)}]$ is a success and, therefore, the methods developed have to be evaluated according to their ability to lead to a success.

### 6.1.1 The Exclusion Procedure as a Decision Procedure

Exclusion of $H_0$ means making one of two possible decisions:

1. Reject the null hypothesis $H_0$ (i.e., exclude the set $[\underline{p}^{(0)}, \overline{p}^{(0)}]$).
2. Do not reject the null hypothesis $H_0$ (i.e., don't exclude the set $[\underline{p}^{(0)}, \overline{p}^{(0)}]$).

The second decision means keeping (accepting) the whole parameter space $[\underline{p}, \overline{p}]$.

Evidently, the two decisions may also be formulated in the following way:

1. Assume $[\underline{p}, \overline{p}] \setminus [\underline{p}^{(0)}, \overline{p}^{(0)}]$.
2. Keep the original parameter space $[\underline{p}, \overline{p}]$.

They may be formulated as positive decisions as follows:

1. Decide for $p \in [\underline{p}, \overline{p}] \setminus [\underline{p}^{(0)}, \overline{p}^{(0)}]$.
2. Decide for $p \in [\underline{p}, \overline{p}]$.

The decision procedure is formally given by a measurable function $\phi_{H_0}$ with

$$\phi_{H_0} : \{0,1\}^n \to \{0,1\}, \tag{6.2}$$

where $\phi_{H_0}(\vec{x}) = 0$ means a decision in favor of $[\underline{p}, \overline{p}]$, and $\phi_{H_0}(\vec{x}) = 1$ means a decision in favor of $[\underline{p}, \overline{p}] \setminus [\underline{p}^{(0)}, \overline{p}^{(0)}]$.

The above alternative formulations will clarify the situation in the following sense: A significance test (called an exclusion procedure here) does not decide between a given null hypothesis $H_0$ and its complement, but it indicates in a scientifically correct way whether the range $[\underline{p},\overline{p}]$ may be reduced by $H_0$ or whether such a reduction is not possible.

A necessary condition for a decision procedure to be looked upon as scientifically justified is that its reliability is given. Clearly, an error may occur only if the procedure rejects $H_0$. Such a decision is wrong if the unknown value $p$ is actually an element of $[\underline{p}^{(0)},\overline{p}^{(0)}]$.

On the other hand, deciding for $p \in [\underline{p},\overline{p}]$ does not constitute an error, as it is true with certainty. Hence, it is only necessary to take precautions against false decisions in favor of the first alternative. As shown earlier, the reliability of an exclusion procedure is stated by the significance level $\alpha$, which is an upper bound for committing an exclusion error:

$$\mathbf{P}_{\vec{X}}^{(p)}\Big(\phi_{H_0}^{-1}(\{1\})\Big) \leq \alpha \quad \text{for any } p \in [\underline{p}^{(0)},\overline{p}^{(0)}], \qquad (6.3)$$

where $\phi_{H_0}^{-1}(\{1\}) = \Big\{\vec{x}\,\Big|\,\phi_{H_0}(\vec{x}) = 1\Big\}$.

Any exclusion procedure for $H_0$ meeting the reliability requirement (6.3) is called an $\alpha$-exclusion procedure. In the case that $H_0$ is not excluded, the aim of applying the procedure is not reached, and the procedure has been applied unsuccessfully. Therefore, one looks for a procedure that maximizes the success probability and meets the side condition with respect to the reliability given by the significance level $\alpha$. Clearly, before the problem can be solved, the question of how to calculate the success probability of an exclusion procedure has to be answered.

## 6.2 $\alpha$-Exclusion Space

Let $H_0 : [\underline{p}^{(0)},\overline{p}^{(0)}]$ be the null hypothesis. The problem of rejecting $H_0$ on the basis of $\vec{X}$ (or $X_S$) may be solved preliminary by means of a $(1-\alpha)$-measurement & prediction space. The following considerations are restricted to the case in which the sample $\vec{X}$ is used, but can easily be extended to the case of $X_S$

by exchanging the two symbols and the respective probability distributions.

- Let $\beta = 1 - \alpha$; then, determine for each element $p \in [\underline{p}^{(0)}, \overline{p}^{(0)}]$, a $\beta$-prediction region $A_{\vec{X}}^{(\beta,n)}(p)$ implying that the conditional probability $\mathbf{P}_{\vec{X}}^{(p)}\big(A_{\vec{X}}^{(\beta,n)}(p)\big)$ is equal to or larger than $\beta$ for any $p \in [\underline{p}^{(0)}, \overline{p}^{(0)}]$.

- Next, define the decision function $\phi_{H_0}$ by

$$\begin{aligned}\phi_{H_0}^{-1}\big(\{0\}\big) &= \bigcup_{p\in[\underline{p}^{(0)},\overline{p}^{(0)}]} A_{\vec{X}}^{(\beta,n)}(p)\\ \phi_{H_0}^{-1}\big(\{1\}\big) &= \big\{0,1\big\}^n \setminus \bigcup_{p\in[\underline{p}^{(0)},\overline{p}^{(0)}]} A_{\vec{X}}^{(\beta,n)}(p).\end{aligned} \tag{6.4}$$

The decision function defined above meets the reliability requirement (6.3) because

$$\begin{aligned}\mathbf{P}_{\vec{X}}^{(p)}\big(\phi_{H_0}^{-1}(\{1\})\big) &= 1 - \mathbf{P}_{\vec{X}}^{(p)}\Big(\bigcup_{p\in[\underline{p}^{(0)},\overline{p}^{(0)}]} A_{\vec{X}}^{(\beta,n)}(p)\Big)\\ &\leq 1 - \mathbf{P}_{\vec{X}}^{(p)}\big(A_{\vec{X}}^{(\beta,n)}(\tilde{p})\big)\\ &\leq 1 - \beta = \alpha\\ &\qquad \text{for any fixed } \tilde{p} \in [\underline{p}^{(0)}, \overline{p}^{(0)}].\end{aligned} \tag{6.5}$$

If $p \in [\underline{p}^{(0)}, \overline{p}^{(0)}]$, then the event $\bigcup_{p\in[\underline{p}^{(0)},\overline{p}^{(0)}]} A_{\vec{X}}^{(\beta,n)}(p)$ occurs with a high probability ($\geq \beta$). Thus, in case of its occurrence, the null hypothesis $H_0$ may not be rejected. In the case of the complementary event, however, one may exclude $[\underline{p}^{(0)}, \overline{p}^{(0)}]$, as the complementary event is not compatible (on the basis of the significance level $\alpha$) with any pair $(p, \vec{X})$ where $p \in [\underline{p}^{(0)}, \overline{p}^{(0)}]$.

In summary, an $\alpha$-exclusion procedure for $H_0$ may be given by the set of $(1-\alpha)$-prediction regions for $p \in [\underline{p}^{(0)}, \overline{p}^{(0)}]$ or, more precisely, by the union of the $(1-\alpha)$-prediction regions:

$$A_{\vec{X}}^{(1-\alpha,n)}([\underline{p}^{(0)}, \overline{p}^{(0)}]) = \bigcup_{p\in[\underline{p}^{(0)},\overline{p}^{(0)}]} A_{\vec{X}}^{(1-\alpha,n)}(p). \tag{6.6}$$

The set $A_{\vec{X}}^{(1-\alpha,n)}([\underline{p}^{(0)},\overline{p}^{(0)}])$ is called a *local* $(1-\alpha)$*-prediction region.*

A local $(1-\alpha)$-prediction region together with $[\underline{p}^{(0)},\overline{p}^{(0)}]$ determines a *local* $(1-\alpha)$*-prediction procedure*, which is also completely given by the following subset of the corresponding measurement & prediction space:

$$[\underline{p}^{(0)},\overline{p}^{(0)}] \times A_{\vec{X}}^{(1-\alpha,n)}([\underline{p}^{(0)},\overline{p}^{(0)}]). \tag{6.7}$$

The set (6.7) is called a *local* $(1-\alpha)$*-prediction space.* Its projection on the $p$-axis equals $[\underline{p}^{(0)},\overline{p}^{(0)}]$, and its projection on the $\vec{x}$-axis equals the local $(1-\alpha)$-prediction region.

Any local $(1-\alpha)$-prediction space determines a set denoted by $E_{\check{p},\vec{X}}^{(\alpha,n)}\big([\underline{p}^{(0)},\overline{p}^{(0)}]\big)$:

$$\begin{aligned} E_{\check{p},\vec{X}}^{(\alpha,n)}\big([\underline{p}^{(0)},\overline{p}^{(0)}]\big) &= \big([\underline{p},\overline{p}] \setminus [\underline{p}^{(0)},\overline{p}^{(0)}]\big) \\ &\quad \times \big(\{0,1\}^n \setminus A_{\vec{X}}^{(1-\alpha,n)}([\underline{p}^{(0)},\overline{p}^{(0)}])\big). \end{aligned} \tag{6.8}$$

$E_{\check{p},\vec{X}}^{(1-\alpha,n)}\big([\underline{p}^{(0)},\overline{p}^{(0)}]\big)$ determines completely the exclusion procedure, and therefore it is called an $\alpha$*-exclusion space.* Its projection on the $p$-axis equals the set $[\underline{p},\overline{p}]\setminus[\underline{p}^{(0)},\overline{p}^{(0)}]$, and the projection on the $\vec{x}$-axis equals $\{0,1\}^n \setminus A_{\vec{X}}^{(1-\alpha,n)}([\underline{p}^{(0)},\overline{p}^{(0)}])$, which allows one to reject $H_0$ and accept its complement.

Similar to the duality between corresponding $\beta$-prediction and $\beta$-measurement procedures, any local $(1-\alpha)$-prediction procedure is dual to a corresponding $\alpha$-exclusion procedure. A local $(1-\alpha)$-prediction procedure determines uniquely an $\alpha$-exclusion procedure and vice versa.

It is easy to see that the reliability of the exclusion procedure given by (6.6) is generally larger than the required $\alpha$. Thus, the problem arises of how to improve (i.e., reduce) a given local prediction region $A_{\vec{X}}^{(1-\alpha,n)}([\underline{p}^{(0)},\overline{p}^{(0)}])$ in a way that the probability of exclusion (i.e., the success probability) of the corresponding $\alpha$-exclusion procedure is maximized.

The reliability requirement for an exclusion procedure is specified by the significance level $\alpha$ or by its complement $1-\alpha$, where the latter is a lower bound for the probability of not committing an exclusion error. In view of the local $(1-\alpha)$-prediction region (6.6) for the null hypothesis given by $[\underline{p}^{(0)}, \overline{p}^{(0)}]$, the reliability requirement may be formulated in the following way:

$$\min_{p\in[\underline{p}^{(0)},\overline{p}^{(0)}]} \mathbf{P}_{\vec{X}}^{(p)}\Big(A_{\vec{X}}^{(1-\alpha,n)}([\underline{p}^{(0)}, \overline{p}^{(0)}])\Big) \geq 1-\alpha. \tag{6.9}$$

In the continuous case, it is always possible to exactly meet the reliability requirement [i.e., obtain an equality sign in (6.9)]. However, in the discrete case, this is not possible in general, and, therefore, the left-hand side of (6.9) is often larger than $1-\alpha$. Analogous to the reliability excess of a measurement procedure, we call

$$\min_{p\in[\underline{p}^{(0)},\overline{p}^{(0)}]} \mathbf{P}_{\vec{X}}^{(p)}\Big(A_{\vec{X}}^{(1-\alpha,n)}([\underline{p}^{(0)}, \overline{p}^{(0)}])\Big) - (1-\alpha) \tag{6.10}$$

the reliability excess of the local $(1-\alpha)$-prediction procedure. The reliability excess of the dual $\alpha$-exclusion procedure is obtained by

$$\alpha - \max_{p\in[\underline{p}^{(0)},\overline{p}^{(0)}]} \mathbf{P}_{\vec{X}}^{(p)}\Big(\{0,1\}\setminus A_{\vec{X}}^{(1-\alpha,n)}([\underline{p}^{(0)}, \overline{p}^{(0)}])\Big)$$
$$= \alpha - \left(1 - \min_{p\in[\underline{p}^{(0)},\overline{p}^{(0)}]} \mathbf{P}_{\vec{X}}^{(p)}\Big(A_{\vec{X}}^{(1-\alpha,n)}([\underline{p}^{(0)}, \overline{p}^{(0)}])\Big)\right)$$

which, of course, is equal to (6.10).

## 6.3 Quality of an $\alpha$-Exclusion Procedure

Clearly, from two exclusion procedures with the same reliability $\alpha = 1-\beta$ and the same expense $n$, the one which more frequently leads to a success (i.e., excludes $[\underline{p}^{(0)}, \overline{p}^{(0)}]$) will be called better. Calling the event of not rejecting $H_0$ a failure, the quality of an $\alpha$-exclusion procedure is given by its success probability or, equivalently, by its failure probability.

The procedure results in a failure if the event

$$A_{\vec{X}}^{(1-\alpha,n)}([\underline{p}^{(0)},\overline{p}^{(0)}]) \subset \{0,1\}^n \tag{6.11}$$

occurs. The probability distribution on $\left(\{0,1\}^n, \mathcal{P}_{\{0,1\}^n}\right)$ is given the marginal distribution $Q_{\{0,1\}^n}$ defined by (4.13). Thus, the failure probability (cf. (4.53)) is given by

$$Q_{\{0,1\}^n}\left(A_{\vec{X}}^{(1-\alpha,n)}([\underline{p}^{(0)},\overline{p}^{(0)}])\right) \tag{6.12}$$

and the success probability by:

$$1 - Q_{\{0,1\}^n}\left(A_{\vec{X}}^{(1-\alpha,n)}([\underline{p}^{(0)},\overline{p}^{(0)}])\right). \tag{6.13}$$

Note that in contrast to a $(1-\alpha)$-prediction procedure determined by a $(1-\alpha)$-measurement & prediction space, a local $(1-\alpha)$-prediction procedure consists of one single prediction region only, which is valid for the whole null hypothesis $H_0$.

With (6.12) or (6.13), the quality of any $\alpha$-exclusion procedure is defined and enables the comparison of two exclusion procedures (with respect to their success probability). Note that the success probability does not depend on the actual and unknown value $p$ (i.e., on a special application), but is a property of the procedure developed to be applied to the specified measurement space.

**Definition 6.1**
*Let the two local* $(1-\alpha)$*-prediction regions* $A_{\vec{X}}^{(1-\alpha,n)}([\underline{p}^{(0)},\overline{p}^{(0)}])$ *and* $\tilde{A}_{\vec{X}}^{(1-\alpha,n)}([\underline{p}^{(0)},\overline{p}^{(0)}])$ *define two* $\alpha$*-exclusion procedures for the null hypothesis* $H_0 : [\underline{p}^{(0)},\overline{p}^{(0)}]$*. Then,*

- $A_{\vec{X}}^{(1-\alpha,n)}([\underline{p}^{(0)},\overline{p}^{(0)}])$ *is superior to* $\tilde{A}_{\vec{X}}^{(1-\alpha,n)}([\underline{p}^{(0)},\overline{p}^{(0)}])$ *if*
  $$Q_{\{0,1\}^n}\left(A_{\vec{X}}^{(1-\alpha,n)}([\underline{p}^{(0)},\overline{p}^{(0)}])\right) < Q_{\{0,1\}^n}\left(\tilde{A}_{\vec{X}}^{(1-\alpha,n)}([\underline{p}^{(0)},\overline{p}^{(0)}])\right).$$

- $A_{\vec{X}}^{(1-\alpha,n)}([\underline{p}^{(0)},\overline{p}^{(0)}])$ *and* $\tilde{A}_{\vec{X}}^{(1-\alpha,n)}([\underline{p}^{(0)},\overline{p}^{(0)}])$ *are equally appropriate, if*
$$Q_{\{0,1\}^n}\left(A_{\vec{X}}^{(1-\alpha,n)}([\underline{p}^{(0)},\overline{p}^{(0)}])\right) = Q_{\{0,1\}^n}\left(\tilde{A}_{\vec{X}}^{(1-\alpha,n)}([\underline{p}^{(0)},\overline{p}^{(0)}])\right).$$

•

## 6.4 Neyman Exclusion Procedure

The definition of a quality indicator for an $\alpha$-exclusion procedure enables the definition of an $\alpha$-exclusion procedure which is optimal with respect to the success probability.

**Definition 6.2**
*An $\alpha$-exclusion procedure given by the local $(1-\alpha)$-prediction region* ${}^*A_{\vec{X}}^{(1-\alpha,n)}([\underline{p}^{(0)},\overline{p}^{(0)}])$ *is called a* Neyman $\alpha$-exclusion procedure *if*

$$Q_{\{0,1\}^n}\left({}^*A_{\vec{X}}^{(1-\alpha,n)}([\underline{p}^{(0)},\overline{p}^{(0)}])\right) \leq Q_{\{0,1\}^n}\left(A_{\vec{X}}^{(1-\alpha,n)}([\underline{p}^{(0)},\overline{p}^{(0)}])\right) \tag{6.14}$$

*for any $\alpha$-exclusion procedure given by a local $(1-\alpha)$-prediction region* $A_{\vec{X}}^{(1-\alpha,n)}([\underline{p}^{(0)},\overline{p}^{(0)}])$. •

Note that for any exclusion procedure calculation of the success probability and thus determination of its quality constitutes no problem. Moreover, the existence of Neyman $\alpha$-exclusion procedures is, of course, guaranteed, although not uniquely.

## 6.5 Determination of Neyman Exclusion Procedures

For determining a Neyman $\alpha$-exclusion procedure, the following optimization problem has to be solved:

**Problem 6.1**

$$Q_{\{0,1\}^n}\Big(A_{\vec{X}}^{(1-\alpha,n)}([\underline{p}^{(0)},\overline{p}^{(0)}])\Big) \stackrel{!}{=} \min \tag{6.15}$$

$$\mathbf{P}_{\vec{X}}^{(p)}\Big(A_{\vec{X}}^{(1-\alpha,n)}([\underline{p}^{(0)},\overline{p}^{(0)}])\Big) \geq 1-\alpha \quad \textit{for any } p \in [\underline{p}^{(0)},\overline{p}^{(0)}]. \tag{6.16}$$

•

The problem may also be formulated in the following way:

- For any $p \in [\underline{p}^{(0)},\overline{p}^{(0)}]$, select an appropriate element of $\mathcal{P}_{\{0,1\}^n}$ in such a way that its measure with respect to $\mathbf{P}_{\vec{X}}^{(p)}$ is not less than $1-\alpha$, and
- its measure with respect to $Q_{\{0,1\}^n}$ is minimum.

The solution of the problem is to determine the local $(1-\alpha)$-prediction region ${}^*A_{\vec{X}}^{(1-\alpha,n)}([\underline{p}^{(0)},\overline{p}^{(0)}])$ according to the following rule.

- For $p = \underline{p}^{(0)}$ and $p = \overline{p}^{(0)}$, order the elements $\vec{x} \in \{0,1\}^n$ according to the *likelihood ratio* of the two measure functions concerned:

$$\frac{\mathbf{P}_{\vec{X}}^{(p)}(\vec{x})}{Q_{\{0,1\}^n}(\vec{x})}. \tag{6.17}$$

Next, define the following sets, which are filled according to the order of the likelihood ratios:

$$A(r) = \left\{\vec{x} \,\middle|\, \frac{\mathbf{P}_{\vec{X}}^{(\underline{p}^{(0)})}(\vec{x})}{Q_{\{0,1\}^n}(\vec{x})} \geq r \text{ and } \frac{\mathbf{P}_{\vec{X}}^{(\overline{p}^{(0)})}(\vec{x})}{Q_{\{0,1\}^n}(\vec{x})} \geq r\right\}. \tag{6.18}$$

Then, an almost optimal local $(1-\alpha)$-prediction region ${}^*A_{\vec{X}}^{(1-\alpha,n)}([\underline{p}^{(0)},\overline{p}^{(0)}])$ is obtained by

$${}^*A_{\vec{X}}^{(1-\alpha,n)}([\underline{p}^{(0)},\overline{p}^{(0)}]) = A\Big(r^*_{1-\alpha,n}([\underline{p}^{(0)},\overline{p}^{(0)}])\Big), \tag{6.19}$$

where $r^*_{1-\alpha,n}([\underline{p}^{(0)}, \overline{p}^{(0)}])$ is given by

$$r^*_{1-\alpha,n}([\underline{p}^{(0)}, \overline{p}^{(0)}]) = \arg\max_{r>0} \Big\{ r \,\Big|\, \mathbf{P}_{\vec{X}}^{(\underline{p}^{(0)})}(A(r)) \geq 1-\alpha \text{ and } \mathbf{P}^{(\overline{p}^{(0)})}(A(r)) \geq 1-\alpha \Big\}. \quad (6.20)$$

The local $(1-\alpha)$-prediction space is filled on the basis of the likelihood ratio order until its conditional probability volume (under the condition of the null hypothesis) for the first time is equal to or exceeds the given reliability bound $1-\alpha$.

The resulting local $(1-\alpha)$-prediction region may not be optimal because of the discreteness of the binomial distribution. This can be taken into account by additional rules, analogous to the case of the determination of Neyman $\beta$-prediction procedures. •

If a $(1-\alpha)$-measurement & prediction space is available, then one can start the search for ${}^*A_{\vec{X}}^{(1-\alpha,n)}([\underline{p}^{(0)}, \overline{p}^{(0)}])$ with the union

$$\bigcup_{p \in [\underline{p}^{(0)}, \overline{p}^{(0)}]} A_{\vec{X}}^{(1-\alpha,n)}(p)$$

and reduce it according to the likelihood ratio order of the elements, until the reliability bound is reached.

**Remark 6.1**
*If, instead of the sample $\vec{X}$, the sum $X_S$ is used, then the above derivations are easily transferred by exchanging the relevant measure functions.* •

## 6.6 Illustrative Example for an $\alpha$-Exclusion Procedure

To illustrate the determination of $\alpha$-exclusion procedures based on a sample $\vec{X}$, the already well-known example

$$\beta = 1-\alpha = 0.90, \; n = 2, \; \text{and } [\underline{p}, \overline{p}] = [0.0, 1.0]$$

is used. For this example, a (0.9)-measurement and prediction space is available and displayed in Figure 4.1.

In the following, different ways of proceeding are exemplified by means of different null hypotheses.

- As a first example, a simple null hypothesis is investigated:

$$H_0^{(1)} : [\underline{p}^{(0)}, \overline{p}^{(0)}] = \{0.1\}. \tag{6.21}$$

The ordered likelihood ratios and the conditional probabilities of the realizations of $\vec{X}$ are given in Table 6.1.

Selecting the realization according to the size of its likelihood ratio means to start with $(0,0)$ and proceed to $(0,1)$. As the probability for $\{(0,0),(0,1)\}$ sums up to $1-\alpha = 0.9$, a Neyman 0.1-exclusion procedure for $H_0^{(1)}$ is defined by the local prediction region

$$^*A_{\vec{X}}^{(0.90,2)}\big(\{0.1\}\big) = \big\{(0,0),(0,1)\big\}. \tag{6.22}$$

Clearly, another equally appropriate local prediction region is

$$^*A_{\vec{X}}^{(0.90,2)}\big(\{0.1\}\big) = \big\{(0,0),(1,0)\big\}$$

illustrating the fact that there might be several Neyman $\alpha$-exclusion procedures.

**Table 6.1** *Probabilities and likelihood ratios of the realizations of $\vec{X}$.*

| $\vec{x}$ | $\mathbf{P}_{\vec{X}}^{(0.10)}(\vec{x})$ | $Q_{\{0,1\}^n}(\vec{x})$ | $\frac{\mathbf{P}_{\vec{X}}^{(0.10)}(\vec{x})}{Q_{\{0,1\}^2}(\vec{x})}$ |
|---|---|---|---|
| $(0,0)$ | 0.81 | 0.333 | 2.430 |
| $(0,1)$ | 0.09 | 0.167 | 0.540 |
| $(1,0)$ | 0.09 | 0.167 | 0.540 |
| $(1,1)$ | 0.01 | 0.333 | 0.030 |

The above-determined local $(1-\alpha)$-prediction region (6.22) leads to the following decision function:

$$\phi_{H_0^{(1)}}(\vec{x}) = \begin{cases} 0 & \text{if } \vec{x} \in \Big\{(0,0),(0,1)\Big\} \\ \\ 1 & \text{if } \vec{x} \notin \Big\{(1,0),(1,1)\Big\}. \end{cases} \tag{6.23}$$

The success probability for the above $\alpha$-exclusion procedure is given by the probability of rejecting the null hypothesis:

$$Q_{\{0,1\}^2}\Big(\{(1,0),(1,1)\}\Big) = 0.5. \tag{6.24}$$

The failure probability is complementary to the success probability, and, therefore, it adopts the value 0.5 also.

- Next, a null hypothesis with a positive measure is considered:

$$H_0^{(2)} : \; [\underline{p}^{(0)}, \overline{p}^{(0)}] = [0.05, 0.10]. \tag{6.25}$$

The probabilities and the likelihood ratios of the realizations of $\vec{X}$ are given in Table 6.2.

Selecting the realizations in the order of their likelihood ratio means to start with $(0,0)$. Evidently, already $\{(0,0)\}$ meets the reliability requirement of $(1-\alpha)$ for $\underline{p}^{(0)} = 0.05$. However, it is not met for $\overline{p}^{(0)} = 0.10$. The next realization to be selected is $(0,1)$, with a likelihood ratio of 0.540. The set of the two realizations fulfills the reliability requirement for either $\underline{p}^{(0)} = 0.05$ or $\overline{p}^{(0)} = 0.10$. Clearly, the set cannot be further reduced, and, therefore, the optimal $\alpha$-exclusion procedure for $\alpha = 0.1$ and $H_0^{(2)}$ is defined by the local prediction region:

$$^{*}A_{\vec{X}}^{(0.90,2)}\Big([0.05, 0.1]\Big) = \Big\{(0,0),(0,1)\Big\}$$

which is identical to the one for the 0.1-exclusion procedure in the case of $H_0^{(1)}$. Thus, also the decision function is the same:

$$\phi_{H_0^{(2)}}(\vec{x}) = \begin{cases} 0 & \text{if } \vec{x} \in \Big\{(0,0),(0,1)\Big\} \\ 1 & \text{if } \vec{x} \notin \Big\{(1,0),(1,1)\Big\}. \end{cases} \tag{6.26}$$

**Table 6.2** *Probabilities and likelihood ratios of the realizations of $\vec{X}$ for $H_0^{(2)}$.*

| $\vec{x}$ | $\mathbf{P}_{\vec{X}}^{(0.05)}(\vec{x})$ | $Q_{\{0,1\}^n}(\vec{x})$ | $\frac{\mathbf{P}_{\vec{X}}^{(0.05)}(\vec{x})}{Q_{\{0,1\}^2}(\vec{x})}$ |
|---|---|---|---|
| $(0,0)$ | 0.9025 | 0.3333 | 2.7075 |
| $(0,1)$ | 0.0475 | 0.1667 | 0.2850 |
| $(1,0)$ | 0.0475 | 0.1667 | 0.2850 |
| $(1,1)$ | 0.0025 | 0.3333 | 0.0075 |

| $\vec{x}$ | $\mathbf{P}_{\vec{X}}^{(0.10)}(\vec{x})$ | $Q_{\{0,1\}^n}(\vec{x})$ | $\frac{\mathbf{P}_{\vec{X}}^{(0.10)}(\vec{x})}{Q_{\{0,1\}^2}(\vec{x})}$ |
|---|---|---|---|
| $(0,0)$ | 0.8100 | 0.3333 | 2.4300 |
| $(0,1)$ | 0.0900 | 0.1667 | 0.5400 |
| $(1,0)$ | 0.0900 | 0.1667 | 0.5400 |
| $(1,1)$ | 0.0100 | 0.3333 | 0.0300 |

Of course, also the success probability for the exclusion procedure for $H_0^{(2)}$ is the same as in the case of $H_0^{(1)}$ that is,

$$Q\Big((1,0),(0,0)\Big) = 0.5.$$

- Next, consider the example that an $\alpha$-exclusion procedure shall be derived based on the sum $X_S$ instead of the sample $\vec{X}$. The null hypothesis shall be identical to $H_0^{(2)}$:

$$H_0^{(3)} : \; [\underline{p}^{(0)}, \overline{p}^{(0)}] = [0.05, 0.10]. \tag{6.27}$$

The probabilities and likelihood ratios of the realizations of $X_S$ are given in Table 6.3.

In this case, the optimal $\alpha$-exclusion procedure based on $X_S$ for $H_0^{(3)}$ is defined by the local prediction region

$$ {}^*A_{X_S}^{(0.90,2)}\left([0.05, 0.1]\right) = \left\{0, 1\right\} $$

and the decision function

$$ \phi_{H_0^{(3)}}(x_S) = \begin{cases} 0 & \text{if } x_S \in \left\{0, 1\right\} \\ \\ 1 & \text{if } x_S \notin \left\{2\right\}. \end{cases} \qquad (6.28) $$

**Table 6.3** *Likelihood ratios and probabilities of the realizations of $X_S$ for $H_0^{(3)}$.*

| $x_S$ | $\mathbf{P}_{X_S}^{(0.05)}(x_S)$ | $Q_{\{0,1,2\}}(x_S)$ | $\frac{\mathbf{P}_{X_S}^{(0.05)}(x_S)}{Q_{\{0,1,2\}}(x_S)}$ |
|---|---|---|---|
| 0 | 0.9025 | 0.3333 | 2.7075 |
| 1 | 0.0950 | 0.3333 | 0.2850 |
| 2 | 0.0025 | 0.3333 | 0.0075 |
| $x_S$ | $\mathbf{P}_{X_S}^{(0.10)}(x_S)$ | $Q_{\{0,1,2\}}(x_S)$ | $\frac{\mathbf{P}_{X_S}^{(0.10)}(x_S)}{Q_{\{0,1,2\}}(x_S)}$ |
| 0 | 0.8100 | 0.3333 | 2.4300 |
| 1 | 0.1900 | 0.3333 | 0.5400 |
| 2 | 0.0100 | 0.3333 | 0.0300 |

The difference between the two examples becomes obvious when comparing the success probabilities. In the case that the $\alpha$-exclusion procedure is based on the sum $X_S$ instead of the sample $\vec{X}$, we obtain

$$Q_{\{0,1,2\}}\Big(\{2\}\Big) = 0.33 \tag{6.29}$$

which is considerably smaller than the success probability of $Q_{\{0,1\}^2}\Big(\{(1,0),(1,1)\}\Big) = 0.5$ for the procedure based on the sample $\vec{X}$.

### 6.6.1 The Measurement Space and the Exclusion Procedure

This section deals with the question of how the measurement space $[\underline{p}, \overline{p}]$ affects a Neyman exclusion procedure. The effect of the measurement space is illustrated by some examples. In each of them, the same reliability $\beta = 1 - \alpha = 0.8$, the same sample size $n = 40$, and the same null hypothesis

$$H_0 : \ \{0.5\} \tag{6.30}$$

are selected and combined with different measurement spaces.

With the notation introduced above, the null hypothesis is given by $\underline{p}^{(0)} = \overline{p}^{(0)} = 0.5$. In each of the examples, the exclusion procedure is not based on the sample $\vec{X}$, but on the sum of sample elements $X_S$.

**Example 6.1**

*The first example refers to the maximum measurement space* $[0,1]$. *For determining the Neyman exclusion procedure, the conditional probability, the marginal probability, and the likelihood ratio of the realizations* $x_S$ *under the null hypothesis are needed and listed in Table 6.4.*

*A Neyman exclusion procedure for the null hypothesis* $H_0$ *is completely determined by the corresponding local prediction region* $A_{X_S}^{(0.80,40)}\Big([0.5, 0.5]\Big)$, *which is obtained by filling it according to the likelihood ratio order given in Table 6.4 until the reliability requirement given by* $1 - \alpha = 0.8$ *is met.*

*The resulting local $(1-\alpha)$-prediction region is given by*

$$A_{X_S}^{(0.80,40)}\Big([0.5,0.5]\Big) = \Big\{16,17,\ldots,24\Big\}. \tag{6.31}$$

*Note that the likelihood ratio order of the elements is not uniquely defined, as there are elements with the same value of the likelihood ratio. Hence, there is not a unique Neyman exclusion procedure.*

***Table 6.4*** *Conditional probability, marginal probability, and likelihood ratio for the measurement space* $[0,1]$.

| $x_S$ | $\mathbf{P}_{X_S}^{(0.5)}(x_S)$ | $Q_{\{0,1,\ldots,40\}}(x_S)$ | $\frac{\mathbf{P}_{X_S}^{(0.5)}(x_S)}{Q_{\{0,1,\ldots,40\}}(x_S)}$ | *Likelihood ratio order* |
|---|---|---|---|---|
| 9 | 0.000248691 | 0.0243902 | 0.0101964 | 23 |
| 10 | 0.000770943 | 0.0243902 | 0.0316087 | 21 |
| 11 | 0.00210257 | 0.0243902 | 0.0862054 | 19 |
| 12 | 0.00508121 | 0.0243902 | 0.20833 | 17 |
| 13 | 0.0109442 | 0.0243902 | 0.448711 | 15 |
| 14 | 0.0211066 | 0.0243902 | 0.865371 | 13 |
| 15 | 0.0365847 | 0.0243902 | 1.49998 | 11 |
| 16 | 0.0571637 | 0.0243902 | 2.34371 | 9 |
| 17 | 0.0807016 | 0.0243902 | 3.30877 | 7 |
| 18 | 0.103119 | 0.0243902 | 4.22787 | 5 |
| 19 | 0.119401 | 0.0243902 | 4.89543 | 3 |
| 20 | 0.125371 | 0.0243902 | 5.1402 | 1 |
| 21 | 0.119401 | 0.0243902 | 4.89543 | 2 |
| 22 | 0.103119 | 0.0243902 | 4.22787 | 4 |
| 23 | 0.0807016 | 0.0243902 | 3.30877 | 6 |
| 24 | 0.0571637 | 0.0243902 | 2.34371 | 8 |
| 25 | 0.0365847 | 0.0243902 | 1.49998 | 10 |
| 26 | 0.0211066 | 0.0243902 | 0.865371 | 12 |
| 27 | 0.0109442 | 0.0243902 | 0.448711 | 14 |
| 28 | 0.00508121 | 0.0243902 | 0.20833 | 16 |
| 29 | 0.00210257 | 0.0243902 | 0.0862054 | 18 |
| 30 | 0.000770943 | 0.0243902 | 0.0316087 | 20 |
| 31 | 0.000248691 | 0.0243902 | 0.0101964 | 22 |

*The local prediction procedure and the corresponding exclusion procedure have the following properties:*

$$\mathbf{P}_{X_S}^{(0.5)}\left(A_{X_S}^{(0.80,40)}([0.5,0.5])\right) = 0.846 \tag{6.32}$$

*implying that the reliability excess adopts the value of 0.046.*

$$Q_{\{0,1,\ldots,40\}}\left(A_{X_S}^{(0.80,40)}([0.5,0.5])\right) = 0.220 \tag{6.33}$$

*that is,the failure probability amounts to 0.220 and the success probability to 0.780.*

•

**Example 6.2**
*For the second example, the measurement space is selected to be* $[0.3, 0.7]$, *which is still symmetric to 0.5 but limited to 40% of the maximum range. The small values of* $p$ *and the large values of* $p$, *which would lead with high probability to a rejection of* $H_0$, *are not admitted. The Neyman exclusion procedure can be easily obtained by means of Table 6.5.*

*Not surprisingly, the likelihood ratio order does not change and, therefore, the same prediction region*

$$A_{X_S}^{(0.80,40)}\left([0.5,0.5]\right) = \left\{16,\ldots,24\right\}$$

*and the same reliability excess are obtained as in the case of maximum measurement space.*

*However, because of the restricted measurement space, the failure probability is considerably larger than in the case of maximum space:*

$$Q_{\{0,1,\ldots,40\}}\left(A_{X_S}^{(0.80,40)}\left([0.5,0.5]\right)\right) = 0.530. \tag{6.34}$$

*Finally, in this case the success probability is only 0.470 which is quite different than the value of 0.780 obtained for the maximum measurement space.*

***Table 6.5*** *Conditional probability, marginal probability, and likelihood ratio for the measurement space* $[0.3, 0.7]$.

| $x_S$ | $\mathbf{P}_{X_S}^{(0.50)}(x_S)$ | $Q_{\{0,1,\ldots,40\}}(x_S)$ | $\frac{\mathbf{P}_{X_S}^{(0.80,40)}(x_S)}{Q_{\{0,1,\ldots,40\}}(x_S)}$ | *Likelihood ratio order* |
|---|---|---|---|---|
| 9 | 0.000248691 | 0.0103933 | 0.023928 | 23 |
| 10 | 0.000770943 | 0.016762 | 0.0459934 | 21 |
| 11 | 0.00210257 | 0.024454 | 0.0859802 | 19 |
| 12 | 0.00508121 | 0.032696 | 0.155409 | 17 |
| 13 | 0.0109442 | 0.040575 | 0.269727 | 15 |
| 14 | 0.0211066 | 0.047328 | 0.44596 | 13 |
| 15 | 0.0365847 | 0.052538 | 0.696349 | 11 |
| 16 | 0.0571637 | 0.056164 | 1.01779 | 9 |
| 17 | 0.0807016 | 0.058445 | 1.38081 | 7 |
| 18 | 0.103119 | 0.059733 | 1.72633 | 5 |
| 19 | 0.119401 | 0.060358 | 1.97822 | 3 |
| 20 | 0.125371 | 0.060540 | 2.0708 | 1 |
| 21 | 0.119401 | 0.060358 | 1.97822 | 2 |
| 22 | 0.103119 | 0.059733 | 1.72633 | 4 |
| 23 | 0.0807016 | 0.058445 | 1.38081 | 6 |
| 24 | 0.0571637 | 0.056164 | 1.01779 | 8 |
| 25 | 0.0365847 | 0.052538 | 0.696349 | 10 |
| 26 | 0.0211066 | 0.047328 | 0.44596 | 12 |
| 27 | 0.0109442 | 0.040575 | 0.269727 | 14 |
| 28 | 0.00508121 | 0.032696 | 0.155409 | 16 |
| 29 | 0.00210257 | 0.024454 | 0.0859802 | 18 |
| 30 | 0.000770943 | 0.016762 | 0.0459934 | 20 |
| 31 | 0.000248691 | 0.0103933 | 0.023928 | 22 |

•

**Example 6.3**

*The third example refers to the measurement space* $[0.0, 0.5]$ *which often occurs in reality. In this case, the null hypothesis*

$$H_0 : \quad \{0.5\}$$

*can be looked upon as one sided. In traditional statistics the one sided case has to be treated separately from the two sided one. In our approach there is no separate treatment necessary as the change is expressed in order of the likelihood ratios which turns out to be completely different. However, the local* $(1-\alpha)$*-prediction region changes only slightly, because of the side condition.*

*In Table 6.6, the probabilities and the likelihood ratios (from the largest one downward) are listed so that the local prediction region defining the Neyman exclusion procedure is easily obtained. It has to be noted that the local prediction region must fulfill three requirements:*

- $A_{X_S}^{(0.80,40)}\Big([0.5,0.5]\Big)$ *is connected.*
- $\mathbf{P}_{X_S}^{(0.50)}\Big(A_{X_S}^{(0.80,40)}([0.5,0.5])\Big) \geq 0.80.$
- $Q_{\{0,1,\ldots,40\}}\Big(A_{X_S}^{(0.80,40)}([0.5,0.5])\Big) \stackrel{!}{=} \min.$

*Because of the discreteness of the binomial distribution, it is not possible to arrive at equality in condition (6.16). This fact, in conjunction with the minimum condition (6.15), causes some of the realizations with high likelihood ratio but small probability may be taken off the local prediction region, thus leading to a result which, in some sense, runs counter to intuition.*

*The different measurement space results in a completely different likelihood ratio order as can be seen in Table 6.6. The first element to be selected for the local prediction region according to the likelihood ratio order is the largest realization* $x_S = 40$*, the next is* $x_S = 39$*, and so on until* $x_S = 17$*.*

***Table 6.6*** *Conditional probability, marginal probability, and likelihood ratio for the measurement range* $[0.0, 0.5]$.

| $x_S$ | $\mathbf{P}_{X_S}^{(0.50)}(x_S)$ | $Q_{\{0,1,\ldots,40\}}(x_S)$ | $\frac{\mathbf{P}_{X_S}^{(0.80,40)}(x_S)}{Q_{\{0,1,\ldots,40\}}(x_S)}$ | *Likelihood ratio order* |
|---|---|---|---|---|
| 11 | 0.00210257 | 0.048675 | 0.0431961 | 30 |
| 12 | 0.00508121 | 0.0484998 | 0.104768 | 29 |
| 13 | 0.0109442 | 0.0481089 | 0.227487 | 28 |
| 14 | 0.0211066 | 0.0473272 | 0.445971 | 27 |
| 15 | 0.0365847 | 0.0459201 | 0.796704 | 26 |
| 16 | 0.0571637 | 0.0436336 | 1.31008 | 25 |
| 17 | 0.0807016 | 0.040271 | 2.00396 | 24 |
| 18 | 0.103119 | 0.0357876 | 2.88141 | 23 |
| 19 | 0.119401 | 0.0303603 | 3.93279 | 22 |
| 20 | 0.125371 | 0.0243902 | 5.1402 | 21 |
| 21 | 0.119401 | 0.0184202 | 6.48205 | 20 |
| 22 | 0.103119 | 0.0129929 | 7.93654 | 19 |
| 23 | 0.0807016 | 0.00850948 | 9.48373 | 18 |
| 24 | 0.0571637 | 0.00514692 | 11.1064 | 17 |
| 25 | 0.0365847 | 0.00286037 | 12.7902 | 16 |
| 26 | 0.0211066 | 0.00145327 | 14.5236 | 15 |
| 27 | 0.0109442 | 0.00067154 | 16.2971 | 14 |
| 28 | 0.00508121 | 0.000280678 | 18.1034 | 13 |
| 29 | 0.00210257 | 0.000105464 | 19.9365 | 12 |
| 30 | 0.000770943 | 0.0000353778 | 21.7917 | 11 |
| 31 | 0.000248691 | 0.0000105087 | 23.6653 | 10 |
| 32 | 0.0000699444 | $2.73711 \cdot 10^{-6}$ | 25.5541 | 9 |
| 33 | 0.0000169562 | $6.17581 \cdot 10^{-7}$ | 27.4559 | 8 |
| 34 | $3.49099 \cdot 10^{-6}$ | $1.18868 \cdot 10^{-7}$ | 29.3685 | 7 |
| 35 | $5.98455 \cdot 10^{-7}$ | $1.91258 \cdot 10^{-8}$ | 31.2905 | 6 |
| 36 | $8.31187 \cdot 10^{-8}$ | $2.50204 \cdot 10^{-9}$ | 33.2204 | 5 |
| 37 | $8.98581 \cdot 10^{-9}$ | $2.5559 \cdot 10^{-10}$ | 35.1571 | 4 |
| 38 | $7.09406 \cdot 10^{-10}$ | $1.91216 \cdot 10^{-11}$ | 37.0998 | 3 |
| 39 | $3.63798 \cdot 10^{-11}$ | $9.31677 \cdot 10^{-13}$ | 39.0476 | 2 |
| 40 | $9.09495 \cdot 10^{-13}$ | $2.21828 \cdot 10^{-14}$ | 41.0000 | 1 |

*The reliability requirement is met for the first time for the set* $\{40, 39, \ldots, 17\}$ *, with*

$$\mathbf{P}_{X_S}^{(0.5)}\Big(\{17, \ldots, 40\}\Big) = 0.865908. \tag{6.35}$$

*Unfortunately, the procedure suffers from a rather large reliability excess of 0.065908. In order to reduce the reliability excess without violating the connectedness of the local prediction region, the large realizations with high likelihood ratio, which are extremely unlikely to occur, can be taken off, again yielding a local prediction region which is only slightly different from those of the previous examples:*

$$A_{X_S}^{(0.80,40)}\Big([0.5, 0.5]\Big) = \Big\{17, 18, \ldots, 24, 25\Big\}. \tag{6.36}$$

*The local prediction region* $A_{X_S}^{(0.80,40)}\Big([0.5, 0.5]\Big)$ *defines a 0.20-exclusion procedure for* $H_0$ *with the following properties:*

$$\mathbf{P}_{X_S}^{(0.5)}\Big(A_{X_S}^{(0.80,40)}([0.5, 0.5])\Big) = 0.826 \tag{6.37}$$

$$Q_{\{0,1,\ldots,40\}}\Big(A_{X_S}^{(0.80,40)}([0.5, 0.5])\Big) = 0.181 \tag{6.38}$$

*implying that the success probability adopts the value of 0.819.*

•

Compared to the procedures of the two previous examples, the reliability excess in Example 6.3 is smaller and the success probability larger. If one would have taken the Neyman exclusion procedure of Example 6.1 (assuming maximum measurement space), then the success probability for the measurement range $[0.0, 0.5]$ would be 0.780, which means a decrease of about 5%.

**Example 6.4**

*Example 6.3 refers to the measurement space* $[0.4, 0.8]$*, which is similar to that of Example 6.2, but not symmetric to 0.5. The corresponding probabilities and likelihood ratios are listed in Table 6.7.*

***Table 6.7*** *Conditional probability, marginal probability, and likelihood ratio for the measurement space* $[0.4, 0.8]$.

| $x_S$ | $\mathbf{P}_{X_S}^{(0.50)}(x_S)$ | $Q_{\{0,1,\ldots,40\}}(x_S)$ | $\frac{\mathbf{P}_{X_S}^{(0.80,40)}(x_S)}{Q_{\{0,1,\ldots,40\}}(x_S)}$ | *Likelihood ratio order* |
|---|---|---|---|---|
| 7 | 0.0000169562 | 0.000089634 | 0.189172 | 20 |
| 8 | 0.0000699444 | 0.000271961 | 0.257186 | 19 |
| 9 | 0.000248691 | 0.000717648 | 0.346537 | 18 |
| 10 | 0.000770943 | 0.00166845 | 0.462072 | 16 |
| 11 | 0.00210257 | 0.0034548 | 0.608595 | 14 |
| 12 | 0.00508121 | 0.00643205 | 0.789984 | 13 |
| 13 | 0.0109442 | 0.0108598 | 1.00777 | 11 |
| 14 | 0.0211066 | 0.0167634 | 1.25909 | 10 |
| 15 | 0.0365847 | 0.0238477 | 1.5341 | 8 |
| 16 | 0.0571637 | 0.0315224 | 1.81343 | 6 |
| 17 | 0.0807016 | 0.0390466 | 2.0668 | 5 |
| 18 | 0.103119 | 0.0457348 | 2.25471 | 3 |
| 19 | 0.119401 | 0.0511321 | 2.33514 | 1 |
| 20 | 0.125371 | 0.0550899 | 2.27575 | 2 |
| 21 | 0.119401 | 0.0577271 | 2.06836 | 4 |
| 22 | 0.103119 | 0.0593206 | 1.73833 | 7 |
| 23 | 0.0807016 | 0.0601822 | 1.34095 | 9 |
| 24 | 0.0571637 | 0.0605654 | 0.943833 | 12 |
| 25 | 0.0365847 | 0.0606094 | 0.603615 | 15 |
| 26 | 0.0211066 | 0.0603083 | 0.349978 | 17 |
| 27 | 0.0109442 | 0.0594874 | 0.183974 | 21 |

*Although similar to Example 6.2, the likelihood ratio order is different. The local prediction region is readily obtained:*

$$A_{X_S}^{(0.80,40)}\Big([0.5, 0.5]\Big) = \Big\{15, \ldots, 23\Big\}. \tag{6.39}$$

*From*

$$\mathbf{P}_{X_S}^{(0.5)}\Big(A_{X_S}^{(0.80,40)}([0.5, 0.5])\Big) = 0.8256 \tag{6.40}$$

*we obtain a reliability excess of 0.0256.*

*From the failure probability*

$$Q_{\{0,1,\ldots,40\}}\left(A_{X_S}^{(0.80,40)}([0.5,0.5])\right) = 0.4236 \qquad (6.41)$$

*we obtain the success probability, which is 0.5764.* •

The examples clearly show that the measurement space has to be taken into account in order to determine a good exclusion procedure with maximum success probability.

**Remark 6.2**

*If an $\alpha$-exclusion procedure based on $X_S$ has a large reliability excess, one can reduce it in many cases to almost zero by changing from the sum of sample elements $X_S$ to the sample $\vec{X}$. Changing to $\vec{X}$ would yield local $(1-\alpha)$-prediction procedures, which are not only superior to those based on $X_S$ but also intuitively much more attractive.* •

# Chapter 7

# Comparison Procedures

In many textbooks on statistics, tests are introduced by stating the null hypothesis $H_0$ and the complement as an alternative hypothesis, $H_1$. Often, the impression is given that the two alternatives are in some sense symmetric, and the decision to be made is to accept one of them. As shown in the preceding chapter, the above-described impression is wrong for the case of a significance test, which, therefore, has been called the exclusion procedure here.

A different situation arises, however, whenever there is a clear idea about the actual situation in the sense that only one of two or more distinguishable alternatives can be true. Then, there is often the necessity to decide on the basis of a random sample $\vec{X}$ in favor of one of the alternatives. The decision is made by comparing the sample with the clearly specified alternatives. Because a comparison of alternatives is the basis of the decision-making process, it is called here a *comparison procedure.*

## 7.1 The Alternatives

Sometimes, neither the determination of the actual value nor the exclusion of a part of a given measurement space is of primary interest, but the comparison of two or more alternatives with respect to the probability of interest $\breve{p}$ is. The aim is to make a decision in favor of one of the alternatives:

$$H_i : \quad [\underline{p}^{(i)}, \overline{p}^{(i)}]$$

for $i = 1, 2, \ldots, m$ and $m \geq 2$ with

$$\underline{p}^{(1)} \leq \overline{p}^{(1)} < \underline{p}^{(2)} \leq \overline{p}^{(2)} < \ldots < \underline{p}^{(m)} \leq \overline{p}^{(m)}.$$

The alternatives

$$[\underline{p}^{(i)}, \overline{p}^{(i)}], \quad i = 1, 2, \ldots, m \tag{7.1}$$

have to be looked upon as separate parameter spaces, and the procedure aims at deciding which one of them is the right one.

Evidently, any comparison may lead to one of the following results:

- $p \in [\underline{p}^{(1)}, \overline{p}^{(1)}]$
- $p \in [\underline{p}^{(2)}, \overline{p}^{(2)}]$

$\vdots$

- $p \in [\underline{p}^{(m)}, \overline{p}^{(m)}]$
- no decision

The last result means that it is not possible to distinguish between the alternatives. If it is not possible to distinguish between the alternatives, then it is also not possible to distinguish between any two convex linear combinations of elements of the sets representing the alternatives. Hence, the last result admits the convex hull of $[\underline{p}^{(i)}, \overline{p}^{(i)}]$, $i = 1, 2, \ldots, m$ (i.e., the closed interval $[\underline{p}^{(1)}, \overline{p}^{(m)}]$) as the range of possible values of the probability $\breve{p}$.

The case that no decision is possible represents the decision in favor of $[\underline{p}^{(1)}, \overline{p}^{(m)}]$, which includes all the alternatives, and represents an appropriate measurement space. Thus, the measurement & prediction space of a comparison procedure with $m$ alternatives $[\underline{p}^{(i)}, \overline{p}^{(i)}]$, $i = 1, 2, \ldots, m$, is given by

$$[\underline{p}^{(1)}, \overline{p}^{(m)}] \times \{0, 1\}^n. \tag{7.2}$$

# 7.2 The Decision Function

The comparison of the alternatives yields a decision, where the decision procedure is given by a measurable function:

$$\phi_{H_1 H_2 \ldots H_m} : \{0,1\}^n \to \{0,1,2,\ldots,m\} \tag{7.3}$$

with the following meaning:

For $i = 1, 2, \ldots, m$

$$\phi_{H_1 H_2 \ldots H_m}(\vec{x}) = i \iff \text{decide for } [\underline{p}^{(i)}, \overline{p}^{(i)}]$$

$$\phi_{H_1 H_2 \ldots H_m}(\vec{x}) = 0 \iff \text{decide for } [\underline{p}^{(1)}, \overline{p}^{(m)}].$$

The last decision is correct with certainty, but each of the former decisions may be wrong. Therefore, to make the procedure scientifically meaningful, the respective error probabilities have to be controlled by upper bounds. Let $\alpha_i$ denote the upper bound for the error probability if $p \in [\underline{p}^{(i)}, \overline{p}^{(i)}]$ is true.

An error occurs if $p \in [\underline{p}^{(i)}, \overline{p}^{(i)}]$, but the decision is in favor of $[\underline{p}^{(j)}, \overline{p}^{(j)}]$ for $j \neq i$.

The upper bounds $\alpha_1, \ldots, \alpha_m$ specify the reliability requirements of the comparison procedure.

$$\mathbf{P}_{\vec{X}}^{(p)}\Big( \bigcup_{\substack{j=1 \\ j \neq i}}^{m} \{\vec{x} \,|\, \phi_{H_1 H_2 \ldots H_m}(\vec{x}) = j\} \Big) \le \alpha_i$$

$$\text{for } p \in [\underline{p}^{(i)}, \overline{p}^{(i)}],\ i = 1, \ldots, m. \tag{7.4}$$

We call $\alpha_i$ the *significance level of the ith alternative* with $i = 1, \ldots, m$.

Clearly, the procedure is applied successfully only if a decision for one of the nontrivial alternatives is made. Any decision in favor of $[\underline{p}^{(1)}, \overline{p}^{(m)}]$, although correct, constitutes a failure. Therefore, a procedure is looked for, which maximizes the success probability. The problem of deriving an expression for the success probability is solved based on the results obtained earlier in the case of measurement, prediction, and exclusion procedures.

## 7.3 The $\alpha_1, \ldots, \alpha_m$-Comparison Space

The aim of a comparison procedure is to decide in favor of one of the alternatives given by $[\underline{p}^{(i)}, \overline{p}^{(i)}]$, $i = 1, \ldots, m$. If it is not possible to decide for one, because of the side conditions with respect to the reliability given by $\alpha_1, \ldots, \alpha_m$, then the procedure decides for $[\underline{p}^{(1)}, \overline{p}^{(m)}]$, which means that the procedure has been applied unsuccessfully.

Let $\alpha_1, \ldots, \alpha_m$ and $n$ be given and consider for $i = 1, \ldots, m$ the alternatives

$$[\underline{p}^{(i)}, \overline{p}^{(i)}]. \tag{7.5}$$

Assume further that

$$A_{\vec{X}}^{(1-\alpha_i,n)}\Big([\underline{p}^{(i)}, \overline{p}^{(i)}]\Big) \tag{7.6}$$

is a local $(1-\alpha_i)$-prediction region for $H_i : \; [\underline{p}^{(i)}, \overline{p}^{(i)}]$ with confidence level $1-\alpha_i$, for $i = 1, \ldots, m$.

Then, for $i = 1, \ldots, m$ and $p \in [\underline{p}^{(i)}, \overline{p}^{(i)}]$,

$$\mathbf{P}_{\vec{X}}^{(p)}\Big(A_{\vec{X}}^{(1-\alpha_i,n)}([\underline{p}^{(i)}, \overline{p}^{(i)}])\Big) \geq 1-\alpha_i \tag{7.7}$$

is true.

There are two cases to be distinguished:

$$\text{Case 1:} \quad \bigcup_{i=1}^{m} A_{\vec{X}}^{(1-\alpha_i,n)}([\underline{p}^{(i)}, \overline{p}^{(i)}]) = \{0,1\}^n \tag{7.8}$$

$$\text{Case 2:} \quad \bigcup_{i=1}^{m} A_{\vec{X}}^{(1-\alpha_i,n)}([\underline{p}^{(i)}, \overline{p}^{(i)}]) \neq \{0,1\}^n. \tag{7.9}$$

If the second case is true, then the part of the prediction space $\{0,1\}^n$ not covered by the union of the local prediction regions is divided into disjoint sets $A_1, A_2, \ldots, A_m$ in such a way that for $i = 1, \ldots, m$, the sets

$$A_{\vec{X}}^{(1-\alpha_i,n)}([\underline{p}^{(i)}, \overline{p}^{(i)}]) \cup A_i \tag{7.10}$$

are connected sets and their union is identical to the whole sample space. Clearly, the relations (7.7) remain true for the sets (7.10) implying that the sets given by (7.10) are also local $(1-\alpha_i)$-prediction regions for the alternatives $[\underline{p}^{(i)}, \overline{p}^{(i)}]$. By adding a nonempty $A_i$ to $A_{\vec{X}}^{(1-\alpha_i,n)}([\underline{p}^{(i)}, \overline{p}^{(i)}])$ their probability will increase in general. Hence, Case 2 may easily be reduced to Case 1, and, therefore, in the remainder, Case 1 is assumed to be true.

Assuming Case 1, define for $i = 1, \ldots, m$:

$$A_{\vec{X}}^{(1-\alpha_i,n)}(H_i) = A_{\vec{X}}^{(1-\alpha_i,n)}([\underline{p}^{(i)}, \overline{p}^{(i)}]) \setminus \bigcup_{\substack{j=1 \\ j\neq i}}^{m} A_{\vec{X}}^{(1-\alpha_j,n)}([\underline{p}^{(j)}, \overline{p}^{(j)}]) \tag{7.11}$$

and

$$O_{\vec{X}}^{(\alpha_1,\ldots,\alpha_m,n)}(H_1, \ldots, H_m) = \bigcup_{i=1}^{m} \left(A_{\vec{X}}^{(1-\alpha_i,n)}([\underline{p}^{(i)}, \overline{p}^{(i)}]) \setminus A_{\vec{X}}^{(1-\alpha_i,n)}(H_i)\right) \tag{7.12}$$

Then, because of (7.8),

$$\left\{O_{\vec{X}}^{(\alpha_1,\ldots,\alpha_m,n)}(H_1, \ldots, H_m), A_{\vec{X}}^{(1-\alpha_1,n)}(H_1), \ldots, A_{\vec{X}}^{(1-\alpha_m,n)}(H_m)\right\}$$

is a disjoint partition of the sample space $\{0,1\}^n$. The set $A_{\vec{X}}^{(1-\alpha_1,n)}(H_i)$ refers solely to the alternative $i$, while the set $O_{\vec{X}}^{(\alpha_1,\ldots,\alpha_m,n)}(H_1, \ldots, H_m)$ refers to at least two alternatives if it is not empty.

Now, we are in a position to define a decision function for the comparison procedure:

$$\phi_{H_1 \ldots H_m}(\vec{X}) = \begin{cases} 0 & \text{if } \vec{X} \in O_{\vec{X}}^{(\alpha_1,\ldots,\alpha_m,n)}(H_1, \ldots, H_m) \\ i & \text{if } \vec{X} \in A_{\vec{X}}^{(1-\alpha_i,n)}(H_i) \text{ for } i = 1, \ldots, m. \end{cases} \tag{7.13}$$

For the error probabilities of the decision function (7.13), the following inequalities hold:

For $p \in [\underline{p}^{(i)}, \overline{p}^{(i)}]$, $i = 1, \ldots, m$, we have

$$\begin{aligned}
&\mathbf{P}_{\vec{X}}^{(p)}\Big(\{\vec{x} \,|\, \phi_{H_1 \ldots H_m}(\vec{x}) = j, j \neq i, j \neq 0\}\Big) \\
&= \mathbf{P}_{\vec{X}}^{(p)}\Big(\bigcup_{\substack{j=1 \\ j \neq i}}^{m} A_{\vec{X}}^{(1-\alpha_j, n)}(H_j)\Big) \\
&\leq 1 - \mathbf{P}_{\vec{X}}^{(p)}\Big(A_{\vec{X}}^{(1-\alpha_i, n)}([\underline{p}^{(i)}, \overline{p}^{(i)}])\Big) \\
&\leq 1 - (1 - \alpha_i) = \alpha_i \qquad (7.14)
\end{aligned}$$

With (7.14), it is shown that the decision function (7.13) meets the reliability requirements (7.4) and therefore defines an $\alpha_1, \ldots, \alpha_m$-comparison procedure.

We conclude that any $\alpha_1, \ldots, \alpha_m$-comparison procedure is completely determined by $m$ disjoint subsets of the measurement & prediction space, given by

$$[\underline{p}^{(i)}, \overline{p}^{(i)}] \times A_{\vec{X}}^{(1-\alpha_i, n)}(H_i) \qquad (7.15)$$

for $i = 1, \ldots, m$.

The union of the sets (7.15) is called the $\alpha_1, \ldots, \alpha_m$-*comparison space.* Its projection on the $p$-axis determines the alternatives, and its projections on the $\vec{x}$-axis determines the regions for positive decisions.

**Remark 7.1** *In classical statistics the notions of an error of Type I and an error of Type II are introduced as decision errors. As shown earlier, there is only one decision error to be considered for exclusion procedures. The so-called Type II error in significance testing constitutes a failure to reach the aim, but not a decision error. In the case of a comparison procedure as many errors as there are alternatives are possible. If the number of alternatives are m, then there are m possible errors which, of course, could be called error of Type 1, Type 2,..., Type m.* •

## 7.4 Quality of a Comparison Procedure

A comparison procedure leading to a decision in favor of $H_i$, $i = 1, \ldots, m$, has been applied successfully. A decision in favor of the complete measurement space $[\underline{p}^{(1)}, \overline{p}^{(m)}]$ does not constitute a decision error, as it is always correct, but a failure of the procedure, as the aim of which is not reached. A decision in favor of $[\underline{p}^{(1)}, \overline{p}^{(m)}]$ is made if the event

$$O_{\vec{X}}^{(\alpha_1,\ldots,\alpha_m,n)}(H_1, \ldots, H_m) \in \mathcal{P}_{\{0,1\}^n}$$

occurs. The probability measure on $\mathcal{P}_{\{0,1\}^n}$ is the marginal measure $Q_{\{0,1\}^n}$, and, thus, the failure probability is given by

$$Q_{\{0,1\}^n}\left(O_{\vec{X}}^{(\alpha_1,\ldots,\alpha_m,n)}(H_1, \ldots, H_m)\right). \tag{7.16}$$

The success probability is obtained by the complementary probability or the probability of the complementary event:

$$\begin{aligned} &Q_{\{0,1\}^n}\left(\bigcup_{i=1}^{m} A_{\vec{X}}^{(1-\alpha_i,n)}(H_i)\right) \\ &= 1 - Q_{\{0,1\}^n}\left(O_{\vec{X}}^{(\alpha_1,\ldots,\alpha_m,n)}\big(H_1, \ldots, H_m\big)\right). \end{aligned} \tag{7.17}$$

The smaller the failure probability or the larger the success probability, the better is the comparison procedure. Hence, (7.16) or (7.17) is an appropriate quality criterion for comparison procedures.

## 7.5 Neyman Comparison Procedure

The quality criterion enables a comparison of given $\alpha_1, \ldots, \alpha_m$-comparison procedures and the identification of procedures which are optimal with respect to the success probability.

For given values of significance levels $\alpha_1, \ldots, \alpha_m$ and of the sample size $n$, the following optimization problem has to be solved.

**Problem 7.1**

$$Q_{\{0,1\}^n}\left(O_{\vec{X}}^{(\alpha_1,\ldots,\alpha_m,n)}(H_1,\ldots,H_m)\right) \stackrel{!}{=} \min$$

*under the side conditions*

$$\mathbf{P}_{\vec{X}}^{(p)}\left(A_{\vec{X}}^{(1-\alpha_i,n)}([\underline{p}^{(i)},\overline{p}^{(i)}])\right) \geq 1-\alpha_i$$

$$\textit{for } p \in [\underline{p}^{(i)},\overline{p}^{(i)}] \textit{ and } i = 1,\ldots,m$$

•

Any solution of Problem 7.1 represents a comparison procedure which is optimal with respect to the success probability.

**Definition 7.1**
*An $\alpha_1,\ldots,\alpha_m$-comparison procedure given by the local $(1-\alpha_i)$-prediction regions*

$$\left({}^*A_{\vec{X}}^{(1-\alpha_1,n)}([\underline{p}^{(1)},\overline{p}^{(1)}]),\ldots,\ {}^*A_{\vec{X}}^{(1-\alpha_m,n)}([\underline{p}^{(m)},\overline{p}^{(m)}])\right) \tag{7.18}$$

*is called a* Neyman $\alpha_1,\ldots,\alpha_m$-comparison procedure *if it is a solution of Problem 7.1.* •

Clearly, any $\alpha_1,\ldots,\alpha_m$-comparison procedure with

$$Q_{\{0,1\}^n}\left(O_{\vec{X}}^{(\alpha_1,\ldots,\alpha_m,n)}(H_1,\ldots,H_m)\right) = 0$$

is a Neyman comparison procedure.

If the comparison procedure is based on the sum of the sample elements $X_S$, then a Neyman comparison procedure is analogously defined by replacing $\vec{X}$ and its realizations $\vec{x}$ by $X_S$ and $x_S$, respectively, the sample space $\{0,1\}^n$ by $\{0,1,\ldots,n\}$, and the probability measure $\mathbf{P}_{\vec{X}}^{(p)}$ by $\mathbf{P}_{X_S}^{(p)}$.

**Remark 7.2** *In classical statistics emphasis is laid on the following special case*

- *There are only two alternatives $m = 2$.*
- *The alternatives have Lebesgue measure zero:* $H_i : \{p_i\}$ *for* $i = 1, 2$.

*In fact the two-alternatives case occurs rather often and therefore it is treated separately and in detail in the next section. Probably, the rather strange notion of a Type I and Type II error used for exclusion procedures, i.e. significance tests, might be derived from comparing two alternatives.* •

## 7.6 The Special Case of an $\alpha_1, \alpha_2$-Comparison Procedure

Traditional test theory concentrates on the case $m = 2$ that is, there are only two alternatives, $H_1$ and $H_2$. For $m = 2$, determination of a Neyman comparison procedure is rather simple and it is described in detail below.

For solving the optimization problem, the $2^n$ elements of the sample space $\{0, 1\}^n$ are numbered consecutively in the following manner taking into account the connectedness:

$$
\begin{aligned}
\vec{x}_1 &= (0, 0, \dots, 0, 0) \\
\vec{x}_2 &= (0, 0, \dots, 0, 1) \\
&\dots \\
\vec{x}_{2^n-1} &= (1, 1, \dots, 1, 0) \\
\vec{x}_{2^n} &= (1, 1, \dots, 1, 1)
\end{aligned}
$$

Because of

$$
\begin{aligned}
&O_{\vec{X}}^{(\alpha_1,\alpha_2,n)}(H_1, H_2) \\
&= A_{\vec{X}}^{(1-\alpha_1,n)}([\underline{p}^{(1)}, \overline{p}^{(1)}]) \cap A_{\vec{X}}^{(1-\alpha_2,n)}([\underline{p}^{(2)}, \overline{p}^{(2)}])
\end{aligned} \tag{7.19}
$$

a first step to solve the original optimization problem consists of solving two optimization problems:

$$1. \quad m_1 \stackrel{!}{=} \min \tag{7.20}$$

where

$$A_{\vec{X}}^{(1-\alpha_1,n)}([\underline{p}^{(1)}, \overline{p}^{(1)}]) = \{\vec{x}_1, \dots, \vec{x}_{m_1}\}$$

and

$$\mathbf{P}_{\vec{X}}^{(p)}\left(\{\vec{x}_1,\ldots,\vec{x}_{m_1}\}\right) \geq 1-\alpha_1$$
for $p \in [\underline{p}^{(1)}, \overline{p}^{(1)}]$.

2. $m_2 \stackrel{!}{=} \max$ (7.21)

where

$$A_{\vec{X}}^{(1-\alpha_2,n)}([\underline{p}^{(2)}, \overline{p}^{(2)}]) = \{\vec{x}_{m_2},\ldots,\vec{x}_{2^n}\}$$

and

$$\mathbf{P}_{\vec{X}}^{(p)}\left(\{\vec{x}_{m_2},\ldots,\vec{x}_{2^n}\}\right) \geq 1-\alpha_2$$
for $p \in [\underline{p}^{(2)}, \overline{p}^{(2)}]$.

The monotony of the binomial distribution function implies

$$\mathbf{P}_{\vec{X}}^{(p)}\left(\{\vec{x}_1,\ldots,\vec{x}_{m_1}\}\right) \geq 1-\alpha_1 \quad \text{for } p \in [\underline{p}^{(1)}, \overline{p}^{(1)}]$$
$$\Longleftrightarrow \quad \mathbf{P}_{\vec{X}}^{(\overline{p}^{(1)})}\left(\{\vec{x}_1,\ldots,\vec{x}_{m_1}\}\right) \geq 1-\alpha_1 \tag{7.22}$$

and

$$\mathbf{P}_{\vec{X}}^{(p)}\left(\{\vec{x}_{m_2},\ldots,\vec{x}_{2^n}\}\right) \geq 1-\alpha_2 \quad \text{for } p \in [\underline{p}^{(2)}, \overline{p}^{(2)}]$$
$$\Longleftrightarrow \quad \mathbf{P}_{\vec{X}}^{(\underline{p}^{(2)})}\left(\{\vec{x}_{m_2},\ldots,\vec{x}_{2^n}\}\right) \geq 1-\alpha_2. \tag{7.23}$$

**Remark 7.3**
*1. Because $H_1$ covers by definition the left-hand boundary of the measurement range and $H_2$ the right-hand boundary, either of the prediction regions refers to the one-sided case, implying that $(0,\ldots,0)$ has a maximum likelihood ratio for $\underline{p}^{(1)}$ and $(1,\ldots,1)$ has a maximum likelihood ratio for $\overline{p}^{(2)}$. Thus, filling the two prediction regions according to the likelihood ratio order yields the above result.*
*2. An $\alpha$-exclusion procedure for $H_0 : [\underline{p}^{(0)}, \overline{p}^{(0)}] \subset [\underline{p}, \overline{p}]$ may be looked at, as a special comparison procedure. Define*

$$H_1 : [\underline{p}^{(0)}, \overline{p}^{(0)}]$$
$$H_2 : [\underline{p}, \underline{p}^{(0)})$$
$$H_3 : (\overline{p}^{(0)}, \overline{p}]$$

*where $H_2$ or $H_3$ may be empty. Then, an $\alpha$-exclusion procedure is equivalent to an $\alpha, 0, 0$-comparison procedure for $H_1, H_2$ and $H_3$. In other words, an exclusion procedure can be looked upon as a boundary case of a comparison procedure, where the alternatives have significant levels $\alpha_j = 0$, $j = 2, 3$, except for the alternative, which shall be rejected. Of course, any local $(1-\alpha_i)$-prediction region for $\alpha_i = 0$ equals the whole prediction space $\{0,1\}^n$, and, therefore, the intersection of the three local prediction spaces is equal to $A_{\vec{X}}^{(1-\alpha,n)}([\underline{p}^{(0)}, \overline{p}^{(0)}])$, and hence*

$$O_{\vec{X}}^{(\alpha,0,0)}(H_1, H_2, H_3) = A_{\vec{X}}^{(1-\alpha,n)}([\underline{p}^{(0)}, \overline{p}^{(0)}]), \tag{7.24}$$

*implying that a Neyman exclusion procedure is obtained by minimizing $A_{\vec{X}}^{(1-\alpha,n)}([\underline{p}^{(0)}, \overline{p}^{(0)}])$ with respect to the marginal measure $Q_{\{0,1\}^n}$.* •

# 7.7 Determination of Neyman Comparison Procedures

The determination of a Neyman $\alpha_1, \alpha_2$-comparison procedure is rather simple. Any $\alpha_1, \alpha_2$-comparison procedure consists of two local prediction regions, the first one containing the realization $\vec{x}_1 = (0, \ldots, 0)$ and the second one containing the realization $\vec{x}_{2^n} = (1, \ldots, 1)$. Thus, starting from $\vec{x}_1$, the first local prediction region is filled without violating the connectedness condition until the reliability condition

$$\mathbf{P}_{\vec{X}}^{(\overline{p}^{(1)})}(\{\vec{x}_1, \ldots, \vec{x}_{m_1}\}) \geq 1 - \alpha_1$$

is met. Then, the second prediction region is filled starting from $\vec{x}_{2^n}$ until

$$\mathbf{P}_{\vec{X}}^{(\underline{p}^{(2)})}(\{\vec{x}_{m_2}, \ldots, \vec{x}_{2^n}\}) \geq 1 - \alpha_2$$

holds for the first time.

Hence, we have to determine

$$m_1 = \min\left\{m \in I\!N_0 \,\middle|\, \mathbf{P}_{\vec{X}}^{(\overline{p}^{(1)})}(\{\vec{x}_1, \ldots, \vec{x}_m\}) \geq 1 - \alpha_1\right\} \tag{7.25}$$

and

$$m_2 = \max\left\{m \in I\!N_0 \,\middle|\, \mathbf{P}_{\vec{X}}^{(\overline{p}^{(2)})}(\{\vec{x}_m, \ldots, \vec{x}_{2^n}\}) \geq 1 - \alpha_2\right\}. \quad (7.26)$$

If

$$m_1 \geq m_2, \quad (7.27)$$

then the two prediction regions have a nonempty intersection, and the probability of a decision in favor of the complete measurement space $[\underline{p}^{(1)}, \overline{p}^{(2)}]$ is positive. The Neyman $\alpha_1, \alpha_2$-comparison procedure is given by

$$\begin{aligned} {}^*A_{\vec{X}}^{(1-\alpha_1,n)}(H_1) &= \{\vec{x}_1, \ldots, \vec{x}_{m_2-1}\} && (7.28)\\ {}^*A_{\vec{X}}^{(1-\alpha_2,n)}(H_2) &= \{\vec{x}_{m_1+1}, \ldots, \vec{x}_{2^n}\} && (7.29) \end{aligned}$$

with failure probability

$$Q_{\{0,1\}^n}\Big(\{\vec{x}_{m_2}, \ldots, \vec{x}_{m_1}\}\Big). \quad (7.30)$$

If, however, the prediction regions have an empty intersection, which is equivalent to

$$m_1 < m_2, \quad (7.31)$$

then $\tilde{m}_1 \in I\!N$ with $m_1 \leq \tilde{m}_1 < m_2$ has to be selected. Next, define $\tilde{m}_2 = \tilde{m}_1 + 1$. Then, a Neyman $\alpha_1, \alpha_2$-comparison procedure is given by

$$\begin{aligned} {}^*A_{\vec{X}}^{(1-\alpha_1,n)}(H_1) &= \{\vec{x}_1, \ldots, \vec{x}_{\tilde{m}_1}\} && (7.32)\\ {}^*A_{\vec{X}}^{(1-\alpha_2,n)}(H_2) &= \{\vec{x}_{\tilde{m}_2}, \ldots, \vec{x}_{2^n}\}. && (7.33) \end{aligned}$$

The union of the sets ${}^*A_{\vec{X}}^{(1-\alpha_1,n)}(H_1)$ and ${}^*A_{\vec{X}}^{(1-\alpha_2,n)}(H_2)$ is equal to the prediction space. Because

$${}^*A_{\vec{X}}^{(1-\alpha_1,n)}([\underline{p}^{(1)}, \overline{p}^{(1)}]) \cap {}^*A_{\vec{X}}^{(1-\alpha_2,n)}([\underline{p}^{(2)}, \overline{p}^{(2)}]) = \emptyset$$

the failure probability is zero:

$$Q_{\{0,1\}^n}\Big(\emptyset\Big) = 0. \quad (7.34)$$

**Remark 7.4**
*If the two local prediction regions do not overlap (7.31), then it is possible to obtain an ideal comparison procedure with success probability 1 or failure probability 0 for even smaller significance levels $\alpha_1$ and $\alpha_2$ than the originally selected ones.* •

Again, the simple example of $n = 2$ serves for the purpose of illustration.

## 7.8 Illustrative Examples for an $\alpha_1, \alpha_2$-Comparison Procedure

The alternatives are given by

- $H_1$: $[0.02, 0.04]$
- $H_2$: $[0.90, 0.95]$

**Example 7.1**
*The procedure shall be based on the sample $\vec{X}$ and meet the reliability requirements given by $\alpha_1 = 0.05$ and $\alpha_2 = 0.1$. The convex hull of the two alternatives $[0.02, 0.04]$ and $[0.90, 0.95]$ and hence the underlying measurement space is the interval $[0.02, 0.95]$.*

*In Table 7.1 and Table 7.2 the corresponding probabilities and likelihood ratios for the right-hand bound of $H_1$ and the left-hand bound of $H_2$ are displayed.*

***Table 7.1*** *Conditional probability, marginal probability, and likelihood ratio for $p = 0.04$ based on $\vec{X}$.*

| $\vec{x}$ | $\mathbf{P}_{\vec{X}}^{(0.04)}(\vec{x})$ | $Q_{\{0,1\}^n}(\vec{x})$ | $\frac{\mathbf{P}_{\vec{X}}^{(0.04)}(\vec{x})}{Q_{\{0,1\}^2}(\vec{x})}$ |
|---|---|---|---|
| (0,0) | 0.9216 | 0.3373 | 2.7323 |
| (0,1) | 0.0384 | 0.1777 | 0.2161 |
| (1,1) | 0.0384 | 0.1777 | 0.2161 |
| (1,1) | 0.0016 | 0.3073 | 0.0052 |

*From Table 7.1, the optimal prediction region for the first alternative, $H_1$, is immediately obtained. The first element* (0,0) *is not sufficient, as its probability is less than the required* $1 - \alpha_1 = 0.95$*; thus, the second is added. The two elements* (0,0) *and* (0,1) *have a probability exceeding* 0.95*, and therefore*

$$^*A_{\vec{X}}^{(0.95,2)}\Big([0.02, 0.04]\Big) = \Big\{(0,0), (0,1)\Big\}. \qquad (7.35)$$

*Note that the optimal prediction region is not unique since the two elements* (0,1) *and* (1,0) *of the prediction space have the same likelihood ratio.*

*In Table 7.2 the corresponding values for* $p = 0.90$ *are displayed.*

***Table 7.2** Conditional probability, marginal probability, and likelihood ratio for* $p = 0.90$ *based on* $\vec{X}$.

| $\vec{x}$ | $\mathbf{P}_{\vec{X}}^{(0.90)}(\vec{x})$ | $Q_{\{0,1\}^n}(\vec{x})$ | $\frac{\mathbf{P}_{\vec{X}}^{(0.90)}(\vec{x})}{Q_{\{0,1\}^2}(\vec{x})}$ |
|---|---|---|---|
| (0,0) | 0.0100 | 0.3373 | 0.0296 |
| (0,1) | 0.0900 | 0.1777 | 0.5065 |
| (1,0) | 0.0900 | 0.1777 | 0.5065 |
| (1,1) | 0.8100 | 0.3073 | 2.6359 |

*From Table 7.2, the optimal local prediction region for the second alternative,* $H_2$*, is immediately obtained:*

$$^*A_{\vec{X}}^{(0.90,2)}\Big([0.90, 0.95]\Big) = \Big\{(1,0), (1,1)\Big\}. \qquad (7.36)$$

*The union of the two local prediction regions (7.35) and (7.36) is the complete sample space. Therefore, the two local prediction regions define a Neyman* $\alpha_1, \alpha_2$*-comparison procedure. Because they are disjoint, we obtain*

$$^*A_{\vec{X}}^{(0.95,2)}(H_1) = {^*A_{\vec{X}}^{(0.95,2)}}\Big([0.02, 0.04]\Big) \qquad (7.37)$$

$$^*A_{\vec{X}}^{(0.90,2)}(H_2) = {^*A_{\vec{X}}^{(0.90,2)}}\Big([0.90, 0.95]\Big) \qquad (7.38)$$

*and the decision function*

$$\phi_{H_1,H_2}(\vec{X}) = \begin{cases} 1 & \text{if } \vec{X} \in \{(0,0),(0,1)\} \\ 2 & \text{if } \vec{X} \in \{(1,0),(1,1)\}. \end{cases} \tag{7.39}$$

*Evidently, the failure probability for the comparison procedure (7.39) is zero. This follows, for instance, from the fact that the two local prediction regions are disjoint. The last columns in the tables illustrate that the elements for the prediction regions are selected according to the likelihood ratio order.* •

The following case serves to illustrate the difference when using the sum of sample elements $X_S$ instead of the sample $\vec{X}$ itself.

**Example 7.2**

*The procedure shall be based on the sum of sample elements* $X_S$ *and meet the same reliability requirements given by* $\alpha_1 = 0.05$ *and* $\alpha_2 = 0.1$. *Again, the measurement space is given by the convex hull of the alternatives* $H_1 : [0.02, 0.04]$ *and* $H_2 : [0.90, 0.95]$, *(i.e., by the interval* $[0.02, 0.95]$*).*

*In Table 7.3 and Table 7.4 the corresponding probabilities and likelihood ratios are displayed.*

***Table 7.3*** *Conditional probability, marginal probability, and likelihood ratio for* $p = 0.04$ *based on* $X_S$.

| $\vec{x}$ | $\mathbf{P}_{X_S}^{(0.04)}(x_S)$ | $Q_{\{0,1,2\}}(x_S)$ | $\frac{\mathbf{P}_{X_S}^{(0.04)}(x_S)}{Q_{\{0,1,2\}}(x_S)}$ |
|---|---|---|---|
| *0* | *0.9216* | *0.3373* | *2.7323* |
| *1* | *0.0768* | *0.3554* | *0.2161* |
| *2* | *0.0016* | *0.3073* | *0.0052* |

*From Table 7.3, the optimal local prediction region for the first alternative, $H_1$, is obtained:*

$$^*A_{X_S}^{(0.95,2)}\Big([0.02, 0.04]\Big) = \Big\{0, 1\Big\}. \tag{7.40}$$

*The second alternative, $H_2$, yields the values presented in Table 7.4.*

***Table 7.4*** *Conditional probability, marginal probability, and likelihood ratio for $p = 0.90$ based on $X_S$.*

| $\vec{x}$ | $\mathbf{P}_{X_S}^{(0.90)}(x_S)$ | $Q_{\{0,1,2\}}(x_S)$ | $\frac{\mathbf{P}_{X_S}^{(0.90)}(x_S)}{Q_{\{0,1,2\}}(x_S)}$ |
|---|---|---|---|
| 0 | 0.0100 | 0.3373 | 0.0296 |
| 1 | 0.1800 | 0.3554 | 0.5065 |
| 2 | 0.8100 | 0.3073 | 2.6359 |

*From Table 7.4 the optimal local prediction region for the second alternative is obtained:*

$$^*A_{X_S}^{(0.90,2)}\Big([0.90, 0.95]\Big) = \Big\{1, 2\Big\}. \tag{7.41}$$

*The intersection of the two local prediction regions (7.40) and (7.41) is not the empty set, and therefore it is not possible to distinguish the two alternatives with certainty. The decision intervals for the alternatives are given by*

$$\begin{aligned} &^*A_{X_S}^{(0.90,2)}(H_1) \\ &= {}^* A_{X_S}^{(0.95,2)}\Big([0.02, 0.04]\Big) \setminus {}^* A_{X_S}^{(0.90,2)}\Big([0.90, 0.95]\Big) \\ &= \{0\} \end{aligned} \tag{7.42}$$

$$\begin{aligned} &^*A_{X_S}^{(0.90,2)}(H_2) \\ &= {}^* A_{X_S}^{(0.90,2)}\Big([0.90, 0.95]\Big) \setminus {}^* A_{X_S}^{(0.95,2)}\Big([0.02, 0.04]\Big) \\ &= \{2\} \end{aligned} \tag{7.43}$$

*and the decision function*

$$\phi_{H_1,H_2}(X_S) = \begin{cases} 0 & \text{for } X_S \in \{1\} \\ 1 & \text{for } X_S \in \{0\} \\ 2 & \text{for } X_S \in \{2\}. \end{cases} \tag{7.44}$$

*Thus, if the comparison procedure is based on the sum of sample elements and not on the sample itself, then*

$$O_{X_S}^{(0.05,0.10,2)}\Big(H_1, H_2\Big) = \{1\} \tag{7.45}$$

*and the failure probability is given by*

$$Q_{\{0,1,2\}}\Big(\{1\}\Big) = 0.3554. \tag{7.46}$$

*Hence, taking into account the reliability requirements, it is not possible to distinguish between the two alternative, $H_1$ and $H_2$, with certainty. There are cases where an application of the procedure will result in a failure.* •

**Example 7.3**
*For the same alternatives, a comparison procedure based on the sample $\vec{X}$ will be developed with a reliability given by the significance levels $\alpha_1 = 0.10$ and $\alpha_2 = 0.20$.*

*The local prediction region for $[0.02, 0.04]$ is taken from Table 7.1, and that for $[0.90, 0.95]$ from Table 7.2:*

$$^*A_{\vec{X}}^{(0.90,2)}\Big([0.02, 0.04]\Big) = \Big\{(0,0)\Big\} \tag{7.47}$$
$$^*A_{\vec{X}}^{(0.80,2)}\Big([0.90, 0.95]\Big) = \Big\{(1,1)\Big\}. \tag{7.48}$$

*In this case, (7.8) is not fulfilled; that is the union of the prediction regions does not fill the whole sample space, and therefore they have to be enlarged to arrive at local prediction regions of the form (7.10). There are three different possibilities $A_{1,k}, A_{2,k}, k = 1, 2, 3,$ of enlarging the local prediction regions.*

1. $A_{1,1} = \emptyset$, $A_{2,1} = \{(0,1),(1,0)\}$ *with the resulting local prediction regions*

$$\tilde{A}_{\vec{X}}^{(0.90,2)}\Big([0.02, 0.04]\Big) = \Big\{(0,0)\Big\}$$
$$\tilde{A}_{\vec{X}}^{(0.80,2)}\Big([0.90, 0.95]\Big) = \Big\{(0,1),(1,0),(1,1)\Big\}.$$

2. $A_{1,2} = \{(0,1)\}$, $A_{2,2} = \{(1,0)\}$ *with the resulting local prediction regions*

$$\tilde{A}_{\vec{X}}^{(0.90,2)}\Big([0.02, 0.04]\Big) = \Big\{(0,0),(0,1)\Big\}$$
$$\tilde{A}_{\vec{X}}^{(0.80,2)}\Big([0.90, 0.95]\Big) = \Big\{(1,0),(1,1)\Big\}.$$

3. $A_{1,3} = \{(0,1),(1,0)\}$, $A_{2,3} = \emptyset\}$ *with the resulting local prediction regions*

$$\tilde{A}_{\vec{X}}^{(0.90,2)}\Big([0.02, 0.04]\Big) = \Big\{(0,0),(0,1),(1,0)\Big\}$$
$$\tilde{A}_{\vec{X}}^{(0.80,2)}\Big([0.90, 0.95]\Big) = \Big\{(1,1)\Big\}.$$

*Clearly, for each of the above possibilities, the failure probability is zero. The difference refers to the reliability, i.e., to the bounds for the error probabilities. The question of which of the possibilities should be selected in a given situation cannot be answered by means of pure statistical reasoning. The answer should take into account the consequences of errors and select the possibility that leads to the most favorable consequences with respect to a specified aim.* •

# Part III

# Introduction to the Tables

# Chapter 8

# Measurement Intervals

## 8.1 Tables for Measurement Intervals

For selecting an appropriate measurement procedure for $\breve{p}$, the measurement space must be chosen in a first step by the user. If there are no experiences with the event of interest, then the whole unit interval has to be selected as an appropriate measurement space.

There are smaller printed tables, which refer to the maximum measurement space $[0,1]$ and covers small and medium sample sizes. These tables are thought as a first tool to be applied in the experimental state of investigations, which is characterized by two features:

- The sample size is small.
- There are only some vague ideas, of which several possible events might be the interesting ones, and therefore, one aim is to identify the relevant event or the relevant events.

The tables on the CD-ROM are devoted particularly to the case that the measurement space is not the whole unit interval. This case occurs after the experimental stage of the investigations has been left and one has some knowledge about what to measure.

The tables are based on the sum of sample elements $X_S$. If it is desirable to have the procedure based on the sample $\vec{X}$

for improving its quality, then one may start from the tabled values and derive the corresponding measurement intervals for the procedures based on $\vec{X}$. Clearly, this pays only for a small or medium sample size.

The entries in the tables are the measurement intervals

$$[{}^*\ell_{\breve{p}}^{(\beta,n)}(x_S), {}^*u_{\breve{p}}^{(\beta,n)}(x_S)]. \tag{8.1}$$

Each entry is identified by the following quantities:

1. Measurement range $[\underline{p}, \overline{p}]$
2. Sample size $n$
3. Confidence level $\beta$
4. Realization $x_S$

For the tables on the CD-ROM, no symmetry relation has to be utilized for obtaining the desired result. The tables contain for each sample size $n$ the complete set of realizations (i.e., $x_S \in \{0, 1, \ldots, n\}$). For the printed tables, however, one symmetry relation has to be used.

## 8.2 The Printed Tables

The printed tables refer only to the maximum measurement space and thus may be used for a preliminary study:

$$\underline{p} = 0 \quad \text{and} \quad \overline{p} = 1. \tag{8.2}$$

The sample size $n$ is listed in the first column of the printed tables and covers smaller and medium sizes:

$$n \in \{1, 2, \ldots, 100\}. \tag{8.3}$$

There are five different confidence levels listed in the second column of the printed tables:

$$\begin{aligned} \beta_1 &= 0.80 \\ \beta_2 &= 0.90 \\ \beta_3 &= 0.95 \\ \beta_4 &= 0.98 \\ \beta_5 &= 0.99 \end{aligned} \tag{8.4}$$

As for the tabulated realizations of $X_S$ in the printed tables, they are listed in the first row and the entries cover, for each $n$, the range $1 \leq n \leq n_1$, where $n_1$ is an integer with $\frac{n}{2} < n_1 \leq n$.

In more than half of the cases, the realizations of the measurement intervals (8.1) can be found directly in the tables. If this is not the case, then by the symmetry of the binomial distribution, the bounds of the measurement intervals can be easily computed as described below.

Assume that for given sample size $n$ and confidence level $\beta$, there is no entry for the observed value $x_S$ (i.e., $x_S > n_1$), then there is an entry for $n - x_S$, as $n - x_S \leq n_1$. In this case, the bounds of the measurement interval are computed as given by (8.5) and (8.6) using the values obtained for ${}^*\ell_{\breve{p}}^{(\beta,n)}(n - x_S)$ and ${}^*u^{(\beta,n)}(n - x_S)$.

$$
\begin{aligned}
{}^*\ell_{\breve{p}}^{(\beta,n)}(x_S) &= 1 - {}^*u_{\breve{p}}^{(\beta,n)}(n - x_S) &\quad (8.5)\\
{}^*u_{\breve{p}}^{(\beta,n)}(x_S) &= 1 - {}^*\ell_{\breve{p}}^{(\beta,n)}(n - x_S). &\quad (8.6)
\end{aligned}
$$

The bounds of the measurement intervals are given by four decimal digits in the row identified by a pair $n$ and $\beta$ and in the column identified by $n$ for $n \leq n_1$.

## 8.3 Illustrative Examples for the Use of the Printed Tables

The first refers to the case that the observed realization $x_S$ is contained in the table, and the second example to the case that there is no entry of $x_S$ in the table.

**Example 8.1**

*Let the measurement space be maximum,* $[0, 1]$, *implying that the printed tables may be used. Consider the sample size* $n = 24$, *confidence level* $\beta = 0.90$, *and realization of the sum of sample elements* $x_S = 9$.

*These values for the sample size, confidence level, and realization determine the corresponding table, which is given in Table 8.1.*

***Table 8.1*** *Detail from the printed tables.*

| $n$ | $\beta$ | 7 | 8 | **9** | 10 | 11 | 12 | 13 |
|---|---|---|---|---|---|---|---|---|
| | 0.99 | 0.1056 | 0.1332 | 0.1625 | 0.1901 | 0.2163 | 0.2573 | 0.2795 |
| | | 0.5666 | 0.6115 | 0.6380 | 0.6768 | 0.7205 | 0.7427 | 0.7837 |
| | 0.98 | 0.1206 | 0.1501 | 0.1729 | 0.2073 | 0.2395 | 0.2738 | 0.3040 |
| | | 0.5436 | 0.5874 | 0.6156 | 0.6514 | 0.6960 | 0.7262 | 0.7605 |
| **24** | 0.95 | 0.1391 | 0.1690 | 0.2037 | 0.2339 | 0.2665 | 0.3100 | 0.3388 |
| | | 0.5000 | 0.5424 | 0.5849 | 0.6276 | 0.6612 | 0.6900 | 0.7335 |
| | **0.90** | 0.1647 | 0.1860 | **0.2298** | 0.2590 | 0.2911 | 0.3334 | 0.3754 |
| | | 0.4791 | 0.5210 | **0.5527** | 0.5867 | 0.6246 | 0.6666 | 0.7089 |
| | 0.80 | 0.1910 | 0.2117 | 0.2537 | 0.2953 | 0.3338 | 0.3584 | 0.3971 |
| | | 0.4384 | 0.4795 | 0.5206 | 0.5617 | 0.6029 | 0.6416 | 0.6662 |

*From the table, the measurement interval is immediately obtained:*

$$^*\ell_{\breve{p}}^{(0.90,24)}(9) = 0.2298 \qquad \textit{and} \qquad ^*u_{\breve{p}}^{(0.90,24)}(9) = 0.5527$$

*with the length of 0.3229.* •

**Remark 8.1**

*The traditional measurement interval (taken from [29]) has the bounds*

$$\ell_{\breve{p}}^{(0.90,24)}(9) = 0.2116 \qquad \textit{and} \qquad u_{\breve{p}}^{(0.90,24)}(9) = 0.5629$$

*and the length 0.3513. Hence, in this example the length of the traditional confidence interval exceeds by 8.8% the length of the optimal one.* •

**Example 8.2**

*Again consider the maximum measurement space* $[0, 1]$, *and let* $n = 60$, $\beta = 0.98$, *and* $x_S = 50$.

*There is no entry for* $x_S = 50$ *in the printed tables. However, there is an entry for* $n - x_S = 10$. *The corresponding part of the printed tables is presented in Table 8.2.*

***Table 8.2*** *Detail from the printed tables.*

| $n$ | $\beta$ | 7 | 8 | 9 | **10** | 11 | 12 | 13 |
|---|---|---|---|---|---|---|---|---|
| | 0.99 | 0.0400 | 0.0501 | 0.0607 | 0.0718 | 0.0820 | 0.0898 | 0.1053 |
| | | 0.2612 | 0.2787 | 0.3038 | 0.3211 | 0.3383 | 0.3634 | 0.3806 |
| | **0.98** | 0.0460 | 0.0569 | 0.0682 | **0.0755** | 0.0893 | 0.1024 | 0.1098 |
| | | 0.2461 | 0.2634 | 0.2883 | **0.3054** | 0.3225 | 0.3475 | 0.3646 |
| **60** | 0.95 | 0.0560 | 0.0584 | 0.0780 | 0.0888 | 0.1013 | 0.1141 | 0.1272 |
| | | 0.2234 | 0.2430 | 0.2652 | 0.2821 | 0.2991 | 0.3238 | 0.3407 |
| | 0.90 | 0.0623 | 0.0751 | 0.0859 | 0.1001 | 0.1132 | 0.1248 | 0.1421 |
| | | 0.2083 | 0.2253 | 0.2421 | 0.2667 | 0.2835 | 0.3002 | 0.3170 |
| | 0.80 | 0.0712 | 0.0859 | 0.1013 | 0.1104 | 0.1274 | 0.1435 | 0.1518 |
| | | 0.1856 | 0.2100 | 0.2267 | 0.2434 | 0.2601 | 0.2767 | 0.3012 |

*For the set of input data* ($n = 60, \beta = 0.98, n - x_S = 10$), *the corresponding measurement interval is determined as in the preceding example with*

$$^*\ell_{\breve{p}}^{(0.98,60)}(10) = 0.0755 \qquad \textit{and} \qquad ^*u_{\breve{p}}^{(0.98,60)}(10) = 0.3054.$$

*Hence, the measurement interval is found to be*

$$^*\ell_{\breve{p}}^{(0.98,60)}(50) = 1 - {}^*u_{\breve{p}}^{(0.98,60)}(10) = 0.6946 \qquad (8.7)$$

$$^*u_{\breve{p}}^{(0.98,60)}(50) = 1 - {}^*\ell_{\breve{p}}^{(0.98,60)}(10) = 0.9245. \qquad (8.8)$$

•

**Remark 8.2**
*The traditional measurement interval (taken from [29]) has the bounds*

$$\ell_{\breve{p}}^{(0.98,60)}(50) = 0.6922 \qquad \textit{and} \qquad u_{\breve{p}}^{(0.98,60)}(10) = 0.9282$$

•

## 8.4 The Tables on CD-ROM

The primary label of the tables on the CD-ROM is the measurement space $[\underline{p}, \overline{p}]$.

### Measurement Space

Bounds and lengths of the measurement spaces are the multiples of 0.1. Thus, the smallest length of the tabulated measurement spaces is 0.1. There are 10 possible measurement spaces of length 0.1:

1. $[\underline{p}, \overline{p}] = [0.0, 0.1]$
2. $[\underline{p}, \overline{p}] = [0.1, 0.2]$
3. $[\underline{p}, \overline{p}] = [0.2, 0.3]$
4. $[\underline{p}, \overline{p}] = [0.3, 0.4]$
5. $[\underline{p}, \overline{p}] = [0.4, 0.5]$
6. $[\underline{p}, \overline{p}] = [0.5, 0.6]$
7. $[\underline{p}, \overline{p}] = [0.6, 0.7]$
8. $[\underline{p}, \overline{p}] = [0.7, 0.8]$
9. $[\underline{p}, \overline{p}] = [0.8, 0.9]$
10. $[\underline{p}, \overline{p}] = [0.9, 1.0]$

The next smallest length considered in the tables is 0.2. There are 9 measurement spaces with length 0.2:

1. $[\underline{p}, \overline{p}] = [0.0, 0.2]$
2. $[\underline{p}, \overline{p}] = [0.1, 0.3]$
3. $[\underline{p}, \overline{p}] = [0.2, 0.4]$
4. $[\underline{p}, \overline{p}] = [0.3, 0.5]$
5. $[\underline{p}, \overline{p}] = [0.4, 0.6]$
6. $[\underline{p}, \overline{p}] = [0.5, 0.7]$
7. $[\underline{p}, \overline{p}] = [0.6, 0.8]$

8. $[\underline{p}, \overline{p}] = [0.7, 0.9]$

9. $[\underline{p}, \overline{p}] = [0.8, 1.0]$

For length 0.3, there are 8 measurement spaces:

1. $[\underline{p}, \overline{p}] = [0.0, 0.3]$
2. $[\underline{p}, \overline{p}] = [0.1, 0.4]$
3. $[\underline{p}, \overline{p}] = [0.2, 0.5]$
4. $[\underline{p}, \overline{p}] = [0.3, 0.6]$
5. $[\underline{p}, \overline{p}] = [0.4, 0.7]$
6. $[\underline{p}, \overline{p}] = [0.5, 0.8]$
7. $[\underline{p}, \overline{p}] = [0.6, 0.9]$
8. $[\underline{p}, \overline{p}] = [0.7, 1.0]$

For length 0.4, there are 7 measurement spaces:

1. $[\underline{p}, \overline{p}] = [0.0, 0.4]$
2. $[\underline{p}, \overline{p}] = [0.1, 0.5]$
3. $[\underline{p}, \overline{p}] = [0.2, 0.6]$
4. $[\underline{p}, \overline{p}] = [0.3, 0.7]$
5. $[\underline{p}, \overline{p}] = [0.4, 0.8]$
6. $[\underline{p}, \overline{p}] = [0.5, 0.9]$
7. $[\underline{p}, \overline{p}] = [0.6, 1.0]$

For length 0.5, there are 6 measurement spaces:

1. $[\underline{p}, \overline{p}] = [0.0, 0.5]$
2. $[\underline{p}, \overline{p}] = [0.1, 0.6]$

3. $[\underline{p}, \overline{p}] = [0.2, 0.7]$

4. $[\underline{p}, \overline{p}] = [0.3, 0.8]$

5. $[\underline{p}, \overline{p}] = [0.4, 0.9]$

6. $[\underline{p}, \overline{p}] = [0.5, 1.0]$

For length 0.6, there are 5 measurement spaces:

1. $[\underline{p}, \overline{p}] = [0.0, 0.6]$

2. $[\underline{p}, \overline{p}] = [0.1, 0.7]$

3. $[\underline{p}, \overline{p}] = [0.2, 0.8]$

4. $[\underline{p}, \overline{p}] = [0.3, 0.9]$

5. $[\underline{p}, \overline{p}] = [0.4, 1.0]$

For length 0.7, there are 4 measurement spaces:

1. $[\underline{p}, \overline{p}] = [0.0, 0.7]$

2. $[\underline{p}, \overline{p}] = [0.1, 0.8]$

3. $[\underline{p}, \overline{p}] = [0.2, 0.9]$

4. $[\underline{p}, \overline{p}] = [0.3, 1.0]$

For length 0.8, there are 3 measurement spaces:

1. $[\underline{p}, \overline{p}] = [0.0, 0.8]$

2. $[\underline{p}, \overline{p}] = [0.1, 0.9]$

3. $[\underline{p}, \overline{p}] = [0.2, 1.0]$

For length 0.9, there are 2 measurement spaces:

1. $[\underline{p}, \overline{p}] = [0.0, 0.9]$

2. $[\underline{p}, \overline{p}] = [0.1, 1.0]$

Finally, there is the maximum measurement space with length 1.0:

$$[\underline{p}, \overline{p}] = [0.0, 1.0]$$

### Sample Size and Realization

For each of the above-given measurement spaces $[\underline{p}, \overline{p}]$, there are two different cases. For sample sizes between $n = 1$ and $n = 400$, the bounds given in the tables are computed by means of an optimal $\beta$-measurement & prediction space. For sample sizes $400 < n \leq 1000$, the bounds given in the tables are computed by the approximate solutions (4.80).

As for the realizations $x_S$, the tabulation is complete; that is, any exploitation of symmetry relations is unnecessary. The desired result can be taken directly from the tables.

### Confidence Level

The confidence levels considered in the tables on CD-ROM are the same as in the case of the printed tables:

1. $\beta_1 = 0.80$
2. $\beta_2 = 0.90$
3. $\beta_3 = 0.95$
4. $\beta_4 = 0.98$
5. $\beta_5 = 0.99$

## 8.4.1 Input Mask I for Measurement Intervals

The CD-ROM which runs under Windows and Linux generates an input mask for determining a measurement interval or a prediction region. The first questionnaire appears after the CD-ROM has been installed (Figure 8.1).

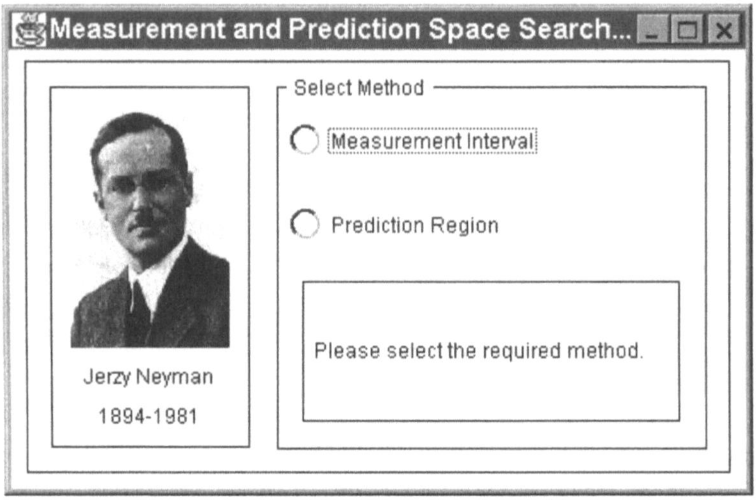


**Figure 8.1** *The questionnaire for measurement or prediction.*

Once the button "Measurement Interval" is pressed, the mask for the corresponding input data appears and the values have to be selected by means of a pull-down menu.

The data to be inserted are the following.

- The confidence level $\beta$ referring to the required procedure's reliability
- The lower and upper bounds $\underline{p}$ and $\overline{p}$ of the measurement space to be selected by the user according to the special proportion or probability of interest
- The sample size $n$ reflecting the expense of the procedure
- The realization $x_S$ which is the observed result of the random experiment

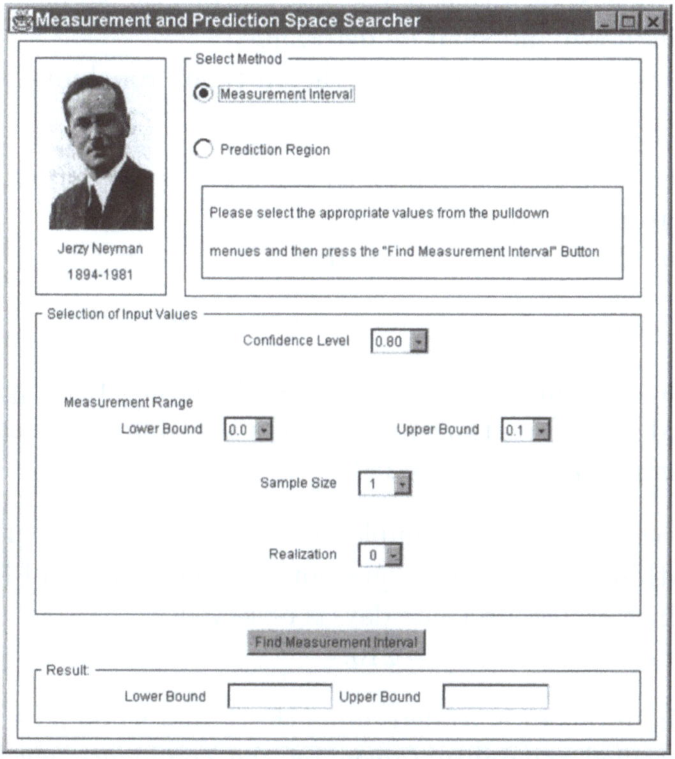


**Figure 8.2** *The input mask for obtaining measurement intervals.*

By pressing the button "Find Measurement Interval," the lower and upper bounds are displayed at the bottom of the mask. (See Figure 8.2.)

## 8.5 Illustrative Example for the Use of the CD-ROM Tables

Consider the event that a letter arrives one day after its posting at the addressee. The probability of this event is called *on-time probability* by the postal organizations and will be denoted by $\breve{p}$. The actual value $p$ of $\breve{p}$ is of interest. For many national postal services, it is known that $0.8 \leq p \leq 1.0$, but the actual value $p$ of $\breve{p}$ is unknown and shall be determined.

The measurement space for the on-time probability $\breve{p}$ is selected as follows:

> The measurement reliability $\beta$ is fixed by the national regulatory office:
> *confidence level:* $\beta = 0.9$.
>
> *measurement space:* $\underline{p} = 0.8$ and $\overline{p} = 1.0$.
>
> The measurement procedure shall be not too expensive and, therefore, the agency selects a medium-sized sample:
> *sample size:* $n = 250$.
>
> According to the sample size, 250 test letters are sent within the country in question and the number of letters, which arrive after one day, is observed:
> *realization:* $x_S = 210$.
>
> The problem is to determine the actual value $p$ of $\breve{p}$ with specified reliability (given by the confidence level) and stated precision.
>
> From the tables on the CD-ROM, for the measurement space $[0.8, 1.0]$, confidence level $\beta = 0.90$, sample size $n = 250$, and realization $x_S = 210$, the following measurement interval is obtained:
>
> *measurement interval:*
>
> $$\begin{aligned} \ell_{\breve{p}}^{(0.90,250)}(210) &= 0.8062 \\ u_{\breve{p}}^{(0.90,250)}(210) &= 0.8783. \end{aligned}$$
>
> Generally, for publishing the result, a point estimate is needed. The simple midpoint $\beta$-estimator yields
>
> *midpoint $\beta$-estimator:*
>
> $$TMP_{\breve{p}}^{(0.90,250)}(210) = 0.8423.$$

**Result:**
The procedure arrives at a correct result in 90% of all cases. The measurement result for the actual value $p$ of the on-time probability is

$$p = 0.8423 \pm 0.0361 \tag{8.9}$$

or in a more popular way,

> The on-time probability measured by means of a procedure with reliability of 90% is equal to 84.23% with measurement precision given by ±3.61%.

Even more important for European postal services than the point estimate would be the statement that the on-time probability exceeds the value of 80% which is required as a lower specification limit by the Commission of the European Union.

**Remark 8.3**
*The much more sophisticated minimum MSE-$\beta$-estimator yields an estimate of*

$$TMS_{\breve{p}}^{(0.90,250)}(210) = 0.8403$$

*and the relative frequency is*

$$TF_{\breve{p}}^{(250)}(210) = 0.8400$$

. *Thus from the viewpoint of the postal organizations, the simple midpoint $\beta$-estimator yields the "best" estimate.* •

## 8.6 Effects of a Restricted Measurement Space

Traditional methods assume generally either maximum measurement space (classical approach) or the existence of prior information about the situation allowing one to select a *prior distribution* with respect to the parameter of interest (Bayes approach). The approach developed here does not start with a given situation, but develops methods for different situations each of them specified by a bounded measurement space. In a case of application, one must select an appropriate measurement space in a first step. In a second step, the actual value of the parameter of interest is measured with a measurement procedure referring to the measurement space. From the viewpoint of stochastics a bounded measurement space or range is equivalent to selecting a bounded support of the relevant random variable. As often the only reliable information about the random phenomenon in question is some knowledge about the possible outcomes, the significance of a realistic support for statistical procedures should be reconsidered. Neglecting this information could indeed lead to inferior methods.

The approach proposed here always yields results that are in conformance with the given measurement space and, thus, with the situation in question. Moreover, there are numerical benefits of the methods and tables recommended in this book, which are indicated by the following example.

**Example 8.3**

*For illustrating the benefits on the resulting measurement interval when using the tables that are based on an appropriately selected measurement space, consider the set of input data given below.*

*Set of input data:*

| | |
|---|---|
| *Measurement space I* | $[\underline{p}, \overline{p}] = [0.0, 1.0]$ |
| *Measurement space II* | $[\underline{p}, \overline{p}] = [0.0, 0.1]$ |
| *Sample size* | $n = 250$ |
| *Confidence level* | $\beta = 0.9$ |

*Let us assume that measurement space II is appropriate, that is, it is known with certainty that the actual value $p$ of the probability $\breve{p}$ of interest is smaller than 0.1. Situations in which small probabilities have to be determined occur and, therefore, the assumption is justified.*

*If the measurement space* $[0.0, 0.1]$ *is appropriate, then the actual value p of the probability* $\breve{p}$ *lies between 0 and 0.1. Therefore, realizations around 15 are not unlikely to appear. In Tables 8.3 and 8.4, the measurement intervals for some values of* $x_S$ *around 12 are displayed.*

*Table 8.3 refers to the traditionally assumed maximum measurement space* $[0.0, 1.0]$, *and Table 8.4 refers to the appropriate measurement space* $[0.0, 0.1]$. *Table 8.3 contains the respective interval lengths in addition to the measurement interval.*

***Table 8.3*** *Measurement intervals based on the maximum measurement space* $[0.0, 1.0]$.

| $[\underline{p}, \overline{p}] = [0.0, 1.0]$ | | |
|---|---|---|
| $x_S$ | *measurement interval* | *length* |
| *8* | [0.01816; 0.05794] | *0.03978* |
| *9* | [0.01978; 0.06206] | *0.04228* |
| *10* | [0.02416; 0.06794] | *0.04378* |
| *11* | [0.02585; 0.07204] | *0.04619* |
| *12* | [0.03014; 0.07612] | *0.04598* |
| *13* | [0.03183; 0.08201] | *0.05018* |
| *14* | [0.03607; 0.08608] | *0.05001* |
| *15* | [0.03915; 0.09014] | *0.05099* |
| *16* | [0.04199; 0.09603] | *0.05404* |
| *17* | [0.04615; 0.10009] | *0.05394* |
| *18* | [0.04790; 0.10414] | *0.05624* |

*Table 8.4 refers to the restricted measurement space* $[0.0, 0.1]$. *In addition to the measurement intervals and their lengths in an additional column it contains the reduction in length of the measurement intervals for realizations around* $x_S = 12$.

*Each of the measurement intervals for those values of* $X_S$ *that are rather likely to occur is shorter if the restricted measurement space* $[0.0, 0.1]$ *is selected instead of the maximum measurement space. The reduction goes up to 13% for* $x_S = 16$, *which is a rather large value. Thus, we may conclude that selecting an appropriate measurement space for a given situation pays, even if the sample size is not small, as in the above example.*

***Table 8.4*** *Measurement intervals based on the appropriate measurement space* [0.0, 0.1].

| $[\underline{p}, \overline{p}] = [0.0, 0.1]$ | | | |
|---|---|---|---|
| $x_S$ | *measurement interval* | *length* | *reduction* |
| *8* | [0.01816; 0.05598] | *0.03782* | *4.9%* |
| *9* | [0.01912; 0.06048] | *0.04136* | *2.2%* |
| *10* | [0.02416; 0.06500] | *0.04084* | *6.7%* |
| *11* | [0.02584; 0.06838] | *0.04254* | *7.9%* |
| *12* | [0.03013; 0.07235] | *0.04222* | *8.2%* |
| *13* | [0.03183; 0.07586] | *0.04403* | *12.3%* |
| *14* | [0.03604; 0.07984] | *0.04380* | *12.4%* |
| *15* | [0.03915; 0.08411] | *0.04496* | *11.8%* |
| *16* | [0.04188; 0.08864] | *0.04676* | *13.5%* |
| *17* | [0.04595; 0.09317] | *0.04722* | *12.5%* |
| *18* | [0.04784; 0.09768] | *0.04984* | *11.4%* |

•

There is another noteworthy effect of the measurement space approach. The smaller a measurement space for a given sample size, the larger is the number of degenerate measurement intervals (i.e., measurement intervals of length zero consisting of one of the bounds of the measurement space). If this happens, then either of three cases has occurred:

- The measurement is correct and the actual value of the probability is indeed one of the boundary values of the measurement space.
- A measurement error has occurred, which is to be expected with probability approximately given by $1 - \beta$.
- A wrong measurement space has been selected.

Thus, whenever a measurement interval of zero length occurs, it is recommended that the measurement be verified by repeating it, in case of doubt, with a larger measurement space.

The result in the case of an inappropriate measurement space is in accordance with the experiences with physical measurement devices. If, for example, the weight of a letter exceeds the measurement range of a given letter balance, then the result will be the upper boundary and repeated measures will always show the same value without any variation.

# Chapter 9

# Prediction Regions

## 9.1 Tables of Prediction Regions on CD-ROM

The tables referring to exclusion and comparison procedures consist of prediction regions for hypotheses containing one element. They can be found on the CD-ROM. Similar to the case of the measurement procedure, each entry, i.e., each prediction region, is based on the sum of sample elements $X_S$ and therefore is represented by a connected subset of $\{0, 1, \cdots, n\}$, where $n$ is the sample size. Thus the prediction regions are completely specified by their smallest and largest element, denoted by

$$\ell_{X_S}^{(\beta,n)}(p_0) \text{ and } u_{X_S}^{(\beta,n)}(p_0)$$

The prediction regions given in the tables are identified by the following set of input data:

1. Measurement space $[\underline{p}, \overline{p}]$
2. Sample size $n$
3. Significance level $\alpha$
4. Null hypothesis $H_0 : \{p_0\}$

The measurement spaces and sample sizes considered in the tables of prediction regions are the same as in the case of the tables of measurement intervals. There are five significance levels which correspond to the confidence levels by the relation $\alpha_i = 1 - \beta_i$ for $i = 1, 2, \ldots, 5$.

For $H_0$, only simple null hypotheses are considered in the tables. However, if a local prediction region is needed for a hypothesis with a positive Lebesgue measure, it can be found using the tabulated prediction regions as the starting value.

For the simple hypotheses, the multitudes of 0.05 contained in each of the measurement ranges are considered in the tables. For example, for the measurement range $[0.2, 0.4]$, there are entries for the hypotheses $(0.20, 0.25, 0.30, 0.35, 0.40)$.

For each of these null hypotheses, the tables contain the lower and upper bound of the corresponding prediction region. If the realization $x_S$ falls within the prediction region, the null hypothesis cannot be rejected, since the realization represents an event which is expected (i.e., predicted) if the null hypotheses is correct. If the realization $x_S$ is an element of the complement of the prediction region, the null hypothesis can be rejected. Thus, the complement of the prediction region is identical to the so-called critical region of a classical significance test. However, arguing by means of the prediction region instead of the critical region has two advantages:

- The prediction region is a connected set and, therefore, topologically simpler.
- The prediction region seems to be more appropriate from a didactic point of view.

### 9.1.1 Input Mask II for Prediction Regions

If the button "Prediction Region" has been pressed on the first questionnaire, an input mask referring to the input data for an prediction region appears and the values have to be selected by means of a pull-down menu. (See Figure 9.1.) The data to be selected are the confidence level, the lower and upper bounds of the measurement space, the sample size, and, finally, the null hypothesis.

By pressing the button "Find Prediction Region," the lower and upper bounds are displayed at the bottom of the mask.

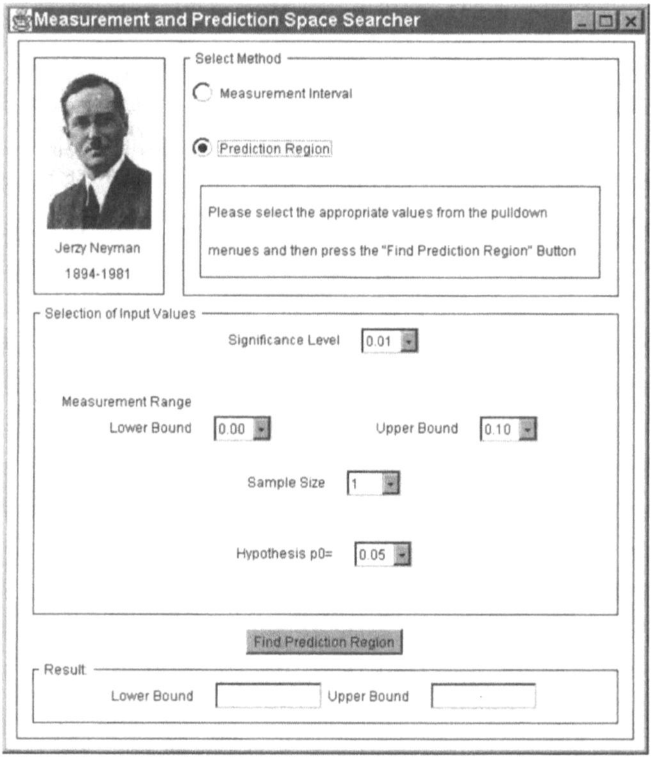


**Figure 9.1** *The input mask for obtaining prediction intervals.*

Note that no manipulation of formula or data is necessary to obtain the prediction region or equivalently the critical region of the test. The user has to insert only those data given by the experimental design and the result of the experiment. Everything else is done by the "statistical procedure."

Thus, application is as simple as the application of many technical devices and errors are almost impossible. The result is a prediction region, which is an intuitively easy to understand quantity. Thus, misunderstanding the result of the statistical procedure is almost impossible.

## 9.2 Illustrative Examples for Determining a Prediction Region

In case of an exclusion procedure, the null hypothesis is rejected if the realization does not fall into the prediction region for the set representing $H_0$. If the realization is an element of the prediction region, then it is clear that the null hypothesis cannot be rejected as the event occurs, which is to be expected if the null hypothesis is correct.

**Example 9.1**
*Let us again consider the example of the on-time probability* $\breve{p}$ *of mailed letters. Assume that the appropriate measurement space is* $[0.8, 1.0]$ *and the aim is to exclude the smallest admissible value, namely* $p_0 = 0.8$ *with a reliability given by the significance level* $\alpha = 0.05$. *Thus, we have the following set of input data:*

- *Measurement space* $[\underline{p}, \overline{p}] = [0.8, 1.0]$
- *Sample size* $n = 250$
- *Significance level* $\alpha = 0.05$
- *Null hypothesis* $H_0\colon\ p_0 = 0.8$

*For the above set of input parameters, the lower and upper bounds of the prediction region are immediately obtained from the CD-ROM tables:*

$$\begin{aligned} \ell_{X_S}^{(0.95,250)}(0.80) &= 183 \\ u_{X_S}^{(0.95,250)}(0.80) &= 210 \end{aligned}$$

*If the observed value $x_S$ exceeds 210, then the null hypothesis $H_0 : p = 0.80$ can be rejected, and it is proved that the on-time probability is larger than 0.80, where the reliability of the procedure is given by $1 - \alpha = 0.05$.*

*The case of a realization $x_S$ smaller than 183 should give rise to the following considerations: Applying the likelihood ratio selection principle for determining the local prediction region would lead to the result*

$$\begin{aligned} {}^{*}\ell_{X_S}^{(0.95,250)}(0.80) &= 0 \\ {}^{*}u_{X_S}^{(0.95,250)}(0.80) &= 210 \end{aligned}$$

*with a comparatively large reliability excess caused by the discreteness of the binomial distribution. In order to decrease the reliability excess, the very unlikely realizations*

$$x_S \in \{0, 1, \ldots, 182\} \tag{9.1}$$

*are taken off the prediction region, without violating the reliability requirement. Therefore, any observed value smaller than 183 should not be used for rejecting the null hypothesis $H_0 : p = 0.80$, but may be looked upon as an indication that an inappropriate measurement space has been selected and that the actual value of the on-time probability may well be smaller than 0.8. Therefore, in the case $x_S leq 182$, it is advisable to select a larger measurement space and thus to admit actual values of $\breve{p}$ which are smaller than 0.80.* •

The CD-ROM tables may be used also for the maximum measurement space as illustrated by Example 9.2.

**Example 9.2**

*Next, consider the same example except for the measurement space, which is selected to be the unit interval yielding the following set of input parameters:*

- *Measurement space* $[\underline{p}, \overline{p}] = [0.0, 1.0]$
- *Sample size* $n = 250$

- *Significance level* $\alpha = 0.05$
- *null hypothesis* $H_0$: $p_0 = 0.8$

*The prediction interval can be taken directly from the tables:*

$$\begin{aligned} \ell_{X_S}^{(0.95,250)}(0.80) &= 188 \\ u_{X_S}^{(0.95,250)}(0.80) &= 212 \end{aligned}$$

*If the whole unit interval is taken as measurement space, then the hypothesis that the on-time probability equals 0.80 can be rejected if the observed value* $x_S$ *exceeds 212 or is less than 188.* •

## 9.3 $H_0$ with Positive Lebesgue Measure

As mentioned earlier, excluding a subset of the measurement space with zero Lebesgue measure does not make much sense. Therefore, emphasis should be laid on the so-called composite null hypotheses, which are represented here by sets of non-zero Lebesgue measure.

In the following, a method of how an exclusion procedure may be obtained in the case of a null hypothesis with positive Lebesgue measure is briefly outlined.

If the null hypothesis is an interval $H_0 : [p_1, p_2]$, then in a first step, the prediction regions for $p \in [p_1, p_2]$ are taken from the CD-ROM tables. Their union is the first preliminary local prediction region for $H_0$.

The second step consists of an attempt to reduce the preliminary local prediction region by taking off the most extreme elements of the two prediction regions (for $p_1$ and $p_2$, respectively) in the order of the size of their likelihood ratio, beginning with the element with smallest likelihood ratio. The aim is to reduce the reliability excess computed for the boundaries of the null hypothesis (i.e., $p_1$ and $p_2$). Because of the discreteness of the problem, the above-outlined algorithm may not necessarily lead to the optimal local prediction region, but it will always lead to a local prediction region close to the optimal one.

For illustration, consider Example 1 with the measurement space $[\underline{p}, \overline{p}] = [0.8, 1.0]$, the sample size $n = 250$, the significance level $\alpha = 0.05$, but now with the null hypothesis

$$H_0\colon\ [p_1, p_2] = [0.8, 0.9].$$

From the CD-ROM tables, we immediately obtain the following preliminary local prediction region for $H_0$, where the lower bound refers to the lower bound of the prediction region for $p = 0.8$, and the upper bound refers to upper bound of the prediction region for $p = 0.9$:

$$\begin{aligned} \ell_{X_S}^{(0.95,250)}([0.8, 0.9]) &= \ell_{X_S}^{(0.95,250)}(0.8) = 183 \\ u_{X_S}^{(0.95,250)}([0.8, 0.9]) &= u_{X_S}^{(0.95,250)}(0.9) = 234 \end{aligned}$$

Generally, the preliminary local prediction region has a high reliability excess and, therefore, the attempt should be made to improve it. How to proceed is in principle clear: The lower bound should be increased and the upper bound decreased. The problem is to answer the question at which end to start and to stop.

For reducing the preliminary local prediction region, the likelihood ratios

$$\frac{\mathbf{P}_{X_S}^{(p)}(x_S)}{Q_{\{0,1,\cdots,250\}}(x_S)}$$

with

$$\mathbf{P}_{X_S}^{(p)}(x_S) = \binom{n}{x_S} p^{x_s}(1-p)^{n-x_S}$$

$$Q_{\{0,1,\cdots,250\}}(x_S) = \frac{\int_{\underline{p}}^{\overline{p}} \mathbf{P}_{X_S}^{(p)}(x_S)dp}{\overline{p} - \underline{p}}$$

of the elements at the lower and upper bounds of the preliminary prediction region are calculated. The results are listed in Table 9.1.

**Table 9.1** *Likelihood ratios for the boundary elements of the preliminary local prediction region.*

| $x_S$ | $\frac{\mathbf{P}_{X_S}^{(0.80)}(x_S)}{Q_{\{0,1,\cdots,250\}}(x_S)}$ | $\frac{\mathbf{P}_{X_S}^{(0.90)}(x_S)}{Q_{\{0,1,\cdots,250\}}(x_S)}$ |
|---|---|---|
| 183 | 24.330 | — |
| 184 | 23.178 | — |
| 185 | 22.034 | — |
| 186 | 20.898 | — |
| 187 | 19.771 | — |
| 188 | 18.654 | — |
| 189 | 17.549 | — |
| 190 | 16.457 | — |
| 232 | — | 1.477 |
| 233 | — | 1.027 |
| 234 | — | 0.671 |

The prediction regions are filled starting with those realizations with the largest likelihood ratios. The reduction of a prediction region is performed in the reverse order; that is the realizations with smallest likelihood ratios are taken off first.

By means of Table 9.1, the boundary values $x_S = 183$ and $x_S = 234$ are compared with respect to their likelihood ratios

$$\frac{\mathbf{P}_{X_S}^{(0.80)}(183)}{Q_{\{0,1,\ldots,250\}}(183)} \text{ and } \frac{\mathbf{P}_{X_S}^{(0.90)}(234)}{Q_{\{0,1,\ldots,250\}}(234)} \tag{9.2}$$

The realization $x_S = 234$ has a smaller value and is therefore the first candidate to be taken off the local prediction region; the next candidate would be $x_S = 233$ and so forth. Thus, the likelihood ratios determine the order of the elements to be taken out of the prediction region.

For determining the number of realizations which can be taken out of the prediction region without violating the reliability requirement given by the confidence level $\beta$ one has to calculate the corresponding probabilities. The prediction region is reduced as long as its probability is at least equal to $\beta$ for every $p \in [0.80, 0.90]$ The probabilities are given in Table 9.2.

**Table 9.2** *Probabilities of the reduced prediction regions.*

| $[\ell, u]$ | $\sum_{x_S=\ell}^{u} \mathbf{P}_{X_S}^{(0.80)}(x_S)$ | $\sum_{x_S=\ell}^{u} \mathbf{P}_{X_S}^{(0.90)}(x_S)$ |
|---|---|---|
| $[183, 234]$ | 0.996291 | 0.982492 |
| $[183, 233]$ | 0.996291 | 0.969119 |
| $[183, 232]$ | 0.996291 | **0.948664** |
| $[184, 233]$ | 0.994301 | 0.969119 |
| $[185, 233]$ | 0.991402 | 0.969119 |
| $[186, 233]$ | 0.987265 | 0.969119 |
| $[187, 233]$ | 0.981482 | 0.969119 |
| $[188, 233]$ | 0.973565 | 0.969119 |
| $[189, 233]$ | 0.962953 | 0.969119 |
| $[190, 233]$ | **0.949028** | 0.969119 |

From Table 9.2 it is seen that the prediction region reduced by 234 still meets the reliability requirements. Therefore, the reduction may be continued by the next candidate $x_S = 233$; and again the reliability requirement is still met. However, removing $x_S = 232$ leads to a prediction region violating the reliability requirement for $p = 0.90$. Therefore, the upper boundary of the reduced prediction region is found to be 233.

It remains to check whether the lower boundary may be increased. As can be seen from Table 9.2 this is possible, up to 189. Thus, the optimal prediction region is found to be

$$\begin{aligned} {}^*\ell_{X_S}^{(0.95,250)}([0.8, 0.9]) &= 189 \\ {}^*u_{X_S}^{(0.95,250)}([0.8, 0.9]) &= 233 \end{aligned}$$

The necessary values for the likelihood ratios and the binomial probabilities may be obtained using any of the standard mathematical software programs, for instance, *Mathematica.*

Note that the development of exclusion procedures for null hypotheses with positive Lebesgue measure does not make any principle difficulties, but may be solved by applying the same methods derived for simple null hypotheses.

# Part IV

# Application

# Chapter 10

# Measuring a Probability

In the preceding chapters the approach based on a restricted measurement space and a detailed analysis of the situation with respect to the corresponding measure functions is given. This theoretical introduction to approach and procedures addresses primarily academic and professional statisticians. In contrast, the present chapter is devoted to non-statisticians and statisticians who are only interested in applying statistical methods without bothering about the theoretical details.

Different situations, in which the methods proposed here may be applied, are briefly outlined by means of examples. The application of the appropriate procedure is represented in the form of a recipe.

In many of the following examples the quantity of interest is a certain proportion, and the sample is drawn without replacement from the corresponding population. Therefore, the correct model is given by the hypergeometric distribution. The used binomial model has to be looked upon as a conservative approximation in the sense that the results (measurement intervals and prediction regions) obtained by using the binomial model instead of the hypergeometric model always meet the reliability requirement, but may be unnecessarily wide.

In the case that the sample size $n$ is not small compared with the size of the population, it is advisable to use the hypergeometric model and derive measurement procedures analogously to those presented in this handbook.

## 10.1 The Actual Value of a Probability

### 10.1.1 Algorithm

The algorithm for determining the actual value of a probability of interest is extremely simple. It assumes that there is a random sample of size $n$ representing $n$ independent Bernoulli experiments and that the measurement procedure shall be based on $X_S$. The algorithm consists of only two steps:

Step 1: List the set of input parameters:

- Measurement space $[\underline{p}, \overline{p}]$
- Sample size $n$
- Confidence level $\beta$
- Realization $x_S$

Step 2: Enter the table, identify the relevant entry by means of the input parameters and, thus, find the lower limit $\ell_{\check{p}}^{(\beta,n)}(x_S)$ and the upper limit $u_{\check{p}}^{(\beta,n)}(x_S)$ of the measurement interval.

Thus, application of the tables is possible without any cumbersome calculations or transformations, which, indeed, assume some often not existing familarity with mathematics and statistics.

### 10.1.2 Examples

The task of determining the actual value of a proportion of interest is one of the most frequently arising statistical problems in almost any field of application.

The following problems (including their formulation) are taken from [32], except for the sample size, which has been changed to enable the use and illustration of the printed tables.

**Example 10.1**

*First the problem from [32] is cited and then some additional assumptions are made in order to better illustrate the approach developed here.*

> *Out of a group of 10,000 degree candidates of The University of North Carolina at Chapel Hill, a random sample of 400 showed that 20 per cent of the students have an earning potential exceeding $30,000 annually. Establish a .95 confidence-interval*[1] *estimate of the number of students with a $30,000 plus earning potential.*

*Additionally, we assume the following:*

> *From past experience, it is known that a range of* [0.1, 0.4] *is appropriate for measuring the actual value of the proportion of students having an earning potential exceeding $30,000 annually.*

*Thus, the set of input parameters is the following:*

- *Measurement space:* $[\underline{p}, \overline{p}] = [0.1, 0.4]$
- *Sample size:* $n = 400$
- *Confidence level:* $\beta = 0.95$
- *Realization:* $x_S = 80$ *that is,* 20% *of 400 students*

*From the tables on the CD-ROM, the lower and upper bounds of the measurement interval are immediately obtained:*

$$\ell_{\breve{p}}^{(0.95,400)}(80) = 0.16345$$
$$u_{\breve{p}}^{(0.95,400)}(80) = 0.24230$$

[1]The examples are citations from [32] and, therefore, the traditional term *confidence interval* is used instead of measurement interval.

*There are 10,000 students and, therefore, the statistical measurement procedure yields the following result with respect to the actual number of students with a $30,000 earning potential.*

**Result:**
*There are between 1,634 and 2,423 students with a $30,000 earning potential, where the reliability of the procedure used is given by* 95%. •

**Example 10.2**
*In this example the printed tables may be used, as the sample size is comparatively small and there is no prior knowledge about the unknown value of the proportion of interest.*

> *Harvey of Brooklyn surveyed a random sample of 100 students*[2] *at SUNY-Stony Brook. Being a premedical student, he hoped that most students would major in the social sciences rather than the natural sciences, thus providing him with less competition. To Harvey's dismay, 60% of the students he surveyed were majoring in the natural sciences. Construct a 95% confidence interval for p, the population proportion of students majoring in the natural sciences.*

*Obviously, Harvey had no idea about the actual value of the proportion, therefore the maximum measurement space is appropriate.*

*The set of input parameter is as follows:*

- *Measurement space:* $[\underline{p}, \overline{p}] = [0.0, 1.0]$
- *Sample size:* $n = 100$
- *Confidence level:* $\beta = 0.95$
- *Realization:* $x_S = 60$

[2] The sample size in [32] is given as 625.

*Because of the maximum measurement space and the comparatively small sample size $n = 100$, the printed tables may be used. There is no entry for the realization $x_S = 60$ in the corresponding table, but the value of the complementary realization $n - x_S = 40$ is contained in the table:*

$$\begin{aligned} \ell_{\breve{p}}^{(0.95,100)}(40) &= 0.3055 \\ u_{\breve{p}}^{(0.95,100)}(40) &= 0.5000 \end{aligned}$$

*From these values, we immediately obtain the required bounds of the measurement interval:*

$$\begin{aligned} \ell_{\breve{p}}^{(0.95,100)}(60) &= 1 - u_{\breve{p}}^{(0.95,100)}(40) = 0.5000 \\ u_{\breve{p}}^{(0.95,100)}(60) &= 1 - \ell_{\breve{p}}^{(0.95,100)}(40) = 0.6945 \end{aligned}$$

**Result:**
*The proportion of students majoring in the natural sciences lies between 50.0% and 69.5%. The procedure's reliability is 95%.* •

**Example 10.3**
*In this example the sample size n exceeds 100 and, moreover, the measurement space is not maximum. Therefore, the CD-ROM tables have to be used.*

> *In assessing the desirability of windowless schools, officials asked 144 elementary school children whether or not the children preferred windows. Thirty percent of the children preferred windows. Establish a 0.95% confidence interval estimate of the proportion of elementary school children who like windows in their classroom.*

*Additionally, we assume the following:*

> *There is some knowledge about the actual value of the proportion to be determined according to which a measurement space of $[0.2, 0.8]$ is appropriate.*

*Thus, the following set of input parameters is obtained:*

- *Measurement space:* $[\underline{p}, \overline{p}] = [0.2, 0.8]$
- *Sample size:* $n = 144$
- *Confidence level:* $\beta = 0.95$
- *Realization:* $x_S = 43$ *(about 30% of 144)*

*Because* $n > 100$, *the printed tables cannot be used in this example, but the CD-ROM tables can. The following values are obtained:*

$$\begin{aligned} \ell_{\breve{p}}^{(0.95,144)}(43) &= 0.2360 \\ u_{\breve{p}}^{(0.95,144)}(43) &= 0.3782 \end{aligned}$$

**Result:**
*More than 23.6% but less than 37.8% of the children prefer schools with windows. The reliability of the procedure applied is 95%*[3]. •

**Example 10.4**
*Very often it is known that a proportion is smaller or larger than 0.50. Of course, this information should be utilized.*

> *A random sample of 300 Oberlin students were asked if they regularly smoke marijuana. Sixty replied in the affirmative. Find an approximate 95% confidence interval on the true proportion, p, of Oberlin marijuana smokers.*

*Additionally, we assume:*

---

[3]The approximate method proposed in [32] is based on the normal distribution. The methods derived here give not only an exact (with respect to the binomial model) but also a most precise measurement interval. The approximate measurement interval given in [32] is [0.225, 0.375], which is a 5.5% increase compared with the Neyman measurement interval.

> *From experience and from other sources, it is known with certainty that the actual value of the proportion is less than 0.50.*

*Thus, we arrive at the following restricted measurement space:*

- *Measurement space:* $[\underline{p}, \overline{p}] = [0.0, 0.5]$

*For the other input parameter, we have*

- *Sample size:* $n = 300$
- *Confidence level:* $\beta = 0.95$
- *Realization:* $x_S = 60$

*From the CD-ROM tables, we immediately obtain the bounds of the required measurement interval:*

$$\begin{aligned} \ell_{\breve{p}}^{(0.95,100)}(60) &= 0.15798 \\ u_{\breve{p}}^{(0.95,100)}(60) &= 0.24972 \end{aligned}$$

**Result:**
*The actual proportion of Oberlin students who regularly smoke marijuana is at least 15.8% and at most 25.0%, where the reliability of the procedure is 95%.* •

**Example 10.5**
*Frequently, it is known that the actual value of a proportion of interest is larger (or smaller) than 0.5 and smaller than 1.0 (larger than 0.0). Again, this information should be utilized to obtained more precise measurement results.*

> *A random sample of 225 students at a college showed that 135 had used and benefited from problem solvers. Make a .95 confidence interval estimate of the exact proportion.*

*Additionally, we assume the following:*

> *The teachers have some knowledge about the actual value of the proportion by experience and are able to chose an appropriate measurement space:*

- *Measurement space:* $[\underline{p}, \overline{p}] = [0.5, 0.8]$

*The other input parameters are as follows:*

- *Sample size:* $n = 225$
- *Confidence level:* $\beta = 0.95$
- *Realization:* $x_S = 135$

*From the CD-ROM tables, we immediately obtain the bounds of the measurement interval:*

$$\begin{aligned} \ell_{\breve{p}}^{(0.95,225)}(135) &= 0.5334 \\ u_{\breve{p}}^{(0.95,225)}(135) &= 0.6624 \end{aligned}$$

**Result:**
*The actual value of the proportion of students using problem-solvers lies between 53.3% and 66.2%. The procedure's reliability is 95%.* •

## 10.2 The Difference of Two Probabilities

Sometimes, there are two proportions or probabilities and not the single probabilities, but their difference is of interest. In such a case, the tables can be used for finding a measurement interval for the difference.

Let

$$X_S \sim Bi(n_1, p_1) \tag{10.1}$$
$$Y_S \sim Bi(n_2, p_2) \tag{10.2}$$

be two independent random variables. There are two parameters $\breve{p}_1$ and $\breve{p}_2$ with actual values $p_1$ and $p_2$, respectively. The parameter of interest $\breve{d}$, however, is the difference of the two parameters $\breve{p}_1$ and $\breve{p}_2$ with actual value $d = p_1 - p_2$.

Moreover, let

$$\left(L_{\breve{p}_1}^{(\beta_1,n_1)}(X_S), U_{\breve{p}_1}^{(\beta_1,n_1)}(X_S)\right)$$

and

$$\left(L_{\breve{p}_2}^{(\beta_2,n_2)}(Y_S), U_{\breve{p}_2}^{(\beta_2,n_2)}(Y_S)\right)$$

be two measurement procedures for $\breve{p}_1$ and $\breve{p}_2$, respectively, and

$$A_{X_S}^{(\beta_1,n_1)}\left(p_1\right)$$

and

$$A_{Y_S}^{(\beta_2,n_2)}\left(p_2\right)$$

be two prediction regions based on $X_S$ and $Y_S$ for $p_1$ and $p_2$, respectively.

Because

$$\begin{aligned} &\left.\begin{array}{c} L_{\breve{p}_1}^{(\beta_1,n_1)}(X_S) \le p_1 \le U_{\breve{p}_1}^{(\beta_1,n_1)}(X_S) \\ L_{\breve{p}_2}^{(\beta_2,n_2)}(Y_S) \le p_2 \le U_{\breve{p}_2}^{(\beta_2,n_2)}(Y_S) \end{array}\right\} \\ \Rightarrow \quad & L_{\breve{p}_1}^{(\beta_1,n_1)}(X_S) - U_{\breve{p}_2}^{(\beta_2,n_2)}(Y_S) \le p_1 - p_2 \\ & \le U_{\breve{p}_1}^{(\beta_1,n_1)}(X_S) - L_{\breve{p}_2}^{(\beta_2,n_2)}(Y_S) \end{aligned} \tag{10.3}$$

it is clear that

$$\left(L_{\breve{p}_1}^{(\beta_1,n_1)}(X_S) - U_{\breve{p}_2}^{(\beta_2,n_2)}(Y_S), U_{\breve{p}_1}^{(\beta_1,n_1)}(X_S) - L_{\breve{p}_2}^{(\beta_2,n_2)}(Y_S)\right) \tag{10.4}$$

is a measurement procedure for the difference $\breve{d}$ with confidence level $\beta = \beta_1\beta_2$.

**Example 10.6**

*It should be noted that the above-derived measurement procedure is not a Neyman measurement procedure.*

> *Tom and Joe*[4] *like to throw darts. Tom throws 100 times and hits the target 54 times; Joe throws 100 times and hits the target 49 times. Find a 95 % confidence interval for* $p_1 - p_2$*, where* $p_1$ *represents the probability*[5] *of a hit in Tom's tosses, and* $p_2$ *represents the probability of a hit in Joe's tosses.*

*Additionally the following is assumed:*

> *It is known that for either Tom or Joe, the hit probability is larger than* 0.4 *and smaller than* 0.6.

*Because of* $0.98^2 = 0.960 > 0.95 = \beta$*, we have the following sets of input parameters for* $\breve{p}_1$ *and* $\breve{p}_2$*:*

- $[\underline{p}_1, \overline{p}_1] = [\underline{p}_2, \overline{p}_2] = [0.4, 0.6]$
- $\beta_1 = \beta_2 = 0.98$
- $n_1 = n_2 = 100$
- $x_S = 54$ *and* $y_S = 49$.

*As the measurement space is not maximum, the CD-ROM tables have to be used. The following measurement intervals are obtained:*

$$\begin{aligned} \ell_{\breve{p}_1}^{(0.98,100)}(54) &= 0.43280 \\ u_{\breve{p}_1}^{(0.98,100)}(54) &= 0.60000 \end{aligned}$$

$$\begin{aligned} \ell_{\breve{p}_2}^{(0.98,100)}(49) &= 0.40000 \\ u_{\breve{p}_2}^{(0.98,100)}(49) &= 0.59652 \end{aligned}$$

*Using (10.4), the following measurement interval for the difference* $\breve{d}$ *is obtained,*[6] *with* $x_S - y_S = 5$*:*

$$\begin{aligned} \ell_{\breve{d}}^{(0.95,100,100)}(5) &= -0.1672 \\ u_{\breve{d}}^{(0.95,100,100)}(5) &= 0.1965 \end{aligned}$$

---

[4] Taken from [32], Problem 14-72.

[5] In [32] the term *proportion* is used instead of *probability.*

[6] The approximate measurement interval given in [32] is $[-0.88, 0.188]$ which is three times longer than the measurement interval obtained here.

**Result:**
*The actual value of the difference of the hit probabilities of Tom and Joe lies between −16.7% and +19.7%. The procedure's reliability is* $0.98^2 \cdot 100\,\%$*=96%.*

## 10.3 Ratio of Poisson Parameters

Consider the case of two independent random variables $X$ and $Y$, where $X$ is distributed according to a Poisson distribution with actual value $\theta_1$ of the distribution parameter $\breve{\theta}_1$ and $Y$ has also a Poisson distribution with actual value $\theta_2$ of the distribution parameter $\breve{\theta}_2$.

Assume that the quantity of interest is the ratio of the actual values $\frac{\theta_1}{\theta_2}$ of the distribution parameters instead of the single values themselves. In this case, the so-called Chapman-Test (see [5]) can be used.

Consider the conditional random variable

$$X|X+Y = n \in \{0, 1, \ldots, n\} \tag{10.5}$$

The probability distribution of $X|X+Y = n$ is easily obtained by calculating the probabilities of the events $\{x\}$ with $x \in \{0, 1, \ldots, n\}$:

$$\begin{aligned}
\mathbf{P}^{(\frac{\theta_1}{\theta_2})}_{X|X+Y=n}\Big(\{x\}\Big) &= \frac{\mathbf{P}^{(\theta_1,\theta_2)}_{X,Y}\Big(\{x, n-x\}\Big)}{\mathbf{P}^{(\theta_1,\theta_2)}_{X+Y}\Big(\{n\}\Big)} \\
&= \frac{\frac{\theta_1^x}{x!}e^{-\theta_1}\frac{\theta_2^{n-x}}{(n-x)!}e^{-\theta_2}}{\frac{(\theta_1+\theta_2)^n}{n!}e^{-(\theta_1+\theta_2)}} \\
&= \binom{n}{x}\left(\frac{\theta_1}{\theta_1+\theta_2}\right)^x\left(\frac{\theta_2}{\theta_1+\theta_2}\right)^{(n-x)}
\end{aligned} \tag{10.6}$$

The expression (10.6) shows that the conditional random variable $X|X+Y = n$ is distributed according to the binomial distribution $Bi\Big(n, \frac{\theta_1}{\theta_1+\theta_2}\Big)$ or $Bi\Big(n, 1 - \frac{1}{1+\theta}\Big)$, with $\theta = \frac{\theta_1}{\theta_2}$ being the actual value of the distribution parameter $\breve{\theta} = \frac{\breve{\theta}_1}{\breve{\theta}_2}$, with measurement space, say $\Theta = [\underline{\theta}, \overline{\theta}]$.

The situation is described in the following diagram:

| quantity | symbol | value | image space |
|---|---|---|---|
| conditional phenomenon | $X \| X+Y=n$ | $x$ | $\{0,1,\ldots,n\}$ |
| parameter | $\breve{\theta}$ | $\theta$ | $\Theta$ |
| connection | $\mathbf{P}^{(\theta)}_{X\|X+Y=n}$ | | |

The problem is to measure the actual value $\theta$ of $\breve{\theta}$ as precisely as possible.

In order to make a direct application of the tables possible, the given problem is substituted by a problem referring to the quantity

$$\breve{p} = 1 - \frac{1}{1+\breve{\theta}} \tag{10.7}$$

and aiming at determining the acutal value $p$ of $\breve{p}$ as precisely as possible. The new problem is described in the following diagram:

| quantity | symbol | value | image space |
|---|---|---|---|
| conditional phenomenon | $X \| X+Y=n$ | $x$ | $\{0,1,\ldots,n\}$ |
| parameter | $\breve{p}$ | $p$ | $[\underline{p}, \overline{p}]$ |
| connection | $\mathbf{P}^{(p)}_{X\|X+Y=n}$ | | |

where

$$\underline{p} = 1 - \frac{1}{1+\underline{\theta}} \quad \text{and} \quad \overline{p} = 1 - \frac{1}{1+\overline{\theta}} \tag{10.8}$$

Once a solution with respect to $\breve{p}$ is determined, a solution for the original problem is obtained by using the transformation

$$\breve{\theta} = \frac{\breve{p}}{1-\breve{p}} \tag{10.9}$$

**Example 10.7**

*The determination of element concentrations in analytical chemistry [33] by means of quantitative electron-probe microanalysis is important for evaluation of geochemical processes. The calculation of concentration is based on the generation of X-rays, which results in a number of impulses at the position of the line maximum. The number of these impulses within a time unit is generally modeled by a Poisson-distributed random variable. Let* $X$ *be the number of impulses of one specimen of material and* $Y$ *the number of impulses of another specimen. Let* $X \sim Po(\theta_1)$ *and* $Y \sim Po(\theta_2)$, *then* $\theta_1$ *and* $\theta_2$ *can be regarded as measures of concentration; thus, their ratio may be used for comparing the element's concentration in the two specimens.*

*Let the confidence level be fixed at* $\beta = 0.99$ *and the measurement space for the ratio be selected as* $\Theta = [0.25, 1.00]$. *There are* $x = 113$ *counts with respect to the first specimen and* $y = 231$ *counts for the second specimen. Thus, the following input parameters are obtained:*

- $[\underline{p}, \overline{p}] = [0.2, 0.5]$
- $\beta = 0.99$
- $n = 344$
- $x = 113$

*From the tables on the CD-ROM, the lower and upper bounds of the measurement interval with respect to* $\breve{p}$ *are obtained:*

$$
\begin{aligned}
\ell_{\breve{p}}^{(0.99,344)}(113) &= 0.32059 \\
u_{\breve{p}}^{(0.99,344)}(113) &= 0.45072
\end{aligned}
$$

*which immediately lead to the following bounds with respect to* $\breve{\theta}$:

$$
\begin{aligned}
\ell_{\breve{\theta}}^{(0.99,344)}(113) &= 0.47187 \\
u_{\breve{\theta}}^{(0.99,344)}(113) &= 0.82057
\end{aligned}
$$

**Result:**
*The concentration of the element of interest in the first specimen is between 47.1% and 82.1% of the concentration of the element in the second specimen. The procedure's reliability is 99%.* •

**Remark 10.1**
*Note that the two problems (i.e., determining the actual value p and the actual value θ) are not equivalent, as the two measure functions $\Lambda_\Theta$ and $\Lambda_{[\underline{p},\overline{p}]}$ are not equivalent.* •

## 10.4 Point Estimator

Without a statement about reliability and precision of the used procedure, any point estimator is more or less meaningless. With a corresponding statement, however, it is not only useful but even indispensable for many calculations and implications. Therefore, any measurement procedure should be completed by a point estimator.

In other words, each measurement interval of a measurement procedure should be supplemented by a corresponding point estimate. The estimator $T_{\breve{p}}(\vec{X})$ or $T_{\breve{p}}(X_S)$ should be selected so as to minimize the distance to the actual but unknown value $p$. Of course, this concept makes only sense if no measurement error has occurred (i.e., $\ell_{\breve{p}}^{(\beta,n)} \leq p \leq u_{\breve{p}}^{(\beta,n)}$).

In accordance with the tables, only the case of procedures based on the sum of sample elements $X_S$ is considered here.

There are three estimators for a probability $\breve{p}$: the traditional relative frequency, $TF_{\breve{p}}^{(n)}(X_S)$, the minimum MSE-$\beta$-estimator $TMS_{\breve{p}}^{(\beta,n)}(X_S)$, and, finally, the midpoint estimator $TMP_{\breve{p}}^{(\beta,n)}(X_S)$. As the three estimators are asymptotically equivalent, there should not be a considerable difference for a large sample size $n$.

For given measurement procedure $\left(L_{\breve{p}}^{(\beta,n)}(X_S), U_{\breve{p}}^{(\beta,n)}(X_S)\right)$, the three point estimators are defined in the following way:

$$TF_{\breve{p}}^{(n)}(X_S) = \frac{1}{n} X_S \qquad (10.10)$$

$$TMS_{\breve{p}}^{(\beta,n)}(X_S) = \frac{\int\limits_{L_{\breve{p}}^{(\beta,n)}(X_S)}^{U_{\breve{p}}^{(\beta,n)}(X_S)} \binom{n}{X_S} p^{X_S+1}(1-p)^{n-X_S}\,dp}{\int\limits_{L_{\breve{p}}^{(\beta,n)}(X_S)}^{U_{\breve{p}}^{(\beta,n)}(X_S)} \binom{n}{X_S} p^{X_S}(1-p)^{n-X_S}\,dp} \tag{10.11}$$

$$TMP_{\breve{p}}^{(\beta,n)}(X_S) = \frac{U_{\breve{p}}^{(\beta,n)}(X_S) + L_{\breve{p}}^{(\beta,n)}(X_S)}{2} \tag{10.12}$$

Clearly, there are no computational difficulties with the relative frequency (10.10) and the midpoint $\beta$-estimator (10.12), at least for given measurement interval. In contrast to these two simple estimators, computation of the estimate for the minimum MSE-$\beta$-estimator is rather sophisticated. However, as mentioned earlier, the computations can easily be performed using a mathematical software package.

In Table 10.1, the values of the point estimators for the examples of the section "Determination of the Actual Value of a Probability" are listed.

**Table 10.1** *Measurement intervals and point estimates for Examples 1-5.*

| Example | $[\ell, u]$ | $TF$ | $TMS$ | $TMP$ |
|---|---|---|---|---|
| No. 1 | $[0.1635, 0.2423]$ | 0.2000 | 0.2014 | 0.2029 |
| No. 2 | $[0.5000, 0.6945]$ | 0.6000 | 0.5984 | 0.5972 |
| No. 3 | $[0.2000, 0.3134]$ | 0.2639 | 0.2617 | 0.2567 |
| No. 4 | $[0.1580, 0.2497]$ | 0.2000 | 0.2018 | 0.2038 |
| No. 5 | $[0.5334, 0.6624]$ | 0.6000 | 0.5991 | 0.5979 |

It appears that the actual values of the three estimators are almost the same, and one could be inclined to conclude that there is no essential difference between the traditional relative frequency, and the minimum MSE-$\beta$-estimator and the midpoint $\beta$-estimator. However, such a conclusion would miss the point. The differences and advantages of the approach developed here become clear in the case of small sample sizes.

Asymptotically, the three estimators are equivalent; therefore, the differences become small with increasing sample size. Therefore, if the sample size is sufficiently large, then the traditional approach yields more or less the same results as obtained by means of procedures developed here.

In many instances, however, samples of a size below 50 are typical; for example, in clinical studies and studies in the natural sciences. Moreover, often the measurement space may be selected comparatively small, as, for example, for high quality manufacturing processes where the nonconforming probability is far below 1%. In these cases, one should not rely on the traditional procedures, but take those introduced here.

For illustration, the results in case of a small sample size are given below. The set of input parameters for the example is as follows:

- Measurement space: $[\underline{p}, \overline{p}] = [0.1, 0.4]$
- Sample size: $n = 20$
- Confidence level: $\beta = 0.80$

From Table 10.2, it is clear that the relative frequency results in many cases in useless estimates, that is, in values which are neither elements of the corresponding measurement interval nor of the selected measurement space and, therefore, have just to be discarded, as they do not add to the already available knowledge. In contrast to the relative frequency, any $\beta$-estimator yields results that are reasonable with certainty and, therefore, increase the available knowledge.

**Remark 10.2**

*The above-outlined difficulties with the relative frequency are, of course, well-known and, therefore, the fact is surprising that a majority of statisticians still supports the frequency concept in statistics.* •

**Table 10.2** *Complete measurement procedure for the small sample size $n = 20$.*

| $x_S$ | $[\ell, u]$ | $TF$ | $TMS$ | $TMP$ |
|---|---|---|---|---|
| 0 | $[0.1000, 0.1584]$ | 0.00 | 0.1229 | 0.1292 |
| 1 | $[0.1000, 0.1975]$ | 0.05 | 0.1372 | 0.1488 |
| 2 | $[0.1000, 0.2313]$ | 0.10 | 0.1536 | 0.1656 |
| 3 | $[0.1000, 0.2693]$ | 0.15 | 0.1763 | 0.1846 |
| 4 | $[0.1170, 0.3135]$ | 0.20 | 0.2123 | 0.2153 |
| 5 | $[0.1584, 0.3665]$ | 0.25 | 0.2601 | 0.2624 |
| 6 | $[0.1975, 0.4000]$ | 0.30 | 0.3002 | 0.2988 |
| 7 | $[0.2313, 0.4000]$ | 0.35 | 0.3230 | 0.3156 |
| 8 | $[0.2615, 0.4000]$ | 0.40 | 0.3403 | 0.3307 |
| 9 | $[0.2693, 0.4000]$ | 0.45 | 0.3484 | 0.3346 |
| 10 | $[0.3135, 0.4000]$ | 0.50 | 0.3642 | 0.3567 |
| 11 | $[0.3135, 0.4000]$ | 0.55 | 0.3667 | 0.3567 |
| 12 | $[0.3135, 0.4000]$ | 0.60 | 0.3690 | 0.3567 |
| 13 | $[0.3135, 0.4000]$ | 0.65 | 0.3712 | 0.3567 |
| 14 | $[0.3665, 0.4000]$ | 0.70 | 0.3857 | 0.3832 |
| 15 | $[0.3665, 0.4000]$ | 0.75 | 0.3861 | 0.3832 |
| 16 | $[0.3665, 0.4000]$ | 0.80 | 0.3864 | 0.3832 |
| 17 | $[0.3665, 0.4000]$ | 0.85 | 0.3868 | 0.3832 |
| 18 | $[0.4000, 0.4000]$ | 0.90 | 0.4000 | 0.4000 |
| 19 | $[0.4000, 0.4000]$ | 0.95 | 0.4000 | 0.4000 |
| 20 | $[0.4000, 0.4000]$ | 1.00 | 0.4000 | 0.4000 |

In order to quantify the advantage of the two $\beta$-estimators over the asymptotic estimator given by the relative frequency, the conditional mean squared error is used.

The probability volume of the $\beta$-measurement space $M_{X_S,\breve{p}}^{(0.80,20)}$ defined by the measurement intervals given in Table 10.2 adopts the value

$$V_{prob}\Big(M_{X_S,\breve{p}}^{(0.80,20)}\Big) = 0.842987 \tag{10.13}$$

Thus, the reliability excess amounts to about 4%.

The conditional mean squared errors of the three point estimators are given in Table 10.3.

**Table 10.3** *The conditional mean squared error of the three point estimators.*

| $TF$ | $TMS$ | $TMP$ |
|---|---|---|
| 0.005532 | 0.002079 | 0.002140 |

It turns out that the conditional mean squared error of the relative frequency is about 2.6 times larger than the minimum given by the conditional mean squared error of the minimum MSE-$\beta$-estimator. The efficiency of the simple midpoint estimator compared to the minimum MSE-$\beta$-estimator is surprisingly good, as the increase of the conditional mean squared error amounts to about 3% only. Thus, it is strongly recommended for small and medium sample sizes not to use the relative frequency as point estimator, but one of the $\beta$-estimators.

## 10.5 Determination of Point Estimates

As mentioned earlier, the calculation starts from a measurement intervals for a $\beta$ measurement procedure based on the sample $\vec{X}$ or on the sample sum $X_S$. As measurement procedures based on $\vec{X}$ are primarily of theoretic interest, the following recipe will be formulated for the case $X_S$ only.

The measurement interval is given by

$$\left[\ell_{\breve{p}}^{(\beta,n)}(x_S), u_{\breve{p}}^{(\beta,n)}(x_S)\right] \tag{10.14}$$

Then, the midpoint $\beta$-estimate is defined by

$$TMP_{\breve{p}}^{(\beta,n)}(x_S) = \frac{\ell_{\breve{p}}^{(\beta,n)}(x_S) + u_{\breve{p}}^{(\beta,n)}(x_S)}{2} \tag{10.15}$$

In contrast to the midpoint $\beta$-estimates which can be calculated without any difficulties, the minimum MSE-$\beta$-estimates is the ratio of two integrals and its calculation by hand would be too troublesome. For a given measurement interval

$$[\ell_{\breve{p}}^{(\beta,n)}(x_S), u_{\breve{p}}^{(\beta,n)}(x_S)],$$

the two integrals may easily be calculated by means of a mathematical program package like *Mathematica*. The minimum MSE-$\beta$-estimate based on $X_S$ is defined by

$$TMS_{\breve{p}}^{(\beta,n)}(x_S) = \frac{\int_{\ell_{\breve{p}}^{(\beta,n)}(x_S)}^{u_{\breve{p}}^{(\beta,n)}(x_S)} p\mathbf{P}_{X_S}^{(p)}(x_S)dp}{\int_{\ell_{\breve{p}}^{(\beta,n)}(x_S)}^{u_{\breve{p}}^{(\beta,n)}(x_S)} \mathbf{P}_{X_S}^{(p)}(x_S)dp}$$

$$= \frac{\int_{\ell_{\breve{p}}^{(\beta,n)}(x_S)}^{u_{\breve{p}}^{(\beta,n)}(x_S)} p\binom{n}{x_S}p^{x_S}(1-p)^{n-x_S}dp}{\int_{\ell_{\breve{p}}^{(\beta,n)}(x_S)}^{u_{\breve{p}}^{(\beta,n)}(x_S)} \binom{n}{x_S}p^{x_S}(1-p)^{n-x_S}dp} \qquad (10.16)$$

The denominator of the quotient in (10.16) is the probability volume of the underlying $\beta$-measurement space. Hence, the reliability excess of the procedure is obtained by subtracting the confidence level from the value of the denominator.

The simple midpoint $\beta$-estimator $TMP_{\breve{p}}(X_S)$ should be used in preliminary studies, where one wants to quickly get some knowledge about a phenomenon of interest. Later, one can also calculate the slightly better (with respect to the conditional mean squared error) minimum MSE-$\beta$-estimator $TMS_{\breve{p}}(X_S)$.

It should be noted that the point estimators proposed here are the beginning of a theory, which basically differs from the classical theory of point estimators. Emphasis is laid on the properties of the estimator in the case of finite sample size, whereas asymptotic properties are completely omitted in the argumentation. For more details, see [11].

# Chapter 11

# Excluding a Probability

In this chapter, various problems are listed which are solved by means of a significance test, or, as it is called here, an exclusion procedure. Any exclusion procedure for a probability refers to a given interval $[\underline{p}^{(0)}, \overline{p}^{(0)}] \subset [\underline{p}, \overline{p}]$ with $\underline{p} \leq \underline{p}^{(0)} \leq \overline{p}^{(0)} \leq \overline{p}$ and $\overline{p}^{(0)} - \underline{p}^{(0)} < \overline{p} - \underline{p}$. There are two possible decisions that are specified by the decision function $\phi_{H_0}$:

$$\phi_{H_0}(X_S) = \begin{cases} 0 & \text{if } X_S \in A_{X_S}^{(1-\alpha,n)}\Big([\underline{p}^{(0)}, \overline{p}^{(0)}]\Big) \\ 1 & \text{if } X_S \notin A_{X_S}^{(1-\alpha,n)}\Big([\underline{p}^{(0)}, \overline{p}^{(0)}]\Big) \end{cases} \tag{11.1}$$

where

$$\phi_{H_0}(X_S) = 1$$

means a decision in favor of $[\underline{p}, \overline{p}] \setminus [\underline{p}^{(0)}, \overline{p}^{(0)}]$, and

$$\phi_{H_0}\Big(X_S\Big) = 0$$

a decision in favor of $[\underline{p}, \overline{p}]$.

The connected set

$$A_{X_S}^{(1-\alpha,n)}\Big([\underline{p}^{(0)}, \overline{p}^{(0)}]\Big) \subset \{0, 1, \ldots, n\} \tag{11.2}$$

is a local $(1-\alpha)$-prediction region for $H_0$ meeting the reliability requirement given by the confidence level $1-\alpha$:

$$\mathbf{P}_{X_S}^{(p)}\left(A_{X_S}^{(1-\alpha,n)}\Big([\underline{p}^{(0)}, \overline{p}^{(0)}]\Big)\right) \geq 1-\alpha \tag{11.3}$$

Thus, $A_{X_S}^{(1-\alpha,n)}\big([\underline{p}^{(0)},\overline{p}^{(0)}]\big)$ contains those realizations of $X_S$ that are to be expected if the actual value $p$ is an element of $[\underline{p}^{(0)},\overline{p}^{(0)}]$ (i.e., if $H_0$ is correct). Hence, in case the event given by the local prediction region occurs, $H_0$ should not be rejected.

The aim of applying an exclusion procedure is, of course, to exclude the values $p$ given by the null hypothesis $H_0$ with stated reliability given by $\alpha$. An exclusion by means of (11.1) is equivalent to the occurrence of the event complementary to the local prediction region. This complementary event is often called the *critical region*, and tests are given by means of the corresponding critical region. However, from a more pedagogical point of view, the local prediction region seems to be more appropriate, as it is a connected set consisting of those sample realizations that are to be expected if the null hypothesis is correct. Thus, if the event given by the local prediction region occurs, then it is clear that $H_0$ cannot be rejected.

## 11.1 Special Exclusion Procedures

Exclusion procedures based on the binomial distribution are applied in many situations. Four of these tests, the so-called binomial test, the sign test, the Test of McNemar, and the Chapman Test, will be used to illustrate the application of the tables.

### 11.1.1 Binomial Test

Consider an event of interest and its probability denoted by $\breve{p}$. The problem is to show that the actual value $p$ of $\breve{p}$ is different from any value in a certain interval $[\underline{p}^{(0)},\overline{p}^{(0)}] \subset [\underline{p},\overline{p}]$, where the reliability of the exclusion procedure is specified by the significance level $\alpha$.

A direct use of the tables is restricted to the simple null hypothesis $H_0 : \{p^{(0)}\}$. Therefore, we start with the case of a simple null hypothesis:

$$H_0 : p \in \{p^{(0)}\}$$

To use the tables, the following input parameters must be listed:

- Measurement space $[\underline{p}, \overline{p}]$
- Sample size $n$
- Significance level $\alpha$
- Null hypothesis $H_0 : \{p^{(0)}\}$

If $p^{(0)} = k \cdot 0.05$, $k \in \{1, 2, \ldots, 19\}$, then the prediction region for $H_0$

$$A_{X_S}^{(1-\alpha,n)}\left(p^{(0)}\right) = \left\{\ell_{X_S}^{(1-\alpha,n)}(p^{(0)}), \ldots, u_{X_S}^{(1-\alpha,n)}(p^{(0)})\right\}$$

given by its smallest $\ell_{X_S}^{(1-\alpha,n)}(p^{(0)})$ and largest $u_{X_S}^{(1-\alpha,n)}(p^{(0)})$ integers can be taken directly from the tables.

Once the prediction region is known, the decision function is given by

$$\phi_{H_0}(x_S) = \begin{cases} 0 & \text{if } \ell_{X_S}^{(1-\alpha,n)}(p^{(0)}) \leq x_S \leq u_{X_S}^{(1-\alpha,n)}(p^{(0)}) \\ 1 & \text{otherwise} \end{cases}$$

If $\phi_{H_0}(x_S) = 1$, the aim of exclusion has been reached; otherwise the attempt to exclude $p^{(0)}$ was unsuccessful.

The following examples (except the solutions) are taken from [32].

**Example 11.1**

*Remember that a exclusion procedure aims at excluding certain probability distributions from the set of admissible ones.*

> *A sports magazine reports that the people who watch Monday night football games on television are evenly divided between men and women. Out of a random sample of 400 people who regularly watch the Monday night game, 220 are men. Using a .10 level of significance, can it be concluded that the report is false?*

*The aim is to falsify the report, which claims that the actual value of the proportion of men watching the Monday-night football games is equal to* $p^{(0)} = 0.5$.

*Additionally, the following is assumed:*

> *There is evidence that the actual value is at least 0.3 and at most 0.7.*

*Thus, we obtain the following set of input data:*

- *Measurement space* $[\underline{p}, \overline{p}] = [0.3, 0.7]$
- *Sample size* $n = 400$
- *Significance level* $\alpha = 0.1$
- *Null hypothesis* $H_0 : p^{(0)} = 0.5$

*For this set of input data, we immediately obtain the prediction region for* $H_0$ *from the tables:*

$$\begin{aligned} A_{X_S}^{(0.90,400)}(0.5) &= \left\{\ell_{X_S}^{(0.90,400)}(0.5), \ldots, u_{X_S}^{(0.90,400)}(0.5)\right\} \\ &= \{184, \ldots, 216\} \end{aligned}$$

*Because* $x_S = 220$ *is not contained in* $A_{X_S}^{(0.90,400)}(0.5)$, *the value* $p^{(0)}$ *is excluded from the set of possible values.*

**Result:**
*By means of the exclusion procedure, it is proved that the report is wrong. The reliability of the procedure is 90%.* •

**Example 11.2**
*This example refers to a so-called one-sided null hypothesis.*

> *In 1964, 40% of shipments of U.S. cotton to Germany were sent to arbitration because of complaints about the quality being sub-standard, according to*

> *Time Magazine. Would it signify a worsening of the situation if 20 out of the first 40 shipment in 1965 were likewise the cause of complaint, or might this difference be attributed to chance?*

*Additionally, the following is assumed:*

> *Because of the past experience, it is decided that the actual value p is for sure larger than 0.2 and less than 0.6.*

*Setting the reliability of the procedure to* $1-\alpha = 0.95$*, we have the following:*

- *Measurement space* $[\underline{p}, \overline{p}] = [0.2, 0.6]$
- *Sample size* $n = 40$
- *Significance level* $\alpha = 0.05$
- *Null hypothesis* $H_0: \; [\underline{p}^{(0)}, \overline{p}^{(0)}] = [0.2, 0.4]$

*Because the null hypothesis* $H_0$ *includes the left limit of the measurement range* $[\underline{p}, \overline{p}]$*, the problem is solved by means of the following auxiliary problem with the simple null hypothesis:*

- *Measurement space* $[\underline{p}, \overline{p}] = [0.4, 0.6]$
- *Sample size* $n = 40$
- *Significance level* $\alpha = 0.05$
- *Null hypothesis* $H_0: \; p^{(0)} = 0.4$

*The corresponding prediction region* $A_{X_S}^{(0.95,40)}(0.4)$ *is taken from the CD-ROM tables:*

$$\tilde{A}_{X_S}^{(0.95,40)}(0.4) = \Big\{11, \ldots, 28\Big\} \tag{11.4}$$

*To obtain a solution of the given problem, the values smaller than 11 are added and we obtain*

$$A_{X_S}^{(0.95,40)}(0.4) = \left\{0, 1, \ldots, 28\right\} \tag{11.5}$$

*The observation* $x_S = 20$ *is an element of the prediction region, and, therefore, the null hypothesis cannot be rejected.*

**Result:**
*Application of the exclusion procedure shows that the statement about a worsening of quality is not justified. The procedure's reliability is 95%.* •

**Example 11.3**
*The formulation of this problem in [32] calls the decision of not rejecting* $H_0$ *"acceptance" which is misleading and, therefore, should be avoided.*

> *An electrical company claimed that at least 95% of the parts which they supplied on a government contract conformed to specifications. A sample of 400 parts was tested, and 45 did not meet specifications. Can we accept the company's claim at a 0.05 level of significance?*

*Additionally, the following is assumed:*

> *The conformance probability is for sure larger than 0.8.*

*Hence, the following set of input parameters is obained:*

- *Measurement space* $[\underline{p}, \overline{p}] = [0.8, 1.0]$
- *Sample size* $n = 400$
- *Significance level* $\alpha = 0.05$
- *Null hypothesis* $H_0 :\ [\underline{p}^{(0)}, \overline{p}^{(0)}] = [0.95, 1.00]$

*In this example, the null hypothesis* $H_0$ *includes the right limit of the measurement range* $[\underline{p}, \overline{p}]$, *and we obtain the following auxiliary problem with a simple null hypothesis:*

- *Measurement space* $[\underline{p}, \overline{p}] = [0.80, 0.95]$
- *Sample size* $n = 400$
- *Significance level* $\alpha = 0.05$
- *Null hypothesis* $H_0 : \; p^{(0)} = 0.95$

*From the CD-ROM tables, we obtain*

$$\tilde{A}_{X_S}^{(0.95,400)}(0.95) = \left\{372, \ldots, 388\right\} \tag{11.6}$$

*To get a solution of the given problem, the values larger than 388 are added and we obtain*

$$A_{X_S}^{(0.95,400)}(0.95) = \left\{372, \ldots, 400\right\} \tag{11.7}$$

*The observation* $x_S = 400 - 45 = 355$ *is not an element of the prediction region; therefore, the null hypothesis can be rejected.*

**Result:**
*It is proved that the company's claim is wrong. The procedure's reliability is 95%.* •

### 11.1.2 Sign Test

Let $Y$ be a random variable of continuous type with distribution function $F_Y^{(M)}(y)$, where the model parameter is the median $M$. Then,

$$F_Y^{(M)}(M) = 1 - F_Y^{(M)}(M) = \frac{1}{2} \tag{11.8}$$

Therefore, the following statements are equivalent:

$$M = M_0 \tag{11.9}$$

$$F_Y^{(M)}(M_0) = 1 - F_Y^{(M)}(M_0) \tag{11.10}$$

Let $(Y_1, \ldots, Y_n)$ be independent random variables identically distributed like $Y$, and define

$$X_i = \begin{cases} 0 & \text{if } Y_i \leq M_0 \\ 1 & \text{if } Y_i > M_0 \end{cases} \tag{11.11}$$

Then, $(X_1, \ldots, X_n)$ are independent random variables identically distributed like a Bernoulli variable with parameter $p = 1 - F_Y(M)$, and the following assertion with respect to $p$ is equivalent to (11.9) and (11.10) :

$$p = \frac{1}{2} \tag{11.12}$$

Thus, a simple null hypothesis with respect to the median of a random variable $Y$ of continuous type is equivalent to the null hypothesis

$$H_0: \quad p = \frac{1}{2}$$

for a Bernoulli variable $X$.

The following numerical examples are again taken from [32].

**Example 11.4**

*Just as the preceding examples, the null hypothesis in this example is of Lebesgue measure zero. It would be more appropriate to select a null hypothesis with positive Lebesgue measure.*

> *Consider the following data obtained from testing the breaking strength of ceramic tile manufactured by a new cheaper process: 20, 42, 18, 21, 22, 35, 19, 18, 26, 20, 21, 32, 22, 20, 24. Suppose that experience with the old process produced a median of 25. Then test the hypothesis $H_0: M = 25$ against $H_1: M < 25$.*

*The two alternatives $H_0$ and $H_1$ mean that the median of the breaking strength is at most 25. Setting $M_0 = 25$ implies that $p = F_Y^{(M)}(M_0)$ is at most $\frac{1}{2}$, thus leading to a measurement space with an upper bound of $\frac{1}{2}$. Having no further information, we select the lower bound as 0. Requiring a significance level of 0.05, we obtain the following set of input parameters:*

- *Measurement space* $[\underline{p}, \overline{p}] = [0.0, 0.5]$
- *Sample size:* $n = 15$

- *Significance level* $\alpha = 0.05$
- *Null hypothesis* $H_0 : p = 0.5$

*From the CD-ROM tables, we immediately obtain the prediction region*

$$\tilde{A}_{X_S}^{(0.95,15)}(0.5) = \{3, 4, \ldots, 12\} \tag{11.13}$$

*As mentioned earlier, the prediction region obtained by solely applying the likelihood ratio selection principle is often larger than the tabled ones. In this example, one would get the following prediction region:*

$$A_{X_S}^{(0.95,15)}(0.5) = \{3, 4, \ldots, 15\} \tag{11.14}$$

*The reduced prediction region (11.13) given in the tables is obtained by removing the extremely unlikely realization from (11.14) in order to make the reliability excess smaller. Thus, to avoid a wrong decision, one should either use (11.14) or interpret a realization* $x_S > 12$ *as an indication that the measurement space has been selected inappropriately.*

*The realization* $x_S$ *in the example is obtained by from the data and the definition (11.11) of* $X_i$*. There are 4 observations larger than 25 and 11 observations smaller than 25. Hence, we have* $x_S = 4$*, which is an element of* $A_{X_S}^{(0.95,15)}(0.5)$*, implying that the null hypothesis cannot be rejected.*

**Result:**
*The experiments do not prove that the cheaper process yields ceramic tiles with smaller breaking strength. The procedure's reliability is 95%.* •

**Example 11.5**
*This example illustrates the fact that a small sample size in conjunction with a high reliability requirement generally leads to a failure of a exclusion procedure.*

*The following table gives scores on equivalent forms of the same test taken before (Y) and after (Z) a special audio/visual instructional program. Determine whether these data support the conclusion that this special program improved test performance.*

| $i$ | $Y_i$ | $Z_i$ | $i$ | $Y_i$ | $Z_i$ |
|---|---|---|---|---|---|
| 1 | 87 | 98 | 12 | 72 | 79 |
| 2 | 89 | 93 | 13 | 80 | 78 |
| 3 | 83 | 90 | 14 | 73 | 77 |
| 4 | 93 | 89 | 15 | 77 | 76 |
| 5 | 85 | 87 | 16 | 70 | 75 |
| 6 | 81 | 86 | 17 | 66 | 73 |
| 7 | 82 | 85 | 18 | 69 | 71 |
| 8 | 78 | 84 | 19 | 71 | 70 |
| 9 | 79 | 83 | 20 | 58 | 67 |
| 10 | 75 | 81 | 21 | 64 | 61 |
| 11 | 74 | 80 | | | |

*Assuming that each pair $(Y_i, Z_i)$ consists of independent and identically distributed random variables of continuous type implies*

$$\mathbf{P}(Y_i \geq Z_i) = \mathbf{P}(Y_i < Z_i) = \frac{1}{2} \tag{11.15}$$

*or*

$$\mathbf{P}(Y_i - Z_i \geq 0) = \frac{1}{2} \tag{11.16}$$

*Let*

$$X_i = \begin{cases} 1 & \text{if } Y_i - Z_i < 0 \\ 0 & \text{if } Y_i - Z_i \geq 0 \end{cases} \tag{11.17}$$

*Under the above assumption, we have*

$$X_S = \sum_{i=1}^{n} X_i \sim Bi\left(n, \frac{1}{2}\right)$$

*Thus, the assumption of identical distribution of Y and Z can be excluded by excluding $p = \frac{1}{2}$ as the actual value $p$ of the distribution parameter $\breve{p}$ of $X_S$.*

*Selecting as measurement space* $[0.3, 0.8]$ *and requiring a significance level of* $\alpha = 0.02$ *results in the following input data:*

- *Measurement space* $[\underline{p}, \overline{p}] = [0.3, 0.8]$
- *Sample size* $n = 21$
- *Significance level* $\alpha = 0.02$
- *Null hypothesis* $H_0 : p = 0.5$

*For these input data, we obtain the following prediction region from the CD-ROM tables:*

$$A_{X_S}^{(0.98,21)}(0.5) = \left\{5, \ldots, 15\right\} \tag{11.18}$$

*The observed value of* $X_S$ *is given by* $x_S = 5$, *which is an element of the prediction region and, therefore, the null hypothesis cannot be rejected.*

**Result:**
*The scores do not prove that the special program improves the performance. The procedure's reliability is 98%.* •

### 11.1.3 Test of McNemar

Let $(Y, Z)$ be a pair of independent dichotomic random variables with joint distribution function $F_{Y,Z}$ and marginal distributions $Y \sim Bi(1, p_1)$ and $Z \sim Bi(1, p_2)$.

Define the conditional random variable

$$U = U|Y + Z = 1 = \begin{cases} 0 & \text{if } Y = 0 \\ 1 & \text{if } Z = 0 \end{cases} \tag{11.19}$$

then, $U \sim Bi(1, p)$ with some unknown parameter $p$.
If $p_1 = p_2 = \tilde{p}$, then $Y + Z \sim Bi(2, \tilde{p})$, and

$$\mathbf{P}(U = 0) = \mathbf{P}(Y = 0|Y + Z = 1) = \frac{\tilde{p}(1 - \tilde{p})}{2\tilde{p}(1 - \tilde{p})} = \frac{1}{2}$$

Hence, $p_1 = p_2$ is equivalent to $p = \frac{1}{2}$, and instead of excluding $p_1 = p_2$, we might also exclude $p = \frac{1}{2}$.

The resulting exclusion procedure may be described as follows:

For fixed $n$ and significance level $\alpha$, let $\big((Y_1, Z_1), \ldots, (Y_n, Z_n)\big)$ be a random sample for $(Y, Z)$. Let $Y_{i_j} + Z_{i_j} = 1$ for $j = 1, \ldots, N_U$, where $N_U$ is a random variable, which may adopt any value from $\{0, 1, \ldots, n\}$.

For fixed $N_U = m$, define the random variables

$$U_{i_j} = \begin{cases} 0 & \text{if } Y = 0 \\ 1 & \text{if } Z = 0 \end{cases}$$

Then $(U_{i_1}, \ldots, U_{i_m})$ represents a random sample of size $m$ for a Bernoulli variable with parameter $p$, and the problem of excluding the equality of $p_1$ and $p_2$ reduces to a simple sign test based on

$$V_S = \sum_{j=1}^{m} U_{i_j}$$

with sample size given by $m$.

For significance level $\alpha$, let $A_{V_S}^{(1-\alpha,m)}(0.5)$ be a prediction region for $p = 0.5$. Then, the following relation holds for the error probability of the exclusion procedure:

$$\begin{aligned}
&\mathbf{P}^{(0.5)}\Big(V_S \notin A_{V_S}^{(1-\alpha,m)}(0.5)\Big) \\
&= \sum_{m=0}^{n} \mathbf{P}^{(0.5)}\Big(\sum_{j=1}^{N_U} U_{i_j} \notin A_{V_S}^{(1-\alpha,N_U)}(0.5)\Big)\Big|N_U = m\Big)\mathbf{P}^{(\tilde{p})}(N_U = m) \\
&\leq \alpha \sum_{m=0}^{n} \mathbf{P}^{(\tilde{p})}(N_U = m) = \alpha \qquad (11.20)
\end{aligned}$$

It should be noted that this is a so-called conditional test (conditioned by fixing $m$), and, therefore, the actual significance level may be smaller than the required significance level $\alpha$.

**Example 11.6**

*This example (except for the solution) is taken from [19], Vol. 5, p. 362.*

*Suppose we want to compare two treatments, chemotherapy and surgery, for a particular type of cancer with regard to three-year survival rate. In order to utilize smaller sample sizes a matched pairs design is chosen. Possible participants are matched on stage of cancer, age of patient, and the patient's mental outlook. These variables are thought to most influence the binomial outcome of survival after three years of treatment. Within the matched pair, patients are randomly assigned to receive one of the two treatments. If there are 100 matched pairs, the data may be put in the following table:*

| | | Chemotherapy Alive | Chemotherapy Dead | |
|---|---|---|---|---|
| Surgery | Alive | 5 | 20 | 25 |
| | Dead | 15 | 60 | 75 |
| | | 20 | 80 | 100 |

*From the table it is seen that 25 of the 100 patients treated with surgery are alive after three years while the chemotherapy three years survival rate is 20/100.*

*Let* $Y$ *describe the effect of surgery with*

$$Y = \begin{cases} 0, & \textit{participant dead} \\ 1, & \textit{participant alive} \end{cases} \tag{11.21}$$

*and* $Z$ *the effect of chemotherapy with*

$$Z = \begin{cases} 0, & \textit{participant dead} \\ 1, & \textit{participant alive} \end{cases} \tag{11.22}$$

*We have* $m = 15 + 20 = 35$ *and* $v_S = \sum_{j=1}^{35} U_{i_j} = 15$.

*Selecting* $[0.2, 0.8]$ *as the measurement space and* $\alpha = 0.05$ *as the significance level, we obtain the following set of input data.*

- *Measurement space* $[\underline{p}, \overline{p}] = [0.2, 0.8]$
- *Sample size* $n = 35$
- *Significance level* $\alpha = 0.05$
- *Null hypothesis* $H_0 : p = 0.5$

*From the CD-ROM tables, we immediately obtain the corresponding prediction region:*

$$A_{V_S}^{(0.95,35)}(0.5) = \{12, 13, \ldots, 23\} \tag{11.23}$$

*As* $v_S = 15$ *is an element of the prediction region, the null hypothesis cannot be rejected.*

**Result:**
*The clinical study does not prove that there is a difference in the success probability between surgery and chemotherapy. The procedure's reliability is 95%.* •

### 11.1.4 Chapman Test

Let $X$ and $Y$ be a Poisson random variables with $X \sim Po(\theta_1)$ and $Y \sim Po(\theta_2)$. Consider the problem of excluding the possibility that the two random variables are identical distributed; that is, of rejecting the null hypothesis

$$H_0^{(1)} : \quad \theta_1 = \theta_2 \tag{11.24}$$

This problem can be solved by means of the Chapman test [5] using the tables of the handbook.

Let $\theta = \frac{\theta_1}{\theta_1 + \theta_2}$; then, the conditional random variable $X|X + Y = n$ is binomially distributed with parameters $n$ and $p = \theta$.

The null hypothesis (11.24) with respect to $\theta_1$ and $\theta_2$ is equivalent to the following null hypothesis with respect to $p$:

$$H_0^{(2)} : p = 0.5 \tag{11.25}$$

Thus, the question of whether or not to reject the null hypothesis (11.24) referring to two Poisson-distributed random variables may be answered by means of the null hypothesis (11.25) referring to a binomial random variable.

**Example 11.7**
*Consider the case that* $0 \leq \theta_2 \leq \theta_1$ *is true with certainty, implying that* $0.5 \leq p \leq 1.0$. *The reliability of the procedure is fixed by the significance level* $\alpha = 0.1$. *The realizations of the two Poisson random variables are given by* $x = 188$ *and* $y = 134$, *implying the following set of input data:*

- *Measurement space:* $[\underline{p}, \overline{p}] = [0.5, 1.0]$
- *Sample size:* $n = x + y = 322$
- *Significance level:* $\alpha = 0.1$
- *Null hypothesis:* $H_0 : p_0 = 0.5$

*From the CD-ROM tables, the following prediction region is obtained:*

$$\tilde{A}_{X|X+Y}^{(0.90,322)}(0.5) = \{127, 128, \ldots, 172\} \tag{11.26}$$

*The null hypothesis refers to the smallest value of the measurement space; therefore, the set* $\{0, 1, \ldots, 126\}$, *which was removed in order to reduce the reliability excess, is again added to* $\tilde{A}_{X|X+Y}^{(0.90,322)}(0.5)$, *thus arriving at*

$$A_{X|X+Y}^{(0.90,322)}(0.5) = \{0, 1, 2, \ldots, 172\} \tag{11.27}$$

**Result:**
*The observed value of* $X$ *is given by* $x = 188$, *which is not an element of the prediction region, and therefore the value* $p = \theta = 0.5$ *or* $\theta_1 = \theta_2$ *may be excluded from the measurement space. The procedure's reliability is given by 90%.* •

# Chapter 12

# Comparing Probabilities

Besides measuring and excluding probabilities, the problem of comparing different probabilities, or, more generally said, alternatives, is a frequently occurring task in a variety of fields. In this chapter, it is shown how a comparison procedure can be designed using the tables on the CD-ROM. Similar to that in the case of exclusion procedures, the resulting methods are optimal or nearly optimal with respect to success probability. Additionally, they are, in a certain sense, also optimal with respect to reliability. The determination is comparatively simple, as the corresponding prediction regions can be taken directly from the tables as soon as the set of input parameters is specified.

## 12.1 Comparisons by Means of the Tables

Consider two hypotheses, $H_1$ and $H_2$, concerning the actual value $p$ of a probability $\breve{p}$ given by two disjoint intervals $[\underline{p}^{(1)}, \overline{p}^{(1)}]$ and $[\underline{p}^{(2)}, \overline{p}^{(2)}]$, respectively. The procedure aims at deciding either for the first alternative or for the second alternative, of course, with stated reliability given by the significance level $\alpha_1$ and $\alpha_2$, where $\alpha_i$ is an upper bound for the probability of a decision in favor of $H_j$, $j \neq i$, in the case that the alternative $H_i$ is the true one.

The solution of the problem is determined by two local prediction regions, the first one referring to $H_1$ and the second one to $H_2$, which can be taken from the CD-ROM tables just as in the case of a prediction region for excluding a null hypothesis.

The appropriate measurement space is the convex hull $[\underline{p}^{(1)}, \overline{p}^{(2)}]$ of the two alternatives $[\underline{p}^{(1)}, \overline{p}^{(1)}]$ and $[\underline{p}^{(2)}, \overline{p}^{(2)}]$. Evidently, the sets representing the two alternatives are located at the left boundary and the right boundary of the measurement space, implying that the local prediction region

$$A_{X_S}^{(1-\alpha-1,n)}\Big([\underline{p}^{(1)}, \overline{p}^{(1)}]\Big) \tag{12.1}$$

for $H_1$ should be filled beginning with the smallest realization (i.e., 0), until the reliability requirement is met for all values $p \in [\underline{p}^{(1)}, \overline{p}^{(1)}]$:

$$\mathbf{P}_{X_S}^{(p)}\Big(A_{X_S}^{(1-\alpha_1,n)}\Big([\underline{p}^{(1)}, \overline{p}^{(1)}]\Big)\Big) \geq 1-\alpha_1 \qquad \text{for } p \in [\underline{p}^{(1)}, \overline{p}^{(1)}] \tag{12.2}$$

This is true if (12.2) is fulfilled for the largest possible value $p = \overline{p}^{(1)}$.

Analogously, the local prediction region

$$A_{X_S}^{(1-\alpha_2,n)}\Big([\underline{p}^{(2)}, \overline{p}^{(2)}]\Big) \tag{12.3}$$

for $H_2$ is filled starting with the largest realization (i.e. $n$), until the reliability requirement is met for all values $p \in [\underline{p}^{(2)}, \overline{p}^{(2)}]$:

$$\mathbf{P}_{X_S}^{(p)}\Big(A_{X_S}^{(1-\alpha_2,n)}\Big([\underline{p}^{(2)}, \overline{p}^{(2)}]\Big)\Big) \geq 1-\alpha_2 \qquad \text{for } p \in [\underline{p}^{(2)}, \overline{p}^{(2)}] \tag{12.4}$$

This is true if (12.4) is fulfilled for the smallest possible value $p = \underline{p}^{(2)}$.

Therefore, we may reduce the actual measurement space to the range $[\overline{p}^{(1)}, \underline{p}^{(2)}]$, yielding the same result with respect to the local prediction regions. Hence, we obtain the following two sets of input data for the determination of the two local prediction regions:

| | $H_1$ | $H_2$ |
|---|---|---|
| Measurement space | $[\overline{p}^{(1)}, \underline{p}^{(2)}]$ | $[\overline{p}^{(1)}, \underline{p}^{(2)}]$ |
| Sample size | $n$ | $n$ |
| Significance level | $\alpha_1$ | $\alpha_2$ |
| Hypotheses | $\overline{p}^{(1)}$ | $\underline{p}^{(2)}$ |

With the above sets of input data, the corresponding prediction regions $A_{X_S}^{(1-\alpha_1,n)}(\overline{p}^{(1)})$ and $A_{X_S}^{(1-\alpha_2,n)}(\underline{p}^{(2)})$ can be taken from the CD-ROM tables if the bounds $\overline{p}^{(1)}$ and $\underline{p}^{(2)}$ are multiples of 0.1; thus, the measurement space $[\overline{p}^{(1)}, \underline{p}^{(2)}]$ is contained in the tables.

If the two prediction regions have a nonempty intersection

$$\begin{aligned} & O_{X_S}^{(\alpha_1,\alpha_2,n)}(H_1, H_2) \\ & = A_{X_S}^{(1-\alpha_1,n)}\left([\underline{p}^{(1)}, \overline{p}^{(1)}]\right) \cap A_{X_S}^{(1-\alpha-2,n)}\left([\underline{p}^{(2)}, \overline{p}^{(2)}]\right) \\ & \neq \emptyset \end{aligned} \tag{12.5}$$

then the decision function is given by

$$\phi_{H_1,H_2}(x_S) = \begin{cases} 0 & \text{for } x_S \in O_{X_S}^{(\alpha_1,\alpha_2,n)}(H_1, H_2) \\ 1 & \text{for } x_S \in A_{X_S}^{(1-\alpha_1,n)}(H_1) \\ 2 & \text{for } x_S \in A_{X_S}^{(1-\alpha_2,n)}(H_2) \end{cases} \tag{12.6}$$

and the comparison procedure fails with probability

$$Q_{\{0,1,\ldots,n\}}\left(O_{X_S}^{(\alpha_1,\alpha_2,n)}(H_1, H_2).\right) \tag{12.7}$$

If the intersection is the empty set and at the same time the union is equal to the prediction space $\{0, 1, \ldots, n\}$, then the comparison is successful with probability 1, and we have

$$\phi_{H_1,H_2}(x_S) = \begin{cases} 1 & \text{for } x_S \in A_{X_S}^{(1-\alpha_1,n)}(\overline{p}^{(1)}) \\ 2 & \text{for } x_S \in A_{X_S}^{(1-\alpha_2,n)}(\underline{p}^{(2)}). \end{cases} \tag{12.8}$$

If the intersection is the empty set, but the union is not equal to the prediction space $\{0,1,\ldots,n\}$, then the set difference

$$\Big\{0,1,\ldots,n\Big\} \setminus \Big\{A_{X_S}^{(1-\alpha_1,n)}(\overline{p}^{(1)}) \cup A_{X_S}^{(1-\alpha_2,n)}(\underline{p}^{(2)})\Big\} \qquad (12.9)$$

is divided into two disjoint parts which are added to $A_{X_S}^{(1-\alpha_1,n)}(\overline{p}^{(1)})$ and $A_{X_S}^{(1-\alpha_2,n)}(\underline{p}^{(2)})$, respectively, thus increasing the required reliability given by $\alpha_1$ and $\alpha_2$.

A comparison procedure may be applied to decide which of two distinct and known distributions or families of distributions is the actual one. Clearly, the main problem of designing a comparison procedure is to select the sample size $n$ and the significance levels $\alpha_1$ and $\alpha_2$ in a way that the procedure is successful with certainty. Assuming that the two significance levels are chosen, because of substantial and non-statistical reasons, the only way to guarantee a success is to have the sample size $n$ selected sufficiently large. Any increase of the sample size means increasing expenses, and therefore the main problem of developing comparison tests with success probability 1 is to determine a procedure with minimum sample size.

**Remark 12.1** *Although the above considerations refer to the case of two alternatives, it is easy to see how to treat the case of more than two alternatives. Any comparison procedure can be reduced to the problem of determining appropriate local prediction regions for the different alternatives. These prediction regions must be selected so as to meet the corresponding reliability requirement on the one hand and to maximize the success probability on the other hand.*

*If for $i = 1,\ldots,m$ there are $(1-\alpha_i)$-measurement & prediction spaces already available, then they can be used at least as a preliminary comparison procedure. If the resulting local prediction regions are disjoint sets with union equal to the entire prediction space, then a Neyman $\alpha_1,\ldots,\alpha_m$-comparison procedure has already been found. If the local prediction regions are disjoint sets but their union is not equal to the entire prediction space, then a Neyman $\alpha_1,\ldots,\alpha_m$-comparison procedure is obtained by* *appropriately increasing the local prediction regions until their union fills the prediction space. In case the local prediction regions have a non-empty intersection, then they can be used as a starting point for deriving a Neyman $\alpha_1, \ldots, \alpha_m$-comparison procedure.* •

## 12.2 Illustrative Examples for $\alpha_1, \alpha_2$-Comparison Procedures

**Example 12.1**

*A manufacturing process for an advanced semiconductor may produce in two different modes. The first one is characterized by a nonconformance probability of $p_1 = 0.10$, and the second one, which is adopted after a certain not observable disturbance, by a nonconformance probability of $p_2 = 0.20$. Because of different costs, the significant levels are selected to be unequal: $\alpha_1 = 0.05$ and $\alpha_2 = 0.02$. The problem is to determine the minimum sample size $n$ with which a positive decision for one of the two modes can be made with certainty meeting the reliability requirements.*

*The bounds $p_1 = 0.10$ and $p_2 = 0.20$ are multiples of 0.1, and therefore the CD-ROM tables may be used to solve the problem of determining the minimum sample size by means of which a decision in favor of $H_1$ or $H_2$ can be made with certainty.*

1. *In the first step, the larger significance level is replaced by the smaller one. In the example, $\alpha_1 = 0.05$ is replaced by $\tilde{\alpha}_1 = 0.02$. For this simplified problem (the two significance levels are equal!), the minimum sample size $n_1$ is determined by means of the CD-ROM tables such that the upper bound $u_1^{(p_1,\tilde{\alpha}_1)}(n_1)$ of the prediction region for $p_1$ and the lower bound $\ell_1^{(p_2,\alpha_2)}(n_1)$ of the prediction region for $p_2$ are consecutive integers.*

2. *In the second step, the original reliability given by $\alpha_1 = 0.05$ is used to obtain the upper bound $u_2^{(p_1,\alpha_1)}(n_1)$*

*of the prediction region for* $p_1$, $\alpha_1 = 0.05$, *and sample size* $n_1$ *from the tables. Generally,* $u_2^{(p_1,\alpha_1)}(n_1)$ *is smaller than* $u_1^{(p_1,\tilde{\alpha}_1)}(n_1)$.

3. *Next, the minimum sample size* $n_2$ *is taken from the tables, for which the prediction region for* $p_2$ *has a lower bound* $\ell_2^{(p_2,0.02)}(n_2) = u_1^{(p_1,0.05)}(n_1) + 1$.

4. *The determination of the upper bound* $u_i(p_1, 0.05)(n_i)$ *of the prediction region for* $p_1$ *and, subsequently, the minimum sample size* $n_{i+1}$ *still yielding a prediction region for* $p_2$ *having a lower bound* $\ell_i^{(p_2,0.02)}(n_{i+1})$ *larger than* $u_i(p_1, 0.05)(n_i)$ *is continued until for the first time* $u_{k+1}(p_1, 0.05)(n_{k+1}) = u_k(p_1, 0.05)(n_k)$ *holds.*

5. *Then,* $n^{(0.05,0.02)} = n_k$ *and*

$$A^{(0.95,n_k)} = \{0, \ldots, u_k^{(p_1,0.05)}(n_k)\}$$

*and*

$$A^{(0.98,n_k)} = \{\ell_k^{(p_2,0.02)}(n_k), \ldots, n_k\}$$

*The set of input parameters for the example is*

| | $H_1$ | $H_2$ |
|---|---|---|
| *Measurement space* | $[\overline{p}^{(1)}, \underline{p}^{(2)}] =$ $[0.10, 0.20]$ | $[\overline{p}^{(1)}, \underline{p}^{(2)}] =$ $[0.10, 0.20]$ |
| *Significance level* | $\alpha_1 = 0.05$ | $\alpha_2 = 0.02$ |
| *Hypotheses* | $p_1 = 0.10$ | $p_2 = 0.20$ |

*The aim with respect to the sample size is*

- *Sample size* $n \stackrel{!}{=} \min$

*According to the above-given algorithm, the following numbers are obtained from the CD-ROM tables. The algorithm stops after* $i = 6$ *steps, since then*

$$u_6(p_1, 0.05)(187) = u_5(p_1, 0.05)(193) = 26$$

***Table 12.1*** *Illustration of the algorithm for determining a Neyman comparison procedure by means of Example 1.*

| | | |
|---|---|---|
| 1 | $u_1(p_1, 0.02)(222) = 32$ | $\ell_1^{(p_2,0.02)}(222) = 33$ |
| 2 | $u_2(p_1, 0.05)(222) = 30$ | $\ell_2^{(p_2,0.02)}(211) = 31$ |
| 3 | $u_3(p_1, 0.05)(211) = 28$ | $\ell_3^{(p_2,0.02)}(199) = 29$ |
| 4 | $u_4(p_1, 0.05)(199) = 27$ | $\ell_4^{(p_2,0.02)}(193) = 28$ |
| 5 | $u_5(p_1, 0.05)(193) = 26$ | $\ell_5^{(p_2,0.02)}(187) = 27$ |
| 6 | $u_6(p_1, 0.05)(187) = 26$ | |

*The last row of Table 12.1 shows the minimum sample size for which the success probability is 1.*

$$n^{(0.05,0.02)} = 187 \tag{12.10}$$

*The decision function for the thus-determined Neyman comparison procedure is the following:*

$$\phi_{H_1,H_2}(X_S) = \begin{cases} 1 & \text{for } X_S \in \{0,\ldots,26\} \\ 2 & \text{for } X_S \in \{27,\ldots,187\} \end{cases} \tag{12.11}$$

*The required reliability is given by $\alpha_1 = 0.05$ and $\alpha_2 = 0.02$; the actually achieved reliability is given by the values 0.03352 with respect to $H_1$ and 0.01983 with respect to $H_2$.* •

The tables may also be useful in order to determine single sampling plans for acceptance sampling as used in quality control.

**Example 12.2**

*Suppose that an automobile assembly plant is interested in a single-sampling plan such that lots of axles that do not contain more than 10% nonconforming will be accepted at least 95% of the time. On the other hand, the plant would like lots of axles that contain at least 30% nonconforming not to be accepted more than 1% of the time. In other words, the plant wants a plan with* $\alpha_1 = 0.05$, $\alpha_2 = 0.01$, $\overline{p}^{(1)} = 0.10$, *and* $\underline{p}^{(2)} = 0.30$.

*From the above, we obtain the following sets of input parameters:*

| | $H_1$ | $H_2$ |
|---|---|---|
| *Measurement space* | $[\overline{p}^{(1)}, \underline{p}^{(2)}] = [0.10, 0.30]$ | $[\overline{p}^{(1)}, \underline{p}^{(2)}] = [0.10, 0.30]$ |
| *Sample size* | $n \overset{!}{=} \min$ | $n \overset{!}{=} \min$ |
| *Significance level* | $\alpha_1 = 0.05$ | $\alpha_2 = 0.01$ |
| *Hypotheses* | $\overline{p}_1 = 0.10$ | $\underline{p}_2 = 0.30$ |

*Applying the above-stated algorithm yields the result in Table 12.2.*

***Table 12.2*** *Illustration of the algorithm for determining a Neyman comparison procedure by means of Example 2.*

| | | |
|---|---|---|
| *1* | $u_1(p_1, 0.01)(83) = 15$ | $\ell_1^{(p_2,0.01)}(83) = 16$ |
| *2* | $u_2(p_1, 0.05)(83) = 13$ | $\ell_2^{(p_2,0.01)}(75) = 14$ |
| *3* | $u_3(p_1, 0.05)(75) = 12$ | $\ell_3^{(p_2,0.01)}(71) = 13$ |
| *4* | $u_4(p_1, 0.05)(71) = 11$ | $\ell_4^{(p_2,0.01)}(67) = 12$ |
| *5* | $u_5(p_1, 0.05)(67) = 11$ | |

*Thus, we arrive at the following Neyman comparison procedure:*

- *Minimum sample size for obtaining a comparison procedure with zero failure probability:*

$$n^{(0.05,0.01)} = 67.$$

- *Decision function:*

$$\phi_{H_1,H_2}\big(X_S\big) = \begin{cases} 1 & \text{for } X_S \in \{0, 1, \ldots, 11\} \\ 2 & \text{for } X_S \in \{12, 13, \ldots, 67\}. \end{cases}$$

- *Achieved reliabilities:*

$$\mathbf{P}_{X_S}^{(0.10)}\Big(\{0,1,\ldots,11\}\Big) = 0.9673$$
$$\mathbf{P}_{X_S}^{(0.30)}\Big(\{12,13,\ldots,67\}\Big) = 0.991702.$$

*In the above example the reliability requirements were rather high, leading to a comparatively large minimum sample size of* $n^{(0.05,0.01)} = 67$*. Next, consider the case that the reliability requirements are reduced to* $\alpha_1 = 0.10$ *and* $\alpha_2 = 0.02$*. Then the algorithm yields the results given in Table 12.3.*

***Table 12.3*** *Illustration of the algorithm for determining a Neyman comparison procedure by means of Example 2 for reduced reliability requirements.*

| | | |
|---|---|---|
| *1* | $u_1(p_1, 0.02)(67) = 12$ | $\ell_1^{(p_2,0.02)}(67) = 13$ |
| *2* | $u_2(p_1, 0.10)(67) = 10$ | $\ell_2^{(p_2,0.02)}(59) = 11$ |
| *3* | $u_3(p_1, 0.10)(59) = 9$ | $\ell_3^{(p_2,0.02)}(54) = 10$ |
| *4* | $u_4(p_1, 0.10)(54) = 8$ | $\ell_4^{(p_2,0.02)}(50) = 9$ |
| 5 | $u_5(p_1, 0.10)(50) = 8$ | |

*The Neyman comparison procedure for the reduced reliability requirements is given by the following:*

- *Minimum sample size for obtaining a comparison procedure with zero failure probability:*

$$n^{(0.10,0.02)} = 50.$$

- *Decision function:*

$$\phi_{H_1,H_2}(X_S) = \begin{cases} 1 & \text{for } X_S \in \{0,1,\ldots,8\} \\ 2 & \text{for } X_S \in \{9,10,\ldots,50\}. \end{cases}$$

- *Achieved reliabilities:*

$$\mathbf{P}_{X_S}^{(0.10)}\Big(\{0, 1, \ldots, 8\}\Big) = 0.9421$$
$$\mathbf{P}_{X_S}^{(0.30)}\Big(\{9, 10, \ldots, 50\}\Big) = 0.9817.$$

*From the results it becomes clear that relaxing the reliability requirements may lead to a large reduction in the sample size (25% in the example) with only a small decrease of the procedure's reliability.*

●

# Part V

# Tables

| $n$ | $\beta$ | 0 | 1 | 2 | 3 | 4 |
|---|---|---|---|---|---|---|
| 1 | 0.99 | 0.0000 | 0.0100 | | | |
| | | 0.9900 | 1.0000 | | | |
| | 0.98 | 0.0000 | 0.0200 | | | |
| | | 0.9800 | 1.0000 | | | |
| | 0.95 | 0.0000 | 0.0500 | | | |
| | | 0.9500 | 1.0000 | | | |
| | 0.90 | 0.0000 | 0.1000 | | | |
| | | 0.9000 | 1.0000 | | | |
| | 0.80 | 0.0000 | 0.2000 | | | |
| | | 0.8000 | 1.0000 | | | |
| 2 | 0.99 | 0.0000 | 0.0050 | 0.1000 | | |
| | | 0.9000 | 0.9950 | 1.0000 | | |
| | 0.98 | 0.0000 | 0.0100 | 0.1414 | | |
| | | 0.8586 | 0.9900 | 1.0000 | | |
| | 0.95 | 0.0000 | 0.0253 | 0.2236 | | |
| | | 0.7764 | 0.9747 | 1.0000 | | |
| | 0.90 | 0.0000 | 0.0513 | 0.3162 | | |
| | | 0.6838 | 0.9487 | 1.0000 | | |
| | 0.80 | 0.0000 | 0.1055 | 0.4472 | | |
| | | 0.5528 | 0.8945 | 1.0000 | | |
| 3 | 0.99 | 0.0000 | 0.0033 | 0.0589 | 0.2154 | |
| | | 0.7846 | 0.9411 | 0.9967 | 1.0000 | |
| | 0.98 | 0.0000 | 0.0067 | 0.0840 | 0.2714 | |
| | | 0.7286 | 0.9160 | 0.9933 | 1.0000 | |
| | 0.95 | 0.0000 | 0.0169 | 0.1353 | 0.3684 | |
| | | 0.6316 | 0.8647 | 0.9831 | 1.0000 | |
| | 0.90 | 0.0000 | 0.0345 | 0.1958 | 0.4641 | |
| | | 0.5359 | 0.8042 | 0.9655 | 1.0000 | |
| | 0.80 | 0.0000 | 0.0716 | 0.2871 | 0.5000 | |
| | | 0.5000 | 0.7129 | 0.9284 | 1.0000 | |
| 4 | 0.99 | 0.0000 | 0.0025 | 0.0419 | 0.1408 | 0.3162 |
| | | 0.6838 | 0.8592 | 0.9581 | 0.9975 | 1.0000 |
| | 0.98 | 0.0000 | 0.0050 | 0.0601 | 0.1794 | 0.3760 |
| | | 0.6240 | 0.8206 | 0.9399 | 0.9950 | 1.0000 |
| | 0.95 | 0.0000 | 0.0127 | 0.0976 | 0.2486 | 0.4728 |
| | | 0.5272 | 0.7514 | 0.9024 | 0.9873 | 1.0000 |
| | 0.90 | 0.0000 | 0.0259 | 0.1425 | 0.3204 | 0.5000 |
| | | 0.5000 | 0.6796 | 0.8575 | 0.9741 | 1.0000 |
| | 0.80 | 0.0000 | 0.0542 | 0.2123 | 0.3864 | 0.6135 |
| | | 0.3865 | 0.6136 | 0.7877 | 0.9458 | 1.0000 |

| n | β | 0 | 1 | 2 | 3 | 4 | 5 | 6 |
|---|---|---|---|---|---|---|---|---|
| | 0.99 | 0.0000 | 0.0020 | 0.0326 | 0.1056 | 0.2220 | 0.3981 | |
| | | 0.6019 | 0.7780 | 0.8944 | 0.9674 | 0.9980 | 1.0000 | |
| | 0.98 | 0.0000 | 0.0040 | 0.0468 | 0.1352 | 0.2670 | 0.4573 | |
| | | 0.5427 | 0.7330 | 0.8648 | 0.9532 | 0.9960 | 1.0000 | |
| 5 | 0.95 | 0.0000 | 0.0102 | 0.0764 | 0.1892 | 0.3425 | 0.5000 | |
| | | 0.5000 | 0.6575 | 0.8108 | 0.9236 | 0.9898 | 1.0000 | |
| | 0.90 | 0.0000 | 0.0208 | 0.1122 | 0.2466 | 0.4007 | 0.5992 | |
| | | 0.4008 | 0.5993 | 0.7534 | 0.8878 | 0.9792 | 1.0000 | |
| | 0.80 | 0.0000 | 0.0436 | 0.1686 | 0.3170 | 0.4780 | 0.6829 | |
| | | 0.3171 | 0.5220 | 0.6830 | 0.8314 | 0.9564 | 1.0000 | |
| | 0.99 | 0.0000 | 0.0016 | 0.0267 | 0.0847 | 0.1730 | 0.2943 | 0.4641 |
| | | 0.5359 | 0.7057 | 0.8270 | 0.9153 | 0.9733 | 0.9984 | 1.0000 |
| | 0.98 | 0.0000 | 0.0033 | 0.0384 | 0.1088 | 0.2093 | 0.3417 | 0.5000 |
| | | 0.5000 | 0.6583 | 0.7907 | 0.8912 | 0.9616 | 0.9967 | 1.0000 |
| 6 | 0.95 | 0.0000 | 0.0085 | 0.0628 | 0.1531 | 0.2713 | 0.4113 | 0.5886 |
| | | 0.4114 | 0.5887 | 0.7287 | 0.8469 | 0.9372 | 0.9915 | 1.0000 |
| | 0.90 | 0.0000 | 0.0174 | 0.0925 | 0.2009 | 0.3331 | 0.4577 | 0.6553 |
| | | 0.3447 | 0.5423 | 0.6669 | 0.7991 | 0.9075 | 0.9826 | 1.0000 |
| | 0.80 | 0.0000 | 0.0365 | 0.1398 | 0.2686 | 0.3650 | 0.5000 | 0.7291 |
| | | 0.2709 | 0.5000 | 0.6350 | 0.7314 | 0.8602 | 0.9635 | 1.0000 |
| | 0.99 | 0.0000 | 0.0014 | 0.0226 | 0.0708 | 0.1422 | 0.2363 | 0.3566 |
| | | 0.5000 | 0.6434 | 0.7637 | 0.8578 | 0.9292 | 0.9774 | 0.9986 |
| | 0.98 | 0.0000 | 0.0028 | 0.0325 | 0.0911 | 0.1727 | 0.2760 | 0.4043 |
| | | 0.4429 | 0.5957 | 0.7240 | 0.8273 | 0.9089 | 0.9675 | 0.9972 |
| 7 | 0.95 | 0.0000 | 0.0073 | 0.0533 | 0.1287 | 0.2253 | 0.3412 | 0.4457 |
| | | 0.3772 | 0.5543 | 0.6588 | 0.7747 | 0.8713 | 0.9467 | 0.9927 |
| | 0.90 | 0.0000 | 0.0149 | 0.0788 | 0.1696 | 0.2786 | 0.3523 | 0.5000 |
| | | 0.3524 | 0.5000 | 0.6477 | 0.7214 | 0.8304 | 0.9212 | 0.9851 |
| | 0.80 | 0.0000 | 0.0313 | 0.1195 | 0.2283 | 0.2913 | 0.4317 | 0.5682 |
| | | 0.2914 | 0.4318 | 0.5683 | 0.7087 | 0.7717 | 0.8805 | 0.9687 |
| | 0.99 | 0.0000 | 0.0012 | 0.0196 | 0.0608 | 0.1209 | 0.1982 | 0.2932 |
| | | 0.4507 | 0.5900 | 0.7068 | 0.8018 | 0.8791 | 0.9392 | 0.9804 |
| | 0.98 | 0.0000 | 0.0025 | 0.0282 | 0.0784 | 0.1471 | 0.2323 | 0.3344 |
| | | 0.4263 | 0.5738 | 0.6656 | 0.7677 | 0.8529 | 0.9216 | 0.9718 |
| 8 | 0.95 | 0.0000 | 0.0063 | 0.0463 | 0.1111 | 0.1929 | 0.2892 | 0.3646 |
| | | 0.3647 | 0.5000 | 0.6354 | 0.7108 | 0.8071 | 0.8889 | 0.9537 |
| | 0.90 | 0.0000 | 0.0130 | 0.0686 | 0.1468 | 0.2396 | 0.3089 | 0.4376 |
| | | 0.3090 | 0.4377 | 0.5624 | 0.6911 | 0.7604 | 0.8532 | 0.9314 |
| | 0.80 | 0.0000 | 0.0275 | 0.1043 | 0.1985 | 0.2569 | 0.3807 | 0.5000 |
| | | 0.2570 | 0.3808 | 0.5000 | 0.6193 | 0.7431 | 0.8015 | 0.8957 |

| $n$ | $\beta$ | 0 | 1 | 2 | 3 | 4 | 5 | 6 |
|---|---|---|---|---|---|---|---|---|
| 9 | 0.99 | 0.0000 | 0.0011 | 0.0173 | 0.0533 | 0.1052 | 0.1709 | 0.2500 |
| | | 0.4318 | 0.5683 | 0.6564 | 0.7500 | 0.8291 | 0.8948 | 0.9467 |
| | 0.98 | 0.0000 | 0.0022 | 0.0249 | 0.0688 | 0.1283 | 0.2010 | 0.2862 |
| | | 0.3748 | 0.5064 | 0.6253 | 0.7138 | 0.7990 | 0.8717 | 0.9312 |
| | 0.95 | 0.0000 | 0.0056 | 0.0410 | 0.0977 | 0.1687 | 0.2513 | 0.3233 |
| | | 0.3234 | 0.4435 | 0.5583 | 0.6767 | 0.7487 | 0.8313 | 0.9023 |
| | 0.90 | 0.0000 | 0.0116 | 0.0607 | 0.1294 | 0.2103 | 0.2754 | 0.3896 |
| | | 0.2755 | 0.3910 | 0.5152 | 0.6104 | 0.7246 | 0.7897 | 0.8706 |
| | 0.80 | 0.0000 | 0.0244 | 0.0926 | 0.1757 | 0.2298 | 0.3407 | 0.4457 |
| | | 0.2299 | 0.3408 | 0.4491 | 0.5543 | 0.6593 | 0.7702 | 0.8243 |
| 10 | 0.99 | 0.0000 | 0.0010 | 0.0155 | 0.0475 | 0.0932 | 0.1504 | 0.2183 |
| | | 0.3833 | 0.5124 | 0.6168 | 0.7029 | 0.7817 | 0.8496 | 0.9068 |
| | 0.98 | 0.0000 | 0.0020 | 0.0223 | 0.0613 | 0.1138 | 0.1772 | 0.2506 |
| | | 0.3376 | 0.5000 | 0.5881 | 0.6657 | 0.7494 | 0.8228 | 0.8862 |
| | 0.95 | 0.0000 | 0.0051 | 0.0367 | 0.0872 | 0.1500 | 0.2224 | 0.2908 |
| | | 0.2909 | 0.4465 | 0.5536 | 0.6195 | 0.7092 | 0.7776 | 0.8500 |
| | 0.90 | 0.0000 | 0.0104 | 0.0545 | 0.1158 | 0.1875 | 0.2486 | 0.3408 |
| | | 0.2487 | 0.3980 | 0.5000 | 0.6021 | 0.6592 | 0.7514 | 0.8125 |
| | 0.80 | 0.0000 | 0.0220 | 0.0832 | 0.1576 | 0.2080 | 0.3085 | 0.3868 |
| | | 0.2081 | 0.3086 | 0.4512 | 0.5489 | 0.6132 | 0.6915 | 0.7920 |
| 11 | 0.99 | 0.0000 | 0.0009 | 0.0140 | 0.0428 | 0.0836 | 0.1343 | 0.1939 |
| | | 0.3592 | 0.5000 | 0.5928 | 0.6605 | 0.7378 | 0.8061 | 0.8657 |
| | 0.98 | 0.0000 | 0.0018 | 0.0202 | 0.0553 | 0.1022 | 0.1585 | 0.2232 |
| | | 0.3457 | 0.4499 | 0.5502 | 0.6544 | 0.7042 | 0.7768 | 0.8415 |
| | 0.95 | 0.0000 | 0.0046 | 0.0333 | 0.0788 | 0.1350 | 0.1995 | 0.2645 |
| | | 0.2646 | 0.4045 | 0.5000 | 0.5956 | 0.6672 | 0.7355 | 0.8005 |
| | 0.90 | 0.0000 | 0.0095 | 0.0494 | 0.1047 | 0.1692 | 0.2266 | 0.3023 |
| | | 0.2267 | 0.3620 | 0.4544 | 0.5457 | 0.6381 | 0.6977 | 0.7734 |
| | 0.80 | 0.0000 | 0.0200 | 0.0756 | 0.1429 | 0.1900 | 0.2820 | 0.3379 |
| | | 0.1901 | 0.2821 | 0.4114 | 0.5000 | 0.5887 | 0.6621 | 0.7180 |
| 12 | 0.99 | 0.0000 | 0.0008 | 0.0128 | 0.0389 | 0.0758 | 0.1214 | 0.1746 |
| | | 0.3546 | 0.4528 | 0.5473 | 0.6455 | 0.6976 | 0.7652 | 0.8254 |
| | 0.98 | 0.0000 | 0.0016 | 0.0185 | 0.0504 | 0.0928 | 0.1435 | 0.2012 |
| | | 0.3153 | 0.4106 | 0.5109 | 0.5903 | 0.6848 | 0.7344 | 0.7988 |
| | 0.95 | 0.0000 | 0.0042 | 0.0304 | 0.0718 | 0.1228 | 0.1810 | 0.2426 |
| | | 0.2427 | 0.3702 | 0.4572 | 0.5429 | 0.6299 | 0.7061 | 0.7574 |
| | 0.90 | 0.0000 | 0.0087 | 0.0452 | 0.0956 | 0.1541 | 0.2083 | 0.2712 |
| | | 0.2084 | 0.3322 | 0.4169 | 0.5000 | 0.5832 | 0.6679 | 0.7288 |
| | 0.80 | 0.0000 | 0.0184 | 0.0692 | 0.1307 | 0.1749 | 0.2565 | 0.2921 |
| | | 0.1750 | 0.2720 | 0.3783 | 0.4596 | 0.5405 | 0.6218 | 0.7079 |

| $n$ | $\beta$ | 0 | 1 | 2 | 3 | 4 | 5 | 6 |
|---|---|---|---|---|---|---|---|---|
| | 0.99 | 0.0000 | 0.0007 | 0.0118 | 0.0357 | 0.0694 | 0.1108 | 0.1588 |
| | | 0.3252 | 0.4290 | 0.5234 | 0.5938 | 0.6749 | 0.7272 | 0.7872 |
| | 0.98 | 0.0000 | 0.0015 | 0.0170 | 0.0462 | 0.0850 | 0.1310 | 0.1833 |
| | | 0.2901 | 0.4150 | 0.5000 | 0.5851 | 0.6340 | 0.7100 | 0.7589 |
| 13 | 0.95 | 0.0000 | 0.0039 | 0.0280 | 0.0660 | 0.1126 | 0.1656 | 0.2239 |
| | | 0.2252 | 0.3416 | 0.4340 | 0.5197 | 0.5866 | 0.6585 | 0.7396 |
| | 0.90 | 0.0000 | 0.0080 | 0.0416 | 0.0879 | 0.1416 | 0.1927 | 0.2455 |
| | | 0.1928 | 0.3071 | 0.3853 | 0.4697 | 0.5446 | 0.6148 | 0.6930 |
| | 0.80 | 0.0000 | 0.0170 | 0.0638 | 0.1204 | 0.1620 | 0.2346 | 0.2735 |
| | | 0.1621 | 0.2736 | 0.3502 | 0.4254 | 0.5000 | 0.5747 | 0.6499 |
| | 0.99 | 0.0000 | 0.0007 | 0.0109 | 0.0330 | 0.0640 | 0.1019 | 0.1456 |
| | | 0.3006 | 0.4193 | 0.5000 | 0.5808 | 0.6365 | 0.6995 | 0.7513 |
| | 0.98 | 0.0000 | 0.0014 | 0.0157 | 0.0427 | 0.0784 | 0.1206 | 0.1683 |
| | | 0.2688 | 0.3836 | 0.4616 | 0.5385 | 0.6165 | 0.6698 | 0.7313 |
| 14 | 0.95 | 0.0000 | 0.0036 | 0.0259 | 0.0611 | 0.1040 | 0.1527 | 0.2060 |
| | | 0.2382 | 0.3172 | 0.4257 | 0.5000 | 0.5744 | 0.6290 | 0.6829 |
| | 0.90 | 0.0000 | 0.0074 | 0.0386 | 0.0814 | 0.1309 | 0.1794 | 0.2238 |
| | | 0.1795 | 0.2856 | 0.3651 | 0.4640 | 0.5361 | 0.5945 | 0.6452 |
| | 0.80 | 0.0000 | 0.0158 | 0.0592 | 0.1116 | 0.1509 | 0.2161 | 0.2547 |
| | | 0.1510 | 0.2548 | 0.3261 | 0.3961 | 0.4654 | 0.5347 | 0.6040 |
| | 0.99 | 0.0000 | 0.0006 | 0.0101 | 0.0307 | 0.0593 | 0.0943 | 0.1345 |
| | | 0.2796 | 0.3892 | 0.4635 | 0.5366 | 0.6109 | 0.6725 | 0.7205 |
| | 0.98 | 0.0000 | 0.0013 | 0.0147 | 0.0397 | 0.0728 | 0.1117 | 0.1556 |
| | | 0.2505 | 0.3570 | 0.4293 | 0.5056 | 0.5708 | 0.6431 | 0.7000 |
| 15 | 0.95 | 0.0000 | 0.0034 | 0.0242 | 0.0568 | 0.0966 | 0.1416 | 0.1908 |
| | | 0.2223 | 0.3021 | 0.3968 | 0.4658 | 0.5343 | 0.6033 | 0.6677 |
| | 0.90 | 0.0000 | 0.0069 | 0.0360 | 0.0758 | 0.1217 | 0.1677 | 0.2053 |
| | | 0.1678 | 0.2670 | 0.3654 | 0.4331 | 0.5000 | 0.5670 | 0.6347 |
| | 0.80 | 0.0000 | 0.0147 | 0.0553 | 0.1040 | 0.1412 | 0.2003 | 0.2382 |
| | | 0.1413 | 0.2383 | 0.3052 | 0.3706 | 0.4446 | 0.5137 | 0.5711 |
| | 0.99 | 0.0000 | 0.0006 | 0.0095 | 0.0286 | 0.0553 | 0.0878 | 0.1250 |
| | | 0.2639 | 0.3634 | 0.4514 | 0.5246 | 0.5792 | 0.6367 | 0.7047 |
| | 0.98 | 0.0000 | 0.0012 | 0.0137 | 0.0371 | 0.0679 | 0.1041 | 0.1447 |
| | | 0.2346 | 0.3339 | 0.4237 | 0.5000 | 0.5652 | 0.6087 | 0.6662 |
| 16 | 0.95 | 0.0000 | 0.0032 | 0.0226 | 0.0531 | 0.0902 | 0.1321 | 0.1777 |
| | | 0.2084 | 0.3055 | 0.3717 | 0.4362 | 0.5000 | 0.5639 | 0.6284 |
| | 0.90 | 0.0000 | 0.0065 | 0.0337 | 0.0709 | 0.1138 | 0.1575 | 0.1892 |
| | | 0.1576 | 0.2507 | 0.3428 | 0.4062 | 0.4688 | 0.5313 | 0.5939 |
| | 0.80 | 0.0000 | 0.0138 | 0.0518 | 0.0974 | 0.1327 | 0.1865 | 0.2238 |
| | | 0.1328 | 0.2239 | 0.2867 | 0.3515 | 0.4386 | 0.5000 | 0.5615 |

| $n$ | $\beta$ | 7 | 8 | 9 | 10 | 11 | 12 | 13 |
|---|---|---|---|---|---|---|---|---|
| | 0.99 | 0.2128 | 0.2728 | 0.3251 | 0.4062 | 0.4766 | 0.5710 | 0.6748 |
| | | 0.8412 | 0.8892 | 0.9306 | 0.9643 | 0.9882 | 0.9993 | 1.0000 |
| | 0.98 | 0.2411 | 0.2900 | 0.3660 | 0.4149 | 0.5000 | 0.5850 | 0.7099 |
| | | 0.8167 | 0.8690 | 0.9150 | 0.9538 | 0.9830 | 0.9985 | 1.0000 |
| 13 | 0.95 | 0.2604 | 0.3415 | 0.4134 | 0.4803 | 0.5660 | 0.6584 | 0.7748 |
| | | 0.7761 | 0.8344 | 0.8874 | 0.9340 | 0.9720 | 0.9961 | 1.0000 |
| | 0.90 | 0.3070 | 0.3852 | 0.4554 | 0.5303 | 0.6147 | 0.6929 | 0.8072 |
| | | 0.7545 | 0.8073 | 0.8584 | 0.9121 | 0.9584 | 0.9920 | 1.0000 |
| | 0.80 | 0.3501 | 0.4253 | 0.5000 | 0.5746 | 0.6498 | 0.7264 | 0.8379 |
| | | 0.7265 | 0.7654 | 0.8380 | 0.8796 | 0.9362 | 0.9830 | 1.0000 |
| | 0.99 | 0.1947 | 0.2487 | 0.3005 | 0.3635 | 0.4192 | 0.5000 | 0.5807 |
| | | 0.8053 | 0.8544 | 0.8981 | 0.9360 | 0.9670 | 0.9891 | 0.9993 |
| | 0.98 | 0.2209 | 0.2687 | 0.3302 | 0.3835 | 0.4615 | 0.5384 | 0.6164 |
| | | 0.7791 | 0.8317 | 0.8794 | 0.9216 | 0.9573 | 0.9843 | 0.9986 |
| 14 | 0.95 | 0.2381 | 0.3171 | 0.3710 | 0.4256 | 0.5000 | 0.5743 | 0.6828 |
| | | 0.7619 | 0.7940 | 0.8473 | 0.8960 | 0.9389 | 0.9741 | 0.9964 |
| | 0.90 | 0.2855 | 0.3548 | 0.4055 | 0.4639 | 0.5360 | 0.6349 | 0.7144 |
| | | 0.7145 | 0.7762 | 0.8206 | 0.8691 | 0.9186 | 0.9614 | 0.9926 |
| | 0.80 | 0.3260 | 0.3960 | 0.4653 | 0.5346 | 0.6039 | 0.6739 | 0.7452 |
| | | 0.6740 | 0.7453 | 0.7839 | 0.8491 | 0.8884 | 0.9408 | 0.9842 |
| | 0.99 | 0.1794 | 0.2287 | 0.2795 | 0.3275 | 0.3891 | 0.4634 | 0.5365 |
| | | 0.7713 | 0.8206 | 0.8655 | 0.9057 | 0.9407 | 0.9693 | 0.9899 |
| | 0.98 | 0.2038 | 0.2504 | 0.3000 | 0.3569 | 0.4292 | 0.4944 | 0.5707 |
| | | 0.7496 | 0.7962 | 0.8444 | 0.8883 | 0.9272 | 0.9603 | 0.9853 |
| 15 | 0.95 | 0.2222 | 0.2939 | 0.3323 | 0.3967 | 0.4657 | 0.5342 | 0.6032 |
| | | 0.7061 | 0.7778 | 0.8092 | 0.8584 | 0.9034 | 0.9432 | 0.9758 |
| | 0.90 | 0.2669 | 0.3251 | 0.3653 | 0.4330 | 0.5000 | 0.5669 | 0.6346 |
| | | 0.6749 | 0.7331 | 0.7947 | 0.8323 | 0.8783 | 0.9242 | 0.9640 |
| | 0.80 | 0.3051 | 0.3705 | 0.4289 | 0.4863 | 0.5554 | 0.6294 | 0.6948 |
| | | 0.6295 | 0.6949 | 0.7618 | 0.7997 | 0.8588 | 0.8960 | 0.9447 |
| | 0.99 | 0.1664 | 0.2117 | 0.2606 | 0.2953 | 0.3633 | 0.4208 | 0.4754 |
| | | 0.7394 | 0.7883 | 0.8336 | 0.8750 | 0.9122 | 0.9447 | 0.9714 |
| | 0.98 | 0.1892 | 0.2345 | 0.2739 | 0.3338 | 0.3913 | 0.4348 | 0.5000 |
| | | 0.7261 | 0.7655 | 0.8108 | 0.8553 | 0.8959 | 0.9321 | 0.9629 |
| 16 | 0.95 | 0.2083 | 0.2719 | 0.3054 | 0.3716 | 0.4361 | 0.5000 | 0.5638 |
| | | 0.6946 | 0.7281 | 0.7917 | 0.8223 | 0.8679 | 0.9098 | 0.9469 |
| | 0.90 | 0.2506 | 0.2994 | 0.3427 | 0.4061 | 0.4687 | 0.5312 | 0.5938 |
| | | 0.6573 | 0.7006 | 0.7494 | 0.8108 | 0.8425 | 0.8862 | 0.9291 |
| | 0.80 | 0.2866 | 0.3465 | 0.3898 | 0.4385 | 0.5000 | 0.5614 | 0.6485 |
| | | 0.6102 | 0.6535 | 0.7134 | 0.7762 | 0.8135 | 0.8673 | 0.9026 |

| $n$ | $\beta$ | 0 | 1 | 2 | 3 | 4 | 5 | 6 |
|---|---|---|---|---|---|---|---|---|
| 17 | 0.99 | 0.0000 | 0.0005 | 0.0089 | 0.0269 | 0.0518 | 0.0821 | 0.1168 |
| | | 0.2714 | 0.3463 | 0.4354 | 0.5000 | 0.5647 | 0.6196 | 0.6616 |
| | 0.98 | 0.0000 | 0.0011 | 0.0129 | 0.0348 | 0.0636 | 0.0974 | 0.1353 |
| | | 0.2206 | 0.3138 | 0.4057 | 0.4688 | 0.5313 | 0.5944 | 0.6424 |
| | 0.95 | 0.0000 | 0.0030 | 0.0213 | 0.0498 | 0.0846 | 0.1237 | 0.1663 |
| | | 0.1962 | 0.2874 | 0.3498 | 0.4166 | 0.4888 | 0.5442 | 0.5938 |
| | 0.90 | 0.0000 | 0.0061 | 0.0317 | 0.0666 | 0.1068 | 0.1485 | 0.1752 |
| | | 0.1486 | 0.2362 | 0.3229 | 0.3825 | 0.4414 | 0.5000 | 0.5587 |
| | 0.80 | 0.0000 | 0.0130 | 0.0487 | 0.0916 | 0.1252 | 0.1746 | 0.2111 |
| | | 0.1253 | 0.2112 | 0.2705 | 0.3553 | 0.4134 | 0.4712 | 0.5289 |
| 18 | 0.99 | 0.0000 | 0.0005 | 0.0084 | 0.0253 | 0.0487 | 0.0771 | 0.1095 |
| | | 0.2557 | 0.3474 | 0.4096 | 0.4700 | 0.5301 | 0.5905 | 0.6527 |
| | 0.98 | 0.0000 | 0.0011 | 0.0121 | 0.0328 | 0.0598 | 0.0915 | 0.1270 |
| | | 0.2133 | 0.3022 | 0.3823 | 0.4415 | 0.5000 | 0.5586 | 0.6178 |
| | 0.95 | 0.0000 | 0.0028 | 0.0201 | 0.0470 | 0.0796 | 0.1164 | 0.1563 |
| | | 0.1853 | 0.2714 | 0.3303 | 0.4141 | 0.4715 | 0.5286 | 0.5860 |
| | 0.90 | 0.0000 | 0.0058 | 0.0299 | 0.0628 | 0.1006 | 0.1404 | 0.1626 |
| | | 0.1405 | 0.2234 | 0.3052 | 0.3616 | 0.4189 | 0.4823 | 0.5360 |
| | 0.80 | 0.0000 | 0.0123 | 0.0460 | 0.0864 | 0.1185 | 0.1640 | 0.1997 |
| | | 0.1186 | 0.1998 | 0.2559 | 0.3360 | 0.3910 | 0.4456 | 0.5000 |
| 19 | 0.99 | 0.0000 | 0.0005 | 0.0080 | 0.0239 | 0.0460 | 0.0727 | 0.1032 |
| | | 0.2418 | 0.3282 | 0.3869 | 0.4552 | 0.5153 | 0.5644 | 0.6132 |
| | 0.98 | 0.0000 | 0.0010 | 0.0115 | 0.0310 | 0.0565 | 0.0864 | 0.1197 |
| | | 0.2192 | 0.3039 | 0.3616 | 0.4220 | 0.4867 | 0.5390 | 0.5856 |
| | 0.95 | 0.0000 | 0.0026 | 0.0190 | 0.0444 | 0.0752 | 0.1099 | 0.1474 |
| | | 0.1756 | 0.2571 | 0.3158 | 0.3920 | 0.4462 | 0.5000 | 0.5539 |
| | 0.90 | 0.0000 | 0.0055 | 0.0283 | 0.0594 | 0.0951 | 0.1332 | 0.1513 |
| | | 0.1333 | 0.2119 | 0.2893 | 0.3428 | 0.4206 | 0.4736 | 0.5265 |
| | 0.80 | 0.0000 | 0.0116 | 0.0436 | 0.0818 | 0.1124 | 0.1546 | 0.1895 |
| | | 0.1125 | 0.1896 | 0.2429 | 0.3188 | 0.3709 | 0.4227 | 0.4743 |
| 20 | 0.99 | 0.0000 | 0.0005 | 0.0075 | 0.0227 | 0.0436 | 0.0688 | 0.0975 |
| | | 0.2293 | 0.3111 | 0.3748 | 0.4459 | 0.5000 | 0.5542 | 0.6011 |
| | 0.98 | 0.0000 | 0.0010 | 0.0109 | 0.0294 | 0.0535 | 0.0817 | 0.1131 |
| | | 0.2081 | 0.2884 | 0.3431 | 0.4206 | 0.4736 | 0.5265 | 0.5795 |
| | 0.95 | 0.0000 | 0.0025 | 0.0180 | 0.0421 | 0.0713 | 0.1040 | 0.1395 |
| | | 0.1669 | 0.2443 | 0.3200 | 0.3723 | 0.4236 | 0.4746 | 0.5255 |
| | 0.90 | 0.0000 | 0.0052 | 0.0269 | 0.0564 | 0.0902 | 0.1267 | 0.1408 |
| | | 0.1268 | 0.2032 | 0.2751 | 0.3280 | 0.3996 | 0.4499 | 0.5000 |
| | 0.80 | 0.0000 | 0.0110 | 0.0414 | 0.0777 | 0.1070 | 0.1463 | 0.1803 |
| | | 0.1071 | 0.1804 | 0.2311 | 0.3032 | 0.3528 | 0.4021 | 0.4511 |

| $n$ | $\beta$ | 7 | 8 | 9 | 10 | 11 | 12 | 13 |
|---|---|---|---|---|---|---|---|---|
| | 0.99 | 0.1552 | 0.1971 | 0.2422 | 0.2713 | 0.3384 | 0.3804 | 0.4353 |
| | | 0.7287 | 0.7578 | 0.8029 | 0.8448 | 0.8832 | 0.9179 | 0.9482 |
| | 0.98 | 0.1766 | 0.2205 | 0.2505 | 0.3137 | 0.3576 | 0.4056 | 0.4687 |
| | | 0.6863 | 0.7495 | 0.7795 | 0.8234 | 0.8647 | 0.9026 | 0.9364 |
| 17 | 0.95 | 0.1961 | 0.2530 | 0.2873 | 0.3497 | 0.4062 | 0.4558 | 0.5112 |
| | | 0.6503 | 0.7127 | 0.7470 | 0.8039 | 0.8337 | 0.8763 | 0.9154 |
| | 0.90 | 0.2361 | 0.2768 | 0.3228 | 0.3824 | 0.4413 | 0.5000 | 0.5586 |
| | | 0.6176 | 0.6772 | 0.7232 | 0.7639 | 0.8248 | 0.8515 | 0.8932 |
| | 0.80 | 0.2704 | 0.3215 | 0.3552 | 0.4133 | 0.4711 | 0.5288 | 0.5866 |
| | | 0.5867 | 0.6448 | 0.6785 | 0.7296 | 0.7889 | 0.8254 | 0.8748 |
| | 0.99 | 0.1454 | 0.1844 | 0.2262 | 0.2556 | 0.3139 | 0.3473 | 0.4095 |
| | | 0.6861 | 0.7444 | 0.7738 | 0.8156 | 0.8546 | 0.8905 | 0.9229 |
| | 0.98 | 0.1656 | 0.2071 | 0.2279 | 0.2938 | 0.3254 | 0.3822 | 0.4414 |
| | | 0.6746 | 0.7062 | 0.7721 | 0.7929 | 0.8344 | 0.8730 | 0.9085 |
| 18 | 0.95 | 0.1852 | 0.2364 | 0.2713 | 0.3302 | 0.3746 | 0.4140 | 0.4714 |
| | | 0.6254 | 0.6698 | 0.7287 | 0.7636 | 0.8148 | 0.8437 | 0.8836 |
| | 0.90 | 0.2233 | 0.2566 | 0.3051 | 0.3615 | 0.4159 | 0.4640 | 0.5177 |
| | | 0.5841 | 0.6385 | 0.6949 | 0.7434 | 0.7767 | 0.8374 | 0.8596 |
| | 0.80 | 0.2558 | 0.2994 | 0.3359 | 0.3909 | 0.4455 | 0.5000 | 0.5544 |
| | | 0.5545 | 0.6091 | 0.6641 | 0.7006 | 0.7442 | 0.8003 | 0.8360 |
| | 0.99 | 0.1368 | 0.1732 | 0.2123 | 0.2417 | 0.2925 | 0.3281 | 0.3868 |
| | | 0.6719 | 0.7075 | 0.7583 | 0.7877 | 0.8268 | 0.8632 | 0.8968 |
| | 0.98 | 0.1559 | 0.1947 | 0.2191 | 0.2749 | 0.3038 | 0.3615 | 0.4144 |
| | | 0.6385 | 0.6962 | 0.7251 | 0.7809 | 0.8053 | 0.8441 | 0.8803 |
| 19 | 0.95 | 0.1755 | 0.2218 | 0.2570 | 0.3117 | 0.3448 | 0.3919 | 0.4461 |
| | | 0.6081 | 0.6552 | 0.6883 | 0.7430 | 0.7782 | 0.8245 | 0.8526 |
| | 0.90 | 0.2118 | 0.2382 | 0.2892 | 0.3427 | 0.3860 | 0.4205 | 0.4735 |
| | | 0.5795 | 0.6140 | 0.6573 | 0.7108 | 0.7618 | 0.7882 | 0.8487 |
| | 0.80 | 0.2428 | 0.2795 | 0.3187 | 0.3708 | 0.4226 | 0.4742 | 0.5257 |
| | | 0.5258 | 0.5774 | 0.6292 | 0.6813 | 0.7205 | 0.7572 | 0.8105 |
| | 0.99 | 0.1291 | 0.1634 | 0.2000 | 0.2292 | 0.2738 | 0.3110 | 0.3627 |
| | | 0.6373 | 0.6890 | 0.7262 | 0.7708 | 0.8000 | 0.8366 | 0.8709 |
| | 0.98 | 0.1473 | 0.1838 | 0.2080 | 0.2582 | 0.2883 | 0.3430 | 0.3844 |
| | | 0.6156 | 0.6570 | 0.7117 | 0.7418 | 0.7920 | 0.8162 | 0.8527 |
| 20 | 0.95 | 0.1668 | 0.2089 | 0.2442 | 0.2927 | 0.3199 | 0.3722 | 0.4235 |
| | | 0.5765 | 0.6278 | 0.6801 | 0.7073 | 0.7558 | 0.7911 | 0.8332 |
| | 0.90 | 0.2010 | 0.2206 | 0.2750 | 0.3249 | 0.3576 | 0.3995 | 0.4498 |
| | | 0.5502 | 0.6005 | 0.6424 | 0.6751 | 0.7250 | 0.7794 | 0.7990 |
| | 0.80 | 0.2310 | 0.2614 | 0.3031 | 0.3527 | 0.4020 | 0.4510 | 0.5000 |
| | | 0.5000 | 0.5490 | 0.5980 | 0.6473 | 0.6969 | 0.7386 | 0.7690 |

| $n$ | $\beta$ | 0 | 1 | 2 | 3 | 4 | 5 | 6 |
|---|---|---|---|---|---|---|---|---|
| | 0.99 | 0.0000 | 0.0004 | 0.0072 | 0.0215 | 0.0414 | 0.0653 | 0.0924 |
| | | 0.2181 | 0.2957 | 0.3714 | 0.4234 | 0.4746 | 0.5255 | 0.5767 |
| | 0.98 | 0.0000 | 0.0009 | 0.0104 | 0.0279 | 0.0508 | 0.0776 | 0.1073 |
| | | 0.1980 | 0.2744 | 0.3304 | 0.3998 | 0.4501 | 0.5000 | 0.5500 |
| 21 | 0.95 | 0.0000 | 0.0024 | 0.0171 | 0.0400 | 0.0678 | 0.0988 | 0.1324 |
| | | 0.1590 | 0.2327 | 0.3047 | 0.3544 | 0.4033 | 0.4553 | 0.5061 |
| | 0.90 | 0.0000 | 0.0050 | 0.0256 | 0.0536 | 0.0857 | 0.1206 | 0.1303 |
| | | 0.1233 | 0.2123 | 0.2622 | 0.3324 | 0.3807 | 0.4286 | 0.4763 |
| | 0.80 | 0.0000 | 0.0105 | 0.0394 | 0.0739 | 0.1020 | 0.1387 | 0.1720 |
| | | 0.1021 | 0.1721 | 0.2205 | 0.2892 | 0.3364 | 0.3834 | 0.4301 |
| | 0.99 | 0.0000 | 0.0004 | 0.0068 | 0.0205 | 0.0394 | 0.0621 | 0.0878 |
| | | 0.2079 | 0.2819 | 0.3538 | 0.4032 | 0.4537 | 0.5046 | 0.5498 |
| | 0.98 | 0.0000 | 0.0009 | 0.0099 | 0.0266 | 0.0484 | 0.0738 | 0.1020 |
| | | 0.1889 | 0.2617 | 0.3324 | 0.3811 | 0.4290 | 0.4764 | 0.5237 |
| 22 | 0.95 | 0.0000 | 0.0023 | 0.0163 | 0.0382 | 0.0645 | 0.0941 | 0.1260 |
| | | 0.1518 | 0.2222 | 0.2908 | 0.3383 | 0.3890 | 0.4535 | 0.5000 |
| | 0.90 | 0.0000 | 0.0047 | 0.0244 | 0.0511 | 0.0817 | 0.1149 | 0.1335 |
| | | 0.1336 | 0.2028 | 0.2505 | 0.3174 | 0.3635 | 0.4092 | 0.4547 |
| | 0.80 | 0.0000 | 0.0100 | 0.0376 | 0.0705 | 0.0975 | 0.1320 | 0.1644 |
| | | 0.0976 | 0.1645 | 0.2128 | 0.2763 | 0.3215 | 0.3664 | 0.4149 |
| | 0.99 | 0.0000 | 0.0004 | 0.0065 | 0.0196 | 0.0376 | 0.0592 | 0.0837 |
| | | 0.1987 | 0.2693 | 0.3378 | 0.3865 | 0.4534 | 0.5000 | 0.5467 |
| | 0.98 | 0.0000 | 0.0008 | 0.0094 | 0.0254 | 0.0462 | 0.0704 | 0.0973 |
| | | 0.1806 | 0.2501 | 0.3176 | 0.3641 | 0.4117 | 0.4674 | 0.5147 |
| 23 | 0.95 | 0.0000 | 0.0022 | 0.0156 | 0.0365 | 0.0616 | 0.0898 | 0.1202 |
| | | 0.1452 | 0.2126 | 0.2782 | 0.3236 | 0.3888 | 0.4335 | 0.4779 |
| | 0.90 | 0.0000 | 0.0045 | 0.0233 | 0.0489 | 0.0780 | 0.1097 | 0.1278 |
| | | 0.1279 | 0.1941 | 0.2398 | 0.3038 | 0.3479 | 0.3916 | 0.4351 |
| | 0.80 | 0.0000 | 0.0096 | 0.0359 | 0.0674 | 0.0934 | 0.1258 | 0.1574 |
| | | 0.0935 | 0.1575 | 0.2208 | 0.2646 | 0.3079 | 0.3508 | 0.4142 |
| | 0.99 | 0.0000 | 0.0004 | 0.0063 | 0.0187 | 0.0359 | 0.0566 | 0.0799 |
| | | 0.1902 | 0.2590 | 0.3233 | 0.3886 | 0.4335 | 0.4779 | 0.5222 |
| | 0.98 | 0.0000 | 0.0008 | 0.0090 | 0.0243 | 0.0442 | 0.0673 | 0.0929 |
| | | 0.1730 | 0.2396 | 0.3041 | 0.3487 | 0.4127 | 0.4565 | 0.5000 |
| 24 | 0.95 | 0.0000 | 0.0021 | 0.0150 | 0.0349 | 0.0590 | 0.0858 | 0.1149 |
| | | 0.1392 | 0.2038 | 0.2666 | 0.3101 | 0.3725 | 0.4152 | 0.4577 |
| | 0.90 | 0.0000 | 0.0043 | 0.0223 | 0.0468 | 0.0747 | 0.1049 | 0.1225 |
| | | 0.1226 | 0.1861 | 0.2299 | 0.2912 | 0.3335 | 0.3755 | 0.4228 |
| | 0.80 | 0.0000 | 0.0092 | 0.0344 | 0.0646 | 0.0896 | 0.1202 | 0.1510 |
| | | 0.0897 | 0.1511 | 0.2118 | 0.2538 | 0.2954 | 0.3432 | 0.3972 |

| $n$ | $\beta$ | 7 | 8 | 9 | 10 | 11 | 12 | 13 |
|---|---|---|---|---|---|---|---|---|
| | 0.99 | 0.1223 | 0.1546 | 0.1891 | 0.2180 | 0.2571 | 0.2956 | 0.3392 |
| | | 0.6287 | 0.6608 | 0.7044 | 0.7429 | 0.7820 | 0.8109 | 0.8454 |
| | 0.98 | 0.1396 | 0.1740 | 0.1979 | 0.2434 | 0.2743 | 0.3246 | 0.3557 |
| | | 0.6003 | 0.6443 | 0.6754 | 0.7257 | 0.7566 | 0.8021 | 0.8260 |
| 21 | 0.95 | 0.1589 | 0.1973 | 0.2326 | 0.2757 | 0.3046 | 0.3543 | 0.4032 |
| | | 0.5511 | 0.5968 | 0.6457 | 0.6954 | 0.7243 | 0.7674 | 0.8027 |
| | 0.90 | 0.1905 | 0.2122 | 0.2621 | 0.3060 | 0.3323 | 0.3806 | 0.4285 |
| | | 0.5238 | 0.5715 | 0.6194 | 0.6677 | 0.6940 | 0.7379 | 0.7878 |
| | 0.80 | 0.2204 | 0.2446 | 0.2891 | 0.3363 | 0.3833 | 0.4300 | 0.4766 |
| | | 0.4767 | 0.5234 | 0.5700 | 0.6167 | 0.6637 | 0.7109 | 0.7554 |
| | 0.99 | 0.1162 | 0.1467 | 0.1793 | 0.2078 | 0.2421 | 0.2818 | 0.3180 |
| | | 0.5969 | 0.6463 | 0.6820 | 0.7182 | 0.7579 | 0.7922 | 0.8207 |
| | 0.98 | 0.1326 | 0.1652 | 0.1888 | 0.2301 | 0.2616 | 0.3060 | 0.3323 |
| | | 0.5711 | 0.6190 | 0.6677 | 0.6940 | 0.7384 | 0.7699 | 0.8112 |
| 22 | 0.95 | 0.1517 | 0.1869 | 0.2221 | 0.2604 | 0.2907 | 0.3382 | 0.3825 |
| | | 0.5466 | 0.5824 | 0.6175 | 0.6618 | 0.7093 | 0.7396 | 0.7779 |
| | 0.90 | 0.1811 | 0.2027 | 0.2504 | 0.2889 | 0.3173 | 0.3634 | 0.4091 |
| | | 0.5000 | 0.5454 | 0.5909 | 0.6366 | 0.6827 | 0.7111 | 0.7496 |
| | 0.80 | 0.2101 | 0.2281 | 0.2762 | 0.3214 | 0.3663 | 0.4084 | 0.4453 |
| | | 0.4709 | 0.5175 | 0.5547 | 0.5916 | 0.6337 | 0.6786 | 0.7238 |
| | 0.99 | 0.1106 | 0.1396 | 0.1705 | 0.1986 | 0.2286 | 0.2692 | 0.2983 |
| | | 0.5802 | 0.6160 | 0.6623 | 0.7017 | 0.7308 | 0.7714 | 0.8014 |
| | 0.98 | 0.1263 | 0.1573 | 0.1805 | 0.2181 | 0.2500 | 0.2891 | 0.3175 |
| | | 0.5539 | 0.5916 | 0.6360 | 0.6825 | 0.7109 | 0.7500 | 0.7819 |
| 23 | 0.95 | 0.1451 | 0.1776 | 0.2125 | 0.2465 | 0.2781 | 0.3235 | 0.3601 |
| | | 0.5222 | 0.5666 | 0.6113 | 0.6399 | 0.6765 | 0.7219 | 0.7535 |
| | 0.90 | 0.1725 | 0.1940 | 0.2397 | 0.2733 | 0.3037 | 0.3478 | 0.3915 |
| | | 0.4786 | 0.5219 | 0.5650 | 0.6085 | 0.6522 | 0.6963 | 0.7267 |
| | 0.80 | 0.2001 | 0.2207 | 0.2645 | 0.3078 | 0.3507 | 0.3839 | 0.4141 |
| | | 0.4572 | 0.5000 | 0.5429 | 0.5859 | 0.6161 | 0.6493 | 0.6922 |
| | 0.99 | 0.1056 | 0.1332 | 0.1625 | 0.1901 | 0.2163 | 0.2573 | 0.2795 |
| | | 0.5666 | 0.6115 | 0.6380 | 0.6768 | 0.7205 | 0.7427 | 0.7837 |
| | 0.98 | 0.1206 | 0.1501 | 0.1729 | 0.2073 | 0.2395 | 0.2738 | 0.3040 |
| | | 0.5436 | 0.5874 | 0.6156 | 0.6514 | 0.6960 | 0.7262 | 0.7605 |
| 24 | 0.95 | 0.1391 | 0.1690 | 0.2037 | 0.2339 | 0.2665 | 0.3100 | 0.3388 |
| | | 0.5000 | 0.5424 | 0.5849 | 0.6276 | 0.6612 | 0.6900 | 0.7335 |
| | 0.90 | 0.1647 | 0.1860 | 0.2298 | 0.2590 | 0.2911 | 0.3334 | 0.3754 |
| | | 0.4791 | 0.5210 | 0.5527 | 0.5867 | 0.6246 | 0.6666 | 0.7089 |
| | 0.80 | 0.1910 | 0.2117 | 0.2537 | 0.2953 | 0.3338 | 0.3584 | 0.3971 |
| | | 0.4384 | 0.4795 | 0.5206 | 0.5617 | 0.6029 | 0.6416 | 0.6662 |

| $n$ | $\beta$ | 0 | 1 | 2 | 3 | 4 | 5 | 6 |
|---|---|---|---|---|---|---|---|---|
| | 0.99 | 0.0000 | 0.0004 | 0.0060 | 0.0180 | 0.0344 | 0.0542 | 0.0765 |
| | | 0.1825 | 0.2650 | 0.3100 | 0.3724 | 0.4153 | 0.4578 | 0.5000 |
| | 0.98 | 0.0000 | 0.0008 | 0.0087 | 0.0233 | 0.0423 | 0.0644 | 0.0889 |
| | | 0.1660 | 0.2299 | 0.2917 | 0.3361 | 0.3957 | 0.4376 | 0.4793 |
| 25 | 0.95 | 0.0000 | 0.0020 | 0.0144 | 0.0335 | 0.0565 | 0.0822 | 0.1100 |
| | | 0.1337 | 0.1957 | 0.2560 | 0.3032 | 0.3575 | 0.3985 | 0.4392 |
| | 0.90 | 0.0000 | 0.0042 | 0.0214 | 0.0449 | 0.0716 | 0.1006 | 0.1177 |
| | | 0.1178 | 0.1788 | 0.2209 | 0.2797 | 0.3204 | 0.3616 | 0.4198 |
| | 0.80 | 0.0000 | 0.0088 | 0.0330 | 0.0620 | 0.0861 | 0.1150 | 0.1451 |
| | | 0.0862 | 0.1452 | 0.2035 | 0.2439 | 0.2838 | 0.3420 | 0.3816 |
| | 0.99 | 0.0000 | 0.0003 | 0.0058 | 0.0173 | 0.0330 | 0.0520 | 0.0733 |
| | | 0.1753 | 0.2545 | 0.2977 | 0.3575 | 0.3987 | 0.4417 | 0.4866 |
| | 0.98 | 0.0000 | 0.0007 | 0.0083 | 0.0224 | 0.0406 | 0.0618 | 0.0853 |
| | | 0.1596 | 0.2210 | 0.2803 | 0.3394 | 0.3801 | 0.4203 | 0.4602 |
| 26 | 0.95 | 0.0000 | 0.0019 | 0.0138 | 0.0321 | 0.0543 | 0.0789 | 0.1055 |
| | | 0.1286 | 0.1882 | 0.2461 | 0.3038 | 0.3437 | 0.3831 | 0.4222 |
| | 0.90 | 0.0000 | 0.0040 | 0.0206 | 0.0431 | 0.0688 | 0.0966 | 0.1132 |
| | | 0.1133 | 0.1720 | 0.2125 | 0.2691 | 0.3082 | 0.3649 | 0.4037 |
| | 0.80 | 0.0000 | 0.0085 | 0.0318 | 0.0595 | 0.0829 | 0.1103 | 0.1397 |
| | | 0.0830 | 0.1398 | 0.1958 | 0.2348 | 0.2732 | 0.3290 | 0.3672 |
| | 0.99 | 0.0000 | 0.0003 | 0.0055 | 0.0166 | 0.0318 | 0.0499 | 0.0704 |
| | | 0.1688 | 0.2449 | 0.2973 | 0.3439 | 0.3845 | 0.4411 | 0.4804 |
| | 0.98 | 0.0000 | 0.0007 | 0.0080 | 0.0215 | 0.0391 | 0.0594 | 0.0819 |
| | | 0.1536 | 0.2127 | 0.2698 | 0.3265 | 0.3657 | 0.4044 | 0.4462 |
| 27 | 0.95 | 0.0000 | 0.0018 | 0.0133 | 0.0309 | 0.0522 | 0.0759 | 0.1014 |
| | | 0.1239 | 0.1813 | 0.2371 | 0.2925 | 0.3309 | 0.3689 | 0.4148 |
| | 0.90 | 0.0000 | 0.0038 | 0.0198 | 0.0415 | 0.0662 | 0.0929 | 0.1091 |
| | | 0.1092 | 0.1657 | 0.2048 | 0.2592 | 0.2969 | 0.3515 | 0.3888 |
| | 0.80 | 0.0000 | 0.0082 | 0.0306 | 0.0573 | 0.0799 | 0.1059 | 0.1346 |
| | | 0.0800 | 0.1347 | 0.1887 | 0.2263 | 0.2633 | 0.3171 | 0.3539 |
| | 0.99 | 0.0000 | 0.0003 | 0.0053 | 0.0160 | 0.0306 | 0.0480 | 0.0678 |
| | | 0.1627 | 0.2359 | 0.2922 | 0.3312 | 0.3866 | 0.4247 | 0.4624 |
| | 0.98 | 0.0000 | 0.0007 | 0.0077 | 0.0207 | 0.0376 | 0.0572 | 0.0788 |
| | | 0.1481 | 0.2051 | 0.2601 | 0.3147 | 0.3524 | 0.3932 | 0.4443 |
| 28 | 0.95 | 0.0000 | 0.0018 | 0.0128 | 0.0298 | 0.0503 | 0.0731 | 0.0976 |
| | | 0.1195 | 0.1749 | 0.2286 | 0.2821 | 0.3191 | 0.3572 | 0.4092 |
| | 0.90 | 0.0000 | 0.0037 | 0.0191 | 0.0400 | 0.0637 | 0.0894 | 0.1052 |
| | | 0.1053 | 0.1599 | 0.2010 | 0.2501 | 0.2864 | 0.3390 | 0.3750 |
| | 0.80 | 0.0000 | 0.0079 | 0.0295 | 0.0553 | 0.0771 | 0.1019 | 0.1299 |
| | | 0.0772 | 0.1300 | 0.1821 | 0.2183 | 0.2546 | 0.3060 | 0.3415 |

| $n$ | $\beta$ | 7 | 8 | 9 | 10 | 11 | 12 | 13 |
|---|---|---|---|---|---|---|---|---|
| | 0.99 | 0.1010 | 0.1273 | 0.1552 | 0.1824 | 0.2051 | 0.2451 | 0.2649 |
| | | 0.5423 | 0.5848 | 0.6277 | 0.6584 | 0.6901 | 0.7351 | 0.7549 |
| | 0.98 | 0.1154 | 0.1436 | 0.1659 | 0.1975 | 0.2298 | 0.2598 | 0.2916 |
| | | 0.5208 | 0.5625 | 0.6044 | 0.6389 | 0.6664 | 0.7084 | 0.7402 |
| 25 | 0.95 | 0.1336 | 0.1613 | 0.1956 | 0.2223 | 0.2559 | 0.2958 | 0.3170 |
| | | 0.4798 | 0.5203 | 0.5609 | 0.6016 | 0.6426 | 0.6830 | 0.7042 |
| | 0.90 | 0.1575 | 0.1787 | 0.2208 | 0.2457 | 0.2796 | 0.3203 | 0.3600 |
| | | 0.4599 | 0.5000 | 0.5402 | 0.5803 | 0.6107 | 0.6400 | 0.6797 |
| | 0.80 | 0.1826 | 0.2034 | 0.2438 | 0.2837 | 0.3171 | 0.3419 | 0.3815 |
| | | 0.4212 | 0.4606 | 0.5000 | 0.5395 | 0.5789 | 0.6185 | 0.6581 |
| | 0.99 | 0.0968 | 0.1219 | 0.1486 | 0.1752 | 0.1946 | 0.2339 | 0.2544 |
| | | 0.5263 | 0.5624 | 0.6014 | 0.6426 | 0.6779 | 0.7024 | 0.7456 |
| | 0.98 | 0.1106 | 0.1376 | 0.1595 | 0.1885 | 0.2209 | 0.2469 | 0.2802 |
| | | 0.5000 | 0.5399 | 0.5798 | 0.6200 | 0.6607 | 0.6827 | 0.7198 |
| 26 | 0.95 | 0.1285 | 0.1541 | 0.1881 | 0.2116 | 0.2460 | 0.2820 | 0.3037 |
| | | 0.4651 | 0.5056 | 0.5422 | 0.5779 | 0.6170 | 0.6564 | 0.6963 |
| | 0.90 | 0.1510 | 0.1719 | 0.2124 | 0.2332 | 0.2690 | 0.3081 | 0.3423 |
| | | 0.4423 | 0.4808 | 0.5193 | 0.5578 | 0.5964 | 0.6352 | 0.6577 |
| | 0.80 | 0.1750 | 0.1957 | 0.2347 | 0.2731 | 0.3015 | 0.3289 | 0.3671 |
| | | 0.4053 | 0.4432 | 0.4811 | 0.5190 | 0.5569 | 0.5948 | 0.6329 |
| | 0.99 | 0.0929 | 0.1170 | 0.1425 | 0.1687 | 0.1848 | 0.2237 | 0.2448 |
| | | 0.5197 | 0.5590 | 0.5872 | 0.6173 | 0.6562 | 0.7019 | 0.7160 |
| | 0.98 | 0.1062 | 0.1320 | 0.1535 | 0.1802 | 0.2126 | 0.2349 | 0.2697 |
| | | 0.4885 | 0.5258 | 0.5600 | 0.5957 | 0.6344 | 0.6736 | 0.6978 |
| 27 | 0.95 | 0.1238 | 0.1476 | 0.1812 | 0.2017 | 0.2370 | 0.2693 | 0.2924 |
| | | 0.4623 | 0.5000 | 0.5378 | 0.5705 | 0.5982 | 0.6312 | 0.6692 |
| | 0.90 | 0.1450 | 0.1656 | 0.2047 | 0.2212 | 0.2591 | 0.2968 | 0.3257 |
| | | 0.4260 | 0.4630 | 0.5000 | 0.5371 | 0.5741 | 0.6113 | 0.6486 |
| | 0.80 | 0.1679 | 0.1886 | 0.2262 | 0.2632 | 0.2866 | 0.3170 | 0.3538 |
| | | 0.3905 | 0.4271 | 0.4636 | 0.5000 | 0.5365 | 0.5730 | 0.6096 |
| | 0.99 | 0.0893 | 0.1124 | 0.1369 | 0.1626 | 0.1753 | 0.2143 | 0.2358 |
| | | 0.5000 | 0.5377 | 0.5754 | 0.6135 | 0.6360 | 0.6689 | 0.7079 |
| | 0.98 | 0.1022 | 0.1269 | 0.1480 | 0.1726 | 0.2050 | 0.2235 | 0.2600 |
| | | 0.4815 | 0.5186 | 0.5558 | 0.5845 | 0.6125 | 0.6477 | 0.6854 |
| 28 | 0.95 | 0.1194 | 0.1416 | 0.1748 | 0.1924 | 0.2285 | 0.2575 | 0.2820 |
| | | 0.4456 | 0.4819 | 0.5182 | 0.5545 | 0.5909 | 0.6195 | 0.6452 |
| | 0.90 | 0.1394 | 0.1598 | 0.1968 | 0.2084 | 0.2500 | 0.2863 | 0.3098 |
| | | 0.4108 | 0.4466 | 0.4822 | 0.5179 | 0.5535 | 0.5893 | 0.6251 |
| | 0.80 | 0.1614 | 0.1820 | 0.2182 | 0.2538 | 0.2720 | 0.3059 | 0.3414 |
| | | 0.3768 | 0.4121 | 0.4473 | 0.4825 | 0.5176 | 0.5528 | 0.5880 |

| $n$ | $\beta$ | 14 | 15 | 16 | 17 | 18 | 19 | 20 |
|---|---|---|---|---|---|---|---|---|
| | 0.99 | 0.3099 | 0.3416 | 0.3723 | 0.4152 | 0.4577 | 0.5000 | 0.5422 |
| | | 0.7949 | 0.8176 | 0.8448 | 0.8727 | 0.8990 | 0.9235 | 0.9458 |
| | 0.98 | 0.3336 | 0.3611 | 0.3956 | 0.4375 | 0.4792 | 0.5207 | 0.5624 |
| | | 0.7702 | 0.8025 | 0.8341 | 0.8564 | 0.8846 | 0.9111 | 0.9356 |
| 25 | 0.95 | 0.3574 | 0.3984 | 0.4391 | 0.4797 | 0.5202 | 0.5608 | 0.6015 |
| | | 0.7441 | 0.7777 | 0.8044 | 0.8387 | 0.8664 | 0.8900 | 0.9178 |
| | 0.90 | 0.3893 | 0.4197 | 0.4598 | 0.5000 | 0.5401 | 0.5802 | 0.6384 |
| | | 0.7204 | 0.7543 | 0.7792 | 0.8213 | 0.8425 | 0.8823 | 0.8994 |
| | 0.80 | 0.4211 | 0.4605 | 0.5000 | 0.5394 | 0.5788 | 0.6184 | 0.6580 |
| | | 0.6829 | 0.7163 | 0.7562 | 0.7966 | 0.8174 | 0.8549 | 0.8850 |
| | 0.99 | 0.2976 | 0.3221 | 0.3574 | 0.3986 | 0.4376 | 0.4737 | 0.5134 |
| | | 0.7661 | 0.8054 | 0.8248 | 0.8514 | 0.8781 | 0.9032 | 0.9267 |
| | 0.98 | 0.3173 | 0.3393 | 0.3800 | 0.4202 | 0.4601 | 0.5000 | 0.5398 |
| | | 0.7531 | 0.7791 | 0.8115 | 0.8405 | 0.8624 | 0.8894 | 0.9147 |
| 26 | 0.95 | 0.3436 | 0.3830 | 0.4221 | 0.4578 | 0.4944 | 0.5349 | 0.5778 |
| | | 0.7180 | 0.7540 | 0.7884 | 0.8119 | 0.8459 | 0.8715 | 0.8945 |
| | 0.90 | 0.3648 | 0.4036 | 0.4422 | 0.4807 | 0.5192 | 0.5577 | 0.5963 |
| | | 0.6919 | 0.7310 | 0.7668 | 0.7876 | 0.8281 | 0.8490 | 0.8868 |
| | 0.80 | 0.4052 | 0.4431 | 0.4810 | 0.5189 | 0.5568 | 0.5947 | 0.6328 |
| | | 0.6711 | 0.6985 | 0.7269 | 0.7653 | 0.8043 | 0.8250 | 0.8603 |
| | 0.99 | 0.2840 | 0.2981 | 0.3438 | 0.3827 | 0.4128 | 0.4410 | 0.4803 |
| | | 0.7552 | 0.7763 | 0.8152 | 0.8313 | 0.8575 | 0.8830 | 0.9071 |
| | 0.98 | 0.3022 | 0.3264 | 0.3656 | 0.4043 | 0.4400 | 0.4742 | 0.5115 |
| | | 0.7303 | 0.7651 | 0.7874 | 0.8198 | 0.8465 | 0.8680 | 0.8938 |
| 27 | 0.95 | 0.3308 | 0.3688 | 0.4018 | 0.4295 | 0.4622 | 0.5000 | 0.5377 |
| | | 0.7076 | 0.7307 | 0.7630 | 0.7983 | 0.8188 | 0.8524 | 0.8762 |
| | 0.90 | 0.3514 | 0.3887 | 0.4259 | 0.4629 | 0.5000 | 0.5370 | 0.5740 |
| | | 0.6743 | 0.7032 | 0.7409 | 0.7788 | 0.7953 | 0.8344 | 0.8550 |
| | 0.80 | 0.3904 | 0.4270 | 0.4635 | 0.5000 | 0.5364 | 0.5729 | 0.6095 |
| | | 0.6462 | 0.6830 | 0.7134 | 0.7368 | 0.7738 | 0.8114 | 0.8321 |
| | 0.99 | 0.2715 | 0.2921 | 0.3311 | 0.3640 | 0.3865 | 0.4246 | 0.4623 |
| | | 0.7285 | 0.7642 | 0.7857 | 0.8247 | 0.8374 | 0.8631 | 0.8876 |
| | 0.98 | 0.2882 | 0.3146 | 0.3523 | 0.3875 | 0.4155 | 0.4442 | 0.4814 |
| | | 0.7118 | 0.7400 | 0.7765 | 0.7950 | 0.8274 | 0.8520 | 0.8731 |
| 28 | 0.95 | 0.3190 | 0.3548 | 0.3805 | 0.4091 | 0.4455 | 0.4818 | 0.5181 |
| | | 0.6810 | 0.7180 | 0.7425 | 0.7715 | 0.8076 | 0.8252 | 0.8584 |
| | 0.90 | 0.3389 | 0.3749 | 0.4107 | 0.4465 | 0.4821 | 0.5178 | 0.5534 |
| | | 0.6611 | 0.6902 | 0.7137 | 0.7500 | 0.7916 | 0.8032 | 0.8402 |
| | 0.80 | 0.3767 | 0.4120 | 0.4472 | 0.4824 | 0.5175 | 0.5527 | 0.5879 |
| | | 0.6233 | 0.6586 | 0.6941 | 0.7280 | 0.7462 | 0.7818 | 0.8180 |

| $n$ | $\beta$ | 0 | 1 | 2 | 3 | 4 | 5 | 6 |
|---|---|---|---|---|---|---|---|---|
| 29 | 0.99 | 0.0000 | 0.0003 | 0.0051 | 0.0154 | 0.0295 | 0.0463 | 0.0653 |
| | | 0.1600 | 0.2276 | 0.2819 | 0.3195 | 0.3728 | 0.4095 | 0.4458 |
| | 0.98 | 0.0000 | 0.0006 | 0.0075 | 0.0200 | 0.0363 | 0.0551 | 0.0760 |
| | | 0.1430 | 0.1980 | 0.2510 | 0.3036 | 0.3401 | 0.3926 | 0.4286 |
| | 0.95 | 0.0000 | 0.0017 | 0.0123 | 0.0287 | 0.0485 | 0.0704 | 0.0941 |
| | | 0.1154 | 0.1689 | 0.2208 | 0.2724 | 0.3081 | 0.3596 | 0.3950 |
| | 0.90 | 0.0000 | 0.0036 | 0.0184 | 0.0386 | 0.0615 | 0.0863 | 0.1016 |
| | | 0.1017 | 0.1544 | 0.2058 | 0.2416 | 0.2792 | 0.3274 | 0.3622 |
| | 0.80 | 0.0000 | 0.0076 | 0.0285 | 0.0533 | 0.0745 | 0.0982 | 0.1255 |
| | | 0.0746 | 0.1256 | 0.1760 | 0.2110 | 0.2611 | 0.2956 | 0.3299 |
| 30 | 0.99 | 0.0000 | 0.0003 | 0.0050 | 0.0149 | 0.0284 | 0.0447 | 0.0630 |
| | | 0.1653 | 0.2199 | 0.2723 | 0.3103 | 0.3600 | 0.3954 | 0.4304 |
| | 0.98 | 0.0000 | 0.0006 | 0.0072 | 0.0193 | 0.0350 | 0.0532 | 0.0733 |
| | | 0.1382 | 0.1963 | 0.2426 | 0.2934 | 0.3294 | 0.3792 | 0.4140 |
| | 0.95 | 0.0000 | 0.0017 | 0.0119 | 0.0278 | 0.0468 | 0.0680 | 0.0908 |
| | | 0.1116 | 0.1633 | 0.2135 | 0.2633 | 0.2979 | 0.3476 | 0.3817 |
| | 0.90 | 0.0000 | 0.0035 | 0.0178 | 0.0373 | 0.0594 | 0.0833 | 0.0983 |
| | | 0.0984 | 0.1493 | 0.1990 | 0.2336 | 0.2827 | 0.3166 | 0.3502 |
| | 0.80 | 0.0000 | 0.0074 | 0.0275 | 0.0515 | 0.0721 | 0.0947 | 0.1214 |
| | | 0.0722 | 0.1215 | 0.1702 | 0.2041 | 0.2525 | 0.2859 | 0.3191 |
| 31 | 0.99 | 0.0000 | 0.0003 | 0.0048 | 0.0144 | 0.0275 | 0.0432 | 0.0608 |
| | | 0.1599 | 0.2127 | 0.2633 | 0.3134 | 0.3481 | 0.3823 | 0.4255 |
| | 0.98 | 0.0000 | 0.0006 | 0.0070 | 0.0187 | 0.0338 | 0.0514 | 0.0708 |
| | | 0.1337 | 0.1989 | 0.2350 | 0.2838 | 0.3329 | 0.3668 | 0.4003 |
| | 0.95 | 0.0000 | 0.0016 | 0.0115 | 0.0269 | 0.0453 | 0.0657 | 0.0878 |
| | | 0.1080 | 0.1607 | 0.2066 | 0.2548 | 0.2883 | 0.3364 | 0.3694 |
| | 0.90 | 0.0000 | 0.0033 | 0.0172 | 0.0360 | 0.0574 | 0.0805 | 0.0952 |
| | | 0.0953 | 0.1446 | 0.1926 | 0.2262 | 0.2736 | 0.3065 | 0.3390 |
| | 0.80 | 0.0000 | 0.0071 | 0.0266 | 0.0499 | 0.0698 | 0.0915 | 0.1176 |
| | | 0.0699 | 0.1177 | 0.1648 | 0.1976 | 0.2445 | 0.2769 | 0.3090 |
| 32 | 0.99 | 0.0000 | 0.0003 | 0.0047 | 0.0139 | 0.0266 | 0.0417 | 0.0588 |
| | | 0.1548 | 0.2059 | 0.2549 | 0.3033 | 0.3369 | 0.3742 | 0.4182 |
| | 0.98 | 0.0000 | 0.0006 | 0.0067 | 0.0181 | 0.0327 | 0.0497 | 0.0685 |
| | | 0.1295 | 0.1926 | 0.2411 | 0.2749 | 0.3223 | 0.3551 | 0.3876 |
| | 0.95 | 0.0000 | 0.0016 | 0.0112 | 0.0260 | 0.0438 | 0.0636 | 0.0849 |
| | | 0.1047 | 0.1662 | 0.2002 | 0.2469 | 0.2815 | 0.3258 | 0.3578 |
| | 0.90 | 0.0000 | 0.0032 | 0.0167 | 0.0349 | 0.0556 | 0.0780 | 0.0922 |
| | | 0.0923 | 0.1401 | 0.1867 | 0.2192 | 0.2652 | 0.2970 | 0.3285 |
| | 0.80 | 0.0000 | 0.0069 | 0.0258 | 0.0483 | 0.0677 | 0.0884 | 0.1140 |
| | | 0.0678 | 0.1141 | 0.1598 | 0.1916 | 0.2370 | 0.2684 | 0.2995 |

| $n$ | $\beta$ | 7 | 8 | 9 | 10 | 11 | 12 | 13 |
|---|---|---|---|---|---|---|---|---|
| 29 | 0.99 | 0.0860 | 0.1082 | 0.1317 | 0.1565 | 0.1652 | 0.2057 | 0.2275 |
| | | 0.4820 | 0.5181 | 0.5543 | 0.5906 | 0.6273 | 0.6537 | 0.6806 |
| | 0.98 | 0.0984 | 0.1222 | 0.1429 | 0.1656 | 0.1979 | 0.2123 | 0.2509 |
| | | 0.4644 | 0.5000 | 0.5357 | 0.5715 | 0.6075 | 0.6310 | 0.6600 |
| | 0.95 | 0.1153 | 0.1360 | 0.1688 | 0.1836 | 0.2207 | 0.2465 | 0.2723 |
| | | 0.4301 | 0.4651 | 0.5000 | 0.5350 | 0.5700 | 0.6051 | 0.6405 |
| | 0.90 | 0.1342 | 0.1543 | 0.1893 | 0.2057 | 0.2415 | 0.2757 | 0.2936 |
| | | 0.3968 | 0.4313 | 0.4657 | 0.5000 | 0.5344 | 0.5688 | 0.6033 |
| | 0.80 | 0.1553 | 0.1759 | 0.2109 | 0.2438 | 0.2610 | 0.2955 | 0.3298 |
| | | 0.3641 | 0.3981 | 0.4321 | 0.4661 | 0.5006 | 0.5340 | 0.5680 |
| 30 | 0.99 | 0.0829 | 0.1043 | 0.1269 | 0.1507 | 0.1652 | 0.1977 | 0.2198 |
| | | 0.4691 | 0.5052 | 0.5380 | 0.5697 | 0.6047 | 0.6401 | 0.6713 |
| | 0.98 | 0.0949 | 0.1178 | 0.1381 | 0.1591 | 0.1905 | 0.1996 | 0.2425 |
| | | 0.4485 | 0.4829 | 0.5172 | 0.5516 | 0.5861 | 0.6209 | 0.6493 |
| | 0.95 | 0.1115 | 0.1308 | 0.1632 | 0.1750 | 0.2134 | 0.2362 | 0.2632 |
| | | 0.4157 | 0.4495 | 0.4832 | 0.5169 | 0.5506 | 0.5844 | 0.6184 |
| | 0.90 | 0.1294 | 0.1492 | 0.1824 | 0.1989 | 0.2335 | 0.2649 | 0.2826 |
| | | 0.3836 | 0.4170 | 0.4555 | 0.4921 | 0.5243 | 0.5539 | 0.5831 |
| | 0.80 | 0.1497 | 0.1701 | 0.2040 | 0.2345 | 0.2524 | 0.2858 | 0.3190 |
| | | 0.3521 | 0.3851 | 0.4205 | 0.4616 | 0.5000 | 0.5295 | 0.5563 |
| 31 | 0.99 | 0.0801 | 0.1007 | 0.1225 | 0.1454 | 0.1598 | 0.1903 | 0.2126 |
| | | 0.4662 | 0.5000 | 0.5339 | 0.5647 | 0.5891 | 0.6178 | 0.6520 |
| | 0.98 | 0.0917 | 0.1138 | 0.1336 | 0.1530 | 0.1836 | 0.1988 | 0.2346 |
| | | 0.4337 | 0.4669 | 0.5003 | 0.5332 | 0.5664 | 0.5998 | 0.6333 |
| | 0.95 | 0.1079 | 0.1260 | 0.1576 | 0.1658 | 0.2065 | 0.2266 | 0.2547 |
| | | 0.4022 | 0.4349 | 0.4675 | 0.5003 | 0.5326 | 0.5652 | 0.5979 |
| | 0.90 | 0.1250 | 0.1445 | 0.1759 | 0.1925 | 0.2261 | 0.2548 | 0.2735 |
| | | 0.3714 | 0.4128 | 0.4516 | 0.4839 | 0.5162 | 0.5485 | 0.5789 |
| | 0.80 | 0.1445 | 0.1647 | 0.1975 | 0.2259 | 0.2444 | 0.2768 | 0.3089 |
| | | 0.3410 | 0.3730 | 0.4202 | 0.4522 | 0.4841 | 0.5160 | 0.5479 |
| 32 | 0.99 | 0.0774 | 0.0973 | 0.1183 | 0.1404 | 0.1547 | 0.1834 | 0.2058 |
| | | 0.4510 | 0.4837 | 0.5164 | 0.5491 | 0.5819 | 0.6097 | 0.6322 |
| | 0.98 | 0.0886 | 0.1100 | 0.1294 | 0.1474 | 0.1772 | 0.1925 | 0.2261 |
| | | 0.4219 | 0.4615 | 0.5000 | 0.5279 | 0.5546 | 0.5816 | 0.6125 |
| | 0.95 | 0.1046 | 0.1215 | 0.1522 | 0.1661 | 0.2001 | 0.2174 | 0.2468 |
| | | 0.3896 | 0.4234 | 0.4623 | 0.5000 | 0.5274 | 0.5538 | 0.5804 |
| | 0.90 | 0.1208 | 0.1400 | 0.1699 | 0.1866 | 0.2191 | 0.2453 | 0.2651 |
| | | 0.3625 | 0.4061 | 0.4375 | 0.4688 | 0.5000 | 0.5313 | 0.5626 |
| | 0.80 | 0.1396 | 0.1597 | 0.1915 | 0.2178 | 0.2369 | 0.2683 | 0.2994 |
| | | 0.3305 | 0.3763 | 0.4073 | 0.4382 | 0.4691 | 0.5000 | 0.5310 |

| $n$ | $\beta$ | 14 | 15 | 16 | 17 | 18 | 19 | 20 |
|---|---|---|---|---|---|---|---|---|
| | 0.99 | 0.2600 | 0.2818 | 0.3194 | 0.3463 | 0.3727 | 0.4094 | 0.4457 |
| | | 0.7182 | 0.7400 | 0.7725 | 0.7943 | 0.8348 | 0.8435 | 0.8683 |
| | 0.98 | 0.2751 | 0.3035 | 0.3400 | 0.3690 | 0.3925 | 0.4285 | 0.4643 |
| | | 0.6965 | 0.7249 | 0.7491 | 0.7877 | 0.8021 | 0.8344 | 0.8571 |
| 29 | 0.95 | 0.3080 | 0.3391 | 0.3595 | 0.3949 | 0.4300 | 0.4650 | 0.5000 |
| | | 0.6609 | 0.6920 | 0.7277 | 0.7535 | 0.7793 | 0.8164 | 0.8312 |
| | 0.90 | 0.3273 | 0.3621 | 0.3967 | 0.4312 | 0.4656 | 0.5000 | 0.5343 |
| | | 0.6379 | 0.6727 | 0.7064 | 0.7243 | 0.7585 | 0.7943 | 0.8107 |
| | 0.80 | 0.3640 | 0.3980 | 0.4320 | 0.4660 | 0.4994 | 0.5339 | 0.5679 |
| | | 0.6020 | 0.6360 | 0.6702 | 0.7045 | 0.7390 | 0.7562 | 0.7891 |
| | 0.99 | 0.2493 | 0.2722 | 0.3079 | 0.3287 | 0.3599 | 0.3953 | 0.4303 |
| | | 0.6921 | 0.7278 | 0.7507 | 0.7802 | 0.8023 | 0.8348 | 0.8493 |
| | 0.98 | 0.2625 | 0.2933 | 0.3281 | 0.3507 | 0.3791 | 0.4139 | 0.4484 |
| | | 0.6719 | 0.7067 | 0.7375 | 0.7575 | 0.8004 | 0.8095 | 0.8409 |
| 30 | 0.95 | 0.2978 | 0.3241 | 0.3475 | 0.3816 | 0.4156 | 0.4494 | 0.4831 |
| | | 0.6525 | 0.6759 | 0.7022 | 0.7368 | 0.7638 | 0.7866 | 0.8250 |
| | 0.90 | 0.3165 | 0.3501 | 0.3835 | 0.4169 | 0.4461 | 0.4757 | 0.5079 |
| | | 0.6165 | 0.6499 | 0.6835 | 0.7174 | 0.7351 | 0.7665 | 0.8011 |
| | 0.80 | 0.3520 | 0.3850 | 0.4162 | 0.4437 | 0.4705 | 0.5000 | 0.5384 |
| | | 0.5838 | 0.6150 | 0.6480 | 0.6810 | 0.7142 | 0.7476 | 0.7655 |
| | 0.99 | 0.2394 | 0.2632 | 0.2954 | 0.3133 | 0.3480 | 0.3822 | 0.4109 |
| | | 0.6867 | 0.7046 | 0.7368 | 0.7606 | 0.7874 | 0.8097 | 0.8402 |
| | 0.98 | 0.2501 | 0.2837 | 0.3148 | 0.3328 | 0.3667 | 0.4002 | 0.4336 |
| | | 0.6672 | 0.6852 | 0.7163 | 0.7499 | 0.7654 | 0.8012 | 0.8164 |
| 31 | 0.95 | 0.2882 | 0.3097 | 0.3363 | 0.3693 | 0.4021 | 0.4348 | 0.4674 |
| | | 0.6307 | 0.6637 | 0.6903 | 0.7118 | 0.7453 | 0.7734 | 0.7935 |
| | 0.90 | 0.3064 | 0.3389 | 0.3713 | 0.3990 | 0.4211 | 0.4515 | 0.4838 |
| | | 0.6010 | 0.6287 | 0.6611 | 0.6936 | 0.7265 | 0.7452 | 0.7739 |
| | 0.80 | 0.3409 | 0.3727 | 0.3971 | 0.4201 | 0.4521 | 0.4840 | 0.5159 |
| | | 0.5799 | 0.6029 | 0.6273 | 0.6591 | 0.6911 | 0.7232 | 0.7556 |
| | 0.99 | 0.2302 | 0.2548 | 0.2838 | 0.3032 | 0.3368 | 0.3678 | 0.3903 |
| | | 0.6632 | 0.6968 | 0.7162 | 0.7452 | 0.7698 | 0.7942 | 0.8166 |
| | 0.98 | 0.2410 | 0.2748 | 0.3022 | 0.3222 | 0.3550 | 0.3875 | 0.4184 |
| | | 0.6450 | 0.6778 | 0.6978 | 0.7252 | 0.7590 | 0.7739 | 0.8075 |
| 32 | 0.95 | 0.2784 | 0.2950 | 0.3257 | 0.3577 | 0.3895 | 0.4196 | 0.4462 |
| | | 0.6105 | 0.6423 | 0.6743 | 0.7050 | 0.7216 | 0.7532 | 0.7826 |
| | 0.90 | 0.2969 | 0.3284 | 0.3584 | 0.3802 | 0.4060 | 0.4374 | 0.4687 |
| | | 0.5940 | 0.6198 | 0.6416 | 0.6716 | 0.7031 | 0.7349 | 0.7547 |
| | 0.80 | 0.3304 | 0.3577 | 0.3762 | 0.4072 | 0.4381 | 0.4690 | 0.5000 |
| | | 0.5619 | 0.5928 | 0.6238 | 0.6423 | 0.6696 | 0.7006 | 0.7317 |

| $n$ | $\beta$ | 0 | 1 | 2 | 3 | 4 | 5 | 6 |
|---|---|---|---|---|---|---|---|---|
| | 0.99 | 0.0000 | 0.0003 | 0.0045 | 0.0135 | 0.0258 | 0.0404 | 0.0569 |
| | | 0.1500 | 0.1996 | 0.2471 | 0.2939 | 0.3265 | 0.3731 | 0.4051 |
| | 0.98 | 0.0000 | 0.0006 | 0.0065 | 0.0175 | 0.0317 | 0.0482 | 0.0663 |
| | | 0.1256 | 0.1867 | 0.2337 | 0.2665 | 0.3124 | 0.3442 | 0.3757 |
| 33 | 0.95 | 0.0000 | 0.0015 | 0.0108 | 0.0252 | 0.0424 | 0.0616 | 0.0822 |
| | | 0.1015 | 0.1612 | 0.1942 | 0.2394 | 0.2846 | 0.3159 | 0.3469 |
| | 0.90 | 0.0000 | 0.0031 | 0.0162 | 0.0338 | 0.0539 | 0.0755 | 0.0895 |
| | | 0.0896 | 0.1359 | 0.1811 | 0.2126 | 0.2572 | 0.2880 | 0.3186 |
| | 0.80 | 0.0000 | 0.0067 | 0.0250 | 0.0468 | 0.0656 | 0.0856 | 0.1106 |
| | | 0.0657 | 0.1107 | 0.1550 | 0.1859 | 0.2300 | 0.2604 | 0.2906 |
| | 0.99 | 0.0000 | 0.0002 | 0.0044 | 0.0131 | 0.0250 | 0.0392 | 0.0551 |
| | | 0.1456 | 0.1936 | 0.2397 | 0.2851 | 0.3167 | 0.3618 | 0.3928 |
| | 0.98 | 0.0000 | 0.0005 | 0.0063 | 0.0170 | 0.0308 | 0.0467 | 0.0642 |
| | | 0.1219 | 0.1812 | 0.2268 | 0.2586 | 0.3031 | 0.3340 | 0.3724 |
| 34 | 0.95 | 0.0000 | 0.0015 | 0.0105 | 0.0244 | 0.0411 | 0.0597 | 0.0797 |
| | | 0.0985 | 0.1565 | 0.1885 | 0.2324 | 0.2763 | 0.3066 | 0.3367 |
| | 0.90 | 0.0000 | 0.0030 | 0.0157 | 0.0328 | 0.0523 | 0.0732 | 0.0869 |
| | | 0.0870 | 0.1320 | 0.1758 | 0.2064 | 0.2497 | 0.2797 | 0.3094 |
| | 0.80 | 0.0000 | 0.0065 | 0.0243 | 0.0454 | 0.0637 | 0.0829 | 0.1074 |
| | | 0.0638 | 0.1075 | 0.1505 | 0.1805 | 0.2233 | 0.2529 | 0.2822 |
| | 0.99 | 0.0000 | 0.0002 | 0.0042 | 0.0127 | 0.0242 | 0.0380 | 0.0535 |
| | | 0.1414 | 0.1880 | 0.2327 | 0.2768 | 0.3112 | 0.3512 | 0.3813 |
| | 0.98 | 0.0000 | 0.0005 | 0.0062 | 0.0165 | 0.0298 | 0.0453 | 0.0623 |
| | | 0.1184 | 0.1760 | 0.2202 | 0.2511 | 0.2943 | 0.3243 | 0.3675 |
| 35 | 0.95 | 0.0000 | 0.0014 | 0.0102 | 0.0237 | 0.0399 | 0.0580 | 0.0773 |
| | | 0.0958 | 0.1520 | 0.1877 | 0.2258 | 0.2684 | 0.2979 | 0.3271 |
| | 0.90 | 0.0000 | 0.0030 | 0.0152 | 0.0319 | 0.0507 | 0.0711 | 0.0844 |
| | | 0.0845 | 0.1283 | 0.1709 | 0.2006 | 0.2426 | 0.2717 | 0.3020 |
| | 0.80 | 0.0000 | 0.0063 | 0.0236 | 0.0441 | 0.0620 | 0.0804 | 0.1044 |
| | | 0.0621 | 0.1045 | 0.1463 | 0.1755 | 0.2170 | 0.2457 | 0.2743 |
| | 0.99 | 0.0000 | 0.0002 | 0.0041 | 0.0123 | 0.0235 | 0.0369 | 0.0519 |
| | | 0.1374 | 0.1828 | 0.2262 | 0.2690 | 0.3116 | 0.3412 | 0.3704 |
| | 0.98 | 0.0000 | 0.0005 | 0.0060 | 0.0160 | 0.0290 | 0.0440 | 0.0605 |
| | | 0.1151 | 0.1710 | 0.2141 | 0.2453 | 0.2861 | 0.3156 | 0.3572 |
| 36 | 0.95 | 0.0000 | 0.0014 | 0.0099 | 0.0230 | 0.0388 | 0.0563 | 0.0751 |
| | | 0.0931 | 0.1478 | 0.1900 | 0.2195 | 0.2609 | 0.2896 | 0.3203 |
| | 0.90 | 0.0000 | 0.0029 | 0.0148 | 0.0310 | 0.0493 | 0.0691 | 0.0821 |
| | | 0.0822 | 0.1247 | 0.1662 | 0.1951 | 0.2360 | 0.2643 | 0.3052 |
| | 0.80 | 0.0000 | 0.0061 | 0.0229 | 0.0429 | 0.0603 | 0.0781 | 0.1015 |
| | | 0.0604 | 0.1016 | 0.1423 | 0.1707 | 0.2111 | 0.2390 | 0.2668 |

| $n$ | $\beta$ | 7 | 8 | 9 | 10 | 11 | 12 | 13 |
|---|---|---|---|---|---|---|---|---|
| | 0.99 | 0.0749 | 0.0941 | 0.1144 | 0.1357 | 0.1499 | 0.1770 | 0.1995 |
| | | 0.4368 | 0.4685 | 0.5000 | 0.5316 | 0.5633 | 0.5950 | 0.6270 |
| | 0.98 | 0.0858 | 0.1064 | 0.1255 | 0.1421 | 0.1712 | 0.1866 | 0.2182 |
| | | 0.4218 | 0.4531 | 0.4844 | 0.5157 | 0.5470 | 0.5783 | 0.6001 |
| 33 | 0.95 | 0.1014 | 0.1173 | 0.1473 | 0.1611 | 0.1941 | 0.2086 | 0.2393 |
| | | 0.3779 | 0.4232 | 0.4539 | 0.4847 | 0.5154 | 0.5462 | 0.5769 |
| | 0.90 | 0.1169 | 0.1358 | 0.1642 | 0.1810 | 0.2125 | 0.2364 | 0.2571 |
| | | 0.3634 | 0.3939 | 0.4243 | 0.4546 | 0.4849 | 0.5152 | 0.5455 |
| | 0.80 | 0.1350 | 0.1549 | 0.1858 | 0.2102 | 0.2299 | 0.2603 | 0.2905 |
| | | 0.3207 | 0.3650 | 0.3951 | 0.4251 | 0.4551 | 0.4851 | 0.5150 |
| | 0.99 | 0.0726 | 0.0912 | 0.1108 | 0.1314 | 0.1455 | 0.1710 | 0.1935 |
| | | 0.4236 | 0.4542 | 0.4856 | 0.5161 | 0.5459 | 0.5765 | 0.6073 |
| | 0.98 | 0.0831 | 0.1031 | 0.1218 | 0.1372 | 0.1656 | 0.1811 | 0.2108 |
| | | 0.4091 | 0.4395 | 0.4698 | 0.5000 | 0.5303 | 0.5606 | 0.5910 |
| 34 | 0.95 | 0.0984 | 0.1134 | 0.1426 | 0.1564 | 0.1884 | 0.1998 | 0.2323 |
| | | 0.3807 | 0.4106 | 0.4405 | 0.4703 | 0.5000 | 0.5298 | 0.5596 |
| | 0.90 | 0.1132 | 0.1319 | 0.1589 | 0.1757 | 0.2063 | 0.2281 | 0.2496 |
| | | 0.3527 | 0.3823 | 0.4118 | 0.4413 | 0.4707 | 0.5000 | 0.5294 |
| | 0.80 | 0.1307 | 0.1504 | 0.1804 | 0.2031 | 0.2232 | 0.2528 | 0.2821 |
| | | 0.3129 | 0.3544 | 0.3836 | 0.4128 | 0.4419 | 0.4710 | 0.5000 |
| | 0.99 | 0.0704 | 0.0884 | 0.1074 | 0.1273 | 0.1413 | 0.1654 | 0.1879 |
| | | 0.4113 | 0.4484 | 0.4851 | 0.5150 | 0.5400 | 0.5642 | 0.5890 |
| | 0.98 | 0.0806 | 0.1000 | 0.1183 | 0.1325 | 0.1604 | 0.1759 | 0.2039 |
| | | 0.3972 | 0.4267 | 0.4561 | 0.4854 | 0.5147 | 0.5440 | 0.5734 |
| 35 | 0.95 | 0.0957 | 0.1097 | 0.1382 | 0.1519 | 0.1824 | 0.1892 | 0.2257 |
| | | 0.3697 | 0.3988 | 0.4278 | 0.4567 | 0.4856 | 0.5145 | 0.5434 |
| | 0.90 | 0.1098 | 0.1282 | 0.1540 | 0.1708 | 0.2005 | 0.2202 | 0.2425 |
| | | 0.3427 | 0.3715 | 0.4001 | 0.4287 | 0.4573 | 0.4858 | 0.5143 |
| | 0.80 | 0.1267 | 0.1462 | 0.1754 | 0.1963 | 0.2169 | 0.2456 | 0.2742 |
| | | 0.3160 | 0.3444 | 0.3728 | 0.4011 | 0.4294 | 0.4577 | 0.4859 |
| | 0.99 | 0.0683 | 0.0858 | 0.1042 | 0.1235 | 0.1373 | 0.1601 | 0.1827 |
| | | 0.4131 | 0.4421 | 0.4711 | 0.5000 | 0.5290 | 0.5580 | 0.5870 |
| | 0.98 | 0.0783 | 0.0970 | 0.1150 | 0.1281 | 0.1555 | 0.1709 | 0.1974 |
| | | 0.3860 | 0.4146 | 0.4431 | 0.4716 | 0.5004 | 0.5285 | 0.5570 |
| 36 | 0.95 | 0.0930 | 0.1062 | 0.1341 | 0.1477 | 0.1768 | 0.1899 | 0.2194 |
| | | 0.3594 | 0.3877 | 0.4159 | 0.4440 | 0.4720 | 0.5000 | 0.5281 |
| | 0.90 | 0.1066 | 0.1246 | 0.1493 | 0.1661 | 0.1950 | 0.2127 | 0.2359 |
| | | 0.3333 | 0.3612 | 0.3891 | 0.4169 | 0.4446 | 0.4723 | 0.5000 |
| | 0.80 | 0.1229 | 0.1422 | 0.1706 | 0.1900 | 0.2110 | 0.2389 | 0.2667 |
| | | 0.3073 | 0.3350 | 0.3626 | 0.3901 | 0.4177 | 0.4451 | 0.4726 |

| $n$ | $\beta$ | 14 | 15 | 16 | 17 | 18 | 19 | 20 |
|---|---|---|---|---|---|---|---|---|
| | 0.99 | 0.2215 | 0.2470 | 0.2730 | 0.2938 | 0.3264 | 0.3525 | 0.3730 |
| | | 0.6475 | 0.6736 | 0.7062 | 0.7270 | 0.7530 | 0.7785 | 0.8005 |
| | 0.98 | 0.2336 | 0.2664 | 0.2904 | 0.3123 | 0.3441 | 0.3756 | 0.3999 |
| | | 0.6244 | 0.6559 | 0.6877 | 0.7096 | 0.7336 | 0.7664 | 0.7818 |
| 33 | 0.95 | 0.2684 | 0.2845 | 0.3158 | 0.3468 | 0.3776 | 0.4012 | 0.4231 |
| | | 0.5988 | 0.6224 | 0.6532 | 0.6842 | 0.7155 | 0.7316 | 0.7607 |
| | 0.90 | 0.2879 | 0.3185 | 0.3443 | 0.3633 | 0.3938 | 0.4242 | 0.4545 |
| | | 0.5758 | 0.6062 | 0.6367 | 0.6557 | 0.6815 | 0.7121 | 0.7429 |
| | 0.80 | 0.3206 | 0.3432 | 0.3649 | 0.3950 | 0.4250 | 0.4550 | 0.4850 |
| | | 0.5450 | 0.5750 | 0.6050 | 0.6351 | 0.6568 | 0.6794 | 0.7095 |
| | 0.99 | 0.2134 | 0.2396 | 0.2628 | 0.2850 | 0.3166 | 0.3377 | 0.3617 |
| | | 0.6383 | 0.6623 | 0.6834 | 0.7150 | 0.7372 | 0.7604 | 0.7866 |
| | 0.98 | 0.2267 | 0.2585 | 0.2790 | 0.3030 | 0.3339 | 0.3612 | 0.3788 |
| | | 0.6212 | 0.6388 | 0.6661 | 0.6970 | 0.7210 | 0.7415 | 0.7733 |
| 34 | 0.95 | 0.2589 | 0.2762 | 0.3065 | 0.3366 | 0.3629 | 0.3806 | 0.4105 |
| | | 0.5895 | 0.6194 | 0.6371 | 0.6634 | 0.6935 | 0.7238 | 0.7411 |
| | 0.90 | 0.2796 | 0.3093 | 0.3306 | 0.3526 | 0.3822 | 0.4117 | 0.4412 |
| | | 0.5588 | 0.5883 | 0.6178 | 0.6474 | 0.6694 | 0.6907 | 0.7204 |
| | 0.80 | 0.3107 | 0.3284 | 0.3543 | 0.3835 | 0.4127 | 0.4418 | 0.4709 |
| | | 0.5291 | 0.5582 | 0.5873 | 0.6165 | 0.6457 | 0.6716 | 0.6893 |
| | 0.99 | 0.2057 | 0.2326 | 0.2531 | 0.2767 | 0.3060 | 0.3221 | 0.3511 |
| | | 0.6188 | 0.6489 | 0.6779 | 0.6940 | 0.7233 | 0.7469 | 0.7674 |
| | 0.98 | 0.2201 | 0.2510 | 0.2680 | 0.2942 | 0.3242 | 0.3470 | 0.3674 |
| | | 0.6029 | 0.6326 | 0.6530 | 0.6758 | 0.7058 | 0.7320 | 0.7490 |
| 35 | 0.95 | 0.2500 | 0.2683 | 0.2978 | 0.3270 | 0.3486 | 0.3696 | 0.3987 |
| | | 0.5723 | 0.6013 | 0.6304 | 0.6514 | 0.6730 | 0.7022 | 0.7317 |
| | 0.90 | 0.2716 | 0.3000 | 0.3167 | 0.3426 | 0.3714 | 0.4000 | 0.4286 |
| | | 0.5428 | 0.5714 | 0.6000 | 0.6286 | 0.6574 | 0.6833 | 0.7000 |
| | 0.80 | 0.2999 | 0.3159 | 0.3443 | 0.3727 | 0.4010 | 0.4293 | 0.4576 |
| | | 0.5142 | 0.5424 | 0.5707 | 0.5990 | 0.6273 | 0.6557 | 0.6841 |
| | 0.99 | 0.1984 | 0.2261 | 0.2438 | 0.2689 | 0.2953 | 0.3115 | 0.3411 |
| | | 0.6052 | 0.6297 | 0.6589 | 0.6885 | 0.7047 | 0.7311 | 0.7562 |
| | 0.98 | 0.2140 | 0.2437 | 0.2569 | 0.2860 | 0.3149 | 0.3329 | 0.3571 |
| | | 0.5855 | 0.6141 | 0.6429 | 0.6671 | 0.6851 | 0.7140 | 0.7431 |
| 36 | 0.95 | 0.2416 | 0.2608 | 0.2895 | 0.3170 | 0.3339 | 0.3593 | 0.3876 |
| | | 0.5561 | 0.5842 | 0.6124 | 0.6407 | 0.6661 | 0.6830 | 0.7105 |
| | 0.90 | 0.2642 | 0.2899 | 0.3051 | 0.3332 | 0.3611 | 0.3890 | 0.4168 |
| | | 0.5278 | 0.5555 | 0.5832 | 0.6110 | 0.6389 | 0.6668 | 0.6949 |
| | 0.80 | 0.2897 | 0.3072 | 0.3349 | 0.3625 | 0.3900 | 0.4176 | 0.4450 |
| | | 0.5000 | 0.5275 | 0.5550 | 0.5824 | 0.6100 | 0.6375 | 0.6651 |

| $n$ | $\beta$ | 0 | 1 | 2 | 3 | 4 | 5 | 6 |
|---|---|---|---|---|---|---|---|---|
| | 0.99 | 0.0000 | 0.0002 | 0.0040 | 0.0120 | 0.0229 | 0.0359 | 0.0505 |
| | | 0.1336 | 0.1778 | 0.2200 | 0.2616 | 0.3030 | 0.3318 | 0.3625 |
| | 0.98 | 0.0000 | 0.0005 | 0.0058 | 0.0156 | 0.0282 | 0.0428 | 0.0588 |
| | | 0.1120 | 0.1664 | 0.2082 | 0.2495 | 0.2783 | 0.3192 | 0.3474 |
| 37 | 0.95 | 0.0000 | 0.0013 | 0.0096 | 0.0224 | 0.0377 | 0.0547 | 0.0730 |
| | | 0.0906 | 0.1438 | 0.1849 | 0.2136 | 0.2539 | 0.2818 | 0.3220 |
| | 0.90 | 0.0000 | 0.0028 | 0.0144 | 0.0301 | 0.0479 | 0.0672 | 0.0799 |
| | | 0.0800 | 0.1214 | 0.1617 | 0.1899 | 0.2296 | 0.2572 | 0.2970 |
| | 0.80 | 0.0000 | 0.0060 | 0.0223 | 0.0417 | 0.0587 | 0.0759 | 0.0988 |
| | | 0.0588 | 0.0989 | 0.1385 | 0.1662 | 0.2055 | 0.2327 | 0.2597 |
| | 0.99 | 0.0000 | 0.0002 | 0.0039 | 0.0117 | 0.0223 | 0.0349 | 0.0491 |
| | | 0.1301 | 0.1730 | 0.2141 | 0.2546 | 0.2949 | 0.3229 | 0.3631 |
| | 0.98 | 0.0000 | 0.0005 | 0.0057 | 0.0152 | 0.0274 | 0.0416 | 0.0572 |
| | | 0.1090 | 0.1620 | 0.2027 | 0.2429 | 0.2709 | 0.3107 | 0.3381 |
| 38 | 0.95 | 0.0000 | 0.0013 | 0.0094 | 0.0218 | 0.0367 | 0.0533 | 0.0710 |
| | | 0.0882 | 0.1401 | 0.1800 | 0.2080 | 0.2472 | 0.2744 | 0.3135 |
| | 0.90 | 0.0000 | 0.0027 | 0.0140 | 0.0293 | 0.0467 | 0.0654 | 0.0778 |
| | | 0.0779 | 0.1182 | 0.1575 | 0.1849 | 0.2237 | 0.2505 | 0.2892 |
| | 0.80 | 0.0000 | 0.0058 | 0.0217 | 0.0406 | 0.0571 | 0.0738 | 0.0963 |
| | | 0.0572 | 0.0964 | 0.1349 | 0.1619 | 0.2002 | 0.2267 | 0.2530 |
| | 0.99 | 0.0000 | 0.0002 | 0.0038 | 0.0114 | 0.0217 | 0.0340 | 0.0477 |
| | | 0.1267 | 0.1686 | 0.2124 | 0.2480 | 0.2872 | 0.3145 | 0.3536 |
| | 0.98 | 0.0000 | 0.0005 | 0.0055 | 0.0148 | 0.0267 | 0.0405 | 0.0557 |
| | | 0.1062 | 0.1578 | 0.1975 | 0.2366 | 0.2639 | 0.3026 | 0.3294 |
| 39 | 0.95 | 0.0000 | 0.0013 | 0.0091 | 0.0212 | 0.0357 | 0.0519 | 0.0691 |
| | | 0.0860 | 0.1365 | 0.1754 | 0.2027 | 0.2409 | 0.2674 | 0.3055 |
| | 0.90 | 0.0000 | 0.0026 | 0.0137 | 0.0285 | 0.0454 | 0.0636 | 0.0758 |
| | | 0.0759 | 0.1152 | 0.1535 | 0.1803 | 0.2180 | 0.2441 | 0.2819 |
| | 0.80 | 0.0000 | 0.0057 | 0.0211 | 0.0396 | 0.0557 | 0.0718 | 0.0939 |
| | | 0.0558 | 0.0940 | 0.1315 | 0.1578 | 0.1951 | 0.2209 | 0.2466 |
| | 0.99 | 0.0000 | 0.0002 | 0.0037 | 0.0111 | 0.0211 | 0.0331 | 0.0465 |
| | | 0.1235 | 0.1665 | 0.2142 | 0.2417 | 0.2799 | 0.3073 | 0.3445 |
| | 0.98 | 0.0000 | 0.0005 | 0.0054 | 0.0144 | 0.0260 | 0.0394 | 0.0542 |
| | | 0.1036 | 0.1539 | 0.1925 | 0.2306 | 0.2573 | 0.2950 | 0.3210 |
| 40 | 0.95 | 0.0000 | 0.0012 | 0.0089 | 0.0207 | 0.0348 | 0.0505 | 0.0674 |
| | | 0.0839 | 0.1331 | 0.1711 | 0.1977 | 0.2349 | 0.2607 | 0.2979 |
| | 0.90 | 0.0000 | 0.0026 | 0.0133 | 0.0278 | 0.0443 | 0.0620 | 0.0740 |
| | | 0.0741 | 0.1124 | 0.1497 | 0.1758 | 0.2126 | 0.2381 | 0.2749 |
| | 0.80 | 0.0000 | 0.0055 | 0.0206 | 0.0386 | 0.0543 | 0.0699 | 0.0915 |
| | | 0.0544 | 0.0916 | 0.1283 | 0.1539 | 0.1903 | 0.2155 | 0.2418 |

| $n$ | $\beta$ | 7 | 8 | 9 | 10 | 11 | 12 | 13 |
|---|---|---|---|---|---|---|---|---|
| | 0.99 | 0.0663 | 0.0833 | 0.1012 | 0.1199 | 0.1335 | 0.1552 | 0.1777 |
| | | 0.4016 | 0.4298 | 0.4580 | 0.4860 | 0.5141 | 0.5421 | 0.5703 |
| | 0.98 | 0.0760 | 0.0942 | 0.1119 | 0.1240 | 0.1509 | 0.1663 | 0.1912 |
| | | 0.3754 | 0.4032 | 0.4340 | 0.4678 | 0.5000 | 0.5246 | 0.5478 |
| 37 | 0.95 | 0.0905 | 0.1030 | 0.1302 | 0.1437 | 0.1716 | 0.1848 | 0.2135 |
| | | 0.3497 | 0.3772 | 0.4046 | 0.4319 | 0.4592 | 0.4865 | 0.5138 |
| | 0.90 | 0.1035 | 0.1213 | 0.1449 | 0.1616 | 0.1898 | 0.2055 | 0.2295 |
| | | 0.3243 | 0.3515 | 0.3786 | 0.4057 | 0.4332 | 0.4642 | 0.4933 |
| | 0.80 | 0.1193 | 0.1384 | 0.1661 | 0.1840 | 0.2054 | 0.2326 | 0.2596 |
| | | 0.2991 | 0.3261 | 0.3529 | 0.3797 | 0.4065 | 0.4333 | 0.4600 |
| | 0.99 | 0.0645 | 0.0810 | 0.0983 | 0.1165 | 0.1300 | 0.1505 | 0.1729 |
| | | 0.3907 | 0.4182 | 0.4455 | 0.4728 | 0.5000 | 0.5273 | 0.5546 |
| | 0.98 | 0.0739 | 0.0916 | 0.1089 | 0.1201 | 0.1465 | 0.1619 | 0.1855 |
| | | 0.3654 | 0.3950 | 0.4324 | 0.4595 | 0.4865 | 0.5136 | 0.5406 |
| 38 | 0.95 | 0.0881 | 0.0999 | 0.1265 | 0.1400 | 0.1666 | 0.1799 | 0.2079 |
| | | 0.3405 | 0.3672 | 0.3939 | 0.4210 | 0.4530 | 0.4867 | 0.5134 |
| | 0.90 | 0.1006 | 0.1181 | 0.1407 | 0.1574 | 0.1848 | 0.1986 | 0.2236 |
| | | 0.3158 | 0.3423 | 0.3687 | 0.3955 | 0.4341 | 0.4605 | 0.4869 |
| | 0.80 | 0.1159 | 0.1348 | 0.1618 | 0.1783 | 0.2001 | 0.2266 | 0.2529 |
| | | 0.2914 | 0.3176 | 0.3438 | 0.3699 | 0.3960 | 0.4220 | 0.4480 |
| | 0.99 | 0.0627 | 0.0788 | 0.0956 | 0.1133 | 0.1266 | 0.1461 | 0.1685 |
| | | 0.3805 | 0.4072 | 0.4338 | 0.4620 | 0.4903 | 0.5165 | 0.5412 |
| | 0.98 | 0.0719 | 0.0891 | 0.1061 | 0.1164 | 0.1424 | 0.1577 | 0.1800 |
| | | 0.3559 | 0.3947 | 0.4211 | 0.4475 | 0.4738 | 0.5000 | 0.5263 |
| 39 | 0.95 | 0.0859 | 0.0970 | 0.1230 | 0.1364 | 0.1619 | 0.1753 | 0.2026 |
| | | 0.3317 | 0.3578 | 0.3838 | 0.4222 | 0.4482 | 0.4741 | 0.5000 |
| | 0.90 | 0.0979 | 0.1151 | 0.1368 | 0.1534 | 0.1802 | 0.1920 | 0.2179 |
| | | 0.3078 | 0.3336 | 0.3593 | 0.3973 | 0.4230 | 0.4487 | 0.4744 |
| | 0.80 | 0.1128 | 0.1314 | 0.1577 | 0.1729 | 0.1950 | 0.2208 | 0.2465 |
| | | 0.2840 | 0.3096 | 0.3351 | 0.3605 | 0.3860 | 0.4113 | 0.4390 |
| | 0.99 | 0.0611 | 0.0767 | 0.0931 | 0.1103 | 0.1234 | 0.1420 | 0.1639 |
| | | 0.3707 | 0.3968 | 0.4276 | 0.4611 | 0.4871 | 0.5130 | 0.5390 |
| | 0.98 | 0.0701 | 0.0868 | 0.1035 | 0.1129 | 0.1385 | 0.1538 | 0.1749 |
| | | 0.3495 | 0.3847 | 0.4104 | 0.4361 | 0.4617 | 0.4873 | 0.5128 |
| 40 | 0.95 | 0.0838 | 0.0942 | 0.1198 | 0.1330 | 0.1575 | 0.1710 | 0.1976 |
| | | 0.3234 | 0.3489 | 0.3797 | 0.4116 | 0.4369 | 0.4622 | 0.4874 |
| | 0.90 | 0.0953 | 0.1123 | 0.1331 | 0.1496 | 0.1757 | 0.1854 | 0.2125 |
| | | 0.3002 | 0.3253 | 0.3515 | 0.3874 | 0.4125 | 0.4375 | 0.4625 |
| | 0.80 | 0.1097 | 0.1282 | 0.1538 | 0.1678 | 0.1902 | 0.2154 | 0.2401 |
| | | 0.2770 | 0.3020 | 0.3268 | 0.3517 | 0.3764 | 0.4042 | 0.4381 |

| $n$ | $\beta$ | 14 | 15 | 16 | 17 | 18 | 19 | 20 |
|---|---|---|---|---|---|---|---|---|
| | 0.99 | 0.1914 | 0.2199 | 0.2349 | 0.2615 | 0.2852 | 0.3029 | 0.3317 |
| | | 0.5985 | 0.6217 | 0.6411 | 0.6683 | 0.6971 | 0.7148 | 0.7385 |
| | 0.98 | 0.2081 | 0.2360 | 0.2494 | 0.2782 | 0.3043 | 0.3191 | 0.3473 |
| | | 0.5713 | 0.5969 | 0.6247 | 0.6527 | 0.6809 | 0.6957 | 0.7218 |
| 37 | 0.95 | 0.2337 | 0.2538 | 0.2817 | 0.3064 | 0.3219 | 0.3496 | 0.3771 |
| | | 0.5409 | 0.5682 | 0.5955 | 0.6229 | 0.6504 | 0.6781 | 0.6936 |
| | 0.90 | 0.2571 | 0.2803 | 0.2969 | 0.3242 | 0.3514 | 0.3785 | 0.4056 |
| | | 0.5196 | 0.5441 | 0.5679 | 0.5944 | 0.6215 | 0.6486 | 0.6758 |
| | 0.80 | 0.2798 | 0.2990 | 0.3260 | 0.3528 | 0.3796 | 0.4064 | 0.4332 |
| | | 0.4867 | 0.5134 | 0.5401 | 0.5668 | 0.5936 | 0.6204 | 0.6472 |
| | 0.99 | 0.1846 | 0.2140 | 0.2260 | 0.2545 | 0.2756 | 0.2948 | 0.3228 |
| | | 0.5819 | 0.6094 | 0.6370 | 0.6537 | 0.6772 | 0.7052 | 0.7244 |
| | 0.98 | 0.2026 | 0.2287 | 0.2428 | 0.2708 | 0.2943 | 0.3106 | 0.3380 |
| | | 0.5677 | 0.5886 | 0.6091 | 0.6347 | 0.6620 | 0.6894 | 0.7057 |
| 38 | 0.95 | 0.2261 | 0.2471 | 0.2743 | 0.2962 | 0.3134 | 0.3404 | 0.3671 |
| | | 0.5348 | 0.5573 | 0.5800 | 0.6062 | 0.6329 | 0.6596 | 0.6866 |
| | 0.90 | 0.2504 | 0.2712 | 0.2891 | 0.3157 | 0.3422 | 0.3686 | 0.3947 |
| | | 0.5132 | 0.5396 | 0.5660 | 0.5842 | 0.6053 | 0.6314 | 0.6578 |
| | 0.80 | 0.2703 | 0.2913 | 0.3175 | 0.3437 | 0.3698 | 0.3959 | 0.4219 |
| | | 0.4740 | 0.5000 | 0.5261 | 0.5521 | 0.5781 | 0.6041 | 0.6302 |
| | 0.99 | 0.1779 | 0.2078 | 0.2157 | 0.2479 | 0.2664 | 0.2871 | 0.3144 |
| | | 0.5663 | 0.5929 | 0.6196 | 0.6465 | 0.6660 | 0.6856 | 0.7129 |
| | 0.98 | 0.1974 | 0.2219 | 0.2365 | 0.2638 | 0.2847 | 0.3025 | 0.3293 |
| | | 0.5526 | 0.5790 | 0.6054 | 0.6232 | 0.6442 | 0.6707 | 0.6975 |
| 39 | 0.95 | 0.2189 | 0.2408 | 0.2673 | 0.2864 | 0.3054 | 0.3316 | 0.3577 |
| | | 0.5260 | 0.5519 | 0.5779 | 0.5952 | 0.6163 | 0.6423 | 0.6684 |
| | 0.90 | 0.2440 | 0.2625 | 0.2818 | 0.3077 | 0.3335 | 0.3592 | 0.3810 |
| | | 0.5000 | 0.5257 | 0.5514 | 0.5771 | 0.6028 | 0.6190 | 0.6408 |
| | 0.80 | 0.2610 | 0.2839 | 0.3095 | 0.3350 | 0.3604 | 0.3859 | 0.4112 |
| | | 0.4687 | 0.4966 | 0.5207 | 0.5431 | 0.5651 | 0.5888 | 0.6141 |
| | 0.99 | 0.1705 | 0.2019 | 0.2141 | 0.2416 | 0.2575 | 0.2798 | 0.3061 |
| | | 0.5601 | 0.5805 | 0.6033 | 0.6294 | 0.6556 | 0.6784 | 0.6939 |
| | 0.98 | 0.1924 | 0.2154 | 0.2305 | 0.2572 | 0.2754 | 0.2949 | 0.3209 |
| | | 0.5384 | 0.5640 | 0.5897 | 0.6154 | 0.6378 | 0.6544 | 0.6791 |
| 40 | 0.95 | 0.2119 | 0.2348 | 0.2606 | 0.2768 | 0.2978 | 0.3233 | 0.3488 |
| | | 0.5127 | 0.5379 | 0.5632 | 0.5885 | 0.6122 | 0.6285 | 0.6512 |
| | 0.90 | 0.2380 | 0.2540 | 0.2748 | 0.3001 | 0.3252 | 0.3498 | 0.3671 |
| | | 0.4876 | 0.5125 | 0.5376 | 0.5626 | 0.5876 | 0.6127 | 0.6329 |
| | 0.80 | 0.2513 | 0.2769 | 0.3019 | 0.3267 | 0.3516 | 0.3763 | 0.3992 |
| | | 0.4629 | 0.4877 | 0.5124 | 0.5372 | 0.5620 | 0.5816 | 0.6008 |

| $n$ | $\beta$ | 0 | 1 | 2 | 3 | 4 | 5 | 6 |
|---|---|---|---|---|---|---|---|---|
| 41 | 0.99 | 0.0000 | 0.0002 | 0.0036 | 0.0108 | 0.0206 | 0.0322 | 0.0453 |
| | | 0.1205 | 0.1705 | 0.2089 | 0.2357 | 0.2730 | 0.3102 | 0.3360 |
| | 0.98 | 0.0000 | 0.0004 | 0.0052 | 0.0140 | 0.0254 | 0.0384 | 0.0528 |
| | | 0.1011 | 0.1501 | 0.1878 | 0.2250 | 0.2510 | 0.2877 | 0.3131 |
| | 0.95 | 0.0000 | 0.0012 | 0.0087 | 0.0202 | 0.0340 | 0.0493 | 0.0657 |
| | | 0.0818 | 0.1299 | 0.1669 | 0.1929 | 0.2291 | 0.2544 | 0.2906 |
| | 0.90 | 0.0000 | 0.0025 | 0.0130 | 0.0271 | 0.0432 | 0.0605 | 0.0722 |
| | | 0.0723 | 0.1097 | 0.1461 | 0.1733 | 0.2074 | 0.2323 | 0.2682 |
| | 0.80 | 0.0000 | 0.0054 | 0.0201 | 0.0376 | 0.0530 | 0.0681 | 0.0893 |
| | | 0.0531 | 0.0894 | 0.1252 | 0.1502 | 0.1857 | 0.2103 | 0.2459 |
| 42 | 0.99 | 0.0000 | 0.0002 | 0.0035 | 0.0105 | 0.0201 | 0.0314 | 0.0442 |
| | | 0.1176 | 0.1664 | 0.2038 | 0.2320 | 0.2664 | 0.3027 | 0.3278 |
| | 0.98 | 0.0000 | 0.0004 | 0.0051 | 0.0137 | 0.0247 | 0.0375 | 0.0515 |
| | | 0.0987 | 0.1465 | 0.1833 | 0.2196 | 0.2450 | 0.2808 | 0.3056 |
| | 0.95 | 0.0000 | 0.0012 | 0.0085 | 0.0197 | 0.0331 | 0.0480 | 0.0640 |
| | | 0.0799 | 0.1268 | 0.1629 | 0.1883 | 0.2237 | 0.2527 | 0.2837 |
| | 0.90 | 0.0000 | 0.0025 | 0.0127 | 0.0265 | 0.0421 | 0.0590 | 0.0705 |
| | | 0.0706 | 0.1071 | 0.1426 | 0.1779 | 0.2025 | 0.2274 | 0.2619 |
| | 0.80 | 0.0000 | 0.0052 | 0.0196 | 0.0367 | 0.0518 | 0.0664 | 0.0873 |
| | | 0.0519 | 0.0874 | 0.1223 | 0.1467 | 0.1814 | 0.2054 | 0.2401 |
| 43 | 0.99 | 0.0000 | 0.0002 | 0.0034 | 0.0103 | 0.0196 | 0.0307 | 0.0431 |
| | | 0.1148 | 0.1625 | 0.1990 | 0.2349 | 0.2601 | 0.2955 | 0.3201 |
| | 0.98 | 0.0000 | 0.0004 | 0.0050 | 0.0134 | 0.0242 | 0.0366 | 0.0503 |
| | | 0.0964 | 0.1431 | 0.1790 | 0.2144 | 0.2424 | 0.2742 | 0.2985 |
| | 0.95 | 0.0000 | 0.0011 | 0.0083 | 0.0192 | 0.0324 | 0.0469 | 0.0625 |
| | | 0.0781 | 0.1238 | 0.1592 | 0.1850 | 0.2185 | 0.2531 | 0.2771 |
| | 0.90 | 0.0000 | 0.0024 | 0.0124 | 0.0259 | 0.0411 | 0.0576 | 0.0688 |
| | | 0.0689 | 0.1046 | 0.1394 | 0.1738 | 0.1979 | 0.2321 | 0.2558 |
| | 0.80 | 0.0000 | 0.0051 | 0.0192 | 0.0358 | 0.0506 | 0.0648 | 0.0853 |
| | | 0.0507 | 0.0854 | 0.1195 | 0.1433 | 0.1772 | 0.2007 | 0.2346 |
| 44 | 0.99 | 0.0000 | 0.0002 | 0.0034 | 0.0100 | 0.0191 | 0.0300 | 0.0421 |
| | | 0.1122 | 0.1587 | 0.1944 | 0.2295 | 0.2541 | 0.2887 | 0.3127 |
| | 0.98 | 0.0000 | 0.0004 | 0.0049 | 0.0131 | 0.0236 | 0.0357 | 0.0491 |
| | | 0.0942 | 0.1398 | 0.1749 | 0.2095 | 0.2439 | 0.2679 | 0.2963 |
| | 0.95 | 0.0000 | 0.0011 | 0.0081 | 0.0188 | 0.0316 | 0.0458 | 0.0610 |
| | | 0.0763 | 0.1210 | 0.1556 | 0.1897 | 0.2136 | 0.2474 | 0.2708 |
| | 0.90 | 0.0000 | 0.0023 | 0.0121 | 0.0253 | 0.0402 | 0.0563 | 0.0673 |
| | | 0.0674 | 0.1023 | 0.1362 | 0.1699 | 0.1934 | 0.2269 | 0.2501 |
| | 0.80 | 0.0000 | 0.0050 | 0.0187 | 0.0350 | 0.0495 | 0.0632 | 0.0833 |
| | | 0.0496 | 0.0834 | 0.1168 | 0.1401 | 0.1732 | 0.1962 | 0.2294 |

| $n$ | $\beta$ | 7 | 8 | 9 | 10 | 11 | 12 | 13 |
|---|---|---|---|---|---|---|---|---|
| | 0.99 | 0.0595 | 0.0747 | 0.0907 | 0.1074 | 0.1204 | 0.1380 | 0.1595 |
| | | 0.3615 | 0.3903 | 0.4242 | 0.4495 | 0.4748 | 0.5000 | 0.5253 |
| | 0.98 | 0.0683 | 0.0846 | 0.1010 | 0.1095 | 0.1348 | 0.1500 | 0.1700 |
| | | 0.3499 | 0.3751 | 0.4002 | 0.4253 | 0.4502 | 0.4751 | 0.5000 |
| 41 | 0.95 | 0.0817 | 0.0916 | 0.1166 | 0.1298 | 0.1533 | 0.1668 | 0.1928 |
| | | 0.3155 | 0.3404 | 0.3767 | 0.4015 | 0.4262 | 0.4508 | 0.4754 |
| | 0.90 | 0.0929 | 0.1096 | 0.1296 | 0.1460 | 0.1712 | 0.1783 | 0.2073 |
| | | 0.2929 | 0.3175 | 0.3535 | 0.3780 | 0.4025 | 0.4269 | 0.4513 |
| | 0.80 | 0.1069 | 0.1251 | 0.1501 | 0.1629 | 0.1856 | 0.2102 | 0.2334 |
| | | 0.2703 | 0.2947 | 0.3190 | 0.3432 | 0.3674 | 0.4033 | 0.4275 |
| | 0.99 | 0.0580 | 0.0728 | 0.0884 | 0.1046 | 0.1175 | 0.1343 | 0.1553 |
| | | 0.3528 | 0.3890 | 0.4138 | 0.4385 | 0.4632 | 0.4878 | 0.5123 |
| | 0.98 | 0.0666 | 0.0825 | 0.0986 | 0.1063 | 0.1313 | 0.1464 | 0.1654 |
| | | 0.3415 | 0.3661 | 0.3906 | 0.4150 | 0.4397 | 0.4672 | 0.4932 |
| 42 | 0.95 | 0.0798 | 0.0891 | 0.1137 | 0.1267 | 0.1493 | 0.1628 | 0.1882 |
| | | 0.3080 | 0.3349 | 0.3677 | 0.3919 | 0.4160 | 0.4400 | 0.4640 |
| | 0.90 | 0.0905 | 0.1070 | 0.1262 | 0.1425 | 0.1667 | 0.1778 | 0.2024 |
| | | 0.2860 | 0.3100 | 0.3451 | 0.3690 | 0.3929 | 0.4168 | 0.4406 |
| | 0.80 | 0.1041 | 0.1222 | 0.1466 | 0.1582 | 0.1813 | 0.2053 | 0.2270 |
| | | 0.2640 | 0.2878 | 0.3115 | 0.3351 | 0.3662 | 0.3938 | 0.4174 |
| | 0.99 | 0.0566 | 0.0710 | 0.0862 | 0.1020 | 0.1147 | 0.1308 | 0.1513 |
| | | 0.3471 | 0.3798 | 0.4040 | 0.4281 | 0.4521 | 0.4761 | 0.5000 |
| | 0.98 | 0.0650 | 0.0804 | 0.0963 | 0.1032 | 0.1279 | 0.1430 | 0.1610 |
| | | 0.3334 | 0.3575 | 0.3814 | 0.4061 | 0.4405 | 0.4643 | 0.4881 |
| 43 | 0.95 | 0.0780 | 0.0867 | 0.1109 | 0.1237 | 0.1455 | 0.1591 | 0.1836 |
| | | 0.3009 | 0.3355 | 0.3591 | 0.3827 | 0.4063 | 0.4297 | 0.4532 |
| | 0.90 | 0.0883 | 0.1045 | 0.1230 | 0.1393 | 0.1625 | 0.1737 | 0.1978 |
| | | 0.2794 | 0.3028 | 0.3371 | 0.3605 | 0.3838 | 0.4071 | 0.4304 |
| | 0.80 | 0.1015 | 0.1194 | 0.1432 | 0.1537 | 0.1771 | 0.2006 | 0.2210 |
| | | 0.2579 | 0.2812 | 0.3043 | 0.3274 | 0.3616 | 0.3847 | 0.4078 |
| | 0.99 | 0.0553 | 0.0693 | 0.0841 | 0.0996 | 0.1121 | 0.1274 | 0.1475 |
| | | 0.3472 | 0.3710 | 0.3946 | 0.4181 | 0.4416 | 0.4669 | 0.4917 |
| | 0.98 | 0.0634 | 0.0785 | 0.0941 | 0.1002 | 0.1248 | 0.1397 | 0.1568 |
| | | 0.3258 | 0.3493 | 0.3726 | 0.4070 | 0.4303 | 0.4536 | 0.4768 |
| 44 | 0.95 | 0.0762 | 0.0845 | 0.1082 | 0.1209 | 0.1419 | 0.1555 | 0.1790 |
| | | 0.2941 | 0.3278 | 0.3510 | 0.3740 | 0.3970 | 0.4199 | 0.4429 |
| | 0.90 | 0.0862 | 0.1022 | 0.1200 | 0.1361 | 0.1585 | 0.1698 | 0.1933 |
| | | 0.2731 | 0.3005 | 0.3295 | 0.3523 | 0.3751 | 0.3979 | 0.4206 |
| | 0.80 | 0.0991 | 0.1167 | 0.1400 | 0.1493 | 0.1731 | 0.1961 | 0.2152 |
| | | 0.2522 | 0.2749 | 0.2975 | 0.3214 | 0.3535 | 0.3761 | 0.3986 |

| $n$ | $\beta$ | 14 | 15 | 16 | 17 | 18 | 19 | 20 |
|---|---|---|---|---|---|---|---|---|
| | 0.99 | 0.1704 | 0.1962 | 0.2088 | 0.2356 | 0.2487 | 0.2729 | 0.2968 |
| | | 0.5506 | 0.5759 | 0.5968 | 0.6151 | 0.6386 | 0.6641 | 0.6899 |
| | 0.98 | 0.1877 | 0.2093 | 0.2249 | 0.2509 | 0.2664 | 0.2876 | 0.3130 |
| | | 0.5250 | 0.5499 | 0.5748 | 0.5999 | 0.6250 | 0.6502 | 0.6654 |
| 41 | 0.95 | 0.2051 | 0.2290 | 0.2543 | 0.2672 | 0.2905 | 0.3154 | 0.3403 |
| | | 0.5000 | 0.5247 | 0.5493 | 0.5739 | 0.5986 | 0.6234 | 0.6410 |
| | 0.90 | 0.2322 | 0.2457 | 0.2681 | 0.2928 | 0.3174 | 0.3388 | 0.3534 |
| | | 0.4757 | 0.5000 | 0.5244 | 0.5488 | 0.5732 | 0.5976 | 0.6221 |
| | 0.80 | 0.2458 | 0.2702 | 0.2946 | 0.3189 | 0.3431 | 0.3673 | 0.3855 |
| | | 0.4517 | 0.4759 | 0.5000 | 0.5242 | 0.5484 | 0.5726 | 0.5968 |
| | 0.99 | 0.1663 | 0.1909 | 0.2037 | 0.2295 | 0.2394 | 0.2663 | 0.2879 |
| | | 0.5369 | 0.5616 | 0.5863 | 0.6111 | 0.6280 | 0.6473 | 0.6723 |
| | 0.98 | 0.1832 | 0.2035 | 0.2195 | 0.2449 | 0.2574 | 0.2807 | 0.3055 |
| | | 0.5170 | 0.5394 | 0.5611 | 0.5851 | 0.6095 | 0.6340 | 0.6586 |
| 42 | 0.95 | 0.1985 | 0.2236 | 0.2473 | 0.2560 | 0.2836 | 0.3079 | 0.3310 |
| | | 0.4880 | 0.5121 | 0.5361 | 0.5601 | 0.5841 | 0.6082 | 0.6324 |
| | 0.90 | 0.2266 | 0.2372 | 0.2618 | 0.2859 | 0.3099 | 0.3282 | 0.3450 |
| | | 0.4644 | 0.4882 | 0.5119 | 0.5357 | 0.5595 | 0.5833 | 0.6072 |
| | 0.80 | 0.2400 | 0.2639 | 0.2877 | 0.3114 | 0.3350 | 0.3561 | 0.3684 |
| | | 0.4410 | 0.4646 | 0.4882 | 0.5119 | 0.5355 | 0.5591 | 0.5827 |
| | 0.99 | 0.1624 | 0.1859 | 0.1989 | 0.2233 | 0.2348 | 0.2600 | 0.2794 |
| | | 0.5240 | 0.5480 | 0.5720 | 0.5961 | 0.6203 | 0.6417 | 0.6569 |
| | 0.98 | 0.1789 | 0.1979 | 0.2143 | 0.2384 | 0.2474 | 0.2741 | 0.2984 |
| | | 0.5120 | 0.5358 | 0.5596 | 0.5762 | 0.5955 | 0.6187 | 0.6426 |
| 43 | 0.95 | 0.1915 | 0.2184 | 0.2406 | 0.2530 | 0.2770 | 0.3008 | 0.3212 |
| | | 0.4766 | 0.5000 | 0.5235 | 0.5469 | 0.5704 | 0.5938 | 0.6174 |
| | 0.90 | 0.2206 | 0.2320 | 0.2557 | 0.2793 | 0.3027 | 0.3174 | 0.3370 |
| | | 0.4536 | 0.4768 | 0.5000 | 0.5233 | 0.5465 | 0.5697 | 0.5930 |
| | 0.80 | 0.2345 | 0.2578 | 0.2811 | 0.3042 | 0.3273 | 0.3450 | 0.3615 |
| | | 0.4309 | 0.4539 | 0.4770 | 0.5000 | 0.5231 | 0.5462 | 0.5692 |
| | 0.99 | 0.1586 | 0.1811 | 0.1943 | 0.2174 | 0.2294 | 0.2540 | 0.2712 |
| | | 0.5148 | 0.5367 | 0.5585 | 0.5820 | 0.6055 | 0.6291 | 0.6529 |
| | 0.98 | 0.1748 | 0.1926 | 0.2094 | 0.2321 | 0.2438 | 0.2678 | 0.2902 |
| | | 0.5000 | 0.5233 | 0.5465 | 0.5698 | 0.5931 | 0.6084 | 0.6275 |
| 44 | 0.95 | 0.1896 | 0.2135 | 0.2342 | 0.2473 | 0.2707 | 0.2940 | 0.3117 |
| | | 0.4679 | 0.4921 | 0.5148 | 0.5362 | 0.5572 | 0.5802 | 0.6031 |
| | 0.90 | 0.2149 | 0.2268 | 0.2500 | 0.2730 | 0.2946 | 0.3050 | 0.3294 |
| | | 0.4433 | 0.4660 | 0.4887 | 0.5114 | 0.5341 | 0.5568 | 0.5795 |
| | 0.80 | 0.2293 | 0.2521 | 0.2748 | 0.2974 | 0.3194 | 0.3334 | 0.3534 |
| | | 0.4212 | 0.4437 | 0.4663 | 0.4888 | 0.5113 | 0.5338 | 0.5564 |

| $n$ | $\beta$ | 21 | 22 | 23 | 24 | 25 | 26 | 27 |
|---|---|---|---|---|---|---|---|---|
| | 0.99 | 0.3101 | 0.3359 | 0.3614 | 0.3849 | 0.4032 | 0.4241 | 0.4494 |
| | | 0.7032 | 0.7271 | 0.7513 | 0.7644 | 0.7912 | 0.8038 | 0.8296 |
| | 0.98 | 0.3346 | 0.3498 | 0.3750 | 0.4001 | 0.4252 | 0.4501 | 0.4750 |
| | | 0.6870 | 0.7124 | 0.7336 | 0.7491 | 0.7751 | 0.7907 | 0.8123 |
| 41 | 0.95 | 0.3590 | 0.3766 | 0.4014 | 0.4261 | 0.4507 | 0.4753 | 0.5000 |
| | | 0.6597 | 0.6846 | 0.7095 | 0.7328 | 0.7457 | 0.7710 | 0.7949 |
| | 0.90 | 0.3779 | 0.4024 | 0.4268 | 0.4512 | 0.4756 | 0.5000 | 0.5243 |
| | | 0.6466 | 0.6612 | 0.6826 | 0.7072 | 0.7319 | 0.7543 | 0.7678 |
| | 0.80 | 0.4032 | 0.4274 | 0.4516 | 0.4758 | 0.5000 | 0.5241 | 0.5483 |
| | | 0.6145 | 0.6327 | 0.6569 | 0.6811 | 0.7054 | 0.7298 | 0.7542 |
| | 0.99 | 0.3026 | 0.3277 | 0.3527 | 0.3720 | 0.3889 | 0.4137 | 0.4384 |
| | | 0.6974 | 0.7121 | 0.7337 | 0.7606 | 0.7705 | 0.7963 | 0.8091 |
| | 0.98 | 0.3238 | 0.3414 | 0.3660 | 0.3905 | 0.4149 | 0.4389 | 0.4606 |
| | | 0.6762 | 0.6945 | 0.7193 | 0.7426 | 0.7551 | 0.7805 | 0.7965 |
| 42 | 0.95 | 0.3458 | 0.3676 | 0.3918 | 0.4159 | 0.4399 | 0.4639 | 0.4879 |
| | | 0.6542 | 0.6690 | 0.6921 | 0.7164 | 0.7440 | 0.7527 | 0.7764 |
| | 0.90 | 0.3689 | 0.3928 | 0.4167 | 0.4405 | 0.4643 | 0.4881 | 0.5118 |
| | | 0.6311 | 0.6550 | 0.6718 | 0.6901 | 0.7141 | 0.7382 | 0.7628 |
| | 0.80 | 0.3937 | 0.4173 | 0.4409 | 0.4645 | 0.4881 | 0.5118 | 0.5354 |
| | | 0.6063 | 0.6316 | 0.6439 | 0.6650 | 0.6886 | 0.7123 | 0.7361 |
| | 0.99 | 0.2954 | 0.3200 | 0.3431 | 0.3583 | 0.3797 | 0.4039 | 0.4280 |
| | | 0.6800 | 0.7046 | 0.7206 | 0.7400 | 0.7652 | 0.7767 | 0.8011 |
| | 0.98 | 0.3130 | 0.3333 | 0.3574 | 0.3813 | 0.4045 | 0.4238 | 0.4404 |
| | | 0.6667 | 0.6870 | 0.7016 | 0.7259 | 0.7526 | 0.7616 | 0.7857 |
| 43 | 0.95 | 0.3354 | 0.3590 | 0.3826 | 0.4062 | 0.4296 | 0.4531 | 0.4765 |
| | | 0.6410 | 0.6646 | 0.6788 | 0.6992 | 0.7230 | 0.7470 | 0.7594 |
| | 0.90 | 0.3604 | 0.3837 | 0.4070 | 0.4303 | 0.4535 | 0.4767 | 0.5000 |
| | | 0.6163 | 0.6396 | 0.6630 | 0.6826 | 0.6973 | 0.7207 | 0.7443 |
| | 0.80 | 0.3846 | 0.4077 | 0.4308 | 0.4538 | 0.4769 | 0.5000 | 0.5230 |
| | | 0.5923 | 0.6154 | 0.6385 | 0.6550 | 0.6727 | 0.6958 | 0.7189 |
| | 0.99 | 0.2886 | 0.3126 | 0.3329 | 0.3471 | 0.3709 | 0.3945 | 0.4180 |
| | | 0.6671 | 0.6874 | 0.7114 | 0.7288 | 0.7460 | 0.7706 | 0.7826 |
| | 0.98 | 0.3006 | 0.3257 | 0.3492 | 0.3725 | 0.3916 | 0.4069 | 0.4302 |
| | | 0.6508 | 0.6743 | 0.6994 | 0.7098 | 0.7322 | 0.7562 | 0.7679 |
| 44 | 0.95 | 0.3277 | 0.3509 | 0.3739 | 0.3969 | 0.4198 | 0.4428 | 0.4638 |
| | | 0.6261 | 0.6491 | 0.6723 | 0.6883 | 0.7060 | 0.7293 | 0.7527 |
| | 0.90 | 0.3522 | 0.3750 | 0.3978 | 0.4205 | 0.4432 | 0.4659 | 0.4886 |
| | | 0.6022 | 0.6250 | 0.6478 | 0.6706 | 0.6950 | 0.7054 | 0.7270 |
| | 0.80 | 0.3760 | 0.3985 | 0.4211 | 0.4436 | 0.4662 | 0.4887 | 0.5112 |
| | | 0.5789 | 0.6015 | 0.6240 | 0.6466 | 0.6666 | 0.6806 | 0.7026 |

| $n$ | $\beta$ | 0 | 1 | 2 | 3 | 4 | 5 | 6 |
|---|---|---|---|---|---|---|---|---|
| 45 | 0.99 | 0.0000 | 0.0002 | 0.0033 | 0.0098 | 0.0187 | 0.0293 | 0.0411 |
| | | 0.1097 | 0.1552 | 0.1901 | 0.2244 | 0.2484 | 0.2822 | 0.3056 |
| | 0.98 | 0.0000 | 0.0004 | 0.0048 | 0.0128 | 0.0231 | 0.0349 | 0.0480 |
| | | 0.0921 | 0.1367 | 0.1710 | 0.2049 | 0.2385 | 0.2619 | 0.2954 |
| | 0.95 | 0.0000 | 0.0011 | 0.0079 | 0.0184 | 0.0309 | 0.0448 | 0.0596 |
| | | 0.0746 | 0.1184 | 0.1521 | 0.1855 | 0.2088 | 0.2419 | 0.2648 |
| | 0.90 | 0.0000 | 0.0023 | 0.0118 | 0.0247 | 0.0393 | 0.0550 | 0.0658 |
| | | 0.0659 | 0.1000 | 0.1332 | 0.1661 | 0.1892 | 0.2219 | 0.2445 |
| | 0.80 | 0.0000 | 0.0049 | 0.0183 | 0.0342 | 0.0484 | 0.0618 | 0.0815 |
| | | 0.0485 | 0.0816 | 0.1143 | 0.1371 | 0.1695 | 0.1919 | 0.2243 |
| 46 | 0.99 | 0.0000 | 0.0002 | 0.0032 | 0.0096 | 0.0183 | 0.0286 | 0.0402 |
| | | 0.1073 | 0.1518 | 0.1859 | 0.2194 | 0.2430 | 0.2760 | 0.2989 |
| | 0.98 | 0.0000 | 0.0004 | 0.0047 | 0.0125 | 0.0225 | 0.0341 | 0.0469 |
| | | 0.0901 | 0.1337 | 0.1673 | 0.2004 | 0.2332 | 0.2562 | 0.2889 |
| | 0.95 | 0.0000 | 0.0011 | 0.0077 | 0.0180 | 0.0302 | 0.0438 | 0.0583 |
| | | 0.0730 | 0.1158 | 0.1488 | 0.1815 | 0.2043 | 0.2366 | 0.2590 |
| | 0.90 | 0.0000 | 0.0022 | 0.0116 | 0.0241 | 0.0384 | 0.0538 | 0.0644 |
| | | 0.0645 | 0.0979 | 0.1304 | 0.1626 | 0.1851 | 0.2171 | 0.2393 |
| | 0.80 | 0.0000 | 0.0048 | 0.0179 | 0.0335 | 0.0473 | 0.0604 | 0.0798 |
| | | 0.0474 | 0.0799 | 0.1118 | 0.1341 | 0.1658 | 0.1878 | 0.2195 |
| 47 | 0.99 | 0.0000 | 0.0002 | 0.0031 | 0.0094 | 0.0179 | 0.0280 | 0.0393 |
| | | 0.1050 | 0.1485 | 0.1819 | 0.2147 | 0.2386 | 0.2700 | 0.2963 |
| | 0.98 | 0.0000 | 0.0004 | 0.0046 | 0.0122 | 0.0220 | 0.0334 | 0.0459 |
| | | 0.0890 | 0.1309 | 0.1638 | 0.1961 | 0.2283 | 0.2507 | 0.2827 |
| | 0.95 | 0.0000 | 0.0010 | 0.0076 | 0.0176 | 0.0295 | 0.0428 | 0.0570 |
| | | 0.0715 | 0.1133 | 0.1457 | 0.1776 | 0.2000 | 0.2316 | 0.2535 |
| | 0.90 | 0.0000 | 0.0022 | 0.0113 | 0.0236 | 0.0376 | 0.0526 | 0.0630 |
| | | 0.0631 | 0.0958 | 0.1276 | 0.1591 | 0.1812 | 0.2125 | 0.2342 |
| | 0.80 | 0.0000 | 0.0047 | 0.0175 | 0.0328 | 0.0464 | 0.0590 | 0.0781 |
| | | 0.0465 | 0.0782 | 0.1095 | 0.1317 | 0.1623 | 0.1838 | 0.2149 |
| 48 | 0.99 | 0.0000 | 0.0002 | 0.0031 | 0.0092 | 0.0175 | 0.0274 | 0.0385 |
| | | 0.1028 | 0.1454 | 0.1781 | 0.2102 | 0.2420 | 0.2643 | 0.2959 |
| | 0.98 | 0.0000 | 0.0004 | 0.0045 | 0.0119 | 0.0216 | 0.0327 | 0.0449 |
| | | 0.0945 | 0.1282 | 0.1603 | 0.1920 | 0.2235 | 0.2455 | 0.2767 |
| | 0.95 | 0.0000 | 0.0010 | 0.0074 | 0.0172 | 0.0289 | 0.0419 | 0.0558 |
| | | 0.0700 | 0.1110 | 0.1426 | 0.1739 | 0.1958 | 0.2268 | 0.2483 |
| | 0.90 | 0.0000 | 0.0021 | 0.0111 | 0.0231 | 0.0368 | 0.0515 | 0.0617 |
| | | 0.0618 | 0.0938 | 0.1250 | 0.1558 | 0.1774 | 0.2081 | 0.2294 |
| | 0.80 | 0.0000 | 0.0046 | 0.0172 | 0.0321 | 0.0454 | 0.0577 | 0.0765 |
| | | 0.0455 | 0.0766 | 0.1072 | 0.1377 | 0.1590 | 0.1801 | 0.2105 |

| $n$ | $\beta$ | 7 | 8 | 9 | 10 | 11 | 12 | 13 |
|---|---|---|---|---|---|---|---|---|
| | 0.99 | 0.0540 | 0.0677 | 0.0822 | 0.0972 | 0.1096 | 0.1242 | 0.1439 |
| | | 0.3394 | 0.3626 | 0.3856 | 0.4086 | 0.4377 | 0.4656 | 0.4886 |
| | 0.98 | 0.0619 | 0.0767 | 0.0920 | 0.0972 | 0.1218 | 0.1366 | 0.1529 |
| | | 0.3185 | 0.3414 | 0.3674 | 0.3978 | 0.4206 | 0.4433 | 0.4660 |
| 45 | 0.95 | 0.0745 | 0.0823 | 0.1056 | 0.1183 | 0.1384 | 0.1520 | 0.1746 |
| | | 0.2875 | 0.3205 | 0.3431 | 0.3657 | 0.3882 | 0.4111 | 0.4403 |
| | 0.90 | 0.0842 | 0.0999 | 0.1171 | 0.1331 | 0.1547 | 0.1660 | 0.1891 |
| | | 0.2671 | 0.2998 | 0.3222 | 0.3446 | 0.3669 | 0.3891 | 0.4113 |
| | 0.80 | 0.0967 | 0.1142 | 0.1370 | 0.1450 | 0.1694 | 0.1918 | 0.2097 |
| | | 0.2466 | 0.2688 | 0.2910 | 0.3236 | 0.3457 | 0.3678 | 0.3899 |
| | 0.99 | 0.0528 | 0.0662 | 0.0803 | 0.0950 | 0.1072 | 0.1212 | 0.1405 |
| | | 0.3318 | 0.3545 | 0.3771 | 0.4068 | 0.4327 | 0.4552 | 0.4776 |
| | 0.98 | 0.0605 | 0.0750 | 0.0900 | 0.0943 | 0.1189 | 0.1336 | 0.1491 |
| | | 0.3115 | 0.3339 | 0.3667 | 0.3890 | 0.4113 | 0.4335 | 0.4557 |
| 46 | 0.95 | 0.0729 | 0.0803 | 0.1032 | 0.1157 | 0.1351 | 0.1487 | 0.1704 |
| | | 0.2826 | 0.3136 | 0.3357 | 0.3577 | 0.3797 | 0.4122 | 0.4342 |
| | 0.90 | 0.0823 | 0.0978 | 0.1144 | 0.1303 | 0.1510 | 0.1625 | 0.1850 |
| | | 0.2613 | 0.2933 | 0.3152 | 0.3371 | 0.3589 | 0.3807 | 0.4040 |
| | 0.80 | 0.0945 | 0.1117 | 0.1340 | 0.1408 | 0.1657 | 0.1877 | 0.2044 |
| | | 0.2413 | 0.2631 | 0.2847 | 0.3166 | 0.3383 | 0.3599 | 0.3815 |
| | 0.99 | 0.0516 | 0.0647 | 0.0785 | 0.0928 | 0.1049 | 0.1183 | 0.1372 |
| | | 0.3247 | 0.3469 | 0.3709 | 0.4013 | 0.4233 | 0.4453 | 0.4672 |
| | 0.98 | 0.0592 | 0.0733 | 0.0880 | 0.0911 | 0.1162 | 0.1308 | 0.1455 |
| | | 0.3048 | 0.3268 | 0.3588 | 0.3806 | 0.4024 | 0.4242 | 0.4459 |
| 47 | 0.95 | 0.0714 | 0.0783 | 0.1009 | 0.1132 | 0.1320 | 0.1456 | 0.1664 |
| | | 0.2851 | 0.3069 | 0.3285 | 0.3501 | 0.3753 | 0.4034 | 0.4249 |
| | 0.90 | 0.0804 | 0.0957 | 0.1117 | 0.1275 | 0.1475 | 0.1590 | 0.1811 |
| | | 0.2558 | 0.2871 | 0.3086 | 0.3300 | 0.3513 | 0.3726 | 0.4042 |
| | 0.80 | 0.0923 | 0.1094 | 0.1312 | 0.1364 | 0.1622 | 0.1837 | 0.1993 |
| | | 0.2363 | 0.2575 | 0.2788 | 0.3099 | 0.3312 | 0.3523 | 0.3735 |
| | 0.99 | 0.0505 | 0.0633 | 0.0767 | 0.0908 | 0.1027 | 0.1155 | 0.1341 |
| | | 0.3178 | 0.3395 | 0.3711 | 0.3928 | 0.4143 | 0.4358 | 0.4572 |
| | 0.98 | 0.0579 | 0.0717 | 0.0861 | 0.0944 | 0.1135 | 0.1281 | 0.1420 |
| | | 0.2984 | 0.3251 | 0.3512 | 0.3726 | 0.3939 | 0.4152 | 0.4376 |
| 48 | 0.95 | 0.0699 | 0.0764 | 0.0987 | 0.1109 | 0.1290 | 0.1425 | 0.1626 |
| | | 0.2792 | 0.3005 | 0.3217 | 0.3428 | 0.3739 | 0.3950 | 0.4160 |
| | 0.90 | 0.0787 | 0.0937 | 0.1092 | 0.1249 | 0.1442 | 0.1557 | 0.1773 |
| | | 0.2505 | 0.2812 | 0.3022 | 0.3232 | 0.3441 | 0.3693 | 0.3958 |
| | 0.80 | 0.0903 | 0.1071 | 0.1283 | 0.1376 | 0.1589 | 0.1800 | 0.1944 |
| | | 0.2314 | 0.2522 | 0.2730 | 0.3036 | 0.3243 | 0.3451 | 0.3658 |

| n | β | 14 | 15 | 16 | 17 | 18 | 19 | 20 |
|---|---|---|---|---|---|---|---|---|
| 45 | 0.99 | 0.1551 | 0.1765 | 0.1900 | 0.2118 | 0.2243 | 0.2483 | 0.2632 |
| | | 0.5115 | 0.5345 | 0.5546 | 0.5723 | 0.5915 | 0.6145 | 0.6375 |
| | 0.98 | 0.1709 | 0.1876 | 0.2048 | 0.2261 | 0.2384 | 0.2618 | 0.2823 |
| | | 0.4887 | 0.5114 | 0.5341 | 0.5568 | 0.5795 | 0.6023 | 0.6223 |
| | 0.95 | 0.1854 | 0.2087 | 0.2281 | 0.2418 | 0.2647 | 0.2874 | 0.3023 |
| | | 0.4664 | 0.4888 | 0.5113 | 0.5337 | 0.5547 | 0.5713 | 0.5899 |
| | 0.90 | 0.2095 | 0.2218 | 0.2444 | 0.2670 | 0.2866 | 0.2997 | 0.3221 |
| | | 0.4348 | 0.4609 | 0.4868 | 0.5095 | 0.5289 | 0.5482 | 0.5675 |
| | 0.80 | 0.2242 | 0.2465 | 0.2687 | 0.2909 | 0.3107 | 0.3235 | 0.3456 |
| | | 0.4119 | 0.4340 | 0.4560 | 0.4780 | 0.5000 | 0.5221 | 0.5441 |
| 46 | 0.99 | 0.1517 | 0.1722 | 0.1858 | 0.2064 | 0.2193 | 0.2429 | 0.2554 |
| | | 0.5000 | 0.5225 | 0.5449 | 0.5674 | 0.5896 | 0.6038 | 0.6230 |
| | 0.98 | 0.1672 | 0.1828 | 0.2003 | 0.2203 | 0.2331 | 0.2561 | 0.2747 |
| | | 0.4779 | 0.5000 | 0.5222 | 0.5444 | 0.5666 | 0.5888 | 0.6111 |
| | 0.95 | 0.1814 | 0.2042 | 0.2222 | 0.2365 | 0.2589 | 0.2807 | 0.2924 |
| | | 0.4562 | 0.4781 | 0.5000 | 0.5220 | 0.5439 | 0.5659 | 0.5879 |
| | 0.90 | 0.2043 | 0.2170 | 0.2392 | 0.2612 | 0.2789 | 0.2932 | 0.3151 |
| | | 0.4348 | 0.4565 | 0.4783 | 0.5000 | 0.5218 | 0.5436 | 0.5653 |
| | 0.80 | 0.2194 | 0.2412 | 0.2630 | 0.2846 | 0.3022 | 0.3165 | 0.3382 |
| | | 0.4031 | 0.4246 | 0.4462 | 0.4677 | 0.4893 | 0.5108 | 0.5324 |
| 47 | 0.99 | 0.1484 | 0.1680 | 0.1818 | 0.2013 | 0.2146 | 0.2374 | 0.2472 |
| | | 0.4891 | 0.5110 | 0.5329 | 0.5548 | 0.5768 | 0.5988 | 0.6168 |
| | 0.98 | 0.1637 | 0.1782 | 0.1960 | 0.2148 | 0.2282 | 0.2506 | 0.2673 |
| | | 0.4676 | 0.4892 | 0.5109 | 0.5325 | 0.5542 | 0.5759 | 0.5977 |
| | 0.95 | 0.1775 | 0.1999 | 0.2166 | 0.2315 | 0.2534 | 0.2736 | 0.2850 |
| | | 0.4464 | 0.4679 | 0.4893 | 0.5108 | 0.5322 | 0.5537 | 0.5752 |
| | 0.90 | 0.1994 | 0.2124 | 0.2341 | 0.2557 | 0.2714 | 0.2870 | 0.3085 |
| | | 0.4255 | 0.4468 | 0.4681 | 0.4894 | 0.5107 | 0.5320 | 0.5533 |
| | 0.80 | 0.2148 | 0.2362 | 0.2574 | 0.2787 | 0.2938 | 0.3098 | 0.3311 |
| | | 0.3946 | 0.4157 | 0.4368 | 0.4579 | 0.4790 | 0.5000 | 0.5211 |
| 48 | 0.99 | 0.1453 | 0.1641 | 0.1780 | 0.1965 | 0.2101 | 0.2316 | 0.2419 |
| | | 0.4786 | 0.5000 | 0.5215 | 0.5429 | 0.5643 | 0.5858 | 0.6073 |
| | 0.98 | 0.1602 | 0.1738 | 0.1919 | 0.2096 | 0.2234 | 0.2454 | 0.2602 |
| | | 0.4623 | 0.4865 | 0.5081 | 0.5273 | 0.5459 | 0.5645 | 0.5849 |
| | 0.95 | 0.1738 | 0.1957 | 0.2113 | 0.2267 | 0.2482 | 0.2667 | 0.2791 |
| | | 0.4371 | 0.4581 | 0.4791 | 0.5000 | 0.5210 | 0.5420 | 0.5630 |
| | 0.90 | 0.1946 | 0.2080 | 0.2293 | 0.2504 | 0.2640 | 0.2811 | 0.3021 |
| | | 0.4167 | 0.4375 | 0.4584 | 0.4792 | 0.5000 | 0.5209 | 0.5417 |
| | 0.80 | 0.2104 | 0.2313 | 0.2521 | 0.2729 | 0.2855 | 0.3035 | 0.3242 |
| | | 0.3865 | 0.4071 | 0.4278 | 0.4484 | 0.4691 | 0.4897 | 0.5104 |

| $n$ | $\beta$ | 21 | 22 | 23 | 24 | 25 | 26 | 27 |
|---|---|---|---|---|---|---|---|---|
| | 0.99 | 0.2821 | 0.3055 | 0.3228 | 0.3393 | 0.3625 | 0.3855 | 0.4085 |
| | | 0.6607 | 0.6772 | 0.6945 | 0.7179 | 0.7368 | 0.7517 | 0.7757 |
| | 0.98 | 0.2953 | 0.3184 | 0.3413 | 0.3626 | 0.3777 | 0.3977 | 0.4205 |
| | | 0.6374 | 0.6587 | 0.6816 | 0.7047 | 0.7177 | 0.7382 | 0.7616 |
| 45 | 0.95 | 0.3204 | 0.3430 | 0.3656 | 0.3881 | 0.4101 | 0.4287 | 0.4453 |
| | | 0.6119 | 0.6344 | 0.6570 | 0.6796 | 0.6977 | 0.7126 | 0.7353 |
| | 0.90 | 0.3445 | 0.3668 | 0.3890 | 0.4112 | 0.4325 | 0.4518 | 0.4711 |
| | | 0.5888 | 0.6110 | 0.6332 | 0.6555 | 0.6779 | 0.7003 | 0.7134 |
| | 0.80 | 0.3677 | 0.3898 | 0.4118 | 0.4339 | 0.4559 | 0.4779 | 0.5000 |
| | | 0.5661 | 0.5882 | 0.6102 | 0.6323 | 0.6544 | 0.6765 | 0.6893 |
| | 0.99 | 0.2759 | 0.2988 | 0.3128 | 0.3317 | 0.3544 | 0.3770 | 0.3962 |
| | | 0.6456 | 0.6683 | 0.6872 | 0.7012 | 0.7241 | 0.7446 | 0.7571 |
| | 0.98 | 0.2888 | 0.3114 | 0.3338 | 0.3520 | 0.3666 | 0.3889 | 0.4112 |
| | | 0.6334 | 0.6480 | 0.6662 | 0.6886 | 0.7112 | 0.7253 | 0.7439 |
| 46 | 0.95 | 0.3135 | 0.3356 | 0.3576 | 0.3796 | 0.3977 | 0.4121 | 0.4341 |
| | | 0.6023 | 0.6204 | 0.6424 | 0.6644 | 0.6865 | 0.7076 | 0.7193 |
| | 0.90 | 0.3370 | 0.3588 | 0.3806 | 0.4013 | 0.4185 | 0.4347 | 0.4564 |
| | | 0.5815 | 0.5987 | 0.6194 | 0.6412 | 0.6630 | 0.6849 | 0.7068 |
| | 0.80 | 0.3598 | 0.3814 | 0.4030 | 0.4245 | 0.4461 | 0.4676 | 0.4892 |
| | | 0.5539 | 0.5755 | 0.5970 | 0.6186 | 0.6402 | 0.6618 | 0.6835 |
| | 0.99 | 0.2699 | 0.2912 | 0.3014 | 0.3246 | 0.3468 | 0.3678 | 0.3832 |
| | | 0.6322 | 0.6532 | 0.6754 | 0.6986 | 0.7088 | 0.7301 | 0.7528 |
| | 0.98 | 0.2826 | 0.3047 | 0.3267 | 0.3412 | 0.3587 | 0.3805 | 0.4023 |
| | | 0.6195 | 0.6413 | 0.6588 | 0.6733 | 0.6953 | 0.7174 | 0.7327 |
| 47 | 0.95 | 0.3068 | 0.3284 | 0.3500 | 0.3697 | 0.3841 | 0.4033 | 0.4248 |
| | | 0.5967 | 0.6159 | 0.6303 | 0.6500 | 0.6716 | 0.6932 | 0.7150 |
| | 0.90 | 0.3299 | 0.3512 | 0.3725 | 0.3894 | 0.4041 | 0.4254 | 0.4467 |
| | | 0.5746 | 0.5959 | 0.6106 | 0.6275 | 0.6488 | 0.6701 | 0.6915 |
| | 0.80 | 0.3522 | 0.3734 | 0.3945 | 0.4156 | 0.4367 | 0.4578 | 0.4789 |
| | | 0.5422 | 0.5633 | 0.5844 | 0.6055 | 0.6266 | 0.6478 | 0.6689 |
| | 0.99 | 0.2642 | 0.2837 | 0.2958 | 0.3177 | 0.3394 | 0.3574 | 0.3710 |
| | | 0.6290 | 0.6426 | 0.6606 | 0.6823 | 0.7042 | 0.7163 | 0.7358 |
| | 0.98 | 0.2766 | 0.2983 | 0.3182 | 0.3284 | 0.3511 | 0.3725 | 0.3938 |
| | | 0.6062 | 0.6275 | 0.6489 | 0.6716 | 0.6818 | 0.7017 | 0.7234 |
| 48 | 0.95 | 0.3004 | 0.3216 | 0.3427 | 0.3592 | 0.3738 | 0.3949 | 0.4159 |
| | | 0.5841 | 0.6051 | 0.6262 | 0.6408 | 0.6573 | 0.6784 | 0.6996 |
| | 0.90 | 0.3231 | 0.3440 | 0.3629 | 0.3759 | 0.3957 | 0.4166 | 0.4374 |
| | | 0.5626 | 0.5834 | 0.6043 | 0.6241 | 0.6371 | 0.6560 | 0.6769 |
| | 0.80 | 0.3450 | 0.3657 | 0.3864 | 0.4070 | 0.4277 | 0.4483 | 0.4690 |
| | | 0.5310 | 0.5517 | 0.5723 | 0.5930 | 0.6136 | 0.6343 | 0.6550 |

| n | β | 0 | 1 | 2 | 3 | 4 | 5 | 6 |
|---|---|---|---|---|---|---|---|---|
| 49 | 0.99 | 0.0000 | 0.0002 | 0.0030 | 0.0090 | 0.0171 | 0.0268 | 0.0376 |
| | | 0.1007 | 0.1424 | 0.1744 | 0.2058 | 0.2370 | 0.2589 | 0.2898 |
| | 0.98 | 0.0000 | 0.0004 | 0.0044 | 0.0117 | 0.0211 | 0.0320 | 0.0439 |
| | | 0.0926 | 0.1255 | 0.1571 | 0.1881 | 0.2189 | 0.2405 | 0.2711 |
| | 0.95 | 0.0000 | 0.0010 | 0.0073 | 0.0168 | 0.0283 | 0.0410 | 0.0546 |
| | | 0.0686 | 0.1087 | 0.1397 | 0.1704 | 0.1919 | 0.2221 | 0.2432 |
| | 0.90 | 0.0000 | 0.0021 | 0.0109 | 0.0227 | 0.0360 | 0.0504 | 0.0605 |
| | | 0.0606 | 0.0919 | 0.1224 | 0.1527 | 0.1739 | 0.2039 | 0.2247 |
| | 0.80 | 0.0000 | 0.0045 | 0.0168 | 0.0314 | 0.0445 | 0.0565 | 0.0749 |
| | | 0.0446 | 0.0750 | 0.1051 | 0.1350 | 0.1558 | 0.1764 | 0.2062 |
| 50 | 0.99 | 0.0000 | 0.0002 | 0.0029 | 0.0088 | 0.0168 | 0.0263 | 0.0369 |
| | | 0.0986 | 0.1395 | 0.1709 | 0.2017 | 0.2322 | 0.2536 | 0.2839 |
| | 0.98 | 0.0000 | 0.0004 | 0.0043 | 0.0115 | 0.0207 | 0.0313 | 0.0430 |
| | | 0.0907 | 0.1230 | 0.1539 | 0.1843 | 0.2145 | 0.2356 | 0.2656 |
| | 0.95 | 0.0000 | 0.0010 | 0.0071 | 0.0165 | 0.0277 | 0.0402 | 0.0535 |
| | | 0.0672 | 0.1066 | 0.1370 | 0.1670 | 0.1880 | 0.2177 | 0.2384 |
| | 0.90 | 0.0000 | 0.0021 | 0.0106 | 0.0222 | 0.0353 | 0.0494 | 0.0593 |
| | | 0.0594 | 0.0901 | 0.1200 | 0.1497 | 0.1704 | 0.1998 | 0.2203 |
| | 0.80 | 0.0000 | 0.0044 | 0.0165 | 0.0308 | 0.0436 | 0.0553 | 0.0735 |
| | | 0.0437 | 0.0736 | 0.1030 | 0.1323 | 0.1527 | 0.1730 | 0.2022 |
| 51 | 0.99 | 0.0000 | 0.0001 | 0.0029 | 0.0086 | 0.0164 | 0.0257 | 0.0361 |
| | | 0.0967 | 0.1368 | 0.1675 | 0.1977 | 0.2276 | 0.2486 | 0.2783 |
| | 0.98 | 0.0000 | 0.0003 | 0.0042 | 0.0112 | 0.0203 | 0.0307 | 0.0421 |
| | | 0.0890 | 0.1206 | 0.1509 | 0.1807 | 0.2103 | 0.2325 | 0.2604 |
| | 0.95 | 0.0000 | 0.0010 | 0.0070 | 0.0162 | 0.0272 | 0.0394 | 0.0524 |
| | | 0.0659 | 0.1045 | 0.1343 | 0.1637 | 0.1844 | 0.2135 | 0.2337 |
| | 0.90 | 0.0000 | 0.0020 | 0.0104 | 0.0218 | 0.0346 | 0.0484 | 0.0581 |
| | | 0.0582 | 0.0884 | 0.1177 | 0.1467 | 0.1671 | 0.1960 | 0.2160 |
| | 0.80 | 0.0000 | 0.0043 | 0.0161 | 0.0302 | 0.0428 | 0.0542 | 0.0720 |
| | | 0.0429 | 0.0721 | 0.1010 | 0.1297 | 0.1498 | 0.1696 | 0.1982 |
| 52 | 0.99 | 0.0000 | 0.0001 | 0.0028 | 0.0085 | 0.0161 | 0.0252 | 0.0354 |
| | | 0.0948 | 0.1341 | 0.1643 | 0.1939 | 0.2232 | 0.2438 | 0.2728 |
| | 0.98 | 0.0000 | 0.0003 | 0.0041 | 0.0110 | 0.0199 | 0.0301 | 0.0413 |
| | | 0.0872 | 0.1183 | 0.1480 | 0.1772 | 0.2062 | 0.2351 | 0.2553 |
| | 0.95 | 0.0000 | 0.0009 | 0.0068 | 0.0159 | 0.0267 | 0.0386 | 0.0514 |
| | | 0.0646 | 0.1025 | 0.1317 | 0.1606 | 0.1808 | 0.2094 | 0.2292 |
| | 0.90 | 0.0000 | 0.0020 | 0.0102 | 0.0213 | 0.0339 | 0.0475 | 0.0570 |
| | | 0.0571 | 0.0867 | 0.1154 | 0.1439 | 0.1639 | 0.1922 | 0.2119 |
| | 0.80 | 0.0000 | 0.0042 | 0.0158 | 0.0296 | 0.0419 | 0.0531 | 0.0707 |
| | | 0.0420 | 0.0708 | 0.0991 | 0.1273 | 0.1469 | 0.1664 | 0.1945 |

| $n$ | $\beta$ | 7 | 8 | 9 | 10 | 11 | 12 | 13 |
|---|---|---|---|---|---|---|---|---|
| | 0.99 | 0.0494 | 0.0619 | 0.0751 | 0.0888 | 0.1006 | 0.1128 | 0.1311 |
| | | 0.3112 | 0.3325 | 0.3634 | 0.3846 | 0.4057 | 0.4267 | 0.4477 |
| | 0.98 | 0.0567 | 0.0702 | 0.0842 | 0.0925 | 0.1110 | 0.1254 | 0.1387 |
| | | 0.2923 | 0.3229 | 0.3440 | 0.3649 | 0.3858 | 0.4083 | 0.4375 |
| 49 | 0.95 | 0.0685 | 0.0746 | 0.0965 | 0.1086 | 0.1262 | 0.1396 | 0.1589 |
| | | 0.2735 | 0.2944 | 0.3151 | 0.3374 | 0.3662 | 0.3869 | 0.4075 |
| | 0.90 | 0.0770 | 0.0918 | 0.1068 | 0.1223 | 0.1410 | 0.1526 | 0.1738 |
| | | 0.2454 | 0.2755 | 0.2961 | 0.3166 | 0.3371 | 0.3673 | 0.3878 |
| | 0.80 | 0.0883 | 0.1050 | 0.1255 | 0.1349 | 0.1557 | 0.1763 | 0.1897 |
| | | 0.2267 | 0.2472 | 0.2695 | 0.2974 | 0.3178 | 0.3381 | 0.3584 |
| | 0.99 | 0.0484 | 0.0606 | 0.0735 | 0.0870 | 0.0985 | 0.1103 | 0.1283 |
| | | 0.3049 | 0.3299 | 0.3560 | 0.3768 | 0.3974 | 0.4180 | 0.4418 |
| | 0.98 | 0.0555 | 0.0687 | 0.0825 | 0.0906 | 0.1086 | 0.1229 | 0.1355 |
| | | 0.2864 | 0.3164 | 0.3370 | 0.3576 | 0.3780 | 0.4082 | 0.4286 |
| 50 | 0.95 | 0.0671 | 0.0729 | 0.0945 | 0.1065 | 0.1234 | 0.1369 | 0.1554 |
| | | 0.2680 | 0.2885 | 0.3088 | 0.3386 | 0.3589 | 0.3791 | 0.3993 |
| | 0.90 | 0.0754 | 0.0900 | 0.1045 | 0.1199 | 0.1379 | 0.1496 | 0.1703 |
| | | 0.2418 | 0.2700 | 0.2902 | 0.3103 | 0.3317 | 0.3600 | 0.3800 |
| | 0.80 | 0.0864 | 0.1029 | 0.1228 | 0.1322 | 0.1526 | 0.1729 | 0.1852 |
| | | 0.2223 | 0.2423 | 0.2716 | 0.2916 | 0.3115 | 0.3314 | 0.3513 |
| | 0.99 | 0.0474 | 0.0594 | 0.0720 | 0.0852 | 0.0966 | 0.1078 | 0.1255 |
| | | 0.2989 | 0.3285 | 0.3489 | 0.3692 | 0.3895 | 0.4142 | 0.4395 |
| | 0.98 | 0.0544 | 0.0673 | 0.0808 | 0.0889 | 0.1063 | 0.1205 | 0.1325 |
| | | 0.2808 | 0.3101 | 0.3304 | 0.3505 | 0.3757 | 0.4001 | 0.4201 |
| 51 | 0.95 | 0.0658 | 0.0712 | 0.0925 | 0.1044 | 0.1208 | 0.1342 | 0.1521 |
| | | 0.2628 | 0.2828 | 0.3028 | 0.3319 | 0.3518 | 0.3717 | 0.3915 |
| | 0.90 | 0.0738 | 0.0883 | 0.1023 | 0.1176 | 0.1350 | 0.1466 | 0.1670 |
| | | 0.2448 | 0.2647 | 0.2845 | 0.3043 | 0.3332 | 0.3529 | 0.3726 |
| | 0.80 | 0.0846 | 0.1009 | 0.1203 | 0.1296 | 0.1497 | 0.1695 | 0.1808 |
| | | 0.2180 | 0.2376 | 0.2663 | 0.2859 | 0.3055 | 0.3250 | 0.3445 |
| | 0.99 | 0.0464 | 0.0582 | 0.0705 | 0.0834 | 0.0947 | 0.1055 | 0.1229 |
| | | 0.2930 | 0.3221 | 0.3421 | 0.3620 | 0.3841 | 0.4111 | 0.4309 |
| | 0.98 | 0.0533 | 0.0660 | 0.0792 | 0.0871 | 0.1041 | 0.1182 | 0.1296 |
| | | 0.2760 | 0.3041 | 0.3240 | 0.3437 | 0.3726 | 0.3923 | 0.4119 |
| 52 | 0.95 | 0.0645 | 0.0696 | 0.0907 | 0.1024 | 0.1183 | 0.1316 | 0.1489 |
| | | 0.2577 | 0.2774 | 0.2970 | 0.3255 | 0.3450 | 0.3645 | 0.3839 |
| | 0.90 | 0.0724 | 0.0866 | 0.1002 | 0.1153 | 0.1322 | 0.1438 | 0.1638 |
| | | 0.2402 | 0.2597 | 0.2791 | 0.2985 | 0.3268 | 0.3462 | 0.3655 |
| | 0.80 | 0.0829 | 0.0990 | 0.1178 | 0.1272 | 0.1468 | 0.1663 | 0.1766 |
| | | 0.2138 | 0.2331 | 0.2612 | 0.2805 | 0.2996 | 0.3188 | 0.3379 |

| $n$ | $\beta$ | 14 | 15 | 16 | 17 | 18 | 19 | 20 |
|---|---|---|---|---|---|---|---|---|
| | 0.99 | 0.1423 | 0.1603 | 0.1743 | 0.1918 | 0.2057 | 0.2261 | 0.2369 |
| | | 0.4688 | 0.4907 | 0.5116 | 0.5316 | 0.5524 | 0.5734 | 0.5944 |
| | 0.98 | 0.1570 | 0.1696 | 0.1880 | 0.2046 | 0.2188 | 0.2404 | 0.2532 |
| | | 0.4584 | 0.4792 | 0.5000 | 0.5209 | 0.5417 | 0.5626 | 0.5779 |
| 49 | 0.95 | 0.1703 | 0.1918 | 0.2061 | 0.2220 | 0.2431 | 0.2601 | 0.2734 |
| | | 0.4281 | 0.4487 | 0.4692 | 0.4898 | 0.5103 | 0.5309 | 0.5514 |
| | 0.90 | 0.1901 | 0.2038 | 0.2246 | 0.2453 | 0.2567 | 0.2754 | 0.2960 |
| | | 0.4082 | 0.4286 | 0.4491 | 0.4695 | 0.4899 | 0.5102 | 0.5306 |
| | 0.80 | 0.2061 | 0.2266 | 0.2471 | 0.2668 | 0.2765 | 0.2973 | 0.3177 |
| | | 0.3787 | 0.3989 | 0.4192 | 0.4394 | 0.4611 | 0.4829 | 0.5036 |
| | 0.99 | 0.1394 | 0.1567 | 0.1708 | 0.1873 | 0.2016 | 0.2208 | 0.2321 |
| | | 0.4692 | 0.4898 | 0.5103 | 0.5309 | 0.5465 | 0.5639 | 0.5821 |
| | 0.98 | 0.1538 | 0.1655 | 0.1842 | 0.1997 | 0.2144 | 0.2355 | 0.2461 |
| | | 0.4491 | 0.4695 | 0.4899 | 0.5102 | 0.5306 | 0.5510 | 0.5715 |
| 50 | 0.95 | 0.1669 | 0.1879 | 0.2011 | 0.2176 | 0.2383 | 0.2536 | 0.2679 |
| | | 0.4195 | 0.4397 | 0.4598 | 0.4799 | 0.5000 | 0.5202 | 0.5403 |
| | 0.90 | 0.1857 | 0.1997 | 0.2202 | 0.2401 | 0.2489 | 0.2699 | 0.2901 |
| | | 0.4001 | 0.4201 | 0.4401 | 0.4601 | 0.4801 | 0.5000 | 0.5200 |
| | 0.80 | 0.2021 | 0.2222 | 0.2422 | 0.2606 | 0.2715 | 0.2915 | 0.3114 |
| | | 0.3712 | 0.3910 | 0.4109 | 0.4354 | 0.4603 | 0.4802 | 0.5000 |
| | 0.99 | 0.1367 | 0.1532 | 0.1674 | 0.1831 | 0.1976 | 0.2157 | 0.2275 |
| | | 0.4597 | 0.4799 | 0.5000 | 0.5202 | 0.5404 | 0.5606 | 0.5783 |
| | 0.98 | 0.1508 | 0.1616 | 0.1806 | 0.1951 | 0.2102 | 0.2305 | 0.2386 |
| | | 0.4401 | 0.4601 | 0.4801 | 0.5000 | 0.5200 | 0.5400 | 0.5600 |
| 51 | 0.95 | 0.1636 | 0.1843 | 0.1963 | 0.2134 | 0.2336 | 0.2474 | 0.2627 |
| | | 0.4113 | 0.4310 | 0.4508 | 0.4724 | 0.4930 | 0.5126 | 0.5313 |
| | 0.90 | 0.1815 | 0.1959 | 0.2159 | 0.2347 | 0.2447 | 0.2646 | 0.2844 |
| | | 0.3923 | 0.4119 | 0.4315 | 0.4511 | 0.4707 | 0.4903 | 0.5098 |
| | 0.80 | 0.1981 | 0.2179 | 0.2375 | 0.2545 | 0.2662 | 0.2858 | 0.3054 |
| | | 0.3640 | 0.3834 | 0.4090 | 0.4319 | 0.4514 | 0.4708 | 0.4903 |
| | 0.99 | 0.1340 | 0.1499 | 0.1642 | 0.1789 | 0.1938 | 0.2109 | 0.2231 |
| | | 0.4507 | 0.4704 | 0.4902 | 0.5099 | 0.5297 | 0.5494 | 0.5692 |
| | 0.98 | 0.1479 | 0.1578 | 0.1771 | 0.1906 | 0.2061 | 0.2254 | 0.2350 |
| | | 0.4316 | 0.4511 | 0.4707 | 0.4903 | 0.5098 | 0.5294 | 0.5490 |
| 52 | 0.95 | 0.1605 | 0.1807 | 0.1916 | 0.2093 | 0.2291 | 0.2413 | 0.2576 |
| | | 0.4033 | 0.4243 | 0.4515 | 0.4710 | 0.4904 | 0.5097 | 0.5291 |
| | 0.90 | 0.1775 | 0.1921 | 0.2118 | 0.2295 | 0.2401 | 0.2596 | 0.2790 |
| | | 0.3847 | 0.4040 | 0.4232 | 0.4424 | 0.4616 | 0.4809 | 0.5000 |
| | 0.80 | 0.1944 | 0.2137 | 0.2330 | 0.2486 | 0.2611 | 0.2804 | 0.2995 |
| | | 0.3570 | 0.3788 | 0.4045 | 0.4237 | 0.4428 | 0.4619 | 0.4810 |

| $n$ | $\beta$ | 21 | 22 | 23 | 24 | 25 | 26 | 27 |
|---|---|---|---|---|---|---|---|---|
| | 0.99 | 0.2588 | 0.2765 | 0.2897 | 0.3111 | 0.3324 | 0.3469 | 0.3633 |
| | | 0.6155 | 0.6367 | 0.6531 | 0.6676 | 0.6889 | 0.7103 | 0.7235 |
| | 0.98 | 0.2710 | 0.2922 | 0.3099 | 0.3228 | 0.3439 | 0.3648 | 0.3857 |
| | | 0.5945 | 0.6143 | 0.6352 | 0.6561 | 0.6772 | 0.6901 | 0.7078 |
| 49 | 0.95 | 0.2943 | 0.3150 | 0.3350 | 0.3482 | 0.3661 | 0.3868 | 0.4074 |
| | | 0.5720 | 0.5926 | 0.6132 | 0.6339 | 0.6518 | 0.6650 | 0.6850 |
| | 0.90 | 0.3165 | 0.3370 | 0.3529 | 0.3672 | 0.3877 | 0.4081 | 0.4285 |
| | | 0.5510 | 0.5715 | 0.5919 | 0.6123 | 0.6328 | 0.6471 | 0.6630 |
| | 0.80 | 0.3380 | 0.3583 | 0.3786 | 0.3988 | 0.4191 | 0.4393 | 0.4583 |
| | | 0.5230 | 0.5417 | 0.5607 | 0.5809 | 0.6012 | 0.6214 | 0.6417 |
| | 0.99 | 0.2535 | 0.2695 | 0.2838 | 0.3048 | 0.3242 | 0.3350 | 0.3559 |
| | | 0.6027 | 0.6233 | 0.6441 | 0.6650 | 0.6758 | 0.6952 | 0.7162 |
| | 0.98 | 0.2655 | 0.2863 | 0.3018 | 0.3163 | 0.3369 | 0.3575 | 0.3779 |
| | | 0.5919 | 0.6061 | 0.6221 | 0.6425 | 0.6631 | 0.6837 | 0.6982 |
| 50 | 0.95 | 0.2884 | 0.3087 | 0.3265 | 0.3385 | 0.3588 | 0.3790 | 0.3992 |
| | | 0.5604 | 0.5806 | 0.6008 | 0.6210 | 0.6412 | 0.6615 | 0.6735 |
| | 0.90 | 0.3102 | 0.3297 | 0.3424 | 0.3599 | 0.3799 | 0.4000 | 0.4200 |
| | | 0.5400 | 0.5600 | 0.5800 | 0.6000 | 0.6201 | 0.6401 | 0.6576 |
| | 0.80 | 0.3313 | 0.3512 | 0.3711 | 0.3909 | 0.4108 | 0.4276 | 0.4432 |
| | | 0.5199 | 0.5398 | 0.5568 | 0.5724 | 0.5892 | 0.6091 | 0.6289 |
| | 0.99 | 0.2485 | 0.2627 | 0.2782 | 0.2988 | 0.3160 | 0.3284 | 0.3488 |
| | | 0.5930 | 0.6106 | 0.6309 | 0.6512 | 0.6716 | 0.6840 | 0.7012 |
| | 0.98 | 0.2603 | 0.2807 | 0.2938 | 0.3100 | 0.3303 | 0.3504 | 0.3683 |
| | | 0.5800 | 0.6000 | 0.6198 | 0.6317 | 0.6496 | 0.6697 | 0.6900 |
| 51 | 0.95 | 0.2827 | 0.3027 | 0.3181 | 0.3318 | 0.3517 | 0.3716 | 0.3914 |
| | | 0.5494 | 0.5691 | 0.5888 | 0.6086 | 0.6284 | 0.6483 | 0.6682 |
| | 0.90 | 0.3042 | 0.3215 | 0.3331 | 0.3528 | 0.3725 | 0.3922 | 0.4118 |
| | | 0.5294 | 0.5490 | 0.5686 | 0.5882 | 0.6078 | 0.6275 | 0.6472 |
| | 0.80 | 0.3249 | 0.3444 | 0.3639 | 0.3833 | 0.3999 | 0.4126 | 0.4318 |
| | | 0.5098 | 0.5293 | 0.5487 | 0.5682 | 0.5874 | 0.6001 | 0.6167 |
| | 0.99 | 0.2437 | 0.2560 | 0.2727 | 0.2929 | 0.3079 | 0.3220 | 0.3420 |
| | | 0.5890 | 0.6052 | 0.6194 | 0.6381 | 0.6580 | 0.6780 | 0.6921 |
| | 0.98 | 0.2552 | 0.2750 | 0.2854 | 0.3040 | 0.3239 | 0.3436 | 0.3584 |
| | | 0.5685 | 0.5882 | 0.6078 | 0.6275 | 0.6416 | 0.6564 | 0.6761 |
| 52 | 0.95 | 0.2773 | 0.2969 | 0.3096 | 0.3254 | 0.3449 | 0.3644 | 0.3838 |
| | | 0.5486 | 0.5623 | 0.5785 | 0.5968 | 0.6162 | 0.6356 | 0.6551 |
| | 0.90 | 0.2984 | 0.3135 | 0.3267 | 0.3461 | 0.3654 | 0.3846 | 0.4039 |
| | | 0.5192 | 0.5385 | 0.5577 | 0.5769 | 0.5961 | 0.6154 | 0.6346 |
| | 0.80 | 0.3187 | 0.3378 | 0.3569 | 0.3747 | 0.3880 | 0.4044 | 0.4236 |
| | | 0.5000 | 0.5191 | 0.5382 | 0.5573 | 0.5764 | 0.5956 | 0.6120 |

| n | β | 0 | 1 | 2 | 3 | 4 | 5 | 6 |
|---|---|---|---|---|---|---|---|---|
| 53 | 0.99 | 0.0000 | 0.0001 | 0.0028 | 0.0083 | 0.0158 | 0.0247 | 0.0347 |
| | | 0.0930 | 0.1316 | 0.1612 | 0.1902 | 0.2189 | 0.2391 | 0.2676 |
| | 0.98 | 0.0000 | 0.0003 | 0.0040 | 0.0108 | 0.0195 | 0.0295 | 0.0405 |
| | | 0.0856 | 0.1161 | 0.1452 | 0.1738 | 0.2023 | 0.2307 | 0.2505 |
| | 0.95 | 0.0000 | 0.0009 | 0.0067 | 0.0156 | 0.0261 | 0.0379 | 0.0504 |
| | | 0.0634 | 0.1006 | 0.1292 | 0.1575 | 0.1774 | 0.2054 | 0.2249 |
| | 0.90 | 0.0000 | 0.0019 | 0.0100 | 0.0209 | 0.0333 | 0.0466 | 0.0559 |
| | | 0.0560 | 0.0851 | 0.1133 | 0.1413 | 0.1608 | 0.1886 | 0.2079 |
| | 0.80 | 0.0000 | 0.0042 | 0.0155 | 0.0290 | 0.0412 | 0.0521 | 0.0694 |
| | | 0.0413 | 0.0695 | 0.0972 | 0.1249 | 0.1442 | 0.1633 | 0.1908 |
| 54 | 0.99 | 0.0000 | 0.0001 | 0.0027 | 0.0081 | 0.0155 | 0.0243 | 0.0340 |
| | | 0.0913 | 0.1291 | 0.1582 | 0.1866 | 0.2149 | 0.2361 | 0.2626 |
| | 0.98 | 0.0000 | 0.0003 | 0.0040 | 0.0106 | 0.0191 | 0.0289 | 0.0397 |
| | | 0.0840 | 0.1139 | 0.1425 | 0.1706 | 0.1985 | 0.2264 | 0.2458 |
| | 0.95 | 0.0000 | 0.0009 | 0.0066 | 0.0153 | 0.0256 | 0.0371 | 0.0495 |
| | | 0.0623 | 0.0987 | 0.1269 | 0.1546 | 0.1742 | 0.2016 | 0.2208 |
| | 0.90 | 0.0000 | 0.0019 | 0.0098 | 0.0205 | 0.0326 | 0.0457 | 0.0549 |
| | | 0.0550 | 0.0835 | 0.1112 | 0.1387 | 0.1579 | 0.1851 | 0.2041 |
| | 0.80 | 0.0000 | 0.0041 | 0.0152 | 0.0285 | 0.0404 | 0.0511 | 0.0681 |
| | | 0.0405 | 0.0682 | 0.0954 | 0.1226 | 0.1415 | 0.1603 | 0.1873 |
| 55 | 0.99 | 0.0000 | 0.0001 | 0.0027 | 0.0080 | 0.0152 | 0.0238 | 0.0334 |
| | | 0.0896 | 0.1268 | 0.1553 | 0.1832 | 0.2109 | 0.2385 | 0.2578 |
| | 0.98 | 0.0000 | 0.0003 | 0.0039 | 0.0104 | 0.0188 | 0.0284 | 0.0390 |
| | | 0.0825 | 0.1118 | 0.1399 | 0.1675 | 0.1949 | 0.2222 | 0.2413 |
| | 0.95 | 0.0000 | 0.0009 | 0.0064 | 0.0150 | 0.0252 | 0.0365 | 0.0485 |
| | | 0.0611 | 0.0969 | 0.1246 | 0.1518 | 0.1710 | 0.1980 | 0.2204 |
| | 0.90 | 0.0000 | 0.0019 | 0.0097 | 0.0202 | 0.0320 | 0.0448 | 0.0539 |
| | | 0.0540 | 0.0820 | 0.1092 | 0.1362 | 0.1550 | 0.1818 | 0.2004 |
| | 0.80 | 0.0000 | 0.0040 | 0.0150 | 0.0280 | 0.0397 | 0.0501 | 0.0669 |
| | | 0.0398 | 0.0670 | 0.0937 | 0.1204 | 0.1390 | 0.1574 | 0.1840 |
| 56 | 0.99 | 0.0000 | 0.0001 | 0.0026 | 0.0078 | 0.0149 | 0.0234 | 0.0328 |
| | | 0.0880 | 0.1245 | 0.1525 | 0.1799 | 0.2071 | 0.2342 | 0.2532 |
| | 0.98 | 0.0000 | 0.0003 | 0.0038 | 0.0102 | 0.0184 | 0.0279 | 0.0383 |
| | | 0.0810 | 0.1098 | 0.1374 | 0.1645 | 0.1914 | 0.2183 | 0.2370 |
| | 0.95 | 0.0000 | 0.0009 | 0.0063 | 0.0147 | 0.0247 | 0.0358 | 0.0477 |
| | | 0.0600 | 0.0952 | 0.1223 | 0.1491 | 0.1687 | 0.1944 | 0.2209 |
| | 0.90 | 0.0000 | 0.0018 | 0.0095 | 0.0198 | 0.0315 | 0.0440 | 0.0530 |
| | | 0.0531 | 0.0806 | 0.1073 | 0.1337 | 0.1523 | 0.1786 | 0.1968 |
| | 0.80 | 0.0000 | 0.0039 | 0.0147 | 0.0275 | 0.0390 | 0.0492 | 0.0657 |
| | | 0.0391 | 0.0658 | 0.0921 | 0.1183 | 0.1366 | 0.1547 | 0.1807 |

| $n$ | $\beta$ | 7 | 8 | 9 | 10 | 11 | 12 | 13 |
|---|---|---|---|---|---|---|---|---|
| | 0.99 | 0.0455 | 0.0570 | 0.0692 | 0.0818 | 0.0929 | 0.1033 | 0.1204 |
| | | 0.2875 | 0.3159 | 0.3356 | 0.3551 | 0.3837 | 0.4032 | 0.4226 |
| | 0.98 | 0.0523 | 0.0647 | 0.0776 | 0.0855 | 0.1020 | 0.1160 | 0.1268 |
| | | 0.2788 | 0.2983 | 0.3178 | 0.3396 | 0.3655 | 0.3848 | 0.4041 |
| 53 | 0.95 | 0.0633 | 0.0680 | 0.0889 | 0.1005 | 0.1159 | 0.1291 | 0.1458 |
| | | 0.2529 | 0.2722 | 0.2933 | 0.3194 | 0.3385 | 0.3576 | 0.3767 |
| | 0.90 | 0.0709 | 0.0850 | 0.0982 | 0.1132 | 0.1295 | 0.1412 | 0.1607 |
| | | 0.2357 | 0.2548 | 0.2739 | 0.2929 | 0.3207 | 0.3397 | 0.3586 |
| | 0.80 | 0.0812 | 0.0971 | 0.1155 | 0.1248 | 0.1441 | 0.1632 | 0.1725 |
| | | 0.2098 | 0.2287 | 0.2564 | 0.2752 | 0.2941 | 0.3129 | 0.3316 |
| | 0.99 | 0.0447 | 0.0559 | 0.0678 | 0.0802 | 0.0912 | 0.1011 | 0.1180 |
| | | 0.2834 | 0.3100 | 0.3293 | 0.3501 | 0.3764 | 0.3956 | 0.4146 |
| | 0.98 | 0.0513 | 0.0634 | 0.0761 | 0.0839 | 0.1000 | 0.1138 | 0.1241 |
| | | 0.2736 | 0.2928 | 0.3119 | 0.3396 | 0.3587 | 0.3776 | 0.3965 |
| 54 | 0.95 | 0.0622 | 0.0666 | 0.0871 | 0.0986 | 0.1135 | 0.1268 | 0.1428 |
| | | 0.2482 | 0.2671 | 0.2946 | 0.3135 | 0.3323 | 0.3510 | 0.3697 |
| | 0.90 | 0.0696 | 0.0834 | 0.0962 | 0.1111 | 0.1269 | 0.1386 | 0.1578 |
| | | 0.2313 | 0.2501 | 0.2688 | 0.2885 | 0.3148 | 0.3334 | 0.3520 |
| | 0.80 | 0.0796 | 0.0953 | 0.1132 | 0.1225 | 0.1414 | 0.1602 | 0.1684 |
| | | 0.2060 | 0.2245 | 0.2517 | 0.2702 | 0.2887 | 0.3071 | 0.3256 |
| | 0.99 | 0.0438 | 0.0549 | 0.0665 | 0.0786 | 0.0895 | 0.0990 | 0.1157 |
| | | 0.2853 | 0.3043 | 0.3232 | 0.3507 | 0.3695 | 0.3882 | 0.4069 |
| | 0.98 | 0.0503 | 0.0622 | 0.0747 | 0.0824 | 0.0981 | 0.1117 | 0.1215 |
| | | 0.2686 | 0.2874 | 0.3062 | 0.3334 | 0.3521 | 0.3707 | 0.3892 |
| 55 | 0.95 | 0.0610 | 0.0651 | 0.0855 | 0.0968 | 0.1113 | 0.1245 | 0.1400 |
| | | 0.2437 | 0.2623 | 0.2893 | 0.3078 | 0.3262 | 0.3446 | 0.3655 |
| | 0.90 | 0.0682 | 0.0819 | 0.0943 | 0.1091 | 0.1244 | 0.1361 | 0.1549 |
| | | 0.2271 | 0.2456 | 0.2640 | 0.2908 | 0.3091 | 0.3274 | 0.3456 |
| | 0.80 | 0.0781 | 0.0936 | 0.1111 | 0.1203 | 0.1389 | 0.1573 | 0.1644 |
| | | 0.2023 | 0.2205 | 0.2471 | 0.2653 | 0.2835 | 0.3016 | 0.3197 |
| | 0.99 | 0.0430 | 0.0539 | 0.0653 | 0.0772 | 0.0879 | 0.0970 | 0.1135 |
| | | 0.2801 | 0.2988 | 0.3174 | 0.3443 | 0.3628 | 0.3812 | 0.3996 |
| | 0.98 | 0.0494 | 0.0611 | 0.0733 | 0.0809 | 0.0962 | 0.1097 | 0.1190 |
| | | 0.2637 | 0.2823 | 0.3007 | 0.3274 | 0.3457 | 0.3640 | 0.3822 |
| 56 | 0.95 | 0.0599 | 0.0637 | 0.0839 | 0.0951 | 0.1092 | 0.1222 | 0.1372 |
| | | 0.2393 | 0.2576 | 0.2841 | 0.3023 | 0.3204 | 0.3385 | 0.3651 |
| | 0.90 | 0.0670 | 0.0805 | 0.0925 | 0.1072 | 0.1220 | 0.1336 | 0.1522 |
| | | 0.2231 | 0.2412 | 0.2593 | 0.2856 | 0.3036 | 0.3216 | 0.3395 |
| | 0.80 | 0.0766 | 0.0920 | 0.1090 | 0.1182 | 0.1365 | 0.1546 | 0.1603 |
| | | 0.1987 | 0.2166 | 0.2428 | 0.2606 | 0.2785 | 0.2963 | 0.3141 |

| $n$ | $\beta$ | 14 | 15 | 16 | 17 | 18 | 19 | 20 |
|---|---|---|---|---|---|---|---|---|
| 53 | 0.99 | 0.1315 | 0.1467 | 0.1611 | 0.1750 | 0.1901 | 0.2062 | 0.2188 |
| | | 0.4420 | 0.4614 | 0.4807 | 0.5000 | 0.5194 | 0.5387 | 0.5581 |
| | 0.98 | 0.1451 | 0.1542 | 0.1737 | 0.1863 | 0.2022 | 0.2204 | 0.2306 |
| | | 0.4233 | 0.4425 | 0.4617 | 0.4809 | 0.5000 | 0.5192 | 0.5384 |
| | 0.95 | 0.1574 | 0.1773 | 0.1870 | 0.2053 | 0.2248 | 0.2352 | 0.2528 |
| | | 0.3968 | 0.4239 | 0.4430 | 0.4620 | 0.4810 | 0.5000 | 0.5191 |
| | 0.90 | 0.1736 | 0.1885 | 0.2078 | 0.2244 | 0.2356 | 0.2547 | 0.2738 |
| | | 0.3775 | 0.3964 | 0.4153 | 0.4341 | 0.4531 | 0.4736 | 0.4933 |
| | 0.80 | 0.1907 | 0.2097 | 0.2286 | 0.2430 | 0.2563 | 0.2751 | 0.2940 |
| | | 0.3504 | 0.3782 | 0.3970 | 0.4157 | 0.4345 | 0.4532 | 0.4720 |
| 54 | 0.99 | 0.1290 | 0.1436 | 0.1581 | 0.1712 | 0.1865 | 0.2017 | 0.2148 |
| | | 0.4336 | 0.4526 | 0.4716 | 0.4906 | 0.5095 | 0.5285 | 0.5475 |
| | 0.98 | 0.1424 | 0.1506 | 0.1705 | 0.1822 | 0.1984 | 0.2157 | 0.2263 |
| | | 0.4154 | 0.4342 | 0.4553 | 0.4765 | 0.4965 | 0.5148 | 0.5320 |
| | 0.95 | 0.1545 | 0.1741 | 0.1825 | 0.2015 | 0.2207 | 0.2290 | 0.2481 |
| | | 0.3973 | 0.4161 | 0.4347 | 0.4534 | 0.4721 | 0.4907 | 0.5094 |
| | 0.90 | 0.1698 | 0.1850 | 0.2040 | 0.2196 | 0.2312 | 0.2500 | 0.2687 |
| | | 0.3706 | 0.3891 | 0.4076 | 0.4279 | 0.4537 | 0.4723 | 0.4908 |
| | 0.80 | 0.1872 | 0.2059 | 0.2244 | 0.2375 | 0.2516 | 0.2701 | 0.2886 |
| | | 0.3459 | 0.3713 | 0.3897 | 0.4081 | 0.4265 | 0.4449 | 0.4633 |
| 55 | 0.99 | 0.1267 | 0.1407 | 0.1552 | 0.1676 | 0.1831 | 0.1974 | 0.2108 |
| | | 0.4256 | 0.4442 | 0.4648 | 0.4848 | 0.5038 | 0.5217 | 0.5388 |
| | 0.98 | 0.1398 | 0.1471 | 0.1674 | 0.1781 | 0.1948 | 0.2112 | 0.2221 |
| | | 0.4078 | 0.4315 | 0.4538 | 0.4723 | 0.4908 | 0.5093 | 0.5278 |
| | 0.95 | 0.1517 | 0.1709 | 0.1780 | 0.1979 | 0.2161 | 0.2211 | 0.2436 |
| | | 0.3901 | 0.4085 | 0.4268 | 0.4451 | 0.4634 | 0.4817 | 0.5000 |
| | 0.90 | 0.1662 | 0.1817 | 0.2003 | 0.2149 | 0.2270 | 0.2455 | 0.2639 |
| | | 0.3639 | 0.3821 | 0.4020 | 0.4273 | 0.4455 | 0.4637 | 0.4819 |
| | 0.80 | 0.1839 | 0.2022 | 0.2204 | 0.2320 | 0.2470 | 0.2652 | 0.2834 |
| | | 0.3465 | 0.3646 | 0.3827 | 0.4007 | 0.4188 | 0.4369 | 0.4549 |
| 56 | 0.99 | 0.1244 | 0.1379 | 0.1524 | 0.1640 | 0.1798 | 0.1932 | 0.2070 |
| | | 0.4186 | 0.4422 | 0.4634 | 0.4817 | 0.5000 | 0.5184 | 0.5367 |
| | 0.98 | 0.1373 | 0.1436 | 0.1644 | 0.1742 | 0.1913 | 0.2068 | 0.2182 |
| | | 0.4061 | 0.4273 | 0.4455 | 0.4637 | 0.4819 | 0.5000 | 0.5182 |
| | 0.95 | 0.1490 | 0.1677 | 0.1732 | 0.1943 | 0.2117 | 0.2208 | 0.2392 |
| | | 0.3831 | 0.4011 | 0.4192 | 0.4371 | 0.4551 | 0.4731 | 0.4911 |
| | 0.90 | 0.1627 | 0.1785 | 0.1967 | 0.2104 | 0.2230 | 0.2411 | 0.2592 |
| | | 0.3574 | 0.3753 | 0.4018 | 0.4197 | 0.4375 | 0.4554 | 0.4733 |
| | 0.80 | 0.1806 | 0.1986 | 0.2165 | 0.2267 | 0.2427 | 0.2605 | 0.2784 |
| | | 0.3403 | 0.3581 | 0.3759 | 0.3936 | 0.4114 | 0.4291 | 0.4469 |

| $n$ | $\beta$ | 21 | 22 | 23 | 24 | 25 | 26 | 27 |
|---|---|---|---|---|---|---|---|---|
| | 0.99 | 0.2390 | 0.2492 | 0.2675 | 0.2874 | 0.2999 | 0.3158 | 0.3355 |
| | | 0.5775 | 0.5969 | 0.6164 | 0.6294 | 0.6450 | 0.6645 | 0.6842 |
| | 0.98 | 0.2504 | 0.2688 | 0.2787 | 0.2982 | 0.3177 | 0.3360 | 0.3477 |
| | | 0.5576 | 0.5768 | 0.5960 | 0.6153 | 0.6346 | 0.6523 | 0.6640 |
| 53 | 0.95 | 0.2721 | 0.2906 | 0.3006 | 0.3193 | 0.3384 | 0.3575 | 0.3766 |
| | | 0.5381 | 0.5571 | 0.5762 | 0.5902 | 0.6051 | 0.6234 | 0.6425 |
| | 0.90 | 0.2928 | 0.3056 | 0.3206 | 0.3396 | 0.3585 | 0.3774 | 0.3963 |
| | | 0.5120 | 0.5299 | 0.5472 | 0.5660 | 0.5848 | 0.6037 | 0.6226 |
| | 0.80 | 0.3128 | 0.3315 | 0.3503 | 0.3651 | 0.3781 | 0.3969 | 0.4156 |
| | | 0.4907 | 0.5094 | 0.5281 | 0.5469 | 0.5656 | 0.5844 | 0.6031 |
| | 0.99 | 0.2342 | 0.2421 | 0.2625 | 0.2815 | 0.2914 | 0.3099 | 0.3292 |
| | | 0.5665 | 0.5855 | 0.6045 | 0.6237 | 0.6400 | 0.6525 | 0.6708 |
| | 0.98 | 0.2457 | 0.2628 | 0.2735 | 0.2927 | 0.3118 | 0.3280 | 0.3395 |
| | | 0.5488 | 0.5659 | 0.5847 | 0.6036 | 0.6225 | 0.6414 | 0.6605 |
| 54 | 0.95 | 0.2670 | 0.2841 | 0.2945 | 0.3134 | 0.3322 | 0.3509 | 0.3696 |
| | | 0.5280 | 0.5467 | 0.5654 | 0.5840 | 0.6028 | 0.6152 | 0.6304 |
| | 0.90 | 0.2870 | 0.2972 | 0.3147 | 0.3333 | 0.3519 | 0.3705 | 0.3890 |
| | | 0.5093 | 0.5278 | 0.5464 | 0.5598 | 0.5753 | 0.5925 | 0.6110 |
| | 0.80 | 0.3070 | 0.3255 | 0.3430 | 0.3548 | 0.3712 | 0.3896 | 0.4080 |
| | | 0.4817 | 0.5000 | 0.5184 | 0.5368 | 0.5552 | 0.5736 | 0.5920 |
| | 0.99 | 0.2292 | 0.2384 | 0.2577 | 0.2753 | 0.2852 | 0.3042 | 0.3231 |
| | | 0.5559 | 0.5745 | 0.5932 | 0.6119 | 0.6306 | 0.6494 | 0.6608 |
| | 0.98 | 0.2412 | 0.2569 | 0.2685 | 0.2873 | 0.3061 | 0.3200 | 0.3333 |
| | | 0.5463 | 0.5632 | 0.5769 | 0.5923 | 0.6109 | 0.6294 | 0.6480 |
| 55 | 0.95 | 0.2622 | 0.2777 | 0.2892 | 0.3077 | 0.3261 | 0.3445 | 0.3617 |
| | | 0.5184 | 0.5367 | 0.5550 | 0.5733 | 0.5916 | 0.6100 | 0.6262 |
| | 0.90 | 0.2807 | 0.2907 | 0.3090 | 0.3273 | 0.3455 | 0.3638 | 0.3820 |
| | | 0.5000 | 0.5182 | 0.5364 | 0.5546 | 0.5728 | 0.5869 | 0.6010 |
| | 0.80 | 0.3015 | 0.3196 | 0.3350 | 0.3464 | 0.3645 | 0.3826 | 0.4006 |
| | | 0.4730 | 0.4910 | 0.5091 | 0.5271 | 0.5452 | 0.5632 | 0.5813 |
| | 0.99 | 0.2244 | 0.2341 | 0.2531 | 0.2692 | 0.2800 | 0.2987 | 0.3173 |
| | | 0.5539 | 0.5674 | 0.5827 | 0.6005 | 0.6189 | 0.6373 | 0.6558 |
| | 0.98 | 0.2369 | 0.2513 | 0.2636 | 0.2822 | 0.3006 | 0.3118 | 0.3273 |
| | | 0.5364 | 0.5546 | 0.5728 | 0.5907 | 0.6023 | 0.6179 | 0.6361 |
| 56 | 0.95 | 0.2575 | 0.2715 | 0.2840 | 0.3022 | 0.3203 | 0.3384 | 0.3530 |
| | | 0.5090 | 0.5270 | 0.5450 | 0.5630 | 0.5809 | 0.5990 | 0.6170 |
| | 0.90 | 0.2746 | 0.2855 | 0.3035 | 0.3215 | 0.3394 | 0.3573 | 0.3752 |
| | | 0.4911 | 0.5090 | 0.5268 | 0.5447 | 0.5626 | 0.5804 | 0.5983 |
| | 0.80 | 0.2962 | 0.3140 | 0.3270 | 0.3402 | 0.3580 | 0.3758 | 0.3935 |
| | | 0.4646 | 0.4823 | 0.5000 | 0.5178 | 0.5355 | 0.5532 | 0.5710 |

| $n$ | $\beta$ | 28 | 29 | 30 | 31 | 32 | 33 | 34 |
|---|---|---|---|---|---|---|---|---|
| | 0.99 | 0.3550 | 0.3706 | 0.3836 | 0.4031 | 0.4225 | 0.4419 | 0.4613 |
| | | 0.7001 | 0.7126 | 0.7325 | 0.7508 | 0.7610 | 0.7812 | 0.7938 |
| | 0.98 | 0.3654 | 0.3847 | 0.4040 | 0.4232 | 0.4424 | 0.4616 | 0.4808 |
| | | 0.6823 | 0.7018 | 0.7213 | 0.7312 | 0.7496 | 0.7694 | 0.7796 |
| 53 | 0.95 | 0.3949 | 0.4098 | 0.4238 | 0.4429 | 0.4619 | 0.4809 | 0.5000 |
| | | 0.6616 | 0.6807 | 0.6994 | 0.7094 | 0.7279 | 0.7472 | 0.7648 |
| | 0.90 | 0.4152 | 0.4340 | 0.4528 | 0.4701 | 0.4880 | 0.5067 | 0.5264 |
| | | 0.6415 | 0.6604 | 0.6794 | 0.6944 | 0.7072 | 0.7262 | 0.7453 |
| | 0.80 | 0.4344 | 0.4531 | 0.4719 | 0.4906 | 0.5093 | 0.5280 | 0.5468 |
| | | 0.6219 | 0.6349 | 0.6497 | 0.6685 | 0.6872 | 0.7060 | 0.7249 |
| | 0.99 | 0.3475 | 0.3600 | 0.3763 | 0.3955 | 0.4145 | 0.4335 | 0.4525 |
| | | 0.6901 | 0.7086 | 0.7185 | 0.7375 | 0.7579 | 0.7658 | 0.7852 |
| | 0.98 | 0.3586 | 0.3775 | 0.3964 | 0.4153 | 0.4341 | 0.4512 | 0.4680 |
| | | 0.6720 | 0.6882 | 0.7073 | 0.7265 | 0.7372 | 0.7543 | 0.7737 |
| 54 | 0.95 | 0.3848 | 0.3972 | 0.4160 | 0.4346 | 0.4533 | 0.4720 | 0.4906 |
| | | 0.6491 | 0.6678 | 0.6866 | 0.7055 | 0.7159 | 0.7330 | 0.7519 |
| | 0.90 | 0.4075 | 0.4247 | 0.4402 | 0.4536 | 0.4722 | 0.4907 | 0.5092 |
| | | 0.6295 | 0.6481 | 0.6667 | 0.6853 | 0.7028 | 0.7130 | 0.7313 |
| | 0.80 | 0.4264 | 0.4448 | 0.4632 | 0.4816 | 0.5000 | 0.5183 | 0.5367 |
| | | 0.6104 | 0.6288 | 0.6452 | 0.6570 | 0.6745 | 0.6930 | 0.7114 |
| | 0.99 | 0.3392 | 0.3506 | 0.3694 | 0.3881 | 0.4068 | 0.4255 | 0.4441 |
| | | 0.6769 | 0.6958 | 0.7148 | 0.7247 | 0.7423 | 0.7616 | 0.7708 |
| | 0.98 | 0.3520 | 0.3706 | 0.3891 | 0.4077 | 0.4231 | 0.4368 | 0.4537 |
| | | 0.6667 | 0.6800 | 0.6939 | 0.7127 | 0.7315 | 0.7431 | 0.7588 |
| 55 | 0.95 | 0.3738 | 0.3900 | 0.4084 | 0.4267 | 0.4450 | 0.4633 | 0.4816 |
| | | 0.6383 | 0.6555 | 0.6739 | 0.6923 | 0.7108 | 0.7223 | 0.7378 |
| | 0.90 | 0.3990 | 0.4131 | 0.4272 | 0.4454 | 0.4636 | 0.4818 | 0.5000 |
| | | 0.6180 | 0.6362 | 0.6545 | 0.6727 | 0.6910 | 0.7093 | 0.7193 |
| | 0.80 | 0.4187 | 0.4368 | 0.4548 | 0.4729 | 0.4909 | 0.5090 | 0.5270 |
| | | 0.5994 | 0.6174 | 0.6355 | 0.6536 | 0.6650 | 0.6804 | 0.6985 |
| | 0.99 | 0.3310 | 0.3442 | 0.3627 | 0.3811 | 0.3995 | 0.4173 | 0.4326 |
| | | 0.6690 | 0.6827 | 0.7013 | 0.7200 | 0.7308 | 0.7469 | 0.7659 |
| | 0.98 | 0.3456 | 0.3639 | 0.3821 | 0.3977 | 0.4093 | 0.4272 | 0.4454 |
| | | 0.6544 | 0.6727 | 0.6882 | 0.6994 | 0.7178 | 0.7364 | 0.7487 |
| 56 | 0.95 | 0.3650 | 0.3830 | 0.4010 | 0.4191 | 0.4370 | 0.4550 | 0.4730 |
| | | 0.6350 | 0.6470 | 0.6616 | 0.6797 | 0.6978 | 0.7160 | 0.7285 |
| | 0.90 | 0.3890 | 0.4017 | 0.4196 | 0.4374 | 0.4553 | 0.4732 | 0.4910 |
| | | 0.6110 | 0.6248 | 0.6427 | 0.6606 | 0.6785 | 0.6965 | 0.7145 |
| | 0.80 | 0.4113 | 0.4290 | 0.4468 | 0.4645 | 0.4822 | 0.5000 | 0.5177 |
| | | 0.5887 | 0.6065 | 0.6242 | 0.6420 | 0.6598 | 0.6730 | 0.6860 |

| $n$ | $\beta$ | 0 | 1 | 2 | 3 | 4 | 5 | 6 |
|---|---|---|---|---|---|---|---|---|
| | 0.99 | 0.0000 | 0.0001 | 0.0026 | 0.0077 | 0.0147 | 0.0229 | 0.0322 |
| | | 0.0865 | 0.1223 | 0.1498 | 0.1767 | 0.2034 | 0.2301 | 0.2487 |
| | 0.98 | 0.0000 | 0.0003 | 0.0037 | 0.0100 | 0.0181 | 0.0274 | 0.0376 |
| | | 0.0796 | 0.1079 | 0.1350 | 0.1616 | 0.1881 | 0.2144 | 0.2328 |
| 57 | 0.95 | 0.0000 | 0.0008 | 0.0062 | 0.0144 | 0.0243 | 0.0352 | 0.0468 |
| | | 0.0590 | 0.0935 | 0.1202 | 0.1465 | 0.1727 | 0.1910 | 0.2171 |
| | 0.90 | 0.0000 | 0.0018 | 0.0093 | 0.0194 | 0.0309 | 0.0432 | 0.0520 |
| | | 0.0521 | 0.0792 | 0.1054 | 0.1314 | 0.1496 | 0.1755 | 0.1934 |
| | 0.80 | 0.0000 | 0.0039 | 0.0144 | 0.0270 | 0.0383 | 0.0483 | 0.0645 |
| | | 0.0384 | 0.0646 | 0.0905 | 0.1162 | 0.1342 | 0.1537 | 0.1776 |
| | 0.99 | 0.0000 | 0.0001 | 0.0025 | 0.0076 | 0.0144 | 0.0225 | 0.0316 |
| | | 0.0850 | 0.1202 | 0.1472 | 0.1737 | 0.1999 | 0.2260 | 0.2443 |
| | 0.98 | 0.0000 | 0.0003 | 0.0037 | 0.0099 | 0.0178 | 0.0269 | 0.0369 |
| | | 0.0782 | 0.1060 | 0.1335 | 0.1588 | 0.1848 | 0.2107 | 0.2288 |
| 58 | 0.95 | 0.0000 | 0.0008 | 0.0061 | 0.0142 | 0.0238 | 0.0345 | 0.0460 |
| | | 0.0580 | 0.0919 | 0.1181 | 0.1440 | 0.1697 | 0.1878 | 0.2133 |
| | 0.90 | 0.0000 | 0.0018 | 0.0092 | 0.0191 | 0.0304 | 0.0425 | 0.0512 |
| | | 0.0513 | 0.0778 | 0.1036 | 0.1292 | 0.1471 | 0.1725 | 0.1901 |
| | 0.80 | 0.0000 | 0.0038 | 0.0142 | 0.0265 | 0.0377 | 0.0474 | 0.0634 |
| | | 0.0378 | 0.0635 | 0.0889 | 0.1142 | 0.1319 | 0.1571 | 0.1746 |
| | 0.99 | 0.0000 | 0.0001 | 0.0025 | 0.0074 | 0.0142 | 0.0221 | 0.0311 |
| | | 0.0835 | 0.1181 | 0.1447 | 0.1707 | 0.1965 | 0.2222 | 0.2402 |
| | 0.98 | 0.0000 | 0.0003 | 0.0036 | 0.0097 | 0.0175 | 0.0264 | 0.0363 |
| | | 0.0769 | 0.1042 | 0.1375 | 0.1561 | 0.1817 | 0.2071 | 0.2249 |
| 59 | 0.95 | 0.0000 | 0.0008 | 0.0060 | 0.0140 | 0.0234 | 0.0339 | 0.0452 |
| | | 0.0570 | 0.0904 | 0.1161 | 0.1416 | 0.1669 | 0.1846 | 0.2097 |
| | 0.90 | 0.0000 | 0.0017 | 0.0090 | 0.0188 | 0.0298 | 0.0417 | 0.0503 |
| | | 0.0504 | 0.0765 | 0.1019 | 0.1270 | 0.1446 | 0.1696 | 0.1869 |
| | 0.80 | 0.0000 | 0.0037 | 0.0139 | 0.0261 | 0.0370 | 0.0466 | 0.0624 |
| | | 0.0371 | 0.0625 | 0.0875 | 0.1123 | 0.1297 | 0.1545 | 0.1716 |
| | 0.99 | 0.0000 | 0.0001 | 0.0024 | 0.0073 | 0.0139 | 0.0218 | 0.0305 |
| | | 0.0821 | 0.1161 | 0.1422 | 0.1678 | 0.1932 | 0.2184 | 0.2361 |
| | 0.98 | 0.0000 | 0.0003 | 0.0036 | 0.0095 | 0.0172 | 0.0260 | 0.0357 |
| | | 0.0756 | 0.1025 | 0.1352 | 0.1535 | 0.1786 | 0.2037 | 0.2212 |
| 60 | 0.95 | 0.0000 | 0.0008 | 0.0059 | 0.0137 | 0.0230 | 0.0334 | 0.0444 |
| | | 0.0561 | 0.0889 | 0.1142 | 0.1392 | 0.1641 | 0.1815 | 0.2062 |
| | 0.90 | 0.0000 | 0.0017 | 0.0088 | 0.0185 | 0.0293 | 0.0410 | 0.0495 |
| | | 0.0496 | 0.0752 | 0.1002 | 0.1249 | 0.1422 | 0.1668 | 0.1838 |
| | 0.80 | 0.0000 | 0.0037 | 0.0137 | 0.0256 | 0.0364 | 0.0458 | 0.0614 |
| | | 0.0365 | 0.0615 | 0.0860 | 0.1105 | 0.1275 | 0.1519 | 0.1688 |

| n | β | 7 | 8 | 9 | 10 | 11 | 12 | 13 |
|---|---|---|---|---|---|---|---|---|
| | 0.99 | 0.0422 | 0.0529 | 0.0641 | 0.0758 | 0.0864 | 0.0951 | 0.1113 |
| | | 0.2751 | 0.2935 | 0.3122 | 0.3382 | 0.3563 | 0.3744 | 0.3924 |
| | 0.98 | 0.0485 | 0.0600 | 0.0719 | 0.0795 | 0.0944 | 0.1078 | 0.1166 |
| | | 0.2591 | 0.2773 | 0.2990 | 0.3216 | 0.3396 | 0.3576 | 0.3775 |
| 57 | 0.95 | 0.0589 | 0.0623 | 0.0823 | 0.0934 | 0.1071 | 0.1201 | 0.1346 |
| | | 0.2351 | 0.2531 | 0.2791 | 0.2970 | 0.3148 | 0.3328 | 0.3586 |
| | 0.90 | 0.0657 | 0.0791 | 0.0908 | 0.1053 | 0.1197 | 0.1313 | 0.1495 |
| | | 0.2192 | 0.2370 | 0.2548 | 0.2806 | 0.2983 | 0.3160 | 0.3336 |
| | 0.80 | 0.0752 | 0.0904 | 0.1070 | 0.1161 | 0.1341 | 0.1517 | 0.1555 |
| | | 0.1953 | 0.2129 | 0.2385 | 0.2561 | 0.2736 | 0.2911 | 0.3095 |
| | 0.99 | 0.0415 | 0.0519 | 0.0629 | 0.0744 | 0.0849 | 0.0933 | 0.1092 |
| | | 0.2703 | 0.2884 | 0.3144 | 0.3323 | 0.3501 | 0.3679 | 0.3912 |
| | 0.98 | 0.0476 | 0.0589 | 0.0706 | 0.0781 | 0.0926 | 0.1059 | 0.1142 |
| | | 0.2546 | 0.2725 | 0.2983 | 0.3160 | 0.3337 | 0.3514 | 0.3772 |
| 58 | 0.95 | 0.0579 | 0.0610 | 0.0808 | 0.0918 | 0.1051 | 0.1180 | 0.1320 |
| | | 0.2311 | 0.2488 | 0.2743 | 0.2919 | 0.3094 | 0.3350 | 0.3524 |
| | 0.90 | 0.0646 | 0.0777 | 0.0891 | 0.1035 | 0.1175 | 0.1291 | 0.1470 |
| | | 0.2155 | 0.2330 | 0.2504 | 0.2758 | 0.2932 | 0.3105 | 0.3278 |
| | 0.80 | 0.0738 | 0.0888 | 0.1050 | 0.1141 | 0.1318 | 0.1488 | 0.1570 |
| | | 0.1919 | 0.2099 | 0.2345 | 0.2517 | 0.2690 | 0.2862 | 0.3115 |
| | 0.99 | 0.0407 | 0.0510 | 0.0618 | 0.0731 | 0.0834 | 0.0915 | 0.1073 |
| | | 0.2657 | 0.2834 | 0.3090 | 0.3266 | 0.3441 | 0.3621 | 0.3872 |
| | 0.98 | 0.0468 | 0.0579 | 0.0694 | 0.0768 | 0.0910 | 0.1041 | 0.1120 |
| | | 0.2503 | 0.2678 | 0.2932 | 0.3106 | 0.3280 | 0.3465 | 0.3708 |
| 59 | 0.95 | 0.0569 | 0.0597 | 0.0794 | 0.0903 | 0.1032 | 0.1160 | 0.1296 |
| | | 0.2272 | 0.2446 | 0.2696 | 0.2869 | 0.3041 | 0.3293 | 0.3465 |
| | 0.90 | 0.0634 | 0.0764 | 0.0875 | 0.1018 | 0.1153 | 0.1269 | 0.1445 |
| | | 0.2118 | 0.2290 | 0.2462 | 0.2712 | 0.2882 | 0.3053 | 0.3223 |
| | 0.80 | 0.0725 | 0.0874 | 0.1032 | 0.1122 | 0.1296 | 0.1461 | 0.1544 |
| | | 0.1887 | 0.2135 | 0.2305 | 0.2475 | 0.2645 | 0.2814 | 0.3063 |
| | 0.99 | 0.0400 | 0.0501 | 0.0607 | 0.0718 | 0.0820 | 0.0898 | 0.1053 |
| | | 0.2612 | 0.2787 | 0.3038 | 0.3211 | 0.3383 | 0.3634 | 0.3806 |
| | 0.98 | 0.0460 | 0.0569 | 0.0682 | 0.0755 | 0.0893 | 0.1024 | 0.1098 |
| | | 0.2461 | 0.2634 | 0.2883 | 0.3054 | 0.3225 | 0.3475 | 0.3646 |
| 60 | 0.95 | 0.0560 | 0.0584 | 0.0780 | 0.0888 | 0.1013 | 0.1141 | 0.1272 |
| | | 0.2234 | 0.2430 | 0.2652 | 0.2821 | 0.2991 | 0.3238 | 0.3407 |
| | 0.90 | 0.0623 | 0.0751 | 0.0859 | 0.1001 | 0.1132 | 0.1248 | 0.1421 |
| | | 0.2083 | 0.2253 | 0.2421 | 0.2667 | 0.2835 | 0.3002 | 0.3170 |
| | 0.80 | 0.0712 | 0.0859 | 0.1013 | 0.1104 | 0.1274 | 0.1435 | 0.1518 |
| | | 0.1856 | 0.2100 | 0.2267 | 0.2434 | 0.2601 | 0.2767 | 0.3012 |

| $n$ | $\beta$ | 14 | 15 | 16 | 17 | 18 | 19 | 20 |
|---|---|---|---|---|---|---|---|---|
| | 0.99 | 0.1222 | 0.1351 | 0.1497 | 0.1606 | 0.1766 | 0.1891 | 0.2033 |
| | | 0.4191 | 0.4371 | 0.4551 | 0.4731 | 0.4911 | 0.5090 | 0.5270 |
| | 0.98 | 0.1349 | 0.1402 | 0.1615 | 0.1704 | 0.1880 | 0.2025 | 0.2143 |
| | | 0.4019 | 0.4198 | 0.4376 | 0.4555 | 0.4733 | 0.4911 | 0.5090 |
| 57 | 0.95 | 0.1464 | 0.1645 | 0.1726 | 0.1909 | 0.2075 | 0.2170 | 0.2350 |
| | | 0.3764 | 0.3941 | 0.4118 | 0.4295 | 0.4471 | 0.4648 | 0.4824 |
| | 0.90 | 0.1593 | 0.1754 | 0.1933 | 0.2060 | 0.2191 | 0.2369 | 0.2547 |
| | | 0.3512 | 0.3736 | 0.3947 | 0.4123 | 0.4299 | 0.4474 | 0.4650 |
| | 0.80 | 0.1775 | 0.1952 | 0.2128 | 0.2214 | 0.2384 | 0.2560 | 0.2735 |
| | | 0.3344 | 0.3519 | 0.3694 | 0.3868 | 0.4042 | 0.4217 | 0.4391 |
| | 0.99 | 0.1201 | 0.1325 | 0.1471 | 0.1573 | 0.1736 | 0.1852 | 0.1998 |
| | | 0.4117 | 0.4294 | 0.4471 | 0.4648 | 0.4824 | 0.5000 | 0.5177 |
| | 0.98 | 0.1325 | 0.1364 | 0.1587 | 0.1667 | 0.1847 | 0.1984 | 0.2106 |
| | | 0.3949 | 0.4124 | 0.4300 | 0.4475 | 0.4650 | 0.4825 | 0.5000 |
| 58 | 0.95 | 0.1439 | 0.1614 | 0.1696 | 0.1877 | 0.2034 | 0.2132 | 0.2310 |
| | | 0.3699 | 0.3873 | 0.4047 | 0.4220 | 0.4394 | 0.4567 | 0.4741 |
| | 0.90 | 0.1560 | 0.1724 | 0.1900 | 0.2018 | 0.2154 | 0.2329 | 0.2503 |
| | | 0.3451 | 0.3707 | 0.3880 | 0.4052 | 0.4225 | 0.4397 | 0.4570 |
| | 0.80 | 0.1745 | 0.1918 | 0.2090 | 0.2157 | 0.2344 | 0.2516 | 0.2689 |
| | | 0.3287 | 0.3459 | 0.3630 | 0.3802 | 0.3973 | 0.4145 | 0.4316 |
| | 0.99 | 0.1180 | 0.1300 | 0.1446 | 0.1542 | 0.1706 | 0.1814 | 0.1964 |
| | | 0.4046 | 0.4220 | 0.4394 | 0.4567 | 0.4741 | 0.4914 | 0.5087 |
| | 0.98 | 0.1300 | 0.1374 | 0.1560 | 0.1630 | 0.1816 | 0.1945 | 0.2070 |
| | | 0.3881 | 0.4054 | 0.4226 | 0.4398 | 0.4571 | 0.4743 | 0.4915 |
| 59 | 0.95 | 0.1415 | 0.1584 | 0.1668 | 0.1845 | 0.1994 | 0.2096 | 0.2271 |
| | | 0.3636 | 0.3807 | 0.3978 | 0.4149 | 0.4319 | 0.4507 | 0.4702 |
| | 0.90 | 0.1528 | 0.1695 | 0.1868 | 0.1977 | 0.2117 | 0.2289 | 0.2461 |
| | | 0.3419 | 0.3644 | 0.3814 | 0.3984 | 0.4153 | 0.4323 | 0.4492 |
| | 0.80 | 0.1715 | 0.1886 | 0.2050 | 0.2134 | 0.2304 | 0.2474 | 0.2644 |
| | | 0.3232 | 0.3401 | 0.3569 | 0.3738 | 0.3907 | 0.4075 | 0.4243 |
| | 0.99 | 0.1160 | 0.1276 | 0.1421 | 0.1511 | 0.1677 | 0.1778 | 0.1931 |
| | | 0.3977 | 0.4148 | 0.4319 | 0.4490 | 0.4660 | 0.4830 | 0.5000 |
| | 0.98 | 0.1277 | 0.1351 | 0.1534 | 0.1593 | 0.1785 | 0.1907 | 0.2036 |
| | | 0.3816 | 0.3985 | 0.4155 | 0.4324 | 0.4494 | 0.4668 | 0.4847 |
| 60 | 0.95 | 0.1391 | 0.1555 | 0.1640 | 0.1814 | 0.1956 | 0.2061 | 0.2233 |
| | | 0.3575 | 0.3744 | 0.3912 | 0.4079 | 0.4286 | 0.4497 | 0.4665 |
| | 0.90 | 0.1497 | 0.1667 | 0.1837 | 0.1936 | 0.2082 | 0.2252 | 0.2420 |
| | | 0.3416 | 0.3584 | 0.3751 | 0.3918 | 0.4084 | 0.4251 | 0.4418 |
| | 0.80 | 0.1687 | 0.1855 | 0.2012 | 0.2099 | 0.2266 | 0.2433 | 0.2600 |
| | | 0.3178 | 0.3345 | 0.3510 | 0.3676 | 0.3842 | 0.4008 | 0.4173 |

| n | β | 21 | 22 | 23 | 24 | 25 | 26 | 27 |
|---|---|---|---|---|---|---|---|---|
| 57 | 0.99 | 0.2198 | 0.2300 | 0.2486 | 0.2634 | 0.2750 | 0.2934 | 0.3114 |
| | | 0.5450 | 0.5630 | 0.5810 | 0.5933 | 0.6077 | 0.6257 | 0.6438 |
| | 0.98 | 0.2327 | 0.2458 | 0.2590 | 0.2772 | 0.2942 | 0.3022 | 0.3215 |
| | | 0.5268 | 0.5446 | 0.5625 | 0.5803 | 0.5982 | 0.6131 | 0.6257 |
| | 0.95 | 0.2530 | 0.2654 | 0.2790 | 0.2969 | 0.3147 | 0.3323 | 0.3441 |
| | | 0.5000 | 0.5177 | 0.5353 | 0.5530 | 0.5706 | 0.5883 | 0.6060 |
| | 0.90 | 0.2687 | 0.2805 | 0.2982 | 0.3159 | 0.3335 | 0.3511 | 0.3667 |
| | | 0.4825 | 0.5000 | 0.5176 | 0.5351 | 0.5527 | 0.5702 | 0.5878 |
| | 0.80 | 0.2910 | 0.3081 | 0.3186 | 0.3343 | 0.3518 | 0.3693 | 0.3867 |
| | | 0.4565 | 0.4739 | 0.4913 | 0.5088 | 0.5262 | 0.5436 | 0.5610 |
| 58 | 0.99 | 0.2154 | 0.2259 | 0.2442 | 0.2577 | 0.2702 | 0.2883 | 0.3047 |
| | | 0.5353 | 0.5530 | 0.5707 | 0.5884 | 0.6064 | 0.6168 | 0.6322 |
| | 0.98 | 0.2287 | 0.2404 | 0.2545 | 0.2724 | 0.2880 | 0.2982 | 0.3159 |
| | | 0.5176 | 0.5351 | 0.5526 | 0.5701 | 0.5877 | 0.6052 | 0.6229 |
| | 0.95 | 0.2487 | 0.2594 | 0.2742 | 0.2918 | 0.3093 | 0.3251 | 0.3349 |
| | | 0.4914 | 0.5087 | 0.5260 | 0.5434 | 0.5607 | 0.5781 | 0.5954 |
| | 0.90 | 0.2629 | 0.2757 | 0.2931 | 0.3104 | 0.3277 | 0.3450 | 0.3579 |
| | | 0.4742 | 0.4914 | 0.5087 | 0.5259 | 0.5431 | 0.5604 | 0.5776 |
| | 0.80 | 0.2861 | 0.3016 | 0.3114 | 0.3286 | 0.3458 | 0.3629 | 0.3801 |
| | | 0.4487 | 0.4658 | 0.4829 | 0.5000 | 0.5172 | 0.5343 | 0.5514 |
| 59 | 0.99 | 0.2111 | 0.2221 | 0.2401 | 0.2521 | 0.2656 | 0.2833 | 0.2982 |
| | | 0.5260 | 0.5434 | 0.5607 | 0.5781 | 0.5955 | 0.6129 | 0.6264 |
| | 0.98 | 0.2248 | 0.2351 | 0.2502 | 0.2677 | 0.2818 | 0.2931 | 0.3105 |
| | | 0.5086 | 0.5258 | 0.5430 | 0.5603 | 0.5775 | 0.5947 | 0.6120 |
| | 0.95 | 0.2445 | 0.2532 | 0.2695 | 0.2868 | 0.3040 | 0.3180 | 0.3292 |
| | | 0.4892 | 0.5066 | 0.5222 | 0.5374 | 0.5525 | 0.5682 | 0.5852 |
| | 0.90 | 0.2571 | 0.2711 | 0.2881 | 0.3052 | 0.3222 | 0.3381 | 0.3482 |
| | | 0.4662 | 0.4831 | 0.5000 | 0.5170 | 0.5339 | 0.5509 | 0.5678 |
| | 0.80 | 0.2813 | 0.2953 | 0.3062 | 0.3231 | 0.3400 | 0.3568 | 0.3737 |
| | | 0.4412 | 0.4580 | 0.4748 | 0.4916 | 0.5085 | 0.5253 | 0.5421 |
| 60 | 0.99 | 0.2070 | 0.2183 | 0.2360 | 0.2466 | 0.2611 | 0.2786 | 0.2917 |
| | | 0.5171 | 0.5341 | 0.5511 | 0.5682 | 0.5853 | 0.6024 | 0.6195 |
| | 0.98 | 0.2211 | 0.2299 | 0.2460 | 0.2633 | 0.2758 | 0.2882 | 0.3053 |
| | | 0.5019 | 0.5184 | 0.5343 | 0.5507 | 0.5677 | 0.5846 | 0.6016 |
| | 0.95 | 0.2398 | 0.2461 | 0.2651 | 0.2820 | 0.2990 | 0.3109 | 0.3237 |
| | | 0.4833 | 0.5000 | 0.5168 | 0.5336 | 0.5504 | 0.5649 | 0.5778 |
| | 0.90 | 0.2514 | 0.2666 | 0.2834 | 0.3001 | 0.3169 | 0.3308 | 0.3415 |
| | | 0.4584 | 0.4751 | 0.4917 | 0.5084 | 0.5250 | 0.5417 | 0.5583 |
| | 0.80 | 0.2766 | 0.2890 | 0.3011 | 0.3177 | 0.3344 | 0.3509 | 0.3675 |
| | | 0.4339 | 0.4504 | 0.4670 | 0.4835 | 0.5001 | 0.5166 | 0.5331 |

| $n$ | $\beta$ | 28 | 29 | 30 | 31 | 32 | 33 | 34 |
|---|---|---|---|---|---|---|---|---|
| | 0.99 | 0.3224 | 0.3381 | 0.3562 | 0.3743 | 0.3923 | 0.4067 | 0.4190 |
| | | 0.6619 | 0.6776 | 0.6886 | 0.7066 | 0.7250 | 0.7366 | 0.7514 |
| | 0.98 | 0.3395 | 0.3575 | 0.3743 | 0.3869 | 0.4018 | 0.4197 | 0.4375 |
| | | 0.6425 | 0.6605 | 0.6785 | 0.6978 | 0.7058 | 0.7228 | 0.7410 |
| 57 | 0.95 | 0.3585 | 0.3763 | 0.3940 | 0.4117 | 0.4294 | 0.4470 | 0.4647 |
| | | 0.6237 | 0.6415 | 0.6559 | 0.6677 | 0.6853 | 0.7031 | 0.7210 |
| | 0.90 | 0.3769 | 0.3946 | 0.4122 | 0.4298 | 0.4473 | 0.4649 | 0.4824 |
| | | 0.6054 | 0.6231 | 0.6333 | 0.6489 | 0.6665 | 0.6841 | 0.7018 |
| | 0.80 | 0.4041 | 0.4216 | 0.4390 | 0.4564 | 0.4738 | 0.4912 | 0.5087 |
| | | 0.5784 | 0.5959 | 0.6133 | 0.6307 | 0.6482 | 0.6657 | 0.6814 |
| | 0.99 | 0.3143 | 0.3322 | 0.3500 | 0.3678 | 0.3832 | 0.3936 | 0.4116 |
| | | 0.6500 | 0.6678 | 0.6857 | 0.6953 | 0.7117 | 0.7298 | 0.7423 |
| | 0.98 | 0.3336 | 0.3513 | 0.3654 | 0.3771 | 0.3948 | 0.4123 | 0.4299 |
| | | 0.6346 | 0.6487 | 0.6664 | 0.6841 | 0.7018 | 0.7120 | 0.7276 |
| 58 | 0.95 | 0.3523 | 0.3698 | 0.3872 | 0.4046 | 0.4219 | 0.4393 | 0.4566 |
| | | 0.6128 | 0.6302 | 0.6477 | 0.6651 | 0.6749 | 0.6907 | 0.7082 |
| | 0.90 | 0.3706 | 0.3879 | 0.4051 | 0.4224 | 0.4396 | 0.4569 | 0.4741 |
| | | 0.5949 | 0.6121 | 0.6294 | 0.6421 | 0.6550 | 0.6723 | 0.6896 |
| | 0.80 | 0.3972 | 0.4144 | 0.4315 | 0.4486 | 0.4657 | 0.4828 | 0.5000 |
| | | 0.5685 | 0.5856 | 0.6028 | 0.6199 | 0.6371 | 0.6542 | 0.6714 |
| | 0.99 | 0.3089 | 0.3265 | 0.3440 | 0.3612 | 0.3736 | 0.3871 | 0.4045 |
| | | 0.6388 | 0.6560 | 0.6735 | 0.6911 | 0.7018 | 0.7167 | 0.7344 |
| | 0.98 | 0.3279 | 0.3447 | 0.3561 | 0.3707 | 0.3880 | 0.4053 | 0.4225 |
| | | 0.6293 | 0.6439 | 0.6553 | 0.6721 | 0.6895 | 0.7069 | 0.7182 |
| 59 | 0.95 | 0.3464 | 0.3635 | 0.3806 | 0.3977 | 0.4148 | 0.4318 | 0.4475 |
| | | 0.6023 | 0.6194 | 0.6365 | 0.6536 | 0.6708 | 0.6820 | 0.6960 |
| | 0.90 | 0.3643 | 0.3813 | 0.3983 | 0.4152 | 0.4322 | 0.4491 | 0.4661 |
| | | 0.5848 | 0.6017 | 0.6187 | 0.6357 | 0.6518 | 0.6619 | 0.6778 |
| | 0.80 | 0.3906 | 0.4074 | 0.4242 | 0.4411 | 0.4579 | 0.4747 | 0.4915 |
| | | 0.5589 | 0.5758 | 0.5926 | 0.6094 | 0.6263 | 0.6432 | 0.6600 |
| | 0.99 | 0.3037 | 0.3210 | 0.3382 | 0.3532 | 0.3633 | 0.3805 | 0.3976 |
| | | 0.6367 | 0.6468 | 0.6618 | 0.6790 | 0.6963 | 0.7083 | 0.7214 |
| | 0.98 | 0.3224 | 0.3373 | 0.3474 | 0.3645 | 0.3815 | 0.3984 | 0.4154 |
| | | 0.6185 | 0.6355 | 0.6526 | 0.6627 | 0.6776 | 0.6947 | 0.7118 |
| 60 | 0.95 | 0.3406 | 0.3574 | 0.3743 | 0.3911 | 0.4078 | 0.4222 | 0.4351 |
| | | 0.5922 | 0.6089 | 0.6257 | 0.6426 | 0.6594 | 0.6763 | 0.6891 |
| | 0.90 | 0.3583 | 0.3750 | 0.3917 | 0.4083 | 0.4250 | 0.4417 | 0.4583 |
| | | 0.5750 | 0.5917 | 0.6083 | 0.6250 | 0.6417 | 0.6585 | 0.6692 |
| | 0.80 | 0.3841 | 0.4007 | 0.4172 | 0.4338 | 0.4503 | 0.4669 | 0.4834 |
| | | 0.5497 | 0.5662 | 0.5828 | 0.5993 | 0.6159 | 0.6325 | 0.6491 |

| $n$ | $\beta$ | 0 | 1 | 2 | 3 | 4 | 5 | 6 |
|---|---|---|---|---|---|---|---|---|
| 61 | 0.99 | 0.0000 | 0.0001 | 0.0024 | 0.0072 | 0.0137 | 0.0214 | 0.0300 |
| | | 0.0808 | 0.1142 | 0.1399 | 0.1651 | 0.1900 | 0.2148 | 0.2322 |
| | 0.98 | 0.0000 | 0.0003 | 0.0035 | 0.0094 | 0.0169 | 0.0256 | 0.0351 |
| | | 0.0743 | 0.1008 | 0.1330 | 0.1516 | 0.1757 | 0.2003 | 0.2179 |
| | 0.95 | 0.0000 | 0.0008 | 0.0058 | 0.0135 | 0.0227 | 0.0328 | 0.0437 |
| | | 0.0551 | 0.0874 | 0.1124 | 0.1369 | 0.1614 | 0.1786 | 0.2029 |
| | 0.90 | 0.0000 | 0.0017 | 0.0087 | 0.0182 | 0.0289 | 0.0404 | 0.0487 |
| | | 0.0488 | 0.0740 | 0.0985 | 0.1229 | 0.1399 | 0.1640 | 0.1808 |
| | 0.80 | 0.0000 | 0.0036 | 0.0135 | 0.0252 | 0.0358 | 0.0450 | 0.0604 |
| | | 0.0359 | 0.0605 | 0.0846 | 0.1087 | 0.1255 | 0.1495 | 0.1661 |
| 62 | 0.99 | 0.0000 | 0.0001 | 0.0024 | 0.0071 | 0.0135 | 0.0210 | 0.0295 |
| | | 0.0795 | 0.1124 | 0.1376 | 0.1624 | 0.1869 | 0.2113 | 0.2291 |
| | 0.98 | 0.0000 | 0.0003 | 0.0034 | 0.0092 | 0.0166 | 0.0251 | 0.0345 |
| | | 0.0731 | 0.0992 | 0.1309 | 0.1555 | 0.1729 | 0.1971 | 0.2212 |
| | 0.95 | 0.0000 | 0.0008 | 0.0057 | 0.0133 | 0.0223 | 0.0323 | 0.0429 |
| | | 0.0545 | 0.0860 | 0.1106 | 0.1347 | 0.1588 | 0.1757 | 0.1996 |
| | 0.90 | 0.0000 | 0.0016 | 0.0086 | 0.0179 | 0.0284 | 0.0397 | 0.0479 |
| | | 0.0480 | 0.0728 | 0.0970 | 0.1209 | 0.1377 | 0.1614 | 0.1779 |
| | 0.80 | 0.0000 | 0.0035 | 0.0133 | 0.0248 | 0.0352 | 0.0443 | 0.0594 |
| | | 0.0353 | 0.0595 | 0.0833 | 0.1069 | 0.1235 | 0.1471 | 0.1634 |
| 63 | 0.99 | 0.0000 | 0.0001 | 0.0023 | 0.0070 | 0.0132 | 0.0207 | 0.0290 |
| | | 0.0782 | 0.1106 | 0.1354 | 0.1598 | 0.1839 | 0.2080 | 0.2319 |
| | 0.98 | 0.0000 | 0.0003 | 0.0034 | 0.0091 | 0.0163 | 0.0247 | 0.0339 |
| | | 0.0720 | 0.0976 | 0.1288 | 0.1530 | 0.1701 | 0.1939 | 0.2177 |
| | 0.95 | 0.0000 | 0.0008 | 0.0056 | 0.0131 | 0.0219 | 0.0317 | 0.0422 |
| | | 0.0596 | 0.0847 | 0.1088 | 0.1326 | 0.1563 | 0.1729 | 0.1964 |
| | 0.90 | 0.0000 | 0.0016 | 0.0084 | 0.0176 | 0.0279 | 0.0391 | 0.0471 |
| | | 0.0472 | 0.0717 | 0.0954 | 0.1190 | 0.1355 | 0.1589 | 0.1751 |
| | 0.80 | 0.0000 | 0.0035 | 0.0131 | 0.0244 | 0.0347 | 0.0435 | 0.0585 |
| | | 0.0348 | 0.0586 | 0.0820 | 0.1053 | 0.1215 | 0.1448 | 0.1609 |
| 64 | 0.99 | 0.0000 | 0.0001 | 0.0023 | 0.0068 | 0.0130 | 0.0204 | 0.0286 |
| | | 0.0770 | 0.1088 | 0.1333 | 0.1573 | 0.1811 | 0.2047 | 0.2283 |
| | 0.98 | 0.0000 | 0.0003 | 0.0033 | 0.0089 | 0.0161 | 0.0243 | 0.0334 |
| | | 0.0709 | 0.0961 | 0.1268 | 0.1506 | 0.1675 | 0.1909 | 0.2143 |
| | 0.95 | 0.0000 | 0.0008 | 0.0055 | 0.0128 | 0.0216 | 0.0312 | 0.0416 |
| | | 0.0587 | 0.0834 | 0.1071 | 0.1306 | 0.1539 | 0.1702 | 0.1934 |
| | 0.90 | 0.0000 | 0.0016 | 0.0083 | 0.0173 | 0.0275 | 0.0384 | 0.0464 |
| | | 0.0465 | 0.0706 | 0.0940 | 0.1171 | 0.1334 | 0.1564 | 0.1724 |
| | 0.80 | 0.0000 | 0.0034 | 0.0128 | 0.0240 | 0.0342 | 0.0428 | 0.0576 |
| | | 0.0343 | 0.0577 | 0.0807 | 0.1036 | 0.1197 | 0.1425 | 0.1584 |

| $n$ | $\beta$ | 7 | 8 | 9 | 10 | 11 | 12 | 13 |
|---|---|---|---|---|---|---|---|---|
| | 0.99 | 0.0394 | 0.0493 | 0.0597 | 0.0706 | 0.0807 | 0.0881 | 0.1035 |
| | | 0.2569 | 0.2741 | 0.2987 | 0.3157 | 0.3327 | 0.3574 | 0.3743 |
| | 0.98 | 0.0452 | 0.0559 | 0.0670 | 0.0742 | 0.0878 | 0.1007 | 0.1077 |
| | | 0.2420 | 0.2590 | 0.2835 | 0.3004 | 0.3172 | 0.3418 | 0.3585 |
| 61 | 0.95 | 0.0550 | 0.0571 | 0.0767 | 0.0873 | 0.0995 | 0.1123 | 0.1249 |
| | | 0.2198 | 0.2440 | 0.2608 | 0.2775 | 0.2946 | 0.3185 | 0.3351 |
| | 0.90 | 0.0613 | 0.0739 | 0.0844 | 0.0984 | 0.1112 | 0.1228 | 0.1398 |
| | | 0.2049 | 0.2216 | 0.2389 | 0.2623 | 0.2788 | 0.2953 | 0.3118 |
| | 0.80 | 0.0700 | 0.0845 | 0.0996 | 0.1086 | 0.1254 | 0.1410 | 0.1494 |
| | | 0.1826 | 0.2066 | 0.2231 | 0.2395 | 0.2559 | 0.2722 | 0.2963 |
| | 0.99 | 0.0387 | 0.0485 | 0.0587 | 0.0694 | 0.0794 | 0.0865 | 0.1017 |
| | | 0.2527 | 0.2696 | 0.2939 | 0.3106 | 0.3286 | 0.3516 | 0.3682 |
| | 0.98 | 0.0445 | 0.0550 | 0.0659 | 0.0730 | 0.0863 | 0.0991 | 0.1057 |
| | | 0.2381 | 0.2548 | 0.2789 | 0.2955 | 0.3122 | 0.3362 | 0.3527 |
| 62 | 0.95 | 0.0542 | 0.0557 | 0.0754 | 0.0859 | 0.0978 | 0.1105 | 0.1227 |
| | | 0.2162 | 0.2401 | 0.2566 | 0.2731 | 0.2970 | 0.3133 | 0.3297 |
| | 0.90 | 0.0602 | 0.0727 | 0.0830 | 0.0969 | 0.1093 | 0.1208 | 0.1376 |
| | | 0.2016 | 0.2180 | 0.2418 | 0.2581 | 0.2744 | 0.2906 | 0.3078 |
| | 0.80 | 0.0688 | 0.0832 | 0.0979 | 0.1068 | 0.1234 | 0.1385 | 0.1470 |
| | | 0.1797 | 0.2033 | 0.2195 | 0.2357 | 0.2518 | 0.2683 | 0.2916 |
| | 0.99 | 0.0381 | 0.0477 | 0.0577 | 0.0682 | 0.0781 | 0.0850 | 0.1000 |
| | | 0.2487 | 0.2677 | 0.2891 | 0.3056 | 0.3295 | 0.3459 | 0.3622 |
| | 0.98 | 0.0437 | 0.0541 | 0.0648 | 0.0719 | 0.0848 | 0.0975 | 0.1037 |
| | | 0.2343 | 0.2525 | 0.2745 | 0.2908 | 0.3145 | 0.3308 | 0.3471 |
| 63 | 0.95 | 0.0533 | 0.0595 | 0.0741 | 0.0846 | 0.0961 | 0.1087 | 0.1206 |
| | | 0.2128 | 0.2363 | 0.2525 | 0.2687 | 0.2923 | 0.3084 | 0.3245 |
| | 0.90 | 0.0592 | 0.0716 | 0.0815 | 0.0953 | 0.1074 | 0.1189 | 0.1354 |
| | | 0.1985 | 0.2146 | 0.2380 | 0.2540 | 0.2700 | 0.2860 | 0.3095 |
| | 0.80 | 0.0676 | 0.0819 | 0.0963 | 0.1052 | 0.1214 | 0.1362 | 0.1447 |
| | | 0.1769 | 0.2001 | 0.2160 | 0.2320 | 0.2478 | 0.2711 | 0.2870 |
| | 0.99 | 0.0375 | 0.0469 | 0.0568 | 0.0671 | 0.0769 | 0.0835 | 0.0983 |
| | | 0.2448 | 0.2683 | 0.2846 | 0.3008 | 0.3243 | 0.3404 | 0.3565 |
| | 0.98 | 0.0430 | 0.0532 | 0.0638 | 0.0708 | 0.0834 | 0.0960 | 0.1018 |
| | | 0.2306 | 0.2540 | 0.2702 | 0.2863 | 0.3096 | 0.3256 | 0.3416 |
| 64 | 0.95 | 0.0524 | 0.0586 | 0.0729 | 0.0833 | 0.0945 | 0.1070 | 0.1185 |
| | | 0.2095 | 0.2326 | 0.2486 | 0.2645 | 0.2877 | 0.3036 | 0.3194 |
| | 0.90 | 0.0583 | 0.0705 | 0.0802 | 0.0939 | 0.1056 | 0.1170 | 0.1333 |
| | | 0.1954 | 0.2113 | 0.2343 | 0.2501 | 0.2658 | 0.2816 | 0.3046 |
| | 0.80 | 0.0665 | 0.0806 | 0.0947 | 0.1035 | 0.1196 | 0.1339 | 0.1424 |
| | | 0.1741 | 0.1970 | 0.2127 | 0.2284 | 0.2440 | 0.2669 | 0.2826 |

| $n$ | $\beta$ | 14 | 15 | 16 | 17 | 18 | 19 | 20 |
|---|---|---|---|---|---|---|---|---|
| 61 | 0.99 | 0.1141 | 0.1252 | 0.1398 | 0.1481 | 0.1650 | 0.1742 | 0.1899 |
| | | 0.3911 | 0.4079 | 0.4247 | 0.4415 | 0.4600 | 0.4784 | 0.4960 |
| | 0.98 | 0.1255 | 0.1329 | 0.1508 | 0.1553 | 0.1756 | 0.1869 | 0.2002 |
| | | 0.3753 | 0.3920 | 0.4086 | 0.4254 | 0.4453 | 0.4667 | 0.4834 |
| | 0.95 | 0.1368 | 0.1527 | 0.1613 | 0.1785 | 0.1919 | 0.2028 | 0.2197 |
| | | 0.3517 | 0.3682 | 0.3847 | 0.4053 | 0.4258 | 0.4423 | 0.4588 |
| | 0.90 | 0.1467 | 0.1639 | 0.1807 | 0.1897 | 0.2048 | 0.2215 | 0.2379 |
| | | 0.3360 | 0.3525 | 0.3689 | 0.3854 | 0.4018 | 0.4182 | 0.4346 |
| | 0.80 | 0.1660 | 0.1825 | 0.1975 | 0.2065 | 0.2230 | 0.2394 | 0.2558 |
| | | 0.3127 | 0.3290 | 0.3453 | 0.3617 | 0.3780 | 0.3943 | 0.4105 |
| 62 | 0.99 | 0.1123 | 0.1229 | 0.1375 | 0.1451 | 0.1623 | 0.1707 | 0.1868 |
| | | 0.3847 | 0.4013 | 0.4181 | 0.4392 | 0.4588 | 0.4753 | 0.4918 |
| | 0.98 | 0.1233 | 0.1308 | 0.1482 | 0.1554 | 0.1728 | 0.1833 | 0.1970 |
| | | 0.3692 | 0.3856 | 0.4020 | 0.4263 | 0.4427 | 0.4591 | 0.4755 |
| | 0.95 | 0.1346 | 0.1500 | 0.1587 | 0.1756 | 0.1884 | 0.1995 | 0.2161 |
| | | 0.3460 | 0.3623 | 0.3801 | 0.4026 | 0.4189 | 0.4351 | 0.4514 |
| | 0.90 | 0.1436 | 0.1613 | 0.1778 | 0.1858 | 0.2015 | 0.2179 | 0.2335 |
| | | 0.3306 | 0.3468 | 0.3630 | 0.3792 | 0.3953 | 0.4114 | 0.4276 |
| | 0.80 | 0.1633 | 0.1796 | 0.1939 | 0.2032 | 0.2194 | 0.2356 | 0.2517 |
| | | 0.3077 | 0.3238 | 0.3398 | 0.3559 | 0.3719 | 0.3880 | 0.4040 |
| 63 | 0.99 | 0.1105 | 0.1208 | 0.1353 | 0.1423 | 0.1597 | 0.1673 | 0.1838 |
| | | 0.3785 | 0.3948 | 0.4188 | 0.4351 | 0.4514 | 0.4676 | 0.4838 |
| | 0.98 | 0.1212 | 0.1287 | 0.1456 | 0.1529 | 0.1700 | 0.1798 | 0.1938 |
| | | 0.3633 | 0.3794 | 0.4010 | 0.4194 | 0.4356 | 0.4517 | 0.4678 |
| | 0.95 | 0.1325 | 0.1474 | 0.1562 | 0.1728 | 0.1849 | 0.1963 | 0.2127 |
| | | 0.3405 | 0.3565 | 0.3802 | 0.3962 | 0.4122 | 0.4282 | 0.4442 |
| | 0.90 | 0.1406 | 0.1588 | 0.1750 | 0.1819 | 0.1984 | 0.2145 | 0.2292 |
| | | 0.3254 | 0.3414 | 0.3573 | 0.3732 | 0.3891 | 0.4049 | 0.4208 |
| | 0.80 | 0.1608 | 0.1768 | 0.1904 | 0.2000 | 0.2159 | 0.2319 | 0.2477 |
| | | 0.3028 | 0.3187 | 0.3345 | 0.3503 | 0.3661 | 0.3818 | 0.3996 |
| 64 | 0.99 | 0.1087 | 0.1186 | 0.1332 | 0.1395 | 0.1572 | 0.1639 | 0.1810 |
| | | 0.3726 | 0.3931 | 0.4122 | 0.4282 | 0.4442 | 0.4602 | 0.4761 |
| | 0.98 | 0.1192 | 0.1267 | 0.1431 | 0.1505 | 0.1674 | 0.1764 | 0.1908 |
| | | 0.3575 | 0.3750 | 0.3969 | 0.4128 | 0.4287 | 0.4446 | 0.4604 |
| | 0.95 | 0.1305 | 0.1449 | 0.1538 | 0.1701 | 0.1816 | 0.1933 | 0.2094 |
| | | 0.3352 | 0.3523 | 0.3742 | 0.3900 | 0.4057 | 0.4215 | 0.4372 |
| | 0.90 | 0.1375 | 0.1563 | 0.1723 | 0.1779 | 0.1953 | 0.2112 | 0.2250 |
| | | 0.3204 | 0.3360 | 0.3517 | 0.3674 | 0.3830 | 0.3986 | 0.4162 |
| | 0.80 | 0.1583 | 0.1740 | 0.1870 | 0.1969 | 0.2126 | 0.2283 | 0.2439 |
| | | 0.2982 | 0.3137 | 0.3293 | 0.3449 | 0.3604 | 0.3760 | 0.3990 |

| $n$ | $\beta$ | 21 | 22 | 23 | 24 | 25 | 26 | 27 |
|---|---|---|---|---|---|---|---|---|
| | 0.99 | 0.2030 | 0.2147 | 0.2321 | 0.2411 | 0.2568 | 0.2740 | 0.2853 |
| | | 0.5124 | 0.5281 | 0.5433 | 0.5586 | 0.5754 | 0.5922 | 0.6090 |
| | 0.98 | 0.2173 | 0.2243 | 0.2419 | 0.2589 | 0.2698 | 0.2834 | 0.3003 |
| | | 0.5000 | 0.5167 | 0.5334 | 0.5467 | 0.5606 | 0.5749 | 0.5915 |
| 61 | 0.95 | 0.2353 | 0.2439 | 0.2607 | 0.2774 | 0.2939 | 0.3035 | 0.3184 |
| | | 0.4753 | 0.4918 | 0.5083 | 0.5248 | 0.5413 | 0.5578 | 0.5743 |
| | 0.90 | 0.2453 | 0.2622 | 0.2787 | 0.2952 | 0.3117 | 0.3235 | 0.3359 |
| | | 0.4509 | 0.4673 | 0.4837 | 0.5000 | 0.5164 | 0.5328 | 0.5492 |
| | 0.80 | 0.2721 | 0.2827 | 0.2962 | 0.3126 | 0.3289 | 0.3452 | 0.3616 |
| | | 0.4268 | 0.4441 | 0.4628 | 0.4815 | 0.5000 | 0.5146 | 0.5290 |
| | 0.99 | 0.1991 | 0.2112 | 0.2282 | 0.2353 | 0.2526 | 0.2695 | 0.2788 |
| | | 0.5083 | 0.5248 | 0.5413 | 0.5563 | 0.5688 | 0.5826 | 0.5988 |
| | 0.98 | 0.2133 | 0.2211 | 0.2380 | 0.2547 | 0.2638 | 0.2788 | 0.2954 |
| | | 0.4919 | 0.5082 | 0.5246 | 0.5410 | 0.5574 | 0.5738 | 0.5848 |
| 62 | 0.95 | 0.2308 | 0.2400 | 0.2565 | 0.2730 | 0.2881 | 0.2969 | 0.3132 |
| | | 0.4676 | 0.4838 | 0.5000 | 0.5163 | 0.5325 | 0.5487 | 0.5650 |
| | 0.90 | 0.2417 | 0.2580 | 0.2743 | 0.2905 | 0.3063 | 0.3158 | 0.3305 |
| | | 0.4437 | 0.4598 | 0.4764 | 0.4930 | 0.5091 | 0.5246 | 0.5403 |
| | 0.80 | 0.2676 | 0.2761 | 0.2915 | 0.3076 | 0.3237 | 0.3397 | 0.3558 |
| | | 0.4223 | 0.4439 | 0.4600 | 0.4760 | 0.4920 | 0.5081 | 0.5241 |
| | 0.99 | 0.1953 | 0.2079 | 0.2240 | 0.2318 | 0.2486 | 0.2645 | 0.2714 |
| | | 0.5000 | 0.5163 | 0.5325 | 0.5487 | 0.5650 | 0.5813 | 0.5921 |
| | 0.98 | 0.2095 | 0.2176 | 0.2342 | 0.2502 | 0.2571 | 0.2744 | 0.2907 |
| | | 0.4839 | 0.5000 | 0.5162 | 0.5323 | 0.5484 | 0.5645 | 0.5807 |
| 63 | 0.95 | 0.2265 | 0.2362 | 0.2524 | 0.2686 | 0.2825 | 0.2922 | 0.3083 |
| | | 0.4601 | 0.4761 | 0.4921 | 0.5080 | 0.5240 | 0.5400 | 0.5559 |
| | 0.90 | 0.2379 | 0.2539 | 0.2699 | 0.2859 | 0.3004 | 0.3094 | 0.3253 |
| | | 0.4378 | 0.4567 | 0.4762 | 0.4921 | 0.5080 | 0.5239 | 0.5372 |
| | 0.80 | 0.2627 | 0.2710 | 0.2869 | 0.3027 | 0.3186 | 0.3344 | 0.3502 |
| | | 0.4211 | 0.4369 | 0.4527 | 0.4685 | 0.4843 | 0.5000 | 0.5158 |
| | 0.99 | 0.1916 | 0.2046 | 0.2200 | 0.2282 | 0.2447 | 0.2596 | 0.2682 |
| | | 0.4921 | 0.5080 | 0.5240 | 0.5399 | 0.5559 | 0.5719 | 0.5879 |
| | 0.98 | 0.2057 | 0.2142 | 0.2305 | 0.2456 | 0.2539 | 0.2701 | 0.2862 |
| | | 0.4763 | 0.4921 | 0.5080 | 0.5238 | 0.5397 | 0.5555 | 0.5714 |
| 64 | 0.95 | 0.2224 | 0.2325 | 0.2485 | 0.2644 | 0.2770 | 0.2876 | 0.3035 |
| | | 0.4529 | 0.4686 | 0.4843 | 0.5000 | 0.5158 | 0.5315 | 0.5472 |
| | 0.90 | 0.2342 | 0.2500 | 0.2657 | 0.2815 | 0.2945 | 0.3045 | 0.3203 |
| | | 0.4375 | 0.4532 | 0.4688 | 0.4844 | 0.5000 | 0.5157 | 0.5313 |
| | 0.80 | 0.2578 | 0.2668 | 0.2825 | 0.2981 | 0.3136 | 0.3292 | 0.3448 |
| | | 0.4146 | 0.4301 | 0.4457 | 0.4612 | 0.4768 | 0.4923 | 0.5078 |

| $n$ | $\beta$ | 28 | 29 | 30 | 31 | 32 | 33 | 34 |
|---|---|---|---|---|---|---|---|---|
| | 0.99 | 0.2986 | 0.3156 | 0.3326 | 0.3452 | 0.3573 | 0.3742 | 0.3910 |
| | | 0.6258 | 0.6427 | 0.6548 | 0.6674 | 0.6844 | 0.7014 | 0.7147 |
| | 0.98 | 0.3171 | 0.3298 | 0.3417 | 0.3584 | 0.3752 | 0.3919 | 0.4085 |
| | | 0.6081 | 0.6248 | 0.6416 | 0.6583 | 0.6702 | 0.6829 | 0.6997 |
| 61 | 0.95 | 0.3350 | 0.3516 | 0.3681 | 0.3846 | 0.3990 | 0.4105 | 0.4257 |
| | | 0.5895 | 0.6010 | 0.6154 | 0.6319 | 0.6484 | 0.6650 | 0.6816 |
| | 0.90 | 0.3524 | 0.3688 | 0.3853 | 0.4017 | 0.4181 | 0.4345 | 0.4508 |
| | | 0.5655 | 0.5819 | 0.5983 | 0.6147 | 0.6312 | 0.6476 | 0.6641 |
| | 0.80 | 0.3779 | 0.3942 | 0.4104 | 0.4267 | 0.4422 | 0.4566 | 0.4710 |
| | | 0.5434 | 0.5578 | 0.5733 | 0.5896 | 0.6058 | 0.6221 | 0.6384 |
| | 0.99 | 0.2938 | 0.3105 | 0.3266 | 0.3367 | 0.3515 | 0.3681 | 0.3846 |
| | | 0.6154 | 0.6319 | 0.6485 | 0.6633 | 0.6734 | 0.6895 | 0.7062 |
| | 0.98 | 0.3119 | 0.3221 | 0.3361 | 0.3526 | 0.3691 | 0.3855 | 0.4019 |
| | | 0.5981 | 0.6145 | 0.6309 | 0.6474 | 0.6639 | 0.6779 | 0.6881 |
| 62 | 0.95 | 0.3296 | 0.3459 | 0.3622 | 0.3776 | 0.3893 | 0.4025 | 0.4188 |
| | | 0.5812 | 0.5975 | 0.6107 | 0.6224 | 0.6378 | 0.6541 | 0.6704 |
| | 0.90 | 0.3467 | 0.3629 | 0.3791 | 0.3952 | 0.4113 | 0.4275 | 0.4436 |
| | | 0.5564 | 0.5725 | 0.5887 | 0.6048 | 0.6209 | 0.6371 | 0.6533 |
| | 0.80 | 0.3718 | 0.3879 | 0.4039 | 0.4184 | 0.4313 | 0.4438 | 0.4599 |
| | | 0.5401 | 0.5562 | 0.5687 | 0.5816 | 0.5961 | 0.6121 | 0.6282 |
| | 0.99 | 0.2890 | 0.3055 | 0.3200 | 0.3294 | 0.3458 | 0.3621 | 0.3784 |
| | | 0.6053 | 0.6216 | 0.6379 | 0.6542 | 0.6706 | 0.6800 | 0.6945 |
| | 0.98 | 0.3058 | 0.3144 | 0.3307 | 0.3470 | 0.3632 | 0.3793 | 0.3931 |
| | | 0.5972 | 0.6069 | 0.6207 | 0.6368 | 0.6530 | 0.6693 | 0.6856 |
| 63 | 0.95 | 0.3244 | 0.3404 | 0.3564 | 0.3694 | 0.3801 | 0.3961 | 0.4121 |
| | | 0.5719 | 0.5879 | 0.6039 | 0.6199 | 0.6306 | 0.6436 | 0.6596 |
| | 0.90 | 0.3413 | 0.3572 | 0.3731 | 0.3890 | 0.4048 | 0.4207 | 0.4357 |
| | | 0.5505 | 0.5643 | 0.5793 | 0.5952 | 0.6110 | 0.6269 | 0.6428 |
| | 0.80 | 0.3660 | 0.3817 | 0.3964 | 0.4083 | 0.4210 | 0.4368 | 0.4526 |
| | | 0.5316 | 0.5474 | 0.5632 | 0.5790 | 0.5917 | 0.6036 | 0.6183 |
| | 0.99 | 0.2845 | 0.3007 | 0.3134 | 0.3242 | 0.3403 | 0.3564 | 0.3725 |
| | | 0.6036 | 0.6136 | 0.6275 | 0.6436 | 0.6597 | 0.6758 | 0.6866 |
| | 0.98 | 0.2997 | 0.3095 | 0.3255 | 0.3415 | 0.3574 | 0.3725 | 0.3837 |
| | | 0.5873 | 0.6032 | 0.6163 | 0.6275 | 0.6426 | 0.6585 | 0.6745 |
| 64 | 0.95 | 0.3193 | 0.3351 | 0.3502 | 0.3607 | 0.3741 | 0.3899 | 0.4056 |
| | | 0.5629 | 0.5786 | 0.5944 | 0.6101 | 0.6259 | 0.6393 | 0.6498 |
| | 0.90 | 0.3359 | 0.3516 | 0.3673 | 0.3829 | 0.3985 | 0.4129 | 0.4253 |
| | | 0.5469 | 0.5626 | 0.5747 | 0.5871 | 0.6015 | 0.6171 | 0.6327 |
| | 0.80 | 0.3603 | 0.3758 | 0.3876 | 0.3989 | 0.4145 | 0.4300 | 0.4456 |
| | | 0.5233 | 0.5389 | 0.5544 | 0.5700 | 0.5855 | 0.6011 | 0.6124 |

| $n$ | $\beta$ | 0 | 1 | 2 | 3 | 4 | 5 | 6 |
|---|---|---|---|---|---|---|---|---|
| | 0.99 | 0.0000 | 0.0001 | 0.0023 | 0.0067 | 0.0128 | 0.0201 | 0.0281 |
| | | 0.0758 | 0.1072 | 0.1313 | 0.1549 | 0.1782 | 0.2015 | 0.2247 |
| | 0.98 | 0.0000 | 0.0003 | 0.0033 | 0.0088 | 0.0158 | 0.0239 | 0.0328 |
| | | 0.0698 | 0.0946 | 0.1248 | 0.1483 | 0.1649 | 0.1879 | 0.2110 |
| 65 | 0.95 | 0.0000 | 0.0007 | 0.0054 | 0.0126 | 0.0212 | 0.0307 | 0.0409 |
| | | 0.0578 | 0.0821 | 0.1055 | 0.1286 | 0.1515 | 0.1676 | 0.1904 |
| | 0.90 | 0.0000 | 0.0016 | 0.0082 | 0.0170 | 0.0271 | 0.0378 | 0.0457 |
| | | 0.0458 | 0.0695 | 0.0925 | 0.1153 | 0.1381 | 0.1540 | 0.1714 |
| | 0.80 | 0.0000 | 0.0034 | 0.0127 | 0.0237 | 0.0336 | 0.0421 | 0.0567 |
| | | 0.0337 | 0.0568 | 0.0795 | 0.1021 | 0.1178 | 0.1403 | 0.1560 |
| | 0.99 | 0.0000 | 0.0001 | 0.0022 | 0.0066 | 0.0126 | 0.0197 | 0.0277 |
| | | 0.0746 | 0.1055 | 0.1293 | 0.1527 | 0.1755 | 0.1984 | 0.2213 |
| | 0.98 | 0.0000 | 0.0003 | 0.0032 | 0.0086 | 0.0156 | 0.0236 | 0.0323 |
| | | 0.0687 | 0.0932 | 0.1229 | 0.1460 | 0.1624 | 0.1851 | 0.2078 |
| 66 | 0.95 | 0.0000 | 0.0007 | 0.0054 | 0.0125 | 0.0209 | 0.0303 | 0.0403 |
| | | 0.0569 | 0.0808 | 0.1039 | 0.1266 | 0.1492 | 0.1651 | 0.1875 |
| | 0.90 | 0.0000 | 0.0015 | 0.0080 | 0.0168 | 0.0266 | 0.0373 | 0.0450 |
| | | 0.0451 | 0.0684 | 0.0911 | 0.1136 | 0.1360 | 0.1517 | 0.1740 |
| | 0.80 | 0.0000 | 0.0033 | 0.0125 | 0.0233 | 0.0331 | 0.0415 | 0.0558 |
| | | 0.0332 | 0.0559 | 0.0783 | 0.1005 | 0.1161 | 0.1382 | 0.1536 |
| | 0.99 | 0.0000 | 0.0001 | 0.0022 | 0.0065 | 0.0124 | 0.0194 | 0.0273 |
| | | 0.0735 | 0.1040 | 0.1273 | 0.1566 | 0.1729 | 0.1955 | 0.2180 |
| | 0.98 | 0.0000 | 0.0003 | 0.0032 | 0.0085 | 0.0153 | 0.0232 | 0.0318 |
| | | 0.0677 | 0.0918 | 0.1211 | 0.1439 | 0.1599 | 0.1823 | 0.2047 |
| 67 | 0.95 | 0.0000 | 0.0007 | 0.0053 | 0.0123 | 0.0206 | 0.0298 | 0.0397 |
| | | 0.0560 | 0.0796 | 0.1023 | 0.1247 | 0.1470 | 0.1626 | 0.1847 |
| | 0.90 | 0.0000 | 0.0015 | 0.0079 | 0.0165 | 0.0262 | 0.0367 | 0.0443 |
| | | 0.0444 | 0.0674 | 0.0898 | 0.1119 | 0.1340 | 0.1494 | 0.1714 |
| | 0.80 | 0.0000 | 0.0033 | 0.0123 | 0.0229 | 0.0326 | 0.0408 | 0.0550 |
| | | 0.0327 | 0.0551 | 0.0771 | 0.0990 | 0.1144 | 0.1362 | 0.1513 |
| | 0.99 | 0.0000 | 0.0001 | 0.0021 | 0.0064 | 0.0123 | 0.0191 | 0.0268 |
| | | 0.0724 | 0.1024 | 0.1316 | 0.1543 | 0.1703 | 0.1926 | 0.2147 |
| | 0.98 | 0.0000 | 0.0002 | 0.0031 | 0.0084 | 0.0151 | 0.0229 | 0.0314 |
| | | 0.0667 | 0.0904 | 0.1193 | 0.1417 | 0.1576 | 0.1796 | 0.2016 |
| 68 | 0.95 | 0.0000 | 0.0007 | 0.0052 | 0.0121 | 0.0203 | 0.0294 | 0.0391 |
| | | 0.0552 | 0.0785 | 0.1008 | 0.1229 | 0.1448 | 0.1602 | 0.1820 |
| | 0.90 | 0.0000 | 0.0015 | 0.0078 | 0.0163 | 0.0259 | 0.0361 | 0.0437 |
| | | 0.0438 | 0.0664 | 0.0885 | 0.1103 | 0.1320 | 0.1473 | 0.1689 |
| | 0.80 | 0.0000 | 0.0032 | 0.0121 | 0.0226 | 0.0322 | 0.0402 | 0.0542 |
| | | 0.0323 | 0.0543 | 0.0760 | 0.0976 | 0.1127 | 0.1342 | 0.1491 |

| $n$ | $\beta$ | 7 | 8 | 9 | 10 | 11 | 12 | 13 |
|---|---|---|---|---|---|---|---|---|
| 65 | 0.99 | 0.0369 | 0.0461 | 0.0559 | 0.0660 | 0.0757 | 0.0820 | 0.0967 |
| | | 0.2410 | 0.2641 | 0.2802 | 0.2961 | 0.3192 | 0.3351 | 0.3510 |
| | 0.98 | 0.0423 | 0.0524 | 0.0628 | 0.0697 | 0.0820 | 0.0945 | 0.0999 |
| | | 0.2271 | 0.2501 | 0.2660 | 0.2818 | 0.3048 | 0.3206 | 0.3363 |
| | 0.95 | 0.0516 | 0.0577 | 0.0718 | 0.0820 | 0.0930 | 0.1054 | 0.1165 |
| | | 0.2063 | 0.2290 | 0.2448 | 0.2605 | 0.2833 | 0.2989 | 0.3145 |
| | 0.90 | 0.0574 | 0.0694 | 0.0789 | 0.0924 | 0.1039 | 0.1152 | 0.1311 |
| | | 0.1924 | 0.2080 | 0.2307 | 0.2463 | 0.2618 | 0.2773 | 0.3000 |
| | 0.80 | 0.0654 | 0.0794 | 0.0932 | 0.1020 | 0.1177 | 0.1317 | 0.1402 |
| | | 0.1715 | 0.1940 | 0.2095 | 0.2249 | 0.2403 | 0.2629 | 0.2782 |
| 66 | 0.99 | 0.0363 | 0.0454 | 0.0550 | 0.0650 | 0.0745 | 0.0806 | 0.0951 |
| | | 0.2373 | 0.2601 | 0.2759 | 0.2922 | 0.3144 | 0.3300 | 0.3456 |
| | 0.98 | 0.0417 | 0.0515 | 0.0618 | 0.0686 | 0.0807 | 0.0931 | 0.0981 |
| | | 0.2236 | 0.2462 | 0.2619 | 0.2775 | 0.3002 | 0.3157 | 0.3312 |
| | 0.95 | 0.0508 | 0.0568 | 0.0706 | 0.0807 | 0.0914 | 0.1038 | 0.1146 |
| | | 0.2032 | 0.2256 | 0.2411 | 0.2565 | 0.2790 | 0.2944 | 0.3097 |
| | 0.90 | 0.0565 | 0.0683 | 0.0776 | 0.0910 | 0.1022 | 0.1135 | 0.1290 |
| | | 0.1895 | 0.2049 | 0.2272 | 0.2425 | 0.2578 | 0.2731 | 0.2954 |
| | 0.80 | 0.0644 | 0.0782 | 0.0917 | 0.1004 | 0.1160 | 0.1295 | 0.1381 |
| | | 0.1689 | 0.1911 | 0.2063 | 0.2215 | 0.2367 | 0.2589 | 0.2741 |
| 67 | 0.99 | 0.0357 | 0.0447 | 0.0542 | 0.0640 | 0.0734 | 0.0793 | 0.0936 |
| | | 0.2337 | 0.2562 | 0.2717 | 0.2941 | 0.3096 | 0.3250 | 0.3404 |
| | 0.98 | 0.0410 | 0.0507 | 0.0608 | 0.0676 | 0.0794 | 0.0917 | 0.0963 |
| | | 0.2203 | 0.2426 | 0.2580 | 0.2734 | 0.2957 | 0.3110 | 0.3262 |
| | 0.95 | 0.0500 | 0.0559 | 0.0695 | 0.0795 | 0.0900 | 0.1022 | 0.1127 |
| | | 0.2001 | 0.2222 | 0.2375 | 0.2527 | 0.2748 | 0.2900 | 0.3051 |
| | 0.90 | 0.0556 | 0.0673 | 0.0764 | 0.0897 | 0.1006 | 0.1118 | 0.1269 |
| | | 0.1867 | 0.2019 | 0.2238 | 0.2389 | 0.2540 | 0.2694 | 0.2911 |
| | 0.80 | 0.0633 | 0.0770 | 0.0903 | 0.0989 | 0.1143 | 0.1274 | 0.1361 |
| | | 0.1664 | 0.1883 | 0.2033 | 0.2182 | 0.2332 | 0.2551 | 0.2700 |
| 68 | 0.99 | 0.0352 | 0.0440 | 0.0533 | 0.0630 | 0.0723 | 0.0779 | 0.0921 |
| | | 0.2303 | 0.2524 | 0.2677 | 0.2898 | 0.3050 | 0.3202 | 0.3378 |
| | 0.98 | 0.0404 | 0.0500 | 0.0599 | 0.0666 | 0.0782 | 0.0903 | 0.0945 |
| | | 0.2170 | 0.2390 | 0.2542 | 0.2723 | 0.2913 | 0.3064 | 0.3214 |
| | 0.95 | 0.0493 | 0.0551 | 0.0685 | 0.0784 | 0.0886 | 0.1007 | 0.1109 |
| | | 0.1972 | 0.2190 | 0.2340 | 0.2512 | 0.2708 | 0.2857 | 0.3006 |
| | 0.90 | 0.0547 | 0.0663 | 0.0752 | 0.0884 | 0.0990 | 0.1102 | 0.1249 |
| | | 0.1840 | 0.1989 | 0.2206 | 0.2354 | 0.2503 | 0.2720 | 0.2868 |
| | 0.80 | 0.0624 | 0.0759 | 0.0889 | 0.0975 | 0.1126 | 0.1254 | 0.1341 |
| | | 0.1640 | 0.1855 | 0.2003 | 0.2151 | 0.2298 | 0.2514 | 0.2661 |

| $n$ | $\beta$ | 14 | 15 | 16 | 17 | 18 | 19 | 20 |
|---|---|---|---|---|---|---|---|---|
| | 0.99 | 0.1071 | 0.1166 | 0.1312 | 0.1367 | 0.1548 | 0.1605 | 0.1781 |
| | | 0.3674 | 0.3900 | 0.4057 | 0.4215 | 0.4372 | 0.4529 | 0.4687 |
| | 0.98 | 0.1172 | 0.1247 | 0.1407 | 0.1482 | 0.1648 | 0.1731 | 0.1878 |
| | | 0.3520 | 0.3751 | 0.3907 | 0.4064 | 0.4220 | 0.4377 | 0.4533 |
| 65 | 0.95 | 0.1285 | 0.1425 | 0.1514 | 0.1675 | 0.1783 | 0.1903 | 0.2062 |
| | | 0.3300 | 0.3529 | 0.3685 | 0.3840 | 0.3995 | 0.4150 | 0.4305 |
| | 0.90 | 0.1380 | 0.1539 | 0.1694 | 0.1733 | 0.1923 | 0.2079 | 0.2210 |
| | | 0.3154 | 0.3309 | 0.3463 | 0.3617 | 0.3771 | 0.3937 | 0.4154 |
| | 0.80 | 0.1559 | 0.1714 | 0.1838 | 0.1939 | 0.2094 | 0.2248 | 0.2402 |
| | | 0.2936 | 0.3089 | 0.3243 | 0.3396 | 0.3549 | 0.3776 | 0.3930 |
| | 0.99 | 0.1054 | 0.1146 | 0.1292 | 0.1339 | 0.1524 | 0.1569 | 0.1754 |
| | | 0.3684 | 0.3840 | 0.3995 | 0.4150 | 0.4305 | 0.4460 | 0.4614 |
| | 0.98 | 0.1153 | 0.1228 | 0.1384 | 0.1459 | 0.1623 | 0.1698 | 0.1850 |
| | | 0.3474 | 0.3693 | 0.3848 | 0.4002 | 0.4156 | 0.4310 | 0.4463 |
| 66 | 0.95 | 0.1265 | 0.1401 | 0.1491 | 0.1650 | 0.1751 | 0.1874 | 0.2031 |
| | | 0.3250 | 0.3476 | 0.3629 | 0.3782 | 0.3934 | 0.4087 | 0.4239 |
| | 0.90 | 0.1359 | 0.1516 | 0.1666 | 0.1739 | 0.1894 | 0.2048 | 0.2171 |
| | | 0.3107 | 0.3259 | 0.3411 | 0.3563 | 0.3714 | 0.3939 | 0.4091 |
| | 0.80 | 0.1535 | 0.1688 | 0.1806 | 0.1910 | 0.2062 | 0.2214 | 0.2366 |
| | | 0.2892 | 0.3043 | 0.3194 | 0.3345 | 0.3496 | 0.3720 | 0.3870 |
| | 0.99 | 0.1039 | 0.1127 | 0.1272 | 0.1311 | 0.1499 | 0.1565 | 0.1728 |
| | | 0.3628 | 0.3782 | 0.3935 | 0.4087 | 0.4240 | 0.4392 | 0.4545 |
| | 0.98 | 0.1134 | 0.1210 | 0.1362 | 0.1438 | 0.1598 | 0.1666 | 0.1822 |
| | | 0.3486 | 0.3638 | 0.3790 | 0.3942 | 0.4093 | 0.4245 | 0.4396 |
| 67 | 0.95 | 0.1246 | 0.1378 | 0.1469 | 0.1625 | 0.1721 | 0.1846 | 0.2000 |
| | | 0.3216 | 0.3424 | 0.3574 | 0.3725 | 0.3875 | 0.4026 | 0.4176 |
| | 0.90 | 0.1339 | 0.1493 | 0.1639 | 0.1713 | 0.1866 | 0.2018 | 0.2132 |
| | | 0.3061 | 0.3211 | 0.3360 | 0.3510 | 0.3685 | 0.3881 | 0.4030 |
| | 0.80 | 0.1512 | 0.1663 | 0.1775 | 0.1882 | 0.2032 | 0.2181 | 0.2331 |
| | | 0.2849 | 0.2998 | 0.3147 | 0.3295 | 0.3481 | 0.3664 | 0.3813 |
| | 0.99 | 0.1023 | 0.1108 | 0.1252 | 0.1315 | 0.1475 | 0.1542 | 0.1702 |
| | | 0.3574 | 0.3725 | 0.3876 | 0.4026 | 0.4177 | 0.4344 | 0.4550 |
| | 0.98 | 0.1117 | 0.1192 | 0.1340 | 0.1416 | 0.1575 | 0.1633 | 0.1795 |
| | | 0.3434 | 0.3584 | 0.3734 | 0.3883 | 0.4033 | 0.4182 | 0.4347 |
| 68 | 0.95 | 0.1228 | 0.1356 | 0.1447 | 0.1601 | 0.1691 | 0.1819 | 0.1971 |
| | | 0.3224 | 0.3373 | 0.3522 | 0.3670 | 0.3818 | 0.3966 | 0.4114 |
| | 0.90 | 0.1319 | 0.1472 | 0.1613 | 0.1688 | 0.1839 | 0.1988 | 0.2095 |
| | | 0.3016 | 0.3164 | 0.3311 | 0.3458 | 0.3676 | 0.3824 | 0.3971 |
| | 0.80 | 0.1490 | 0.1639 | 0.1745 | 0.1854 | 0.2002 | 0.2150 | 0.2297 |
| | | 0.2808 | 0.2954 | 0.3101 | 0.3247 | 0.3464 | 0.3611 | 0.3757 |

| $n$ | $\beta$ | 21 | 22 | 23 | 24 | 25 | 26 | 27 |
|---|---|---|---|---|---|---|---|---|
| | 0.99 | 0.1880 | 0.2014 | 0.2160 | 0.2246 | 0.2409 | 0.2548 | 0.2640 |
| | | 0.4844 | 0.5000 | 0.5157 | 0.5314 | 0.5472 | 0.5629 | 0.5786 |
| | 0.98 | 0.2021 | 0.2109 | 0.2270 | 0.2411 | 0.2500 | 0.2659 | 0.2817 |
| | | 0.4689 | 0.4845 | 0.5000 | 0.5156 | 0.5312 | 0.5468 | 0.5624 |
| 65 | 0.95 | 0.2183 | 0.2289 | 0.2447 | 0.2604 | 0.2716 | 0.2832 | 0.2988 |
| | | 0.4459 | 0.4614 | 0.4769 | 0.4923 | 0.5078 | 0.5232 | 0.5387 |
| | 0.90 | 0.2306 | 0.2462 | 0.2617 | 0.2772 | 0.2887 | 0.2999 | 0.3153 |
| | | 0.4308 | 0.4462 | 0.4616 | 0.4770 | 0.4924 | 0.5077 | 0.5231 |
| | 0.80 | 0.2530 | 0.2628 | 0.2781 | 0.2935 | 0.3088 | 0.3242 | 0.3395 |
| | | 0.4083 | 0.4236 | 0.4389 | 0.4542 | 0.4695 | 0.4848 | 0.5000 |
| | 0.99 | 0.1846 | 0.1983 | 0.2122 | 0.2212 | 0.2372 | 0.2501 | 0.2600 |
| | | 0.4769 | 0.4923 | 0.5078 | 0.5232 | 0.5387 | 0.5541 | 0.5696 |
| | 0.98 | 0.1986 | 0.2077 | 0.2235 | 0.2368 | 0.2461 | 0.2618 | 0.2774 |
| | | 0.4617 | 0.4770 | 0.4924 | 0.5077 | 0.5231 | 0.5384 | 0.5538 |
| 66 | 0.95 | 0.2144 | 0.2255 | 0.2410 | 0.2564 | 0.2661 | 0.2789 | 0.2943 |
| | | 0.4392 | 0.4544 | 0.4696 | 0.4848 | 0.5000 | 0.5153 | 0.5305 |
| | 0.90 | 0.2271 | 0.2424 | 0.2577 | 0.2730 | 0.2828 | 0.2953 | 0.3106 |
| | | 0.4243 | 0.4394 | 0.4546 | 0.4698 | 0.4849 | 0.5000 | 0.5152 |
| | 0.80 | 0.2484 | 0.2588 | 0.2740 | 0.2891 | 0.3042 | 0.3193 | 0.3344 |
| | | 0.4021 | 0.4172 | 0.4323 | 0.4473 | 0.4624 | 0.4775 | 0.4925 |
| | 0.99 | 0.1811 | 0.1954 | 0.2085 | 0.2179 | 0.2336 | 0.2455 | 0.2561 |
| | | 0.4712 | 0.4873 | 0.5028 | 0.5175 | 0.5318 | 0.5458 | 0.5609 |
| | 0.98 | 0.1951 | 0.2046 | 0.2202 | 0.2326 | 0.2425 | 0.2579 | 0.2733 |
| | | 0.4547 | 0.4712 | 0.4872 | 0.5026 | 0.5173 | 0.5315 | 0.5454 |
| 67 | 0.95 | 0.2105 | 0.2221 | 0.2374 | 0.2526 | 0.2606 | 0.2747 | 0.2899 |
| | | 0.4326 | 0.4482 | 0.4651 | 0.4817 | 0.4975 | 0.5121 | 0.5259 |
| | 0.90 | 0.2237 | 0.2388 | 0.2539 | 0.2688 | 0.2767 | 0.2910 | 0.3060 |
| | | 0.4180 | 0.4329 | 0.4478 | 0.4628 | 0.4777 | 0.4926 | 0.5075 |
| | 0.80 | 0.2439 | 0.2550 | 0.2699 | 0.2848 | 0.2997 | 0.3146 | 0.3294 |
| | | 0.3962 | 0.4110 | 0.4259 | 0.4407 | 0.4555 | 0.4704 | 0.4852 |
| | 0.99 | 0.1778 | 0.1925 | 0.2049 | 0.2146 | 0.2302 | 0.2409 | 0.2523 |
| | | 0.4700 | 0.4850 | 0.5000 | 0.5151 | 0.5301 | 0.5451 | 0.5561 |
| | 0.98 | 0.1918 | 0.2015 | 0.2169 | 0.2284 | 0.2389 | 0.2541 | 0.2685 |
| | | 0.4553 | 0.4702 | 0.4851 | 0.5000 | 0.5150 | 0.5299 | 0.5448 |
| 68 | 0.95 | 0.2067 | 0.2189 | 0.2339 | 0.2484 | 0.2542 | 0.2707 | 0.2856 |
| | | 0.4282 | 0.4483 | 0.4631 | 0.4779 | 0.4927 | 0.5074 | 0.5222 |
| | 0.90 | 0.2205 | 0.2353 | 0.2502 | 0.2642 | 0.2719 | 0.2867 | 0.3015 |
| | | 0.4118 | 0.4265 | 0.4413 | 0.4560 | 0.4707 | 0.4854 | 0.5000 |
| | 0.80 | 0.2394 | 0.2513 | 0.2660 | 0.2807 | 0.2953 | 0.3100 | 0.3246 |
| | | 0.3904 | 0.4050 | 0.4196 | 0.4343 | 0.4489 | 0.4635 | 0.4781 |

| $n$ | $\beta$ | 28 | 29 | 30 | 31 | 32 | 33 | 34 |
|---|---|---|---|---|---|---|---|---|
| 65 | 0.99 | 0.2801 | 0.2960 | 0.3068 | 0.3191 | 0.3350 | 0.3509 | 0.3663 |
| | | 0.5944 | 0.6101 | 0.6224 | 0.6337 | 0.6491 | 0.6650 | 0.6809 |
| | 0.98 | 0.2938 | 0.3047 | 0.3205 | 0.3362 | 0.3519 | 0.3647 | 0.3750 |
| | | 0.5781 | 0.5937 | 0.6094 | 0.6250 | 0.6353 | 0.6481 | 0.6638 |
| | 0.95 | 0.3144 | 0.3299 | 0.3432 | 0.3528 | 0.3684 | 0.3839 | 0.3994 |
| | | 0.5542 | 0.5696 | 0.5851 | 0.6006 | 0.6161 | 0.6316 | 0.6472 |
| | 0.90 | 0.3308 | 0.3462 | 0.3616 | 0.3770 | 0.3917 | 0.4035 | 0.4153 |
| | | 0.5385 | 0.5539 | 0.5693 | 0.5847 | 0.5965 | 0.6083 | 0.6230 |
| | 0.80 | 0.3548 | 0.3682 | 0.3775 | 0.3929 | 0.4082 | 0.4235 | 0.4388 |
| | | 0.5153 | 0.5306 | 0.5459 | 0.5612 | 0.5765 | 0.5918 | 0.6071 |
| 66 | 0.99 | 0.2758 | 0.2913 | 0.2999 | 0.3143 | 0.3299 | 0.3455 | 0.3589 |
| | | 0.5851 | 0.6006 | 0.6161 | 0.6317 | 0.6411 | 0.6545 | 0.6701 |
| | 0.98 | 0.2879 | 0.3001 | 0.3156 | 0.3311 | 0.3462 | 0.3565 | 0.3692 |
| | | 0.5691 | 0.5845 | 0.5999 | 0.6153 | 0.6308 | 0.6435 | 0.6538 |
| | 0.95 | 0.3096 | 0.3249 | 0.3360 | 0.3475 | 0.3628 | 0.3781 | 0.3933 |
| | | 0.5457 | 0.5609 | 0.5762 | 0.5914 | 0.6067 | 0.6219 | 0.6372 |
| | 0.90 | 0.3258 | 0.3410 | 0.3562 | 0.3713 | 0.3835 | 0.3938 | 0.4090 |
| | | 0.5303 | 0.5455 | 0.5607 | 0.5758 | 0.5910 | 0.6062 | 0.6165 |
| | 0.80 | 0.3495 | 0.3603 | 0.3719 | 0.3869 | 0.4020 | 0.4171 | 0.4322 |
| | | 0.5076 | 0.5226 | 0.5377 | 0.5528 | 0.5678 | 0.5829 | 0.5980 |
| 67 | 0.99 | 0.2716 | 0.2859 | 0.2940 | 0.3095 | 0.3249 | 0.3403 | 0.3513 |
| | | 0.5761 | 0.5914 | 0.6066 | 0.6219 | 0.6373 | 0.6487 | 0.6597 |
| | 0.98 | 0.2817 | 0.2956 | 0.3109 | 0.3261 | 0.3394 | 0.3485 | 0.3637 |
| | | 0.5605 | 0.5756 | 0.5908 | 0.6059 | 0.6211 | 0.6363 | 0.6515 |
| | 0.95 | 0.3050 | 0.3195 | 0.3284 | 0.3423 | 0.3573 | 0.3724 | 0.3874 |
| | | 0.5395 | 0.5530 | 0.5675 | 0.5825 | 0.5975 | 0.6126 | 0.6276 |
| | 0.90 | 0.3210 | 0.3359 | 0.3509 | 0.3646 | 0.3743 | 0.3880 | 0.4029 |
| | | 0.5224 | 0.5373 | 0.5523 | 0.5672 | 0.5821 | 0.5971 | 0.6120 |
| | 0.80 | 0.3430 | 0.3509 | 0.3663 | 0.3812 | 0.3961 | 0.4109 | 0.4258 |
| | | 0.5000 | 0.5149 | 0.5297 | 0.5446 | 0.5594 | 0.5742 | 0.5891 |
| 68 | 0.99 | 0.2676 | 0.2807 | 0.2897 | 0.3049 | 0.3201 | 0.3342 | 0.3428 |
| | | 0.5687 | 0.5824 | 0.5975 | 0.6125 | 0.6276 | 0.6427 | 0.6572 |
| | 0.98 | 0.2744 | 0.2912 | 0.3063 | 0.3213 | 0.3327 | 0.3433 | 0.3583 |
| | | 0.5556 | 0.5682 | 0.5819 | 0.5968 | 0.6118 | 0.6267 | 0.6417 |
| | 0.95 | 0.3005 | 0.3136 | 0.3223 | 0.3372 | 0.3521 | 0.3669 | 0.3817 |
| | | 0.5370 | 0.5518 | 0.5631 | 0.5752 | 0.5887 | 0.6035 | 0.6183 |
| | 0.90 | 0.3163 | 0.3310 | 0.3457 | 0.3574 | 0.3675 | 0.3823 | 0.3970 |
| | | 0.5147 | 0.5294 | 0.5441 | 0.5588 | 0.5736 | 0.5883 | 0.6030 |
| | 0.80 | 0.3363 | 0.3463 | 0.3610 | 0.3756 | 0.3903 | 0.4049 | 0.4195 |
| | | 0.4927 | 0.5074 | 0.5220 | 0.5366 | 0.5512 | 0.5658 | 0.5805 |

| $n$ | $\beta$ | 0 | 1 | 2 | 3 | 4 | 5 | 6 |
|---|---|---|---|---|---|---|---|---|
| | 0.99 | 0.0000 | 0.0001 | 0.0021 | 0.0063 | 0.0121 | 0.0189 | 0.0265 |
| | | 0.0714 | 0.1009 | 0.1297 | 0.1520 | 0.1679 | 0.1898 | 0.2116 |
| | 0.98 | 0.0000 | 0.0002 | 0.0031 | 0.0083 | 0.0149 | 0.0225 | 0.0309 |
| | | 0.0657 | 0.0891 | 0.1176 | 0.1397 | 0.1553 | 0.1770 | 0.1987 |
| 69 | 0.95 | 0.0000 | 0.0007 | 0.0051 | 0.0119 | 0.0200 | 0.0289 | 0.0385 |
| | | 0.0544 | 0.0773 | 0.0994 | 0.1211 | 0.1427 | 0.1579 | 0.1794 |
| | 0.90 | 0.0000 | 0.0015 | 0.0077 | 0.0160 | 0.0255 | 0.0356 | 0.0430 |
| | | 0.0431 | 0.0655 | 0.0872 | 0.1087 | 0.1301 | 0.1451 | 0.1665 |
| | 0.80 | 0.0000 | 0.0032 | 0.0119 | 0.0223 | 0.0317 | 0.0396 | 0.0534 |
| | | 0.0318 | 0.0535 | 0.0749 | 0.0962 | 0.1111 | 0.1323 | 0.1470 |
| | 0.99 | 0.0000 | 0.0001 | 0.0021 | 0.0063 | 0.0119 | 0.0186 | 0.0261 |
| | | 0.0704 | 0.0995 | 0.1278 | 0.1498 | 0.1655 | 0.1870 | 0.2086 |
| | 0.98 | 0.0000 | 0.0002 | 0.0030 | 0.0081 | 0.0147 | 0.0222 | 0.0304 |
| | | 0.0648 | 0.0879 | 0.1159 | 0.1377 | 0.1543 | 0.1745 | 0.1959 |
| 70 | 0.95 | 0.0000 | 0.0007 | 0.0051 | 0.0117 | 0.0197 | 0.0285 | 0.0379 |
| | | 0.0537 | 0.0762 | 0.0980 | 0.1194 | 0.1407 | 0.1557 | 0.1768 |
| | 0.90 | 0.0000 | 0.0015 | 0.0076 | 0.0158 | 0.0251 | 0.0351 | 0.0424 |
| | | 0.0425 | 0.0646 | 0.0860 | 0.1072 | 0.1283 | 0.1431 | 0.1641 |
| | 0.80 | 0.0000 | 0.0031 | 0.0117 | 0.0220 | 0.0313 | 0.0390 | 0.0527 |
| | | 0.0314 | 0.0528 | 0.0738 | 0.0948 | 0.1095 | 0.1304 | 0.1449 |
| | 0.99 | 0.0000 | 0.0001 | 0.0021 | 0.0062 | 0.0117 | 0.0183 | 0.0257 |
| | | 0.0694 | 0.0981 | 0.1260 | 0.1477 | 0.1632 | 0.1844 | 0.2056 |
| | 0.98 | 0.0000 | 0.0002 | 0.0030 | 0.0080 | 0.0145 | 0.0219 | 0.0300 |
| | | 0.0639 | 0.0866 | 0.1142 | 0.1357 | 0.1570 | 0.1720 | 0.1931 |
| 71 | 0.95 | 0.0000 | 0.0007 | 0.0050 | 0.0116 | 0.0194 | 0.0281 | 0.0374 |
| | | 0.0529 | 0.0752 | 0.0966 | 0.1177 | 0.1387 | 0.1535 | 0.1743 |
| | 0.90 | 0.0000 | 0.0014 | 0.0075 | 0.0156 | 0.0247 | 0.0346 | 0.0418 |
| | | 0.0419 | 0.0637 | 0.0848 | 0.1057 | 0.1265 | 0.1411 | 0.1618 |
| | 0.80 | 0.0000 | 0.0031 | 0.0116 | 0.0216 | 0.0308 | 0.0385 | 0.0519 |
| | | 0.0309 | 0.0520 | 0.0728 | 0.0935 | 0.1080 | 0.1286 | 0.1429 |
| | 0.99 | 0.0000 | 0.0001 | 0.0020 | 0.0061 | 0.0116 | 0.0181 | 0.0253 |
| | | 0.0684 | 0.0967 | 0.1242 | 0.1457 | 0.1668 | 0.1818 | 0.2027 |
| | 0.98 | 0.0000 | 0.0002 | 0.0029 | 0.0079 | 0.0143 | 0.0216 | 0.0296 |
| | | 0.0630 | 0.0856 | 0.1127 | 0.1338 | 0.1548 | 0.1696 | 0.1904 |
| 72 | 0.95 | 0.0000 | 0.0007 | 0.0049 | 0.0114 | 0.0191 | 0.0277 | 0.0369 |
| | | 0.0522 | 0.0741 | 0.0953 | 0.1161 | 0.1368 | 0.1514 | 0.1719 |
| | 0.90 | 0.0000 | 0.0014 | 0.0074 | 0.0154 | 0.0244 | 0.0341 | 0.0413 |
| | | 0.0414 | 0.0628 | 0.0836 | 0.1042 | 0.1247 | 0.1391 | 0.1596 |
| | 0.80 | 0.0000 | 0.0030 | 0.0114 | 0.0213 | 0.0304 | 0.0379 | 0.0512 |
| | | 0.0305 | 0.0513 | 0.0718 | 0.0922 | 0.1065 | 0.1268 | 0.1409 |

| $n$ | $\beta$ | 7 | 8 | 9 | 10 | 11 | 12 | 13 |
|---|---|---|---|---|---|---|---|---|
| | 0.99 | 0.0347 | 0.0434 | 0.0525 | 0.0621 | 0.0713 | 0.0767 | 0.0907 |
| | | 0.2269 | 0.2487 | 0.2638 | 0.2855 | 0.3006 | 0.3155 | 0.3373 |
| | 0.98 | 0.0398 | 0.0492 | 0.0590 | 0.0656 | 0.0770 | 0.0890 | 0.0928 |
| | | 0.2139 | 0.2355 | 0.2505 | 0.2721 | 0.2870 | 0.3019 | 0.3170 |
| 69 | 0.95 | 0.0485 | 0.0543 | 0.0674 | 0.0772 | 0.0872 | 0.0993 | 0.1092 |
| | | 0.1944 | 0.2158 | 0.2306 | 0.2521 | 0.2668 | 0.2816 | 0.2963 |
| | 0.90 | 0.0539 | 0.0654 | 0.0740 | 0.0871 | 0.0974 | 0.1086 | 0.1230 |
| | | 0.1813 | 0.1961 | 0.2174 | 0.2321 | 0.2467 | 0.2681 | 0.2827 |
| | 0.80 | 0.0614 | 0.0748 | 0.0875 | 0.0961 | 0.1110 | 0.1235 | 0.1322 |
| | | 0.1616 | 0.1828 | 0.1974 | 0.2120 | 0.2265 | 0.2478 | 0.2622 |
| | 0.99 | 0.0342 | 0.0427 | 0.0518 | 0.0611 | 0.0703 | 0.0754 | 0.0893 |
| | | 0.2236 | 0.2451 | 0.2600 | 0.2814 | 0.2962 | 0.3110 | 0.3324 |
| | 0.98 | 0.0392 | 0.0485 | 0.0581 | 0.0647 | 0.0758 | 0.0878 | 0.0911 |
| | | 0.2108 | 0.2321 | 0.2469 | 0.2682 | 0.2829 | 0.2976 | 0.3189 |
| 70 | 0.95 | 0.0478 | 0.0536 | 0.0664 | 0.0761 | 0.0858 | 0.0979 | 0.1074 |
| | | 0.1916 | 0.2127 | 0.2273 | 0.2485 | 0.2630 | 0.2776 | 0.2920 |
| | 0.90 | 0.0531 | 0.0645 | 0.0729 | 0.0859 | 0.0959 | 0.1071 | 0.1211 |
| | | 0.1787 | 0.1933 | 0.2143 | 0.2288 | 0.2432 | 0.2642 | 0.2786 |
| | 0.80 | 0.0605 | 0.0737 | 0.0862 | 0.0947 | 0.1094 | 0.1216 | 0.1303 |
| | | 0.1593 | 0.1803 | 0.1946 | 0.2090 | 0.2233 | 0.2442 | 0.2585 |
| | 0.99 | 0.0337 | 0.0421 | 0.0510 | 0.0602 | 0.0693 | 0.0742 | 0.0880 |
| | | 0.2205 | 0.2416 | 0.2563 | 0.2774 | 0.2920 | 0.3069 | 0.3277 |
| | 0.98 | 0.0387 | 0.0478 | 0.0573 | 0.0638 | 0.0747 | 0.0865 | 0.0893 |
| | | 0.2078 | 0.2288 | 0.2434 | 0.2644 | 0.2789 | 0.2934 | 0.3144 |
| 71 | 0.95 | 0.0471 | 0.0528 | 0.0655 | 0.0751 | 0.0846 | 0.0965 | 0.1058 |
| | | 0.1889 | 0.2097 | 0.2241 | 0.2450 | 0.2593 | 0.2737 | 0.2883 |
| | 0.90 | 0.0523 | 0.0636 | 0.0718 | 0.0847 | 0.0945 | 0.1056 | 0.1193 |
| | | 0.1762 | 0.1906 | 0.2113 | 0.2256 | 0.2398 | 0.2605 | 0.2747 |
| | 0.80 | 0.0596 | 0.0727 | 0.0849 | 0.0934 | 0.1079 | 0.1197 | 0.1285 |
| | | 0.1571 | 0.1777 | 0.1919 | 0.2061 | 0.2211 | 0.2408 | 0.2549 |
| | 0.99 | 0.0332 | 0.0415 | 0.0503 | 0.0594 | 0.0683 | 0.0730 | 0.0867 |
| | | 0.2179 | 0.2382 | 0.2527 | 0.2735 | 0.2879 | 0.3088 | 0.3231 |
| | 0.98 | 0.0381 | 0.0471 | 0.0565 | 0.0629 | 0.0736 | 0.0853 | 0.0875 |
| | | 0.2049 | 0.2257 | 0.2400 | 0.2607 | 0.2750 | 0.2893 | 0.3100 |
| 72 | 0.95 | 0.0465 | 0.0521 | 0.0645 | 0.0740 | 0.0833 | 0.0952 | 0.1042 |
| | | 0.1863 | 0.2068 | 0.2210 | 0.2416 | 0.2557 | 0.2699 | 0.2904 |
| | 0.90 | 0.0516 | 0.0627 | 0.0707 | 0.0835 | 0.0931 | 0.1041 | 0.1176 |
| | | 0.1738 | 0.1879 | 0.2084 | 0.2224 | 0.2365 | 0.2569 | 0.2709 |
| | 0.80 | 0.0587 | 0.0717 | 0.0837 | 0.0921 | 0.1064 | 0.1180 | 0.1267 |
| | | 0.1550 | 0.1753 | 0.1893 | 0.2032 | 0.2236 | 0.2375 | 0.2514 |

| n | β | 14 | 15 | 16 | 17 | 18 | 19 | 20 |
|---|---|---|---|---|---|---|---|---|
| 69 | 0.99 | 0.1008 | 0.1090 | 0.1233 | 0.1296 | 0.1451 | 0.1519 | 0.1678 |
| | | 0.3522 | 0.3671 | 0.3819 | 0.3967 | 0.4138 | 0.4335 | 0.4483 |
| | 0.98 | 0.1099 | 0.1175 | 0.1319 | 0.1396 | 0.1552 | 0.1600 | 0.1769 |
| | | 0.3384 | 0.3532 | 0.3679 | 0.3827 | 0.3974 | 0.4143 | 0.4339 |
| | 0.95 | 0.1210 | 0.1335 | 0.1426 | 0.1578 | 0.1661 | 0.1793 | 0.1943 |
| | | 0.3178 | 0.3324 | 0.3471 | 0.3617 | 0.3763 | 0.3909 | 0.4076 |
| | 0.90 | 0.1300 | 0.1450 | 0.1587 | 0.1664 | 0.1812 | 0.1960 | 0.2058 |
| | | 0.2972 | 0.3118 | 0.3263 | 0.3420 | 0.3623 | 0.3769 | 0.3914 |
| | 0.80 | 0.1469 | 0.1615 | 0.1716 | 0.1827 | 0.1973 | 0.2119 | 0.2264 |
| | | 0.2767 | 0.2912 | 0.3056 | 0.3201 | 0.3415 | 0.3559 | 0.3703 |
| 70 | 0.99 | 0.0994 | 0.1072 | 0.1214 | 0.1277 | 0.1429 | 0.1497 | 0.1654 |
| | | 0.3471 | 0.3618 | 0.3764 | 0.3922 | 0.4126 | 0.4272 | 0.4418 |
| | 0.98 | 0.1083 | 0.1158 | 0.1298 | 0.1376 | 0.1528 | 0.1562 | 0.1744 |
| | | 0.3335 | 0.3481 | 0.3626 | 0.3771 | 0.3929 | 0.4131 | 0.4276 |
| | 0.95 | 0.1193 | 0.1314 | 0.1406 | 0.1556 | 0.1633 | 0.1767 | 0.1915 |
| | | 0.3132 | 0.3277 | 0.3421 | 0.3565 | 0.3709 | 0.3861 | 0.4066 |
| | 0.90 | 0.1282 | 0.1430 | 0.1562 | 0.1640 | 0.1786 | 0.1932 | 0.2023 |
| | | 0.2930 | 0.3074 | 0.3217 | 0.3428 | 0.3572 | 0.3715 | 0.3858 |
| | 0.80 | 0.1448 | 0.1592 | 0.1688 | 0.1802 | 0.1945 | 0.2089 | 0.2232 |
| | | 0.2728 | 0.2871 | 0.3013 | 0.3189 | 0.3366 | 0.3509 | 0.3651 |
| 71 | 0.99 | 0.0980 | 0.1055 | 0.1195 | 0.1259 | 0.1407 | 0.1476 | 0.1630 |
| | | 0.3422 | 0.3566 | 0.3710 | 0.3923 | 0.4067 | 0.4211 | 0.4355 |
| | 0.98 | 0.1066 | 0.1141 | 0.1279 | 0.1356 | 0.1505 | 0.1569 | 0.1719 |
| | | 0.3288 | 0.3432 | 0.3575 | 0.3718 | 0.3929 | 0.4073 | 0.4216 |
| | 0.95 | 0.1176 | 0.1294 | 0.1386 | 0.1534 | 0.1605 | 0.1742 | 0.1888 |
| | | 0.3088 | 0.3231 | 0.3373 | 0.3515 | 0.3657 | 0.3867 | 0.4009 |
| | 0.90 | 0.1264 | 0.1410 | 0.1538 | 0.1617 | 0.1761 | 0.1905 | 0.1988 |
| | | 0.2889 | 0.3031 | 0.3172 | 0.3380 | 0.3522 | 0.3663 | 0.3804 |
| | 0.80 | 0.1428 | 0.1570 | 0.1660 | 0.1776 | 0.1918 | 0.2060 | 0.2198 |
| | | 0.2690 | 0.2831 | 0.2971 | 0.3179 | 0.3319 | 0.3459 | 0.3600 |
| 72 | 0.99 | 0.0966 | 0.1039 | 0.1177 | 0.1241 | 0.1385 | 0.1456 | 0.1605 |
| | | 0.3374 | 0.3516 | 0.3681 | 0.3868 | 0.4010 | 0.4151 | 0.4293 |
| | 0.98 | 0.1051 | 0.1126 | 0.1259 | 0.1337 | 0.1482 | 0.1547 | 0.1695 |
| | | 0.3242 | 0.3384 | 0.3525 | 0.3691 | 0.3874 | 0.4015 | 0.4156 |
| | 0.95 | 0.1160 | 0.1274 | 0.1367 | 0.1513 | 0.1577 | 0.1718 | 0.1862 |
| | | 0.3045 | 0.3186 | 0.3326 | 0.3466 | 0.3619 | 0.3813 | 0.3953 |
| | 0.90 | 0.1246 | 0.1390 | 0.1515 | 0.1595 | 0.1737 | 0.1878 | 0.1953 |
| | | 0.2849 | 0.2989 | 0.3131 | 0.3334 | 0.3473 | 0.3612 | 0.3751 |
| | 0.80 | 0.1408 | 0.1549 | 0.1633 | 0.1752 | 0.1892 | 0.2031 | 0.2164 |
| | | 0.2653 | 0.2792 | 0.2930 | 0.3135 | 0.3273 | 0.3412 | 0.3550 |

| $n$ | $\beta$ | 21 | 22 | 23 | 24 | 25 | 26 | 27 |
|---|---|---|---|---|---|---|---|---|
| | 0.99 | 0.1745 | 0.1897 | 0.2014 | 0.2115 | 0.2268 | 0.2365 | 0.2486 |
| | | 0.4631 | 0.4779 | 0.4927 | 0.5074 | 0.5222 | 0.5370 | 0.5518 |
| | 0.98 | 0.1886 | 0.1986 | 0.2138 | 0.2244 | 0.2354 | 0.2504 | 0.2639 |
| | | 0.4486 | 0.4633 | 0.4780 | 0.4927 | 0.5074 | 0.5221 | 0.5368 |
| 69 | 0.95 | 0.2030 | 0.2157 | 0.2305 | 0.2442 | 0.2520 | 0.2667 | 0.2815 |
| | | 0.4271 | 0.4417 | 0.4563 | 0.4709 | 0.4855 | 0.5000 | 0.5146 |
| | 0.90 | 0.2173 | 0.2320 | 0.2466 | 0.2596 | 0.2680 | 0.2826 | 0.2971 |
| | | 0.4059 | 0.4204 | 0.4349 | 0.4494 | 0.4638 | 0.4783 | 0.4928 |
| | 0.80 | 0.2349 | 0.2477 | 0.2621 | 0.2766 | 0.2911 | 0.3055 | 0.3200 |
| | | 0.3848 | 0.3992 | 0.4136 | 0.4280 | 0.4424 | 0.4568 | 0.4712 |
| | 0.99 | 0.1711 | 0.1869 | 0.1980 | 0.2085 | 0.2235 | 0.2321 | 0.2450 |
| | | 0.4564 | 0.4709 | 0.4855 | 0.5000 | 0.5146 | 0.5292 | 0.5437 |
| | 0.98 | 0.1855 | 0.1958 | 0.2107 | 0.2204 | 0.2320 | 0.2468 | 0.2593 |
| | | 0.4421 | 0.4566 | 0.4711 | 0.4856 | 0.5000 | 0.5145 | 0.5290 |
| 70 | 0.95 | 0.1994 | 0.2126 | 0.2272 | 0.2401 | 0.2484 | 0.2629 | 0.2775 |
| | | 0.4210 | 0.4354 | 0.4498 | 0.4641 | 0.4785 | 0.4929 | 0.5072 |
| | 0.90 | 0.2142 | 0.2287 | 0.2431 | 0.2551 | 0.2641 | 0.2785 | 0.2929 |
| | | 0.4001 | 0.4144 | 0.4287 | 0.4430 | 0.4572 | 0.4715 | 0.4858 |
| | 0.80 | 0.2303 | 0.2441 | 0.2584 | 0.2727 | 0.2870 | 0.3012 | 0.3144 |
| | | 0.3793 | 0.3935 | 0.4077 | 0.4219 | 0.4362 | 0.4504 | 0.4646 |
| | 0.99 | 0.1676 | 0.1843 | 0.1947 | 0.2055 | 0.2204 | 0.2276 | 0.2415 |
| | | 0.4498 | 0.4642 | 0.4785 | 0.4929 | 0.5072 | 0.5216 | 0.5359 |
| | 0.98 | 0.1824 | 0.1930 | 0.2077 | 0.2165 | 0.2287 | 0.2433 | 0.2548 |
| | | 0.4358 | 0.4501 | 0.4644 | 0.4787 | 0.4929 | 0.5072 | 0.5214 |
| 71 | 0.95 | 0.1957 | 0.2096 | 0.2240 | 0.2360 | 0.2449 | 0.2592 | 0.2736 |
| | | 0.4151 | 0.4292 | 0.4434 | 0.4576 | 0.4717 | 0.4859 | 0.5000 |
| | 0.90 | 0.2112 | 0.2255 | 0.2397 | 0.2508 | 0.2604 | 0.2746 | 0.2888 |
| | | 0.3945 | 0.4086 | 0.4227 | 0.4367 | 0.4508 | 0.4649 | 0.4789 |
| | 0.80 | 0.2254 | 0.2407 | 0.2548 | 0.2689 | 0.2830 | 0.2970 | 0.3088 |
| | | 0.3740 | 0.3880 | 0.4020 | 0.4160 | 0.4301 | 0.4441 | 0.4581 |
| | 0.99 | 0.1667 | 0.1817 | 0.1915 | 0.2026 | 0.2172 | 0.2229 | 0.2381 |
| | | 0.4435 | 0.4576 | 0.4718 | 0.4859 | 0.5000 | 0.5142 | 0.5283 |
| | 0.98 | 0.1795 | 0.1903 | 0.2048 | 0.2127 | 0.2256 | 0.2399 | 0.2504 |
| | | 0.4297 | 0.4438 | 0.4579 | 0.4719 | 0.4860 | 0.5000 | 0.5141 |
| 72 | 0.95 | 0.1920 | 0.2067 | 0.2209 | 0.2321 | 0.2415 | 0.2556 | 0.2698 |
| | | 0.4093 | 0.4233 | 0.4372 | 0.4512 | 0.4652 | 0.4791 | 0.4931 |
| | 0.90 | 0.2083 | 0.2223 | 0.2364 | 0.2464 | 0.2568 | 0.2708 | 0.2848 |
| | | 0.3890 | 0.4029 | 0.4168 | 0.4307 | 0.4446 | 0.4584 | 0.4725 |
| | 0.80 | 0.2235 | 0.2374 | 0.2513 | 0.2652 | 0.2791 | 0.2929 | 0.3033 |
| | | 0.3688 | 0.3827 | 0.3965 | 0.4103 | 0.4241 | 0.4379 | 0.4517 |

| $n$ | $\beta$ | 28 | 29 | 30 | 31 | 32 | 33 | 34 |
|---|---|---|---|---|---|---|---|---|
| | 0.99 | 0.2637 | 0.2756 | 0.2854 | 0.3005 | 0.3154 | 0.3279 | 0.3372 |
| | | 0.5666 | 0.5786 | 0.5900 | 0.6034 | 0.6182 | 0.6330 | 0.6479 |
| | 0.98 | 0.2720 | 0.2869 | 0.3018 | 0.3165 | 0.3257 | 0.3383 | 0.3531 |
| | | 0.5515 | 0.5662 | 0.5780 | 0.5894 | 0.6027 | 0.6174 | 0.6322 |
| 69 | 0.95 | 0.2962 | 0.3077 | 0.3177 | 0.3323 | 0.3470 | 0.3616 | 0.3762 |
| | | 0.5292 | 0.5438 | 0.5584 | 0.5730 | 0.5849 | 0.5959 | 0.6092 |
| | 0.90 | 0.3117 | 0.3262 | 0.3402 | 0.3496 | 0.3622 | 0.3768 | 0.3913 |
| | | 0.5073 | 0.5218 | 0.5363 | 0.5507 | 0.5652 | 0.5797 | 0.5942 |
| | 0.80 | 0.3295 | 0.3414 | 0.3558 | 0.3702 | 0.3847 | 0.3991 | 0.4135 |
| | | 0.4856 | 0.5000 | 0.5145 | 0.5289 | 0.5433 | 0.5577 | 0.5721 |
| | 0.99 | 0.2599 | 0.2706 | 0.2813 | 0.2961 | 0.3109 | 0.3215 | 0.3323 |
| | | 0.5583 | 0.5729 | 0.5875 | 0.5989 | 0.6098 | 0.6237 | 0.6383 |
| | 0.98 | 0.2681 | 0.2828 | 0.2975 | 0.3109 | 0.3188 | 0.3334 | 0.3480 |
| | | 0.5435 | 0.5580 | 0.5725 | 0.5870 | 0.5983 | 0.6092 | 0.6230 |
| 70 | 0.95 | 0.2919 | 0.3018 | 0.3131 | 0.3276 | 0.3420 | 0.3564 | 0.3708 |
| | | 0.5216 | 0.5360 | 0.5503 | 0.5647 | 0.5791 | 0.5935 | 0.6045 |
| | 0.90 | 0.3073 | 0.3216 | 0.3340 | 0.3427 | 0.3571 | 0.3714 | 0.3857 |
| | | 0.5000 | 0.5143 | 0.5286 | 0.5429 | 0.5571 | 0.5714 | 0.5857 |
| | 0.80 | 0.3212 | 0.3365 | 0.3508 | 0.3650 | 0.3792 | 0.3934 | 0.4076 |
| | | 0.4788 | 0.4930 | 0.5071 | 0.5213 | 0.5355 | 0.5497 | 0.5639 |
| | 0.99 | 0.2562 | 0.2655 | 0.2773 | 0.2919 | 0.3063 | 0.3149 | 0.3276 |
| | | 0.5503 | 0.5646 | 0.5790 | 0.5934 | 0.6078 | 0.6176 | 0.6291 |
| | 0.98 | 0.2643 | 0.2788 | 0.2933 | 0.3053 | 0.3143 | 0.3287 | 0.3431 |
| | | 0.5357 | 0.5500 | 0.5643 | 0.5785 | 0.5928 | 0.6072 | 0.6169 |
| 71 | 0.95 | 0.2877 | 0.2956 | 0.3087 | 0.3230 | 0.3372 | 0.3514 | 0.3656 |
| | | 0.5142 | 0.5284 | 0.5425 | 0.5567 | 0.5709 | 0.5850 | 0.5992 |
| | 0.90 | 0.3030 | 0.3171 | 0.3277 | 0.3379 | 0.3521 | 0.3662 | 0.3803 |
| | | 0.4930 | 0.5071 | 0.5212 | 0.5352 | 0.5493 | 0.5634 | 0.5774 |
| | 0.80 | 0.3178 | 0.3318 | 0.3458 | 0.3599 | 0.3739 | 0.3879 | 0.4019 |
| | | 0.4721 | 0.4861 | 0.5000 | 0.5140 | 0.5280 | 0.5420 | 0.5560 |
| | 0.99 | 0.2526 | 0.2604 | 0.2734 | 0.2878 | 0.3010 | 0.3087 | 0.3230 |
| | | 0.5425 | 0.5566 | 0.5708 | 0.5850 | 0.5991 | 0.6133 | 0.6261 |
| | 0.98 | 0.2606 | 0.2749 | 0.2892 | 0.2997 | 0.3099 | 0.3241 | 0.3383 |
| | | 0.5282 | 0.5422 | 0.5563 | 0.5704 | 0.5845 | 0.5986 | 0.6127 |
| 72 | 0.95 | 0.2828 | 0.2903 | 0.3044 | 0.3185 | 0.3325 | 0.3465 | 0.3598 |
| | | 0.5070 | 0.5210 | 0.5349 | 0.5489 | 0.5629 | 0.5768 | 0.5908 |
| | 0.90 | 0.2988 | 0.3126 | 0.3211 | 0.3333 | 0.3472 | 0.3611 | 0.3750 |
| | | 0.4870 | 0.5011 | 0.5147 | 0.5280 | 0.5417 | 0.5555 | 0.5694 |
| | 0.80 | 0.3134 | 0.3272 | 0.3411 | 0.3549 | 0.3687 | 0.3826 | 0.3964 |
| | | 0.4655 | 0.4793 | 0.4931 | 0.5070 | 0.5208 | 0.5346 | 0.5484 |

| $n$ | $\beta$ | 35 | 36 | 37 | 38 | 39 | 40 | 41 |
|---|---|---|---|---|---|---|---|---|
| 69 | 0.99 | 0.3521 | 0.3670 | 0.3818 | 0.3966 | 0.4100 | 0.4214 | 0.4334 |
| | | 0.6628 | 0.6721 | 0.6846 | 0.6995 | 0.7146 | 0.7244 | 0.7363 |
| | 0.98 | 0.3678 | 0.3826 | 0.3973 | 0.4106 | 0.4220 | 0.4338 | 0.4485 |
| | | 0.6469 | 0.6617 | 0.6743 | 0.6835 | 0.6982 | 0.7131 | 0.7280 |
| | 0.95 | 0.3908 | 0.4041 | 0.4151 | 0.4270 | 0.4416 | 0.4562 | 0.4708 |
| | | 0.6238 | 0.6384 | 0.6530 | 0.6677 | 0.6823 | 0.6923 | 0.7038 |
| | 0.90 | 0.4058 | 0.4203 | 0.4348 | 0.4493 | 0.4637 | 0.4782 | 0.4927 |
| | | 0.6087 | 0.6232 | 0.6378 | 0.6504 | 0.6598 | 0.6738 | 0.6883 |
| | 0.80 | 0.4279 | 0.4423 | 0.4567 | 0.4711 | 0.4855 | 0.5000 | 0.5144 |
| | | 0.5865 | 0.6009 | 0.6153 | 0.6298 | 0.6442 | 0.6586 | 0.6705 |
| 70 | 0.99 | 0.3470 | 0.3617 | 0.3763 | 0.3902 | 0.4011 | 0.4125 | 0.4271 |
| | | 0.6530 | 0.6677 | 0.6785 | 0.6891 | 0.7039 | 0.7187 | 0.7294 |
| | 0.98 | 0.3625 | 0.3770 | 0.3908 | 0.4017 | 0.4130 | 0.4275 | 0.4420 |
| | | 0.6375 | 0.6520 | 0.6666 | 0.6812 | 0.6891 | 0.7025 | 0.7172 |
| | 0.95 | 0.3847 | 0.3955 | 0.4065 | 0.4209 | 0.4353 | 0.4497 | 0.4640 |
| | | 0.6153 | 0.6292 | 0.6436 | 0.6580 | 0.6724 | 0.6869 | 0.6982 |
| | 0.90 | 0.4000 | 0.4143 | 0.4286 | 0.4429 | 0.4571 | 0.4714 | 0.4857 |
| | | 0.6000 | 0.6143 | 0.6286 | 0.6429 | 0.6573 | 0.6660 | 0.6784 |
| | 0.80 | 0.4218 | 0.4361 | 0.4503 | 0.4645 | 0.4787 | 0.4929 | 0.5070 |
| | | 0.5782 | 0.5924 | 0.6066 | 0.6208 | 0.6350 | 0.6492 | 0.6635 |
| 71 | 0.99 | 0.3421 | 0.3565 | 0.3709 | 0.3824 | 0.3922 | 0.4066 | 0.4210 |
| | | 0.6435 | 0.6579 | 0.6724 | 0.6851 | 0.6937 | 0.7081 | 0.7227 |
| | 0.98 | 0.3574 | 0.3717 | 0.3831 | 0.3928 | 0.4072 | 0.4215 | 0.4357 |
| | | 0.6283 | 0.6426 | 0.6569 | 0.6713 | 0.6857 | 0.6947 | 0.7067 |
| | 0.95 | 0.3773 | 0.3866 | 0.4008 | 0.4150 | 0.4291 | 0.4433 | 0.4575 |
| | | 0.6134 | 0.6227 | 0.6344 | 0.6486 | 0.6628 | 0.6770 | 0.6913 |
| | 0.90 | 0.3944 | 0.4085 | 0.4226 | 0.4366 | 0.4507 | 0.4648 | 0.4788 |
| | | 0.5915 | 0.6056 | 0.6197 | 0.6338 | 0.6479 | 0.6621 | 0.6723 |
| | 0.80 | 0.4159 | 0.4300 | 0.4440 | 0.4580 | 0.4720 | 0.4860 | 0.5000 |
| | | 0.5700 | 0.5841 | 0.5981 | 0.6121 | 0.6261 | 0.6401 | 0.6542 |
| 72 | 0.99 | 0.3373 | 0.3515 | 0.3646 | 0.3739 | 0.3867 | 0.4009 | 0.4150 |
| | | 0.6354 | 0.6485 | 0.6627 | 0.6770 | 0.6913 | 0.6990 | 0.7122 |
| | 0.98 | 0.3524 | 0.3653 | 0.3744 | 0.3873 | 0.4014 | 0.4155 | 0.4296 |
| | | 0.6256 | 0.6347 | 0.6476 | 0.6617 | 0.6759 | 0.6901 | 0.7003 |
| | 0.95 | 0.3694 | 0.3812 | 0.3952 | 0.4092 | 0.4232 | 0.4371 | 0.4511 |
| | | 0.6048 | 0.6188 | 0.6306 | 0.6402 | 0.6535 | 0.6675 | 0.6815 |
| | 0.90 | 0.3889 | 0.4028 | 0.4167 | 0.4306 | 0.4445 | 0.4583 | 0.4720 |
| | | 0.5833 | 0.5972 | 0.6111 | 0.6250 | 0.6389 | 0.6528 | 0.6667 |
| | 0.80 | 0.4102 | 0.4240 | 0.4378 | 0.4516 | 0.4654 | 0.4792 | 0.4930 |
| | | 0.5622 | 0.5760 | 0.5898 | 0.6036 | 0.6174 | 0.6313 | 0.6451 |

| $n$ | $\beta$ | 0 | 1 | 2 | 3 | 4 | 5 | 6 |
|---|---|---|---|---|---|---|---|---|
| | 0.99 | 0.0000 | 0.0001 | 0.0020 | 0.0060 | 0.0114 | 0.0178 | 0.0250 |
| | | 0.0675 | 0.0954 | 0.1225 | 0.1437 | 0.1645 | 0.1793 | 0.1999 |
| | 0.98 | 0.0000 | 0.0002 | 0.0029 | 0.0078 | 0.0141 | 0.0213 | 0.0292 |
| | | 0.0621 | 0.0898 | 0.1111 | 0.1320 | 0.1527 | 0.1673 | 0.1878 |
| 73 | 0.95 | 0.0000 | 0.0007 | 0.0048 | 0.0112 | 0.0189 | 0.0273 | 0.0363 |
| | | 0.0515 | 0.0731 | 0.0940 | 0.1145 | 0.1349 | 0.1493 | 0.1696 |
| | 0.90 | 0.0000 | 0.0014 | 0.0073 | 0.0151 | 0.0241 | 0.0336 | 0.0407 |
| | | 0.0408 | 0.0619 | 0.0825 | 0.1028 | 0.1230 | 0.1372 | 0.1574 |
| | 0.80 | 0.0000 | 0.0030 | 0.0113 | 0.0210 | 0.0300 | 0.0374 | 0.0505 |
| | | 0.0301 | 0.0506 | 0.0708 | 0.0910 | 0.1050 | 0.1251 | 0.1390 |
| | 0.99 | 0.0000 | 0.0001 | 0.0020 | 0.0059 | 0.0112 | 0.0176 | 0.0246 |
| | | 0.0665 | 0.0941 | 0.1209 | 0.1417 | 0.1623 | 0.1769 | 0.1972 |
| | 0.98 | 0.0000 | 0.0002 | 0.0029 | 0.0077 | 0.0139 | 0.0210 | 0.0288 |
| | | 0.0613 | 0.0886 | 0.1096 | 0.1302 | 0.1506 | 0.1650 | 0.1853 |
| 74 | 0.95 | 0.0000 | 0.0006 | 0.0048 | 0.0111 | 0.0186 | 0.0269 | 0.0358 |
| | | 0.0508 | 0.0721 | 0.0927 | 0.1130 | 0.1331 | 0.1473 | 0.1673 |
| | 0.90 | 0.0000 | 0.0014 | 0.0072 | 0.0149 | 0.0237 | 0.0332 | 0.0401 |
| | | 0.0402 | 0.0611 | 0.0813 | 0.1014 | 0.1214 | 0.1354 | 0.1553 |
| | 0.80 | 0.0000 | 0.0030 | 0.0111 | 0.0208 | 0.0296 | 0.0369 | 0.0498 |
| | | 0.0297 | 0.0499 | 0.0699 | 0.0898 | 0.1036 | 0.1234 | 0.1372 |
| | 0.99 | 0.0000 | 0.0001 | 0.0019 | 0.0058 | 0.0111 | 0.0173 | 0.0243 |
| | | 0.0656 | 0.0928 | 0.1192 | 0.1398 | 0.1601 | 0.1745 | 0.1946 |
| | 0.98 | 0.0000 | 0.0002 | 0.0028 | 0.0076 | 0.0137 | 0.0207 | 0.0284 |
| | | 0.0605 | 0.0874 | 0.1081 | 0.1285 | 0.1486 | 0.1628 | 0.1828 |
| 75 | 0.95 | 0.0000 | 0.0006 | 0.0047 | 0.0109 | 0.0184 | 0.0266 | 0.0354 |
| | | 0.0501 | 0.0712 | 0.0915 | 0.1115 | 0.1314 | 0.1453 | 0.1651 |
| | 0.90 | 0.0000 | 0.0014 | 0.0071 | 0.0147 | 0.0234 | 0.0327 | 0.0396 |
| | | 0.0397 | 0.0603 | 0.0803 | 0.1001 | 0.1198 | 0.1336 | 0.1532 |
| | 0.80 | 0.0000 | 0.0029 | 0.0110 | 0.0205 | 0.0292 | 0.0364 | 0.0492 |
| | | 0.0293 | 0.0493 | 0.0690 | 0.0886 | 0.1023 | 0.1218 | 0.1353 |
| | 0.99 | 0.0000 | 0.0001 | 0.0019 | 0.0057 | 0.0109 | 0.0171 | 0.0240 |
| | | 0.0648 | 0.0916 | 0.1177 | 0.1380 | 0.1580 | 0.1722 | 0.1920 |
| | 0.98 | 0.0000 | 0.0002 | 0.0028 | 0.0075 | 0.0135 | 0.0204 | 0.0280 |
| | | 0.0597 | 0.0863 | 0.1067 | 0.1268 | 0.1467 | 0.1607 | 0.1804 |
| 76 | 0.95 | 0.0000 | 0.0006 | 0.0046 | 0.0108 | 0.0181 | 0.0262 | 0.0349 |
| | | 0.0494 | 0.0702 | 0.0903 | 0.1100 | 0.1296 | 0.1444 | 0.1629 |
| | 0.90 | 0.0000 | 0.0013 | 0.0070 | 0.0145 | 0.0231 | 0.0323 | 0.0391 |
| | | 0.0392 | 0.0595 | 0.0792 | 0.0988 | 0.1182 | 0.1319 | 0.1512 |
| | 0.80 | 0.0000 | 0.0029 | 0.0108 | 0.0202 | 0.0288 | 0.0359 | 0.0485 |
| | | 0.0289 | 0.0486 | 0.0681 | 0.0874 | 0.1009 | 0.1202 | 0.1336 |

| $n$ | $\beta$ | 7 | 8 | 9 | 10 | 11 | 12 | 13 |
|---|---|---|---|---|---|---|---|---|
| 73 | 0.99 | 0.0327 | 0.0409 | 0.0496 | 0.0585 | 0.0674 | 0.0719 | 0.0854 |
| | | 0.2205 | 0.2349 | 0.2506 | 0.2698 | 0.2839 | 0.3045 | 0.3186 |
| | 0.98 | 0.0376 | 0.0465 | 0.0557 | 0.0620 | 0.0725 | 0.0841 | 0.0897 |
| | | 0.2021 | 0.2226 | 0.2367 | 0.2571 | 0.2712 | 0.2853 | 0.3057 |
| | 0.95 | 0.0458 | 0.0514 | 0.0636 | 0.0730 | 0.0821 | 0.0939 | 0.1026 |
| | | 0.1845 | 0.2040 | 0.2180 | 0.2383 | 0.2522 | 0.2662 | 0.2865 |
| | 0.90 | 0.0509 | 0.0618 | 0.0697 | 0.0824 | 0.0917 | 0.1027 | 0.1158 |
| | | 0.1714 | 0.1854 | 0.2055 | 0.2194 | 0.2332 | 0.2534 | 0.2672 |
| | 0.80 | 0.0579 | 0.0707 | 0.0825 | 0.0909 | 0.1049 | 0.1162 | 0.1250 |
| | | 0.1529 | 0.1729 | 0.1867 | 0.2005 | 0.2206 | 0.2343 | 0.2480 |
| 74 | 0.99 | 0.0323 | 0.0404 | 0.0489 | 0.0577 | 0.0664 | 0.0708 | 0.0842 |
| | | 0.2175 | 0.2317 | 0.2520 | 0.2661 | 0.2801 | 0.3003 | 0.3143 |
| | 0.98 | 0.0371 | 0.0458 | 0.0549 | 0.0612 | 0.0715 | 0.0829 | 0.0885 |
| | | 0.1994 | 0.2195 | 0.2335 | 0.2537 | 0.2676 | 0.2824 | 0.3016 |
| | 0.95 | 0.0452 | 0.0507 | 0.0627 | 0.0720 | 0.0809 | 0.0926 | 0.1011 |
| | | 0.1873 | 0.2012 | 0.2151 | 0.2351 | 0.2488 | 0.2626 | 0.2826 |
| | 0.90 | 0.0502 | 0.0610 | 0.0687 | 0.0812 | 0.0904 | 0.1013 | 0.1142 |
| | | 0.1691 | 0.1829 | 0.2028 | 0.2165 | 0.2301 | 0.2500 | 0.2636 |
| | 0.80 | 0.0570 | 0.0698 | 0.0813 | 0.0897 | 0.1035 | 0.1145 | 0.1233 |
| | | 0.1508 | 0.1706 | 0.1842 | 0.1978 | 0.2176 | 0.2311 | 0.2447 |
| 75 | 0.99 | 0.0318 | 0.0398 | 0.0482 | 0.0569 | 0.0655 | 0.0697 | 0.0830 |
| | | 0.2146 | 0.2286 | 0.2486 | 0.2625 | 0.2763 | 0.2963 | 0.3100 |
| | 0.98 | 0.0366 | 0.0452 | 0.0542 | 0.0604 | 0.0705 | 0.0817 | 0.0873 |
| | | 0.2027 | 0.2166 | 0.2304 | 0.2503 | 0.2640 | 0.2839 | 0.2975 |
| | 0.95 | 0.0446 | 0.0500 | 0.0618 | 0.0711 | 0.0797 | 0.0914 | 0.0996 |
| | | 0.1848 | 0.1986 | 0.2122 | 0.2319 | 0.2455 | 0.2591 | 0.2788 |
| | 0.90 | 0.0495 | 0.0602 | 0.0677 | 0.0802 | 0.0891 | 0.1000 | 0.1126 |
| | | 0.1669 | 0.1813 | 0.2001 | 0.2136 | 0.2271 | 0.2467 | 0.2601 |
| | 0.80 | 0.0562 | 0.0689 | 0.0802 | 0.0885 | 0.1022 | 0.1129 | 0.1217 |
| | | 0.1488 | 0.1683 | 0.1818 | 0.1952 | 0.2147 | 0.2281 | 0.2414 |
| 76 | 0.99 | 0.0314 | 0.0393 | 0.0475 | 0.0561 | 0.0647 | 0.0686 | 0.0819 |
| | | 0.2118 | 0.2256 | 0.2453 | 0.2590 | 0.2727 | 0.2924 | 0.3059 |
| | 0.98 | 0.0361 | 0.0446 | 0.0534 | 0.0596 | 0.0695 | 0.0806 | 0.0862 |
| | | 0.2000 | 0.2137 | 0.2282 | 0.2470 | 0.2605 | 0.2801 | 0.2936 |
| | 0.95 | 0.0440 | 0.0493 | 0.0610 | 0.0701 | 0.0786 | 0.0902 | 0.0982 |
| | | 0.1824 | 0.1959 | 0.2094 | 0.2289 | 0.2423 | 0.2557 | 0.2752 |
| | 0.90 | 0.0488 | 0.0594 | 0.0668 | 0.0791 | 0.0878 | 0.0987 | 0.1110 |
| | | 0.1647 | 0.1841 | 0.1975 | 0.2108 | 0.2254 | 0.2435 | 0.2567 |
| | 0.80 | 0.0555 | 0.0680 | 0.0791 | 0.0873 | 0.1008 | 0.1113 | 0.1201 |
| | | 0.1469 | 0.1661 | 0.1794 | 0.1926 | 0.2119 | 0.2251 | 0.2383 |

| $n$ | $\beta$ | 14 | 15 | 16 | 17 | 18 | 19 | 20 |
|---|---|---|---|---|---|---|---|---|
| 73 | 0.99 | 0.0953 | 0.1023 | 0.1160 | 0.1224 | 0.1365 | 0.1436 | 0.1581 |
| | | 0.3327 | 0.3467 | 0.3674 | 0.3814 | 0.3954 | 0.4094 | 0.4234 |
| | 0.98 | 0.1035 | 0.1110 | 0.1241 | 0.1319 | 0.1460 | 0.1526 | 0.1672 |
| | | 0.3197 | 0.3337 | 0.3476 | 0.3681 | 0.3821 | 0.3960 | 0.4099 |
| | 0.95 | 0.1144 | 0.1255 | 0.1348 | 0.1492 | 0.1549 | 0.1695 | 0.1835 |
| | | 0.3004 | 0.3142 | 0.3281 | 0.3419 | 0.3623 | 0.3761 | 0.3899 |
| | 0.90 | 0.1229 | 0.1371 | 0.1492 | 0.1573 | 0.1713 | 0.1853 | 0.1918 |
| | | 0.2810 | 0.2948 | 0.3150 | 0.3288 | 0.3425 | 0.3563 | 0.3700 |
| | 0.80 | 0.1389 | 0.1528 | 0.1607 | 0.1728 | 0.1866 | 0.2004 | 0.2130 |
| | | 0.2617 | 0.2754 | 0.2890 | 0.3092 | 0.3229 | 0.3365 | 0.3502 |
| 74 | 0.99 | 0.0940 | 0.1007 | 0.1143 | 0.1208 | 0.1345 | 0.1416 | 0.1557 |
| | | 0.3282 | 0.3428 | 0.3623 | 0.3762 | 0.3900 | 0.4038 | 0.4176 |
| | 0.98 | 0.1020 | 0.1095 | 0.1223 | 0.1301 | 0.1438 | 0.1505 | 0.1649 |
| | | 0.3154 | 0.3292 | 0.3440 | 0.3631 | 0.3769 | 0.3906 | 0.4043 |
| | 0.95 | 0.1129 | 0.1236 | 0.1330 | 0.1472 | 0.1522 | 0.1672 | 0.1807 |
| | | 0.2963 | 0.3100 | 0.3236 | 0.3373 | 0.3574 | 0.3710 | 0.3846 |
| | 0.90 | 0.1213 | 0.1353 | 0.1470 | 0.1552 | 0.1690 | 0.1828 | 0.1882 |
| | | 0.2772 | 0.2908 | 0.3108 | 0.3244 | 0.3379 | 0.3515 | 0.3650 |
| | 0.80 | 0.1371 | 0.1507 | 0.1581 | 0.1705 | 0.1841 | 0.1977 | 0.2097 |
| | | 0.2582 | 0.2717 | 0.2866 | 0.3051 | 0.3186 | 0.3320 | 0.3455 |
| 75 | 0.99 | 0.0927 | 0.0992 | 0.1127 | 0.1191 | 0.1325 | 0.1397 | 0.1534 |
| | | 0.3237 | 0.3438 | 0.3574 | 0.3711 | 0.3847 | 0.3983 | 0.4119 |
| | 0.98 | 0.1006 | 0.1080 | 0.1205 | 0.1284 | 0.1417 | 0.1485 | 0.1627 |
| | | 0.3112 | 0.3248 | 0.3447 | 0.3583 | 0.3718 | 0.3854 | 0.3989 |
| | 0.95 | 0.1114 | 0.1218 | 0.1313 | 0.1452 | 0.1494 | 0.1650 | 0.1781 |
| | | 0.2923 | 0.3058 | 0.3193 | 0.3362 | 0.3526 | 0.3660 | 0.3795 |
| | 0.90 | 0.1197 | 0.1335 | 0.1448 | 0.1531 | 0.1668 | 0.1802 | 0.1844 |
| | | 0.2736 | 0.2870 | 0.3067 | 0.3201 | 0.3334 | 0.3468 | 0.3602 |
| | 0.80 | 0.1352 | 0.1487 | 0.1556 | 0.1682 | 0.1817 | 0.1951 | 0.2065 |
| | | 0.2548 | 0.2681 | 0.2877 | 0.3010 | 0.3143 | 0.3276 | 0.3409 |
| 76 | 0.99 | 0.0915 | 0.0977 | 0.1111 | 0.1176 | 0.1306 | 0.1379 | 0.1512 |
| | | 0.3195 | 0.3392 | 0.3527 | 0.3662 | 0.3796 | 0.3930 | 0.4087 |
| | 0.98 | 0.0992 | 0.1066 | 0.1188 | 0.1267 | 0.1397 | 0.1466 | 0.1606 |
| | | 0.3070 | 0.3205 | 0.3401 | 0.3535 | 0.3669 | 0.3803 | 0.3936 |
| | 0.95 | 0.1099 | 0.1201 | 0.1295 | 0.1432 | 0.1462 | 0.1628 | 0.1755 |
| | | 0.2885 | 0.3018 | 0.3151 | 0.3347 | 0.3479 | 0.3612 | 0.3745 |
| | 0.90 | 0.1181 | 0.1318 | 0.1427 | 0.1511 | 0.1646 | 0.1776 | 0.1840 |
| | | 0.2700 | 0.2832 | 0.3027 | 0.3159 | 0.3291 | 0.3423 | 0.3554 |
| | 0.80 | 0.1335 | 0.1468 | 0.1531 | 0.1660 | 0.1793 | 0.1925 | 0.2034 |
| | | 0.2514 | 0.2646 | 0.2840 | 0.2971 | 0.3102 | 0.3234 | 0.3365 |

| $n$ | $\beta$ | 21 | 22 | 23 | 24 | 25 | 26 | 27 |
|---|---|---|---|---|---|---|---|---|
| | 0.99 | 0.1644 | 0.1792 | 0.1883 | 0.1998 | 0.2138 | 0.2204 | 0.2348 |
| | | 0.4373 | 0.4513 | 0.4652 | 0.4791 | 0.4934 | 0.5073 | 0.5210 |
| | 0.98 | 0.1766 | 0.1877 | 0.2020 | 0.2088 | 0.2225 | 0.2366 | 0.2460 |
| | | 0.4238 | 0.4377 | 0.4515 | 0.4654 | 0.4793 | 0.4931 | 0.5070 |
| 73 | 0.95 | 0.1880 | 0.2039 | 0.2179 | 0.2283 | 0.2382 | 0.2521 | 0.2661 |
| | | 0.4037 | 0.4175 | 0.4312 | 0.4450 | 0.4588 | 0.4725 | 0.4863 |
| | 0.90 | 0.2054 | 0.2193 | 0.2331 | 0.2422 | 0.2533 | 0.2671 | 0.2809 |
| | | 0.3837 | 0.3974 | 0.4111 | 0.4248 | 0.4391 | 0.4551 | 0.4726 |
| | 0.80 | 0.2205 | 0.2342 | 0.2479 | 0.2616 | 0.2753 | 0.2889 | 0.2976 |
| | | 0.3638 | 0.3775 | 0.3911 | 0.4047 | 0.4184 | 0.4320 | 0.4456 |
| | 0.99 | 0.1622 | 0.1768 | 0.1853 | 0.1971 | 0.2104 | 0.2174 | 0.2316 |
| | | 0.4313 | 0.4460 | 0.4618 | 0.4781 | 0.4932 | 0.5069 | 0.5200 |
| | 0.98 | 0.1738 | 0.1852 | 0.1993 | 0.2047 | 0.2194 | 0.2334 | 0.2417 |
| | | 0.4180 | 0.4317 | 0.4454 | 0.4591 | 0.4732 | 0.4875 | 0.5014 |
| 74 | 0.95 | 0.1872 | 0.2011 | 0.2150 | 0.2245 | 0.2350 | 0.2487 | 0.2625 |
| | | 0.3982 | 0.4118 | 0.4254 | 0.4390 | 0.4526 | 0.4661 | 0.4797 |
| | 0.90 | 0.2027 | 0.2164 | 0.2300 | 0.2379 | 0.2499 | 0.2635 | 0.2771 |
| | | 0.3785 | 0.3921 | 0.4056 | 0.4205 | 0.4392 | 0.4527 | 0.4663 |
| | 0.80 | 0.2175 | 0.2310 | 0.2446 | 0.2581 | 0.2716 | 0.2846 | 0.2913 |
| | | 0.3590 | 0.3724 | 0.3859 | 0.3993 | 0.4127 | 0.4262 | 0.4401 |
| | 0.99 | 0.1600 | 0.1744 | 0.1822 | 0.1945 | 0.2072 | 0.2145 | 0.2285 |
| | | 0.4276 | 0.4457 | 0.4593 | 0.4729 | 0.4865 | 0.5000 | 0.5136 |
| | 0.98 | 0.1710 | 0.1827 | 0.1963 | 0.2026 | 0.2165 | 0.2303 | 0.2373 |
| | | 0.4124 | 0.4259 | 0.4404 | 0.4564 | 0.4730 | 0.4865 | 0.5000 |
| 75 | 0.95 | 0.1847 | 0.1985 | 0.2121 | 0.2208 | 0.2318 | 0.2454 | 0.2590 |
| | | 0.3929 | 0.4063 | 0.4197 | 0.4331 | 0.4465 | 0.4599 | 0.4736 |
| | 0.90 | 0.2000 | 0.2135 | 0.2270 | 0.2335 | 0.2466 | 0.2600 | 0.2735 |
| | | 0.3735 | 0.3869 | 0.4014 | 0.4200 | 0.4334 | 0.4467 | 0.4601 |
| | 0.80 | 0.2146 | 0.2280 | 0.2413 | 0.2547 | 0.2680 | 0.2800 | 0.2876 |
| | | 0.3542 | 0.3675 | 0.3807 | 0.3940 | 0.4073 | 0.4219 | 0.4403 |
| | 0.99 | 0.1579 | 0.1721 | 0.1792 | 0.1919 | 0.2041 | 0.2117 | 0.2255 |
| | | 0.4263 | 0.4397 | 0.4532 | 0.4666 | 0.4800 | 0.4934 | 0.5067 |
| | 0.98 | 0.1683 | 0.1803 | 0.1934 | 0.1999 | 0.2136 | 0.2271 | 0.2325 |
| | | 0.4069 | 0.4222 | 0.4401 | 0.4534 | 0.4667 | 0.4801 | 0.4934 |
| 76 | 0.95 | 0.1823 | 0.1958 | 0.2093 | 0.2171 | 0.2288 | 0.2422 | 0.2556 |
| | | 0.3877 | 0.4010 | 0.4142 | 0.4274 | 0.4415 | 0.4570 | 0.4736 |
| | 0.90 | 0.1974 | 0.2107 | 0.2237 | 0.2287 | 0.2434 | 0.2566 | 0.2699 |
| | | 0.3686 | 0.3818 | 0.4013 | 0.4145 | 0.4277 | 0.4408 | 0.4540 |
| | 0.80 | 0.2118 | 0.2250 | 0.2382 | 0.2513 | 0.2645 | 0.2755 | 0.2839 |
| | | 0.3496 | 0.3627 | 0.3758 | 0.3889 | 0.4032 | 0.4215 | 0.4346 |

| $n$ | $\beta$ | 28 | 29 | 30 | 31 | 32 | 33 | 34 |
|---|---|---|---|---|---|---|---|---|
| | 0.99 | 0.2488 | 0.2547 | 0.2697 | 0.2838 | 0.2958 | 0.3044 | 0.3185 |
| | | 0.5349 | 0.5488 | 0.5628 | 0.5767 | 0.5907 | 0.6047 | 0.6187 |
| | 0.98 | 0.2570 | 0.2711 | 0.2852 | 0.2941 | 0.3056 | 0.3196 | 0.3336 |
| | | 0.5208 | 0.5347 | 0.5486 | 0.5624 | 0.5763 | 0.5902 | 0.6041 |
| 73 | 0.95 | 0.2781 | 0.2864 | 0.3003 | 0.3141 | 0.3280 | 0.3418 | 0.3533 |
| | | 0.5000 | 0.5138 | 0.5276 | 0.5413 | 0.5551 | 0.5689 | 0.5826 |
| | 0.90 | 0.2947 | 0.3073 | 0.3149 | 0.3287 | 0.3424 | 0.3562 | 0.3699 |
| | | 0.4863 | 0.5000 | 0.5138 | 0.5275 | 0.5384 | 0.5501 | 0.5621 |
| | 0.80 | 0.3091 | 0.3228 | 0.3364 | 0.3501 | 0.3637 | 0.3774 | 0.3910 |
| | | 0.4592 | 0.4728 | 0.4870 | 0.5008 | 0.5142 | 0.5273 | 0.5409 |
| | 0.99 | 0.2448 | 0.2519 | 0.2660 | 0.2800 | 0.2907 | 0.3002 | 0.3142 |
| | | 0.5314 | 0.5435 | 0.5557 | 0.5688 | 0.5825 | 0.5963 | 0.6101 |
| | 0.98 | 0.2536 | 0.2675 | 0.2810 | 0.2881 | 0.3015 | 0.3153 | 0.3291 |
| | | 0.5148 | 0.5278 | 0.5410 | 0.5547 | 0.5684 | 0.5821 | 0.5958 |
| 74 | 0.95 | 0.2734 | 0.2825 | 0.2962 | 0.3099 | 0.3235 | 0.3372 | 0.3465 |
| | | 0.4933 | 0.5068 | 0.5204 | 0.5340 | 0.5475 | 0.5611 | 0.5747 |
| | 0.90 | 0.2907 | 0.3020 | 0.3107 | 0.3243 | 0.3378 | 0.3514 | 0.3649 |
| | | 0.4798 | 0.4933 | 0.5068 | 0.5203 | 0.5338 | 0.5474 | 0.5609 |
| | 0.80 | 0.3050 | 0.3185 | 0.3319 | 0.3454 | 0.3589 | 0.3723 | 0.3858 |
| | | 0.4557 | 0.4732 | 0.4866 | 0.5000 | 0.5135 | 0.5269 | 0.5374 |
| | 0.99 | 0.2409 | 0.2485 | 0.2624 | 0.2762 | 0.2855 | 0.2962 | 0.3099 |
| | | 0.5272 | 0.5408 | 0.5544 | 0.5650 | 0.5760 | 0.5882 | 0.6018 |
| | 0.98 | 0.2502 | 0.2639 | 0.2764 | 0.2838 | 0.2974 | 0.3111 | 0.3247 |
| | | 0.5136 | 0.5271 | 0.5384 | 0.5497 | 0.5615 | 0.5742 | 0.5877 |
| 75 | 0.95 | 0.2688 | 0.2787 | 0.2922 | 0.3057 | 0.3192 | 0.3315 | 0.3382 |
| | | 0.4876 | 0.5011 | 0.5143 | 0.5271 | 0.5402 | 0.5536 | 0.5670 |
| | 0.90 | 0.2869 | 0.2968 | 0.3066 | 0.3200 | 0.3333 | 0.3467 | 0.3601 |
| | | 0.4734 | 0.4867 | 0.5000 | 0.5134 | 0.5267 | 0.5400 | 0.5534 |
| | 0.80 | 0.3009 | 0.3142 | 0.3275 | 0.3408 | 0.3541 | 0.3674 | 0.3806 |
| | | 0.4536 | 0.4669 | 0.4801 | 0.4934 | 0.5067 | 0.5200 | 0.5332 |
| | 0.99 | 0.2371 | 0.2452 | 0.2589 | 0.2726 | 0.2802 | 0.2923 | 0.3058 |
| | | 0.5201 | 0.5335 | 0.5469 | 0.5604 | 0.5738 | 0.5849 | 0.5950 |
| | 0.98 | 0.2469 | 0.2604 | 0.2720 | 0.2800 | 0.2935 | 0.3069 | 0.3204 |
| | | 0.5067 | 0.5200 | 0.5334 | 0.5467 | 0.5600 | 0.5705 | 0.5811 |
| 76 | 0.95 | 0.2642 | 0.2751 | 0.2884 | 0.3017 | 0.3150 | 0.3259 | 0.3346 |
| | | 0.4868 | 0.5000 | 0.5133 | 0.5265 | 0.5373 | 0.5486 | 0.5601 |
| | 0.90 | 0.2831 | 0.2915 | 0.3026 | 0.3158 | 0.3290 | 0.3422 | 0.3553 |
| | | 0.4672 | 0.4803 | 0.4935 | 0.5066 | 0.5198 | 0.5329 | 0.5461 |
| | 0.80 | 0.2970 | 0.3101 | 0.3233 | 0.3364 | 0.3495 | 0.3626 | 0.3757 |
| | | 0.4477 | 0.4608 | 0.4739 | 0.4870 | 0.5000 | 0.5131 | 0.5262 |

| n | β | 35 | 36 | 37 | 38 | 39 | 40 | 41 |
|---|---|---|---|---|---|---|---|---|
| | 0.99 | 0.3326 | 0.3466 | 0.3578 | 0.3673 | 0.3813 | 0.3953 | 0.4093 |
| | | 0.6327 | 0.6422 | 0.6534 | 0.6674 | 0.6815 | 0.6956 | 0.7042 |
| | 0.98 | 0.3475 | 0.3585 | 0.3680 | 0.3820 | 0.3959 | 0.4098 | 0.4237 |
| | | 0.6180 | 0.6320 | 0.6415 | 0.6525 | 0.6664 | 0.6804 | 0.6944 |
| 73 | 0.95 | 0.3622 | 0.3760 | 0.3898 | 0.4036 | 0.4174 | 0.4311 | 0.4449 |
| | | 0.5964 | 0.6102 | 0.6240 | 0.6378 | 0.6467 | 0.6582 | 0.6720 |
| | 0.90 | 0.3836 | 0.3973 | 0.4110 | 0.4247 | 0.4379 | 0.4499 | 0.4616 |
| | | 0.5753 | 0.5890 | 0.6027 | 0.6164 | 0.6301 | 0.6438 | 0.6576 |
| | 0.80 | 0.4046 | 0.4183 | 0.4319 | 0.4455 | 0.4591 | 0.4727 | 0.4858 |
| | | 0.5545 | 0.5681 | 0.5817 | 0.5954 | 0.6090 | 0.6226 | 0.6363 |
| | 0.99 | 0.3281 | 0.3415 | 0.3506 | 0.3622 | 0.3761 | 0.3899 | 0.4037 |
| | | 0.6239 | 0.6378 | 0.6494 | 0.6585 | 0.6719 | 0.6858 | 0.6998 |
| | 0.98 | 0.3423 | 0.3512 | 0.3630 | 0.3768 | 0.3905 | 0.4042 | 0.4179 |
| | | 0.6095 | 0.6232 | 0.6370 | 0.6488 | 0.6577 | 0.6709 | 0.6847 |
| 74 | 0.95 | 0.3573 | 0.3709 | 0.3845 | 0.3981 | 0.4117 | 0.4253 | 0.4389 |
| | | 0.5883 | 0.6019 | 0.6155 | 0.6291 | 0.6427 | 0.6535 | 0.6628 |
| | 0.90 | 0.3784 | 0.3920 | 0.4055 | 0.4180 | 0.4290 | 0.4391 | 0.4526 |
| | | 0.5710 | 0.5820 | 0.5945 | 0.6080 | 0.6216 | 0.6351 | 0.6486 |
| | 0.80 | 0.3992 | 0.4126 | 0.4261 | 0.4391 | 0.4510 | 0.4626 | 0.4731 |
| | | 0.5490 | 0.5609 | 0.5739 | 0.5874 | 0.6008 | 0.6142 | 0.6277 |
| | 0.99 | 0.3236 | 0.3356 | 0.3437 | 0.3573 | 0.3710 | 0.3846 | 0.3982 |
| | | 0.6154 | 0.6290 | 0.6427 | 0.6563 | 0.6644 | 0.6764 | 0.6901 |
| | 0.98 | 0.3364 | 0.3446 | 0.3582 | 0.3717 | 0.3853 | 0.3988 | 0.4123 |
| | | 0.6012 | 0.6147 | 0.6283 | 0.6418 | 0.6554 | 0.6636 | 0.6753 |
| 75 | 0.95 | 0.3525 | 0.3659 | 0.3794 | 0.3928 | 0.4062 | 0.4196 | 0.4330 |
| | | 0.5804 | 0.5938 | 0.6072 | 0.6206 | 0.6341 | 0.6475 | 0.6618 |
| | 0.90 | 0.3734 | 0.3868 | 0.3993 | 0.4097 | 0.4199 | 0.4333 | 0.4466 |
| | | 0.5667 | 0.5801 | 0.5903 | 0.6007 | 0.6132 | 0.6266 | 0.6399 |
| | 0.80 | 0.3939 | 0.4072 | 0.4195 | 0.4304 | 0.4402 | 0.4535 | 0.4668 |
| | | 0.5465 | 0.5598 | 0.5696 | 0.5805 | 0.5928 | 0.6061 | 0.6194 |
| | 0.99 | 0.3194 | 0.3297 | 0.3391 | 0.3526 | 0.3661 | 0.3795 | 0.3929 |
| | | 0.6071 | 0.6205 | 0.6339 | 0.6474 | 0.6609 | 0.6703 | 0.6806 |
| | 0.98 | 0.3304 | 0.3400 | 0.3534 | 0.3668 | 0.3802 | 0.3935 | 0.4068 |
| | | 0.5932 | 0.6065 | 0.6198 | 0.6332 | 0.6466 | 0.6600 | 0.6696 |
| 76 | 0.95 | 0.3478 | 0.3611 | 0.3744 | 0.3876 | 0.4009 | 0.4141 | 0.4273 |
| | | 0.5727 | 0.5859 | 0.5991 | 0.6124 | 0.6256 | 0.6389 | 0.6522 |
| | 0.90 | 0.3685 | 0.3817 | 0.3919 | 0.4012 | 0.4144 | 0.4276 | 0.4407 |
| | | 0.5593 | 0.5724 | 0.5856 | 0.5988 | 0.6081 | 0.6183 | 0.6315 |
| | 0.80 | 0.3888 | 0.4011 | 0.4113 | 0.4214 | 0.4345 | 0.4476 | 0.4607 |
| | | 0.5393 | 0.5524 | 0.5655 | 0.5786 | 0.5887 | 0.5989 | 0.6112 |

| $n$ | $\beta$ | 0 | 1 | 2 | 3 | 4 | 5 | 6 |
|---|---|---|---|---|---|---|---|---|
| | 0.99 | 0.0000 | 0.0001 | 0.0019 | 0.0057 | 0.0108 | 0.0169 | 0.0236 |
| | | 0.0639 | 0.0904 | 0.1161 | 0.1362 | 0.1559 | 0.1700 | 0.1895 |
| | 0.98 | 0.0000 | 0.0002 | 0.0028 | 0.0074 | 0.0133 | 0.0201 | 0.0276 |
| | | 0.0589 | 0.0851 | 0.1053 | 0.1251 | 0.1448 | 0.1586 | 0.1780 |
| 77 | 0.95 | 0.0000 | 0.0006 | 0.0046 | 0.0107 | 0.0179 | 0.0259 | 0.0344 |
| | | 0.0488 | 0.0693 | 0.0891 | 0.1086 | 0.1279 | 0.1473 | 0.1608 |
| | 0.90 | 0.0000 | 0.0013 | 0.0069 | 0.0143 | 0.0228 | 0.0319 | 0.0386 |
| | | 0.0387 | 0.0587 | 0.0782 | 0.0975 | 0.1167 | 0.1302 | 0.1493 |
| | 0.80 | 0.0000 | 0.0028 | 0.0107 | 0.0199 | 0.0284 | 0.0354 | 0.0479 |
| | | 0.0285 | 0.0480 | 0.0672 | 0.0863 | 0.0996 | 0.1187 | 0.1319 |
| | 0.99 | 0.0000 | 0.0001 | 0.0019 | 0.0056 | 0.0107 | 0.0166 | 0.0233 |
| | | 0.0631 | 0.0892 | 0.1146 | 0.1344 | 0.1539 | 0.1678 | 0.1871 |
| | 0.98 | 0.0000 | 0.0002 | 0.0027 | 0.0073 | 0.0131 | 0.0199 | 0.0273 |
| | | 0.0581 | 0.0841 | 0.1040 | 0.1235 | 0.1429 | 0.1566 | 0.1757 |
| 78 | 0.95 | 0.0000 | 0.0006 | 0.0045 | 0.0105 | 0.0177 | 0.0255 | 0.0340 |
| | | 0.0482 | 0.0684 | 0.0880 | 0.1072 | 0.1263 | 0.1454 | 0.1587 |
| | 0.90 | 0.0000 | 0.0013 | 0.0068 | 0.0142 | 0.0225 | 0.0315 | 0.0381 |
| | | 0.0382 | 0.0580 | 0.0772 | 0.0962 | 0.1152 | 0.1285 | 0.1474 |
| | 0.80 | 0.0000 | 0.0028 | 0.0105 | 0.0197 | 0.0281 | 0.0349 | 0.0473 |
| | | 0.0282 | 0.0474 | 0.0663 | 0.0852 | 0.0984 | 0.1171 | 0.1302 |
| | 0.99 | 0.0000 | 0.0001 | 0.0018 | 0.0055 | 0.0105 | 0.0164 | 0.0230 |
| | | 0.0623 | 0.0881 | 0.1132 | 0.1327 | 0.1520 | 0.1656 | 0.1847 |
| | 0.98 | 0.0000 | 0.0002 | 0.0027 | 0.0072 | 0.0130 | 0.0196 | 0.0269 |
| | | 0.0574 | 0.0830 | 0.1027 | 0.1220 | 0.1411 | 0.1546 | 0.1735 |
| 79 | 0.95 | 0.0000 | 0.0006 | 0.0045 | 0.0104 | 0.0174 | 0.0252 | 0.0335 |
| | | 0.0476 | 0.0676 | 0.0868 | 0.1058 | 0.1247 | 0.1435 | 0.1567 |
| | 0.90 | 0.0000 | 0.0013 | 0.0067 | 0.0140 | 0.0222 | 0.0311 | 0.0376 |
| | | 0.0377 | 0.0573 | 0.0762 | 0.0950 | 0.1138 | 0.1269 | 0.1455 |
| | 0.80 | 0.0000 | 0.0028 | 0.0104 | 0.0194 | 0.0277 | 0.0345 | 0.0467 |
| | | 0.0278 | 0.0468 | 0.0655 | 0.0841 | 0.0971 | 0.1157 | 0.1286 |
| | 0.99 | 0.0000 | 0.0001 | 0.0018 | 0.0055 | 0.0104 | 0.0162 | 0.0227 |
| | | 0.0615 | 0.0870 | 0.1118 | 0.1310 | 0.1501 | 0.1640 | 0.1824 |
| | 0.98 | 0.0000 | 0.0002 | 0.0026 | 0.0071 | 0.0128 | 0.0194 | 0.0266 |
| | | 0.0567 | 0.0819 | 0.1014 | 0.1204 | 0.1393 | 0.1527 | 0.1713 |
| 80 | 0.95 | 0.0000 | 0.0006 | 0.0044 | 0.0102 | 0.0172 | 0.0249 | 0.0331 |
| | | 0.0470 | 0.0667 | 0.0858 | 0.1045 | 0.1232 | 0.1417 | 0.1548 |
| | 0.90 | 0.0000 | 0.0013 | 0.0066 | 0.0138 | 0.0219 | 0.0307 | 0.0372 |
| | | 0.0373 | 0.0565 | 0.0753 | 0.0938 | 0.1123 | 0.1253 | 0.1437 |
| | 0.80 | 0.0000 | 0.0027 | 0.0103 | 0.0192 | 0.0274 | 0.0340 | 0.0461 |
| | | 0.0275 | 0.0462 | 0.0647 | 0.0831 | 0.0959 | 0.1142 | 0.1270 |

| $n$ | $\beta$ | 7 | 8 | 9 | 10 | 11 | 12 | 13 |
|---|---|---|---|---|---|---|---|---|
| | 0.99 | 0.0310 | 0.0387 | 0.0469 | 0.0554 | 0.0638 | 0.0676 | 0.0807 |
| | | 0.2090 | 0.2227 | 0.2421 | 0.2556 | 0.2711 | 0.2885 | 0.3019 |
| | 0.98 | 0.0356 | 0.0440 | 0.0527 | 0.0588 | 0.0686 | 0.0795 | 0.0850 |
| | | 0.1974 | 0.2110 | 0.2303 | 0.2437 | 0.2571 | 0.2765 | 0.2898 |
| 77 | 0.95 | 0.0434 | 0.0487 | 0.0602 | 0.0692 | 0.0775 | 0.0890 | 0.0968 |
| | | 0.1800 | 0.1934 | 0.2067 | 0.2259 | 0.2392 | 0.2524 | 0.2716 |
| | 0.90 | 0.0481 | 0.0586 | 0.0658 | 0.0781 | 0.0866 | 0.0974 | 0.1095 |
| | | 0.1626 | 0.1817 | 0.1949 | 0.2081 | 0.2272 | 0.2403 | 0.2534 |
| | 0.80 | 0.0547 | 0.0671 | 0.0780 | 0.0862 | 0.0995 | 0.1097 | 0.1186 |
| | | 0.1450 | 0.1640 | 0.1771 | 0.1901 | 0.2092 | 0.2222 | 0.2352 |
| | 0.99 | 0.0306 | 0.0382 | 0.0463 | 0.0546 | 0.0630 | 0.0666 | 0.0796 |
| | | 0.2063 | 0.2198 | 0.2390 | 0.2523 | 0.2715 | 0.2848 | 0.2980 |
| | 0.98 | 0.0351 | 0.0434 | 0.0520 | 0.0580 | 0.0676 | 0.0785 | 0.0840 |
| | | 0.1949 | 0.2082 | 0.2273 | 0.2406 | 0.2538 | 0.2729 | 0.2860 |
| 78 | 0.95 | 0.0428 | 0.0481 | 0.0594 | 0.0683 | 0.0765 | 0.0879 | 0.0954 |
| | | 0.1777 | 0.1909 | 0.2041 | 0.2230 | 0.2361 | 0.2504 | 0.2681 |
| | 0.90 | 0.0475 | 0.0579 | 0.0649 | 0.0771 | 0.0854 | 0.0961 | 0.1080 |
| | | 0.1605 | 0.1794 | 0.1924 | 0.2054 | 0.2243 | 0.2373 | 0.2502 |
| | 0.80 | 0.0540 | 0.0662 | 0.0770 | 0.0851 | 0.0983 | 0.1082 | 0.1170 |
| | | 0.1431 | 0.1619 | 0.1748 | 0.1877 | 0.2065 | 0.2194 | 0.2322 |
| | 0.99 | 0.0302 | 0.0377 | 0.0457 | 0.0539 | 0.0622 | 0.0656 | 0.0786 |
| | | 0.2037 | 0.2170 | 0.2359 | 0.2491 | 0.2681 | 0.2812 | 0.2942 |
| | 0.98 | 0.0347 | 0.0428 | 0.0513 | 0.0573 | 0.0667 | 0.0774 | 0.0829 |
| | | 0.1924 | 0.2056 | 0.2245 | 0.2376 | 0.2506 | 0.2694 | 0.2824 |
| 79 | 0.95 | 0.0423 | 0.0475 | 0.0586 | 0.0675 | 0.0754 | 0.0867 | 0.0941 |
| | | 0.1755 | 0.1885 | 0.2027 | 0.2202 | 0.2331 | 0.2518 | 0.2647 |
| | 0.90 | 0.0469 | 0.0572 | 0.0641 | 0.0761 | 0.0843 | 0.0949 | 0.1065 |
| | | 0.1585 | 0.1771 | 0.1900 | 0.2028 | 0.2215 | 0.2343 | 0.2470 |
| | 0.80 | 0.0532 | 0.0654 | 0.0760 | 0.0840 | 0.0970 | 0.1068 | 0.1156 |
| | | 0.1414 | 0.1599 | 0.1726 | 0.1854 | 0.2039 | 0.2166 | 0.2293 |
| | 0.99 | 0.0298 | 0.0372 | 0.0451 | 0.0532 | 0.0614 | 0.0647 | 0.0775 |
| | | 0.2011 | 0.2143 | 0.2330 | 0.2460 | 0.2647 | 0.2776 | 0.2905 |
| | 0.98 | 0.0342 | 0.0423 | 0.0507 | 0.0566 | 0.0658 | 0.0764 | 0.0818 |
| | | 0.1900 | 0.2030 | 0.2216 | 0.2346 | 0.2474 | 0.2660 | 0.2789 |
| 80 | 0.95 | 0.0417 | 0.0469 | 0.0578 | 0.0666 | 0.0744 | 0.0857 | 0.0928 |
| | | 0.1733 | 0.1862 | 0.2047 | 0.2175 | 0.2302 | 0.2487 | 0.2614 |
| | 0.90 | 0.0463 | 0.0564 | 0.0632 | 0.0752 | 0.0831 | 0.0937 | 0.1051 |
| | | 0.1565 | 0.1749 | 0.1876 | 0.2003 | 0.2187 | 0.2314 | 0.2440 |
| | 0.80 | 0.0525 | 0.0646 | 0.0750 | 0.0830 | 0.0958 | 0.1053 | 0.1141 |
| | | 0.1396 | 0.1579 | 0.1705 | 0.1831 | 0.2014 | 0.2139 | 0.2265 |

| $n$ | $\beta$ | 14 | 15 | 16 | 17 | 18 | 19 | 20 |
|---|---|---|---|---|---|---|---|---|
| | 0.99 | 0.0903 | 0.0962 | 0.1096 | 0.1160 | 0.1288 | 0.1361 | 0.1491 |
| | | 0.3153 | 0.3347 | 0.3481 | 0.3614 | 0.3746 | 0.3889 | 0.4075 |
| | 0.98 | 0.0978 | 0.1052 | 0.1171 | 0.1250 | 0.1377 | 0.1447 | 0.1585 |
| | | 0.3030 | 0.3167 | 0.3357 | 0.3489 | 0.3621 | 0.3753 | 0.3885 |
| 77 | 0.95 | 0.1085 | 0.1184 | 0.1278 | 0.1412 | 0.1472 | 0.1607 | 0.1729 |
| | | 0.2848 | 0.2979 | 0.3110 | 0.3303 | 0.3434 | 0.3565 | 0.3696 |
| | 0.90 | 0.1166 | 0.1301 | 0.1407 | 0.1492 | 0.1625 | 0.1750 | 0.1816 |
| | | 0.2665 | 0.2805 | 0.2987 | 0.3118 | 0.3248 | 0.3378 | 0.3508 |
| | 0.80 | 0.1318 | 0.1449 | 0.1507 | 0.1639 | 0.1770 | 0.1900 | 0.2003 |
| | | 0.2482 | 0.2612 | 0.2803 | 0.2933 | 0.3062 | 0.3192 | 0.3321 |
| | 0.99 | 0.0891 | 0.0948 | 0.1081 | 0.1145 | 0.1270 | 0.1343 | 0.1470 |
| | | 0.3172 | 0.3304 | 0.3436 | 0.3567 | 0.3698 | 0.3891 | 0.4022 |
| | 0.98 | 0.0965 | 0.1039 | 0.1155 | 0.1234 | 0.1358 | 0.1428 | 0.1565 |
| | | 0.2991 | 0.3183 | 0.3313 | 0.3444 | 0.3574 | 0.3704 | 0.3838 |
| 78 | 0.95 | 0.1071 | 0.1167 | 0.1262 | 0.1392 | 0.1453 | 0.1586 | 0.1705 |
| | | 0.2811 | 0.2941 | 0.3081 | 0.3261 | 0.3390 | 0.3519 | 0.3649 |
| | 0.90 | 0.1151 | 0.1284 | 0.1387 | 0.1473 | 0.1604 | 0.1726 | 0.1793 |
| | | 0.2631 | 0.2820 | 0.2949 | 0.3078 | 0.3207 | 0.3335 | 0.3464 |
| | 0.80 | 0.1301 | 0.1430 | 0.1483 | 0.1618 | 0.1747 | 0.1876 | 0.1974 |
| | | 0.2451 | 0.2579 | 0.2767 | 0.2895 | 0.3023 | 0.3151 | 0.3279 |
| | 0.99 | 0.0880 | 0.0935 | 0.1066 | 0.1131 | 0.1252 | 0.1326 | 0.1449 |
| | | 0.3132 | 0.3262 | 0.3392 | 0.3521 | 0.3667 | 0.3841 | 0.3970 |
| | 0.98 | 0.0952 | 0.1026 | 0.1139 | 0.1219 | 0.1339 | 0.1410 | 0.1545 |
| | | 0.2953 | 0.3142 | 0.3271 | 0.3400 | 0.3529 | 0.3657 | 0.3847 |
| 79 | 0.95 | 0.1057 | 0.1151 | 0.1246 | 0.1373 | 0.1434 | 0.1566 | 0.1681 |
| | | 0.2776 | 0.2904 | 0.3091 | 0.3219 | 0.3347 | 0.3475 | 0.3602 |
| | 0.90 | 0.1137 | 0.1268 | 0.1368 | 0.1454 | 0.1584 | 0.1701 | 0.1770 |
| | | 0.2598 | 0.2785 | 0.2912 | 0.3039 | 0.3166 | 0.3293 | 0.3420 |
| | 0.80 | 0.1285 | 0.1413 | 0.1458 | 0.1598 | 0.1725 | 0.1853 | 0.1944 |
| | | 0.2420 | 0.2546 | 0.2733 | 0.2859 | 0.2985 | 0.3112 | 0.3238 |
| | 0.99 | 0.0869 | 0.0921 | 0.1052 | 0.1117 | 0.1235 | 0.1309 | 0.1429 |
| | | 0.3092 | 0.3221 | 0.3349 | 0.3477 | 0.3665 | 0.3792 | 0.3920 |
| | 0.98 | 0.0939 | 0.1013 | 0.1124 | 0.1203 | 0.1321 | 0.1392 | 0.1526 |
| | | 0.2916 | 0.3103 | 0.3230 | 0.3357 | 0.3485 | 0.3618 | 0.3799 |
| 80 | 0.95 | 0.1044 | 0.1136 | 0.1231 | 0.1355 | 0.1416 | 0.1547 | 0.1657 |
| | | 0.2741 | 0.2867 | 0.3053 | 0.3179 | 0.3305 | 0.3431 | 0.3557 |
| | 0.90 | 0.1122 | 0.1252 | 0.1349 | 0.1436 | 0.1564 | 0.1678 | 0.1748 |
| | | 0.2566 | 0.2750 | 0.2876 | 0.3001 | 0.3127 | 0.3252 | 0.3377 |
| | 0.80 | 0.1269 | 0.1395 | 0.1433 | 0.1578 | 0.1704 | 0.1830 | 0.1916 |
| | | 0.2390 | 0.2518 | 0.2699 | 0.2824 | 0.2948 | 0.3073 | 0.3198 |

| $n$ | $\beta$ | 21 | 22 | 23 | 24 | 25 | 26 | 27 |
|---|---|---|---|---|---|---|---|---|
| | 0.99 | 0.1558 | 0.1699 | 0.1763 | 0.1894 | 0.2010 | 0.2089 | 0.2226 |
| | | 0.4207 | 0.4340 | 0.4472 | 0.4604 | 0.4736 | 0.4868 | 0.5000 |
| | 0.98 | 0.1657 | 0.1779 | 0.1905 | 0.1973 | 0.2109 | 0.2237 | 0.2302 |
| | | 0.4033 | 0.4211 | 0.4343 | 0.4475 | 0.4606 | 0.4738 | 0.4869 |
| 77 | 0.95 | 0.1799 | 0.1933 | 0.2066 | 0.2134 | 0.2258 | 0.2391 | 0.2523 |
| | | 0.3827 | 0.3957 | 0.4088 | 0.4237 | 0.4413 | 0.4544 | 0.4674 |
| | 0.90 | 0.1948 | 0.2080 | 0.2204 | 0.2271 | 0.2402 | 0.2533 | 0.2664 |
| | | 0.3638 | 0.3809 | 0.3961 | 0.4091 | 0.4221 | 0.4351 | 0.4481 |
| | 0.80 | 0.2091 | 0.2221 | 0.2351 | 0.2481 | 0.2611 | 0.2710 | 0.2802 |
| | | 0.3451 | 0.3580 | 0.3709 | 0.3840 | 0.4031 | 0.4160 | 0.4290 |
| | 0.99 | 0.1538 | 0.1677 | 0.1733 | 0.1870 | 0.1980 | 0.2062 | 0.2197 |
| | | 0.4152 | 0.4283 | 0.4414 | 0.4544 | 0.4675 | 0.4805 | 0.4935 |
| | 0.98 | 0.1631 | 0.1756 | 0.1878 | 0.1948 | 0.2081 | 0.2203 | 0.2272 |
| | | 0.4027 | 0.4157 | 0.4287 | 0.4417 | 0.4546 | 0.4676 | 0.4806 |
| 78 | 0.95 | 0.1776 | 0.1908 | 0.2040 | 0.2096 | 0.2229 | 0.2360 | 0.2487 |
| | | 0.3778 | 0.3907 | 0.4053 | 0.4227 | 0.4356 | 0.4485 | 0.4614 |
| | 0.90 | 0.1923 | 0.2053 | 0.2171 | 0.2242 | 0.2372 | 0.2501 | 0.2630 |
| | | 0.3593 | 0.3782 | 0.3911 | 0.4039 | 0.4167 | 0.4296 | 0.4424 |
| | 0.80 | 0.2064 | 0.2193 | 0.2321 | 0.2450 | 0.2578 | 0.2666 | 0.2766 |
| | | 0.3407 | 0.3534 | 0.3662 | 0.3852 | 0.3980 | 0.4107 | 0.4235 |
| | 0.99 | 0.1519 | 0.1655 | 0.1703 | 0.1846 | 0.1951 | 0.2036 | 0.2169 |
| | | 0.4099 | 0.4228 | 0.4357 | 0.4486 | 0.4615 | 0.4743 | 0.4872 |
| | 0.98 | 0.1605 | 0.1734 | 0.1851 | 0.1923 | 0.2055 | 0.2171 | 0.2244 |
| | | 0.3975 | 0.4104 | 0.4232 | 0.4360 | 0.4488 | 0.4616 | 0.4745 |
| 79 | 0.95 | 0.1754 | 0.1884 | 0.2012 | 0.2053 | 0.2201 | 0.2330 | 0.2450 |
| | | 0.3730 | 0.3863 | 0.4046 | 0.4173 | 0.4301 | 0.4428 | 0.4555 |
| | 0.90 | 0.1899 | 0.2027 | 0.2140 | 0.2214 | 0.2342 | 0.2469 | 0.2597 |
| | | 0.3608 | 0.3735 | 0.3861 | 0.3988 | 0.4115 | 0.4241 | 0.4368 |
| | 0.80 | 0.2038 | 0.2165 | 0.2292 | 0.2419 | 0.2545 | 0.2621 | 0.2732 |
| | | 0.3364 | 0.3490 | 0.3621 | 0.3803 | 0.3930 | 0.4056 | 0.4182 |
| | 0.99 | 0.1500 | 0.1634 | 0.1671 | 0.1823 | 0.1923 | 0.2010 | 0.2142 |
| | | 0.4047 | 0.4175 | 0.4302 | 0.4429 | 0.4556 | 0.4683 | 0.4817 |
| | 0.98 | 0.1579 | 0.1712 | 0.1825 | 0.1899 | 0.2029 | 0.2140 | 0.2215 |
| | | 0.3925 | 0.4052 | 0.4179 | 0.4305 | 0.4432 | 0.4558 | 0.4685 |
| 80 | 0.95 | 0.1732 | 0.1861 | 0.1983 | 0.2046 | 0.2174 | 0.2301 | 0.2414 |
| | | 0.3683 | 0.3870 | 0.3995 | 0.4121 | 0.4247 | 0.4373 | 0.4498 |
| | 0.90 | 0.1875 | 0.2002 | 0.2109 | 0.2186 | 0.2313 | 0.2439 | 0.2565 |
| | | 0.3563 | 0.3688 | 0.3813 | 0.3938 | 0.4063 | 0.4188 | 0.4313 |
| | 0.80 | 0.2013 | 0.2138 | 0.2264 | 0.2389 | 0.2513 | 0.2574 | 0.2698 |
| | | 0.3322 | 0.3447 | 0.3632 | 0.3756 | 0.3881 | 0.4005 | 0.4130 |

| n | β | 28 | 29 | 30 | 31 | 32 | 33 | 34 |
|---|---|---|---|---|---|---|---|---|
| | 0.99 | 0.2334 | 0.2420 | 0.2555 | 0.2684 | 0.2741 | 0.2884 | 0.3018 |
| | | 0.5133 | 0.5265 | 0.5397 | 0.5529 | 0.5661 | 0.5794 | 0.5926 |
| | 0.98 | 0.2436 | 0.2570 | 0.2676 | 0.2764 | 0.2897 | 0.3029 | 0.3160 |
| | | 0.5000 | 0.5132 | 0.5263 | 0.5395 | 0.5526 | 0.5658 | 0.5790 |
| 77 | 0.95 | 0.2595 | 0.2715 | 0.2847 | 0.2978 | 0.3109 | 0.3201 | 0.3302 |
| | | 0.4805 | 0.4935 | 0.5066 | 0.5196 | 0.5327 | 0.5457 | 0.5588 |
| | 0.90 | 0.2791 | 0.2858 | 0.2986 | 0.3117 | 0.3247 | 0.3377 | 0.3507 |
| | | 0.4611 | 0.4741 | 0.4871 | 0.5000 | 0.5130 | 0.5260 | 0.5390 |
| | 0.80 | 0.2932 | 0.3061 | 0.3191 | 0.3320 | 0.3450 | 0.3579 | 0.3708 |
| | | 0.4419 | 0.4548 | 0.4677 | 0.4807 | 0.4936 | 0.5065 | 0.5194 |
| | 0.99 | 0.2298 | 0.2389 | 0.2522 | 0.2643 | 0.2714 | 0.2847 | 0.2979 |
| | | 0.5066 | 0.5196 | 0.5326 | 0.5457 | 0.5587 | 0.5718 | 0.5849 |
| | 0.98 | 0.2405 | 0.2537 | 0.2633 | 0.2728 | 0.2859 | 0.2990 | 0.3109 |
| | | 0.4936 | 0.5065 | 0.5195 | 0.5325 | 0.5455 | 0.5584 | 0.5714 |
| 78 | 0.95 | 0.2542 | 0.2680 | 0.2810 | 0.2940 | 0.3065 | 0.3139 | 0.3260 |
| | | 0.4743 | 0.4872 | 0.5000 | 0.5129 | 0.5258 | 0.5387 | 0.5516 |
| | 0.90 | 0.2748 | 0.2819 | 0.2948 | 0.3077 | 0.3206 | 0.3334 | 0.3463 |
| | | 0.4552 | 0.4680 | 0.4808 | 0.4936 | 0.5065 | 0.5193 | 0.5321 |
| | 0.80 | 0.2894 | 0.3022 | 0.3150 | 0.3278 | 0.3406 | 0.3533 | 0.3661 |
| | | 0.4363 | 0.4490 | 0.4618 | 0.4745 | 0.4873 | 0.5000 | 0.5128 |
| | 0.99 | 0.2263 | 0.2358 | 0.2490 | 0.2602 | 0.2680 | 0.2811 | 0.2941 |
| | | 0.5000 | 0.5129 | 0.5258 | 0.5386 | 0.5515 | 0.5644 | 0.5773 |
| | 0.98 | 0.2375 | 0.2505 | 0.2589 | 0.2693 | 0.2823 | 0.2952 | 0.3059 |
| | | 0.4873 | 0.5000 | 0.5128 | 0.5256 | 0.5385 | 0.5513 | 0.5641 |
| 79 | 0.95 | 0.2517 | 0.2646 | 0.2775 | 0.2903 | 0.3017 | 0.3090 | 0.3218 |
| | | 0.4683 | 0.4810 | 0.4937 | 0.5064 | 0.5191 | 0.5318 | 0.5446 |
| | 0.90 | 0.2706 | 0.2784 | 0.2911 | 0.3038 | 0.3165 | 0.3292 | 0.3419 |
| | | 0.4494 | 0.4621 | 0.4748 | 0.4874 | 0.5000 | 0.5127 | 0.5253 |
| | 0.80 | 0.2858 | 0.2984 | 0.3111 | 0.3237 | 0.3363 | 0.3489 | 0.3612 |
| | | 0.4308 | 0.4434 | 0.4560 | 0.4686 | 0.4812 | 0.4938 | 0.5063 |
| | 0.99 | 0.2228 | 0.2329 | 0.2459 | 0.2562 | 0.2646 | 0.2775 | 0.2904 |
| | | 0.4948 | 0.5075 | 0.5198 | 0.5318 | 0.5445 | 0.5572 | 0.5699 |
| | 0.98 | 0.2345 | 0.2473 | 0.2546 | 0.2659 | 0.2788 | 0.2915 | 0.3009 |
| | | 0.4811 | 0.4937 | 0.5064 | 0.5190 | 0.5316 | 0.5443 | 0.5569 |
| 80 | 0.95 | 0.2486 | 0.2613 | 0.2740 | 0.2866 | 0.2970 | 0.3052 | 0.3178 |
| | | 0.4624 | 0.4749 | 0.4875 | 0.5000 | 0.5126 | 0.5252 | 0.5377 |
| | 0.90 | 0.2664 | 0.2749 | 0.2875 | 0.3000 | 0.3126 | 0.3251 | 0.3376 |
| | | 0.4438 | 0.4563 | 0.4688 | 0.4813 | 0.4938 | 0.5063 | 0.5188 |
| | 0.80 | 0.2823 | 0.2947 | 0.3072 | 0.3197 | 0.3321 | 0.3446 | 0.3553 |
| | | 0.4254 | 0.4379 | 0.4503 | 0.4627 | 0.4752 | 0.4876 | 0.5000 |

| n | β | 35 | 36 | 37 | 38 | 39 | 40 | 41 |
|---|---|---|---|---|---|---|---|---|
| 77 | 0.99 | 0.3152 | 0.3235 | 0.3346 | 0.3480 | 0.3613 | 0.3745 | 0.3872 |
| | | 0.6029 | 0.6128 | 0.6255 | 0.6387 | 0.6520 | 0.6654 | 0.6765 |
| | 0.98 | 0.3241 | 0.3356 | 0.3488 | 0.3620 | 0.3752 | 0.3884 | 0.4005 |
| | | 0.5895 | 0.5995 | 0.6116 | 0.6248 | 0.6380 | 0.6512 | 0.6644 |
| | 0.95 | 0.3433 | 0.3564 | 0.3695 | 0.3826 | 0.3956 | 0.4087 | 0.4205 |
| | | 0.5690 | 0.5795 | 0.5913 | 0.6044 | 0.6174 | 0.6305 | 0.6436 |
| | 0.90 | 0.3637 | 0.3751 | 0.3825 | 0.3960 | 0.4090 | 0.4220 | 0.4350 |
| | | 0.5520 | 0.5650 | 0.5780 | 0.5910 | 0.6040 | 0.6175 | 0.6249 |
| | 0.80 | 0.3837 | 0.3937 | 0.4030 | 0.4159 | 0.4289 | 0.4418 | 0.4547 |
| | | 0.5324 | 0.5453 | 0.5582 | 0.5711 | 0.5841 | 0.5970 | 0.6063 |
| 78 | 0.99 | 0.3101 | 0.3171 | 0.3303 | 0.3435 | 0.3566 | 0.3697 | 0.3803 |
| | | 0.5979 | 0.6110 | 0.6197 | 0.6303 | 0.6434 | 0.6565 | 0.6697 |
| | 0.98 | 0.3182 | 0.3312 | 0.3443 | 0.3573 | 0.3703 | 0.3831 | 0.3930 |
| | | 0.5844 | 0.5974 | 0.6070 | 0.6169 | 0.6297 | 0.6427 | 0.6557 |
| | 0.95 | 0.3389 | 0.3518 | 0.3648 | 0.3777 | 0.3906 | 0.4024 | 0.4122 |
| | | 0.5645 | 0.5774 | 0.5878 | 0.5976 | 0.6094 | 0.6223 | 0.6352 |
| | 0.90 | 0.3590 | 0.3683 | 0.3781 | 0.3910 | 0.4038 | 0.4166 | 0.4295 |
| | | 0.5449 | 0.5577 | 0.5705 | 0.5834 | 0.5962 | 0.6090 | 0.6219 |
| | 0.80 | 0.3771 | 0.3851 | 0.3979 | 0.4106 | 0.4234 | 0.4362 | 0.4489 |
| | | 0.5256 | 0.5383 | 0.5511 | 0.5638 | 0.5766 | 0.5894 | 0.6021 |
| 79 | 0.99 | 0.3051 | 0.3131 | 0.3261 | 0.3391 | 0.3520 | 0.3641 | 0.3727 |
| | | 0.5902 | 0.6031 | 0.6160 | 0.6273 | 0.6359 | 0.6480 | 0.6609 |
| | 0.98 | 0.3141 | 0.3270 | 0.3399 | 0.3528 | 0.3656 | 0.3766 | 0.3846 |
| | | 0.5769 | 0.5897 | 0.6026 | 0.6154 | 0.6234 | 0.6344 | 0.6472 |
| | 0.95 | 0.3346 | 0.3474 | 0.3601 | 0.3729 | 0.3853 | 0.3949 | 0.4045 |
| | | 0.5573 | 0.5700 | 0.5828 | 0.5955 | 0.6051 | 0.6147 | 0.6271 |
| | 0.90 | 0.3531 | 0.3607 | 0.3734 | 0.3860 | 0.3987 | 0.4114 | 0.4240 |
| | | 0.5380 | 0.5507 | 0.5633 | 0.5760 | 0.5886 | 0.6013 | 0.6140 |
| | 0.80 | 0.3702 | 0.3802 | 0.3929 | 0.4055 | 0.4181 | 0.4307 | 0.4433 |
| | | 0.5189 | 0.5315 | 0.5441 | 0.5567 | 0.5693 | 0.5819 | 0.5945 |
| 80 | 0.99 | 0.3001 | 0.3091 | 0.3220 | 0.3348 | 0.3476 | 0.3579 | 0.3664 |
| | | 0.5826 | 0.5954 | 0.6081 | 0.6209 | 0.6336 | 0.6421 | 0.6524 |
| | 0.98 | 0.3102 | 0.3229 | 0.3356 | 0.3484 | 0.3607 | 0.3696 | 0.3798 |
| | | 0.5696 | 0.5822 | 0.5949 | 0.6076 | 0.6202 | 0.6304 | 0.6393 |
| | 0.95 | 0.3304 | 0.3430 | 0.3556 | 0.3682 | 0.3787 | 0.3869 | 0.3994 |
| | | 0.5503 | 0.5628 | 0.5754 | 0.5880 | 0.6006 | 0.6131 | 0.6213 |
| | 0.90 | 0.3471 | 0.3562 | 0.3687 | 0.3812 | 0.3937 | 0.4062 | 0.4187 |
| | | 0.5313 | 0.5438 | 0.5563 | 0.5688 | 0.5813 | 0.5938 | 0.6063 |
| | 0.80 | 0.3631 | 0.3755 | 0.3880 | 0.4004 | 0.4129 | 0.4253 | 0.4378 |
| | | 0.5125 | 0.5249 | 0.5374 | 0.5498 | 0.5622 | 0.5747 | 0.5871 |

| $n$ | $\beta$ | 0 | 1 | 2 | 3 | 4 | 5 | 6 |
|---|---|---|---|---|---|---|---|---|
| | 0.99 | 0.0000 | 0.0001 | 0.0018 | 0.0054 | 0.0103 | 0.0160 | 0.0224 |
| | | 0.0608 | 0.0859 | 0.1104 | 0.1294 | 0.1482 | 0.1669 | 0.1801 |
| | 0.98 | 0.0000 | 0.0002 | 0.0026 | 0.0070 | 0.0127 | 0.0191 | 0.0262 |
| | | 0.0560 | 0.0809 | 0.1001 | 0.1190 | 0.1376 | 0.1508 | 0.1692 |
| 81 | 0.95 | 0.0000 | 0.0006 | 0.0044 | 0.0101 | 0.0170 | 0.0246 | 0.0327 |
| | | 0.0464 | 0.0659 | 0.0847 | 0.1032 | 0.1216 | 0.1400 | 0.1529 |
| | 0.90 | 0.0000 | 0.0012 | 0.0065 | 0.0136 | 0.0217 | 0.0303 | 0.0367 |
| | | 0.0368 | 0.0559 | 0.0744 | 0.0927 | 0.1110 | 0.1238 | 0.1419 |
| | 0.80 | 0.0000 | 0.0027 | 0.0101 | 0.0190 | 0.0270 | 0.0336 | 0.0456 |
| | | 0.0271 | 0.0457 | 0.0639 | 0.0821 | 0.0948 | 0.1128 | 0.1254 |
| | 0.99 | 0.0000 | 0.0001 | 0.0018 | 0.0053 | 0.0101 | 0.0158 | 0.0222 |
| | | 0.0600 | 0.0849 | 0.1090 | 0.1278 | 0.1464 | 0.1649 | 0.1779 |
| | 0.98 | 0.0000 | 0.0002 | 0.0026 | 0.0069 | 0.0125 | 0.0189 | 0.0259 |
| | | 0.0553 | 0.0799 | 0.0989 | 0.1175 | 0.1359 | 0.1489 | 0.1672 |
| 82 | 0.95 | 0.0000 | 0.0006 | 0.0043 | 0.0100 | 0.0168 | 0.0243 | 0.0323 |
| | | 0.0458 | 0.0651 | 0.0837 | 0.1020 | 0.1202 | 0.1383 | 0.1510 |
| | 0.90 | 0.0000 | 0.0012 | 0.0065 | 0.0135 | 0.0214 | 0.0299 | 0.0362 |
| | | 0.0363 | 0.0552 | 0.0735 | 0.0916 | 0.1096 | 0.1223 | 0.1402 |
| | 0.80 | 0.0000 | 0.0027 | 0.0100 | 0.0187 | 0.0267 | 0.0332 | 0.0450 |
| | | 0.0268 | 0.0451 | 0.0631 | 0.0811 | 0.0936 | 0.1115 | 0.1239 |
| | 0.99 | 0.0000 | 0.0001 | 0.0017 | 0.0053 | 0.0100 | 0.0156 | 0.0219 |
| | | 0.0593 | 0.0839 | 0.1077 | 0.1263 | 0.1446 | 0.1629 | 0.1757 |
| | 0.98 | 0.0000 | 0.0002 | 0.0025 | 0.0068 | 0.0123 | 0.0187 | 0.0256 |
| | | 0.0546 | 0.0790 | 0.0977 | 0.1161 | 0.1343 | 0.1481 | 0.1651 |
| 83 | 0.95 | 0.0000 | 0.0006 | 0.0042 | 0.0099 | 0.0166 | 0.0240 | 0.0319 |
| | | 0.0453 | 0.0643 | 0.0827 | 0.1008 | 0.1187 | 0.1366 | 0.1492 |
| | 0.90 | 0.0000 | 0.0012 | 0.0064 | 0.0133 | 0.0211 | 0.0295 | 0.0358 |
| | | 0.0359 | 0.0545 | 0.0726 | 0.0905 | 0.1083 | 0.1208 | 0.1385 |
| | 0.80 | 0.0000 | 0.0026 | 0.0099 | 0.0185 | 0.0264 | 0.0328 | 0.0445 |
| | | 0.0265 | 0.0446 | 0.0624 | 0.0801 | 0.0925 | 0.1101 | 0.1224 |
| | 0.99 | 0.0000 | 0.0001 | 0.0017 | 0.0052 | 0.0099 | 0.0154 | 0.0216 |
| | | 0.0586 | 0.0829 | 0.1064 | 0.1248 | 0.1429 | 0.1609 | 0.1736 |
| | 0.98 | 0.0000 | 0.0002 | 0.0025 | 0.0068 | 0.0122 | 0.0184 | 0.0253 |
| | | 0.0540 | 0.0780 | 0.0965 | 0.1147 | 0.1327 | 0.1506 | 0.1632 |
| 84 | 0.95 | 0.0000 | 0.0006 | 0.0042 | 0.0098 | 0.0164 | 0.0237 | 0.0315 |
| | | 0.0447 | 0.0636 | 0.0817 | 0.0996 | 0.1173 | 0.1350 | 0.1474 |
| | 0.90 | 0.0000 | 0.0012 | 0.0063 | 0.0131 | 0.0209 | 0.0292 | 0.0354 |
| | | 0.0355 | 0.0539 | 0.0717 | 0.0894 | 0.1070 | 0.1194 | 0.1369 |
| | 0.80 | 0.0000 | 0.0026 | 0.0098 | 0.0183 | 0.0261 | 0.0324 | 0.0440 |
| | | 0.0262 | 0.0441 | 0.0616 | 0.0792 | 0.0914 | 0.1088 | 0.1210 |

| $n$ | $\beta$ | 7 | 8 | 9 | 10 | 11 | 12 | 13 |
|---|---|---|---|---|---|---|---|---|
| | 0.99 | 0.0294 | 0.0368 | 0.0445 | 0.0526 | 0.0607 | 0.0637 | 0.0765 |
| | | 0.1986 | 0.2116 | 0.2301 | 0.2429 | 0.2614 | 0.2742 | 0.2869 |
| | 0.98 | 0.0338 | 0.0417 | 0.0500 | 0.0559 | 0.0650 | 0.0754 | 0.0808 |
| | | 0.1876 | 0.2005 | 0.2189 | 0.2317 | 0.2449 | 0.2627 | 0.2754 |
| 81 | 0.95 | 0.0412 | 0.0463 | 0.0571 | 0.0658 | 0.0734 | 0.0846 | 0.0916 |
| | | 0.1712 | 0.1839 | 0.2021 | 0.2148 | 0.2274 | 0.2456 | 0.2582 |
| | 0.90 | 0.0457 | 0.0558 | 0.0624 | 0.0743 | 0.0820 | 0.0926 | 0.1037 |
| | | 0.1546 | 0.1728 | 0.1853 | 0.1979 | 0.2160 | 0.2285 | 0.2410 |
| | 0.80 | 0.0519 | 0.0638 | 0.0740 | 0.0820 | 0.0947 | 0.1039 | 0.1127 |
| | | 0.1385 | 0.1560 | 0.1684 | 0.1808 | 0.1989 | 0.2113 | 0.2237 |
| | 0.99 | 0.0290 | 0.0363 | 0.0440 | 0.0519 | 0.0599 | 0.0628 | 0.0755 |
| | | 0.1962 | 0.2090 | 0.2273 | 0.2400 | 0.2582 | 0.2708 | 0.2840 |
| | 0.98 | 0.0334 | 0.0412 | 0.0494 | 0.0552 | 0.0642 | 0.0745 | 0.0798 |
| | | 0.1853 | 0.1981 | 0.2162 | 0.2288 | 0.2470 | 0.2595 | 0.2720 |
| 82 | 0.95 | 0.0407 | 0.0457 | 0.0564 | 0.0650 | 0.0725 | 0.0836 | 0.0903 |
| | | 0.1691 | 0.1816 | 0.1997 | 0.2122 | 0.2246 | 0.2426 | 0.2550 |
| | 0.90 | 0.0451 | 0.0551 | 0.0616 | 0.0734 | 0.0810 | 0.0915 | 0.1024 |
| | | 0.1527 | 0.1707 | 0.1831 | 0.1955 | 0.2134 | 0.2257 | 0.2380 |
| | 0.80 | 0.0512 | 0.0630 | 0.0731 | 0.0810 | 0.0935 | 0.1026 | 0.1114 |
| | | 0.1417 | 0.1541 | 0.1664 | 0.1786 | 0.1965 | 0.2088 | 0.2210 |
| | 0.99 | 0.0287 | 0.0359 | 0.0434 | 0.0512 | 0.0592 | 0.0619 | 0.0746 |
| | | 0.1938 | 0.2065 | 0.2245 | 0.2371 | 0.2551 | 0.2675 | 0.2856 |
| | 0.98 | 0.0330 | 0.0407 | 0.0488 | 0.0545 | 0.0633 | 0.0735 | 0.0789 |
| | | 0.1831 | 0.1957 | 0.2136 | 0.2261 | 0.2440 | 0.2564 | 0.2687 |
| 83 | 0.95 | 0.0402 | 0.0452 | 0.0557 | 0.0642 | 0.0716 | 0.0826 | 0.0891 |
| | | 0.1670 | 0.1795 | 0.1973 | 0.2096 | 0.2219 | 0.2397 | 0.2520 |
| | 0.90 | 0.0446 | 0.0544 | 0.0608 | 0.0725 | 0.0799 | 0.0904 | 0.1011 |
| | | 0.1509 | 0.1686 | 0.1809 | 0.1931 | 0.2108 | 0.2230 | 0.2352 |
| | 0.80 | 0.0506 | 0.0623 | 0.0722 | 0.0800 | 0.0924 | 0.1013 | 0.1100 |
| | | 0.1401 | 0.1522 | 0.1644 | 0.1765 | 0.1942 | 0.2063 | 0.2183 |
| | 0.99 | 0.0283 | 0.0354 | 0.0429 | 0.0506 | 0.0585 | 0.0610 | 0.0736 |
| | | 0.1915 | 0.2042 | 0.2218 | 0.2342 | 0.2520 | 0.2643 | 0.2821 |
| | 0.98 | 0.0326 | 0.0402 | 0.0482 | 0.0539 | 0.0625 | 0.0726 | 0.0779 |
| | | 0.1809 | 0.1933 | 0.2111 | 0.2234 | 0.2411 | 0.2533 | 0.2655 |
| 84 | 0.95 | 0.0397 | 0.0446 | 0.0550 | 0.0635 | 0.0706 | 0.0816 | 0.0880 |
| | | 0.1651 | 0.1773 | 0.1949 | 0.2071 | 0.2193 | 0.2369 | 0.2490 |
| | 0.90 | 0.0440 | 0.0538 | 0.0600 | 0.0716 | 0.0789 | 0.0893 | 0.0998 |
| | | 0.1491 | 0.1666 | 0.1787 | 0.1908 | 0.2083 | 0.2204 | 0.2324 |
| | 0.80 | 0.0499 | 0.0615 | 0.0713 | 0.0791 | 0.0913 | 0.1000 | 0.1087 |
| | | 0.1384 | 0.1504 | 0.1624 | 0.1744 | 0.1919 | 0.2038 | 0.2158 |

| $n$ | $\beta$ | 14 | 15 | 16 | 17 | 18 | 19 | 20 |
|---|---|---|---|---|---|---|---|---|
| | 0.99 | 0.0858 | 0.0908 | 0.1038 | 0.1103 | 0.1219 | 0.1293 | 0.1410 |
| | | 0.3054 | 0.3181 | 0.3307 | 0.3439 | 0.3619 | 0.3745 | 0.3871 |
| | 0.98 | 0.0927 | 0.1000 | 0.1109 | 0.1189 | 0.1303 | 0.1375 | 0.1507 |
| | | 0.2880 | 0.3064 | 0.3190 | 0.3316 | 0.3441 | 0.3626 | 0.3751 |
| 81 | 0.95 | 0.1031 | 0.1120 | 0.1215 | 0.1337 | 0.1399 | 0.1528 | 0.1634 |
| | | 0.2707 | 0.2832 | 0.3015 | 0.3140 | 0.3265 | 0.3389 | 0.3513 |
| | 0.90 | 0.1109 | 0.1237 | 0.1331 | 0.1418 | 0.1545 | 0.1655 | 0.1727 |
| | | 0.2534 | 0.2716 | 0.2840 | 0.2964 | 0.3088 | 0.3212 | 0.3348 |
| | 0.80 | 0.1253 | 0.1377 | 0.1406 | 0.1559 | 0.1683 | 0.1807 | 0.1888 |
| | | 0.2360 | 0.2542 | 0.2666 | 0.2789 | 0.2912 | 0.3035 | 0.3158 |
| | 0.99 | 0.0848 | 0.0895 | 0.1024 | 0.1089 | 0.1203 | 0.1277 | 0.1391 |
| | | 0.3016 | 0.3142 | 0.3267 | 0.3450 | 0.3574 | 0.3699 | 0.3823 |
| | 0.98 | 0.0915 | 0.0988 | 0.1094 | 0.1174 | 0.1286 | 0.1358 | 0.1488 |
| | | 0.2860 | 0.3027 | 0.3151 | 0.3275 | 0.3399 | 0.3581 | 0.3705 |
| 82 | 0.95 | 0.1019 | 0.1105 | 0.1201 | 0.1319 | 0.1382 | 0.1509 | 0.1612 |
| | | 0.2674 | 0.2798 | 0.2978 | 0.3102 | 0.3225 | 0.3348 | 0.3471 |
| | 0.90 | 0.1095 | 0.1222 | 0.1313 | 0.1401 | 0.1526 | 0.1632 | 0.1706 |
| | | 0.2503 | 0.2683 | 0.2806 | 0.2928 | 0.3051 | 0.3173 | 0.3354 |
| | 0.80 | 0.1238 | 0.1359 | 0.1416 | 0.1540 | 0.1663 | 0.1785 | 0.1860 |
| | | 0.2332 | 0.2511 | 0.2633 | 0.2755 | 0.2877 | 0.2999 | 0.3120 |
| | 0.99 | 0.0838 | 0.0883 | 0.1011 | 0.1076 | 0.1187 | 0.1262 | 0.1373 |
| | | 0.2980 | 0.3103 | 0.3227 | 0.3408 | 0.3531 | 0.3654 | 0.3777 |
| | 0.98 | 0.0903 | 0.0976 | 0.1080 | 0.1160 | 0.1269 | 0.1342 | 0.1469 |
| | | 0.2867 | 0.2990 | 0.3113 | 0.3236 | 0.3378 | 0.3538 | 0.3660 |
| 83 | 0.95 | 0.1007 | 0.1091 | 0.1186 | 0.1302 | 0.1365 | 0.1491 | 0.1590 |
| | | 0.2642 | 0.2774 | 0.2942 | 0.3064 | 0.3186 | 0.3307 | 0.3430 |
| | 0.90 | 0.1082 | 0.1207 | 0.1295 | 0.1384 | 0.1508 | 0.1611 | 0.1685 |
| | | 0.2473 | 0.2651 | 0.2772 | 0.2893 | 0.3014 | 0.3135 | 0.3313 |
| | 0.80 | 0.1223 | 0.1342 | 0.1400 | 0.1521 | 0.1643 | 0.1764 | 0.1833 |
| | | 0.2304 | 0.2481 | 0.2602 | 0.2722 | 0.2843 | 0.2963 | 0.3083 |
| | 0.99 | 0.0828 | 0.0871 | 0.0999 | 0.1063 | 0.1172 | 0.1247 | 0.1355 |
| | | 0.2944 | 0.3066 | 0.3189 | 0.3367 | 0.3488 | 0.3610 | 0.3731 |
| | 0.98 | 0.0892 | 0.0964 | 0.1066 | 0.1146 | 0.1253 | 0.1326 | 0.1450 |
| | | 0.2833 | 0.2954 | 0.3076 | 0.3197 | 0.3375 | 0.3496 | 0.3617 |
| 84 | 0.95 | 0.0995 | 0.1077 | 0.1172 | 0.1285 | 0.1349 | 0.1473 | 0.1569 |
| | | 0.2611 | 0.2787 | 0.2907 | 0.3028 | 0.3148 | 0.3268 | 0.3445 |
| | 0.90 | 0.1069 | 0.1193 | 0.1278 | 0.1368 | 0.1490 | 0.1589 | 0.1665 |
| | | 0.2452 | 0.2620 | 0.2739 | 0.2859 | 0.2979 | 0.3099 | 0.3274 |
| | 0.80 | 0.1209 | 0.1324 | 0.1383 | 0.1503 | 0.1623 | 0.1743 | 0.1806 |
| | | 0.2277 | 0.2452 | 0.2571 | 0.2690 | 0.2809 | 0.2928 | 0.3046 |

| $n$ | $\beta$ | 21 | 22 | 23 | 24 | 25 | 26 | 27 |
|---|---|---|---|---|---|---|---|---|
| | 0.99 | 0.1481 | 0.1611 | 0.1668 | 0.1800 | 0.1895 | 0.1985 | 0.2115 |
| | | 0.3997 | 0.4123 | 0.4248 | 0.4375 | 0.4521 | 0.4686 | 0.4812 |
| | 0.98 | 0.1554 | 0.1691 | 0.1799 | 0.1875 | 0.2004 | 0.2109 | 0.2188 |
| | | 0.3877 | 0.4002 | 0.4127 | 0.4252 | 0.4377 | 0.4502 | 0.4626 |
| 81 | 0.95 | 0.1711 | 0.1838 | 0.1956 | 0.2020 | 0.2147 | 0.2273 | 0.2379 |
| | | 0.3649 | 0.3822 | 0.3946 | 0.4070 | 0.4194 | 0.4318 | 0.4443 |
| | 0.90 | 0.1852 | 0.1978 | 0.2079 | 0.2159 | 0.2284 | 0.2409 | 0.2533 |
| | | 0.3519 | 0.3643 | 0.3766 | 0.3890 | 0.4013 | 0.4137 | 0.4260 |
| | 0.80 | 0.1988 | 0.2112 | 0.2236 | 0.2359 | 0.2477 | 0.2541 | 0.2665 |
| | | 0.3282 | 0.3405 | 0.3587 | 0.3710 | 0.3833 | 0.3956 | 0.4079 |
| | 0.99 | 0.1463 | 0.1590 | 0.1648 | 0.1778 | 0.1868 | 0.1961 | 0.2089 |
| | | 0.3948 | 0.4072 | 0.4203 | 0.4380 | 0.4504 | 0.4628 | 0.4752 |
| | 0.98 | 0.1527 | 0.1671 | 0.1774 | 0.1852 | 0.1980 | 0.2078 | 0.2161 |
| | | 0.3829 | 0.3953 | 0.4076 | 0.4200 | 0.4323 | 0.4446 | 0.4575 |
| 82 | 0.95 | 0.1690 | 0.1815 | 0.1929 | 0.1996 | 0.2121 | 0.2245 | 0.2344 |
| | | 0.3652 | 0.3775 | 0.3898 | 0.4020 | 0.4143 | 0.4266 | 0.4388 |
| | 0.90 | 0.1830 | 0.1954 | 0.2049 | 0.2133 | 0.2256 | 0.2379 | 0.2502 |
| | | 0.3476 | 0.3598 | 0.3720 | 0.3843 | 0.3965 | 0.4087 | 0.4208 |
| | 0.80 | 0.1964 | 0.2087 | 0.2209 | 0.2331 | 0.2441 | 0.2510 | 0.2632 |
| | | 0.3242 | 0.3385 | 0.3544 | 0.3665 | 0.3787 | 0.3908 | 0.4030 |
| | 0.99 | 0.1445 | 0.1569 | 0.1628 | 0.1756 | 0.1841 | 0.1937 | 0.2064 |
| | | 0.3900 | 0.4027 | 0.4204 | 0.4327 | 0.4449 | 0.4572 | 0.4694 |
| | 0.98 | 0.1497 | 0.1650 | 0.1750 | 0.1830 | 0.1956 | 0.2049 | 0.2135 |
| | | 0.3783 | 0.3905 | 0.4027 | 0.4149 | 0.4270 | 0.4416 | 0.4574 |
| 83 | 0.95 | 0.1669 | 0.1794 | 0.1902 | 0.1972 | 0.2095 | 0.2218 | 0.2310 |
| | | 0.3608 | 0.3729 | 0.3851 | 0.3972 | 0.4093 | 0.4214 | 0.4335 |
| | 0.90 | 0.1808 | 0.1930 | 0.2020 | 0.2107 | 0.2229 | 0.2351 | 0.2472 |
| | | 0.3434 | 0.3555 | 0.3676 | 0.3796 | 0.3917 | 0.4037 | 0.4158 |
| | 0.80 | 0.1941 | 0.2062 | 0.2182 | 0.2303 | 0.2407 | 0.2480 | 0.2601 |
| | | 0.3203 | 0.3381 | 0.3501 | 0.3621 | 0.3741 | 0.3861 | 0.3981 |
| | 0.99 | 0.1428 | 0.1548 | 0.1608 | 0.1735 | 0.1815 | 0.1914 | 0.2039 |
| | | 0.3853 | 0.4032 | 0.4153 | 0.4274 | 0.4396 | 0.4517 | 0.4638 |
| | 0.98 | 0.1505 | 0.1631 | 0.1726 | 0.1808 | 0.1932 | 0.2020 | 0.2110 |
| | | 0.3737 | 0.3858 | 0.3978 | 0.4100 | 0.4266 | 0.4398 | 0.4519 |
| 84 | 0.95 | 0.1650 | 0.1772 | 0.1877 | 0.1948 | 0.2070 | 0.2192 | 0.2276 |
| | | 0.3565 | 0.3685 | 0.3805 | 0.3925 | 0.4044 | 0.4164 | 0.4284 |
| | 0.90 | 0.1786 | 0.1907 | 0.1992 | 0.2082 | 0.2203 | 0.2323 | 0.2441 |
| | | 0.3394 | 0.3513 | 0.3632 | 0.3751 | 0.3870 | 0.3990 | 0.4109 |
| | 0.80 | 0.1918 | 0.2037 | 0.2157 | 0.2276 | 0.2372 | 0.2451 | 0.2570 |
| | | 0.3165 | 0.3341 | 0.3460 | 0.3578 | 0.3697 | 0.3816 | 0.3934 |

| $n$ | $\beta$ | 28 | 29 | 30 | 31 | 32 | 33 | 34 |
|---|---|---|---|---|---|---|---|---|
| | 0.99 | 0.2193 | 0.2300 | 0.2428 | 0.2523 | 0.2613 | 0.2741 | 0.2868 |
| | | 0.4938 | 0.5063 | 0.5189 | 0.5315 | 0.5410 | 0.5518 | 0.5629 |
| | 0.98 | 0.2316 | 0.2441 | 0.2499 | 0.2626 | 0.2753 | 0.2879 | 0.2957 |
| | | 0.4751 | 0.4876 | 0.5000 | 0.5125 | 0.5250 | 0.5375 | 0.5499 |
| 81 | 0.95 | 0.2455 | 0.2581 | 0.2706 | 0.2831 | 0.2922 | 0.3014 | 0.3139 |
| | | 0.4567 | 0.4691 | 0.4815 | 0.4939 | 0.5062 | 0.5186 | 0.5310 |
| | 0.90 | 0.2623 | 0.2715 | 0.2839 | 0.2963 | 0.3087 | 0.3211 | 0.3330 |
| | | 0.4384 | 0.4507 | 0.4630 | 0.4754 | 0.4877 | 0.5000 | 0.5124 |
| | 0.80 | 0.2788 | 0.2911 | 0.3034 | 0.3157 | 0.3281 | 0.3404 | 0.3493 |
| | | 0.4202 | 0.4325 | 0.4448 | 0.4571 | 0.4693 | 0.4816 | 0.4939 |
| | 0.99 | 0.2158 | 0.2272 | 0.2399 | 0.2484 | 0.2581 | 0.2707 | 0.2831 |
| | | 0.4876 | 0.5000 | 0.5125 | 0.5249 | 0.5373 | 0.5497 | 0.5621 |
| | 0.98 | 0.2287 | 0.2406 | 0.2469 | 0.2594 | 0.2719 | 0.2839 | 0.2899 |
| | | 0.4711 | 0.4843 | 0.4971 | 0.5093 | 0.5209 | 0.5324 | 0.5436 |
| 82 | 0.95 | 0.2425 | 0.2549 | 0.2673 | 0.2797 | 0.2873 | 0.2977 | 0.3101 |
| | | 0.4511 | 0.4633 | 0.4756 | 0.4878 | 0.5000 | 0.5123 | 0.5245 |
| | 0.90 | 0.2582 | 0.2682 | 0.2805 | 0.2927 | 0.3050 | 0.3172 | 0.3278 |
| | | 0.4330 | 0.4452 | 0.4574 | 0.4696 | 0.4818 | 0.4940 | 0.5061 |
| | 0.80 | 0.2754 | 0.2876 | 0.2998 | 0.3119 | 0.3241 | 0.3354 | 0.3423 |
| | | 0.4151 | 0.4272 | 0.4394 | 0.4515 | 0.4637 | 0.4758 | 0.4879 |
| | 0.99 | 0.2123 | 0.2244 | 0.2370 | 0.2444 | 0.2550 | 0.2674 | 0.2789 |
| | | 0.4817 | 0.4939 | 0.5062 | 0.5184 | 0.5307 | 0.5429 | 0.5552 |
| | 0.98 | 0.2260 | 0.2372 | 0.2439 | 0.2563 | 0.2686 | 0.2797 | 0.2866 |
| | | 0.4696 | 0.4818 | 0.4940 | 0.5061 | 0.5183 | 0.5305 | 0.5427 |
| 83 | 0.95 | 0.2396 | 0.2519 | 0.2641 | 0.2760 | 0.2820 | 0.2941 | 0.3063 |
| | | 0.4456 | 0.4577 | 0.4698 | 0.4819 | 0.4940 | 0.5061 | 0.5182 |
| | 0.90 | 0.2539 | 0.2650 | 0.2771 | 0.2892 | 0.3013 | 0.3134 | 0.3225 |
| | | 0.4278 | 0.4399 | 0.4519 | 0.4643 | 0.4773 | 0.4899 | 0.5020 |
| | 0.80 | 0.2721 | 0.2842 | 0.2962 | 0.3082 | 0.3202 | 0.3302 | 0.3380 |
| | | 0.4101 | 0.4221 | 0.4341 | 0.4461 | 0.4581 | 0.4701 | 0.4821 |
| | 0.99 | 0.2086 | 0.2217 | 0.2341 | 0.2405 | 0.2519 | 0.2642 | 0.2748 |
| | | 0.4759 | 0.4880 | 0.5000 | 0.5121 | 0.5242 | 0.5363 | 0.5484 |
| | 0.98 | 0.2233 | 0.2339 | 0.2410 | 0.2532 | 0.2654 | 0.2756 | 0.2832 |
| | | 0.4639 | 0.4760 | 0.4880 | 0.5000 | 0.5121 | 0.5241 | 0.5362 |
| 84 | 0.95 | 0.2368 | 0.2489 | 0.2610 | 0.2720 | 0.2786 | 0.2906 | 0.3027 |
| | | 0.4403 | 0.4523 | 0.4642 | 0.4762 | 0.4885 | 0.5006 | 0.5124 |
| | 0.90 | 0.2494 | 0.2619 | 0.2738 | 0.2858 | 0.2978 | 0.3097 | 0.3171 |
| | | 0.4228 | 0.4350 | 0.4491 | 0.4643 | 0.4762 | 0.4881 | 0.5000 |
| | 0.80 | 0.2689 | 0.2808 | 0.2927 | 0.3045 | 0.3164 | 0.3248 | 0.3340 |
| | | 0.4053 | 0.4171 | 0.4290 | 0.4408 | 0.4527 | 0.4645 | 0.4764 |

| $n$ | $\beta$ | 35 | 36 | 37 | 38 | 39 | 40 | 41 |
|---|---|---|---|---|---|---|---|---|
| | 0.99 | 0.2951 | 0.3053 | 0.3180 | 0.3306 | 0.3430 | 0.3514 | 0.3618 |
| | | 0.5753 | 0.5878 | 0.6004 | 0.6130 | 0.6256 | 0.6382 | 0.6486 |
| | 0.98 | 0.3063 | 0.3189 | 0.3315 | 0.3440 | 0.3548 | 0.3625 | 0.3750 |
| | | 0.5624 | 0.5749 | 0.5874 | 0.5999 | 0.6124 | 0.6250 | 0.6375 |
| 81 | 0.95 | 0.3264 | 0.3388 | 0.3512 | 0.3631 | 0.3716 | 0.3821 | 0.3945 |
| | | 0.5434 | 0.5558 | 0.5683 | 0.5807 | 0.5931 | 0.6055 | 0.6179 |
| | 0.90 | 0.3406 | 0.3518 | 0.3642 | 0.3765 | 0.3889 | 0.4012 | 0.4136 |
| | | 0.5247 | 0.5371 | 0.5494 | 0.5617 | 0.5741 | 0.5864 | 0.5988 |
| | 0.80 | 0.3586 | 0.3709 | 0.3832 | 0.3955 | 0.4078 | 0.4201 | 0.4324 |
| | | 0.5062 | 0.5185 | 0.5308 | 0.5430 | 0.5553 | 0.5676 | 0.5799 |
| | 0.99 | 0.2897 | 0.3015 | 0.3141 | 0.3266 | 0.3376 | 0.3449 | 0.3573 |
| | | 0.5708 | 0.5810 | 0.5929 | 0.6053 | 0.6178 | 0.6302 | 0.6427 |
| | 0.98 | 0.3026 | 0.3150 | 0.3274 | 0.3398 | 0.3487 | 0.3580 | 0.3704 |
| | | 0.5555 | 0.5678 | 0.5801 | 0.5925 | 0.6048 | 0.6172 | 0.6296 |
| 82 | 0.95 | 0.3224 | 0.3347 | 0.3470 | 0.3572 | 0.3651 | 0.3774 | 0.3897 |
| | | 0.5368 | 0.5490 | 0.5613 | 0.5735 | 0.5858 | 0.5981 | 0.6103 |
| | 0.90 | 0.3353 | 0.3475 | 0.3597 | 0.3719 | 0.3842 | 0.3964 | 0.4086 |
| | | 0.5183 | 0.5305 | 0.5427 | 0.5549 | 0.5671 | 0.5793 | 0.5914 |
| | 0.80 | 0.3543 | 0.3664 | 0.3786 | 0.3907 | 0.4029 | 0.4150 | 0.4271 |
| | | 0.5000 | 0.5122 | 0.5243 | 0.5364 | 0.5486 | 0.5607 | 0.5729 |
| | 0.99 | 0.2855 | 0.2979 | 0.3102 | 0.3226 | 0.3321 | 0.3407 | 0.3530 |
| | | 0.5674 | 0.5797 | 0.5885 | 0.5982 | 0.6101 | 0.6224 | 0.6347 |
| | 0.98 | 0.2989 | 0.3112 | 0.3235 | 0.3349 | 0.3419 | 0.3537 | 0.3659 |
| | | 0.5525 | 0.5625 | 0.5731 | 0.5852 | 0.5974 | 0.6096 | 0.6218 |
| 83 | 0.95 | 0.3185 | 0.3306 | 0.3427 | 0.3510 | 0.3607 | 0.3728 | 0.3850 |
| | | 0.5303 | 0.5424 | 0.5545 | 0.5666 | 0.5787 | 0.5908 | 0.6029 |
| | 0.90 | 0.3312 | 0.3433 | 0.3554 | 0.3675 | 0.3795 | 0.3916 | 0.4036 |
| | | 0.5138 | 0.5253 | 0.5365 | 0.5482 | 0.5602 | 0.5723 | 0.5843 |
| | 0.80 | 0.3500 | 0.3620 | 0.3740 | 0.3860 | 0.3980 | 0.4100 | 0.4220 |
| | | 0.4941 | 0.5060 | 0.5180 | 0.5300 | 0.5420 | 0.5540 | 0.5660 |
| | 0.99 | 0.2820 | 0.2943 | 0.3065 | 0.3187 | 0.3264 | 0.3366 | 0.3487 |
| | | 0.5605 | 0.5727 | 0.5848 | 0.5969 | 0.6050 | 0.6148 | 0.6270 |
| | 0.98 | 0.2953 | 0.3075 | 0.3196 | 0.3297 | 0.3374 | 0.3495 | 0.3616 |
| | | 0.5482 | 0.5603 | 0.5726 | 0.5804 | 0.5903 | 0.6023 | 0.6143 |
| 84 | 0.95 | 0.3147 | 0.3267 | 0.3375 | 0.3444 | 0.3564 | 0.3684 | 0.3804 |
| | | 0.5239 | 0.5359 | 0.5478 | 0.5598 | 0.5717 | 0.5837 | 0.5957 |
| | 0.90 | 0.3273 | 0.3393 | 0.3512 | 0.3631 | 0.3750 | 0.3869 | 0.3989 |
| | | 0.5120 | 0.5239 | 0.5358 | 0.5454 | 0.5554 | 0.5657 | 0.5773 |
| | 0.80 | 0.3459 | 0.3577 | 0.3696 | 0.3815 | 0.3933 | 0.4052 | 0.4170 |
| | | 0.4882 | 0.5000 | 0.5119 | 0.5237 | 0.5356 | 0.5474 | 0.5593 |

| $n$ | $\beta$ | 42 | 43 | 44 | 45 | 46 | 47 | 48 |
|---|---|---|---|---|---|---|---|---|
| | 0.99 | 0.3744 | 0.3870 | 0.3996 | 0.4122 | 0.4247 | 0.4371 | 0.4482 |
| | | 0.6570 | 0.6694 | 0.6820 | 0.6947 | 0.7049 | 0.7132 | 0.7259 |
| | 0.98 | 0.3876 | 0.4001 | 0.4126 | 0.4251 | 0.4376 | 0.4501 | 0.4625 |
| | | 0.6452 | 0.6560 | 0.6685 | 0.6811 | 0.6937 | 0.7043 | 0.7121 |
| 81 | 0.95 | 0.4069 | 0.4193 | 0.4317 | 0.4442 | 0.4566 | 0.4690 | 0.4814 |
| | | 0.6284 | 0.6369 | 0.6488 | 0.6612 | 0.6736 | 0.6861 | 0.6986 |
| | 0.90 | 0.4259 | 0.4383 | 0.4506 | 0.4629 | 0.4753 | 0.4876 | 0.5000 |
| | | 0.6111 | 0.6235 | 0.6358 | 0.6482 | 0.6594 | 0.6670 | 0.6789 |
| | 0.80 | 0.4447 | 0.4570 | 0.4692 | 0.4815 | 0.4938 | 0.5061 | 0.5184 |
| | | 0.5922 | 0.6045 | 0.6168 | 0.6291 | 0.6414 | 0.6507 | 0.6596 |
| | 0.99 | 0.3698 | 0.3822 | 0.3947 | 0.4071 | 0.4190 | 0.4292 | 0.4379 |
| | | 0.6551 | 0.6624 | 0.6734 | 0.6859 | 0.6985 | 0.7103 | 0.7169 |
| | 0.98 | 0.3828 | 0.3952 | 0.4075 | 0.4199 | 0.4322 | 0.4445 | 0.4564 |
| | | 0.6420 | 0.6513 | 0.6602 | 0.6726 | 0.6850 | 0.6974 | 0.7101 |
| 82 | 0.95 | 0.4019 | 0.4142 | 0.4265 | 0.4387 | 0.4510 | 0.4632 | 0.4755 |
| | | 0.6226 | 0.6349 | 0.6428 | 0.6530 | 0.6653 | 0.6776 | 0.6899 |
| | 0.90 | 0.4207 | 0.4329 | 0.4451 | 0.4573 | 0.4695 | 0.4817 | 0.4939 |
| | | 0.6036 | 0.6158 | 0.6281 | 0.6403 | 0.6525 | 0.6647 | 0.6722 |
| | 0.80 | 0.4393 | 0.4514 | 0.4636 | 0.4757 | 0.4878 | 0.5000 | 0.5121 |
| | | 0.5850 | 0.5971 | 0.6093 | 0.6214 | 0.6336 | 0.6457 | 0.6577 |
| | 0.99 | 0.3653 | 0.3776 | 0.3899 | 0.4018 | 0.4115 | 0.4203 | 0.4326 |
| | | 0.6470 | 0.6593 | 0.6679 | 0.6774 | 0.6898 | 0.7021 | 0.7145 |
| | 0.98 | 0.3782 | 0.3904 | 0.4026 | 0.4148 | 0.4269 | 0.4375 | 0.4475 |
| | | 0.6341 | 0.6463 | 0.6581 | 0.6651 | 0.6765 | 0.6888 | 0.7011 |
| 83 | 0.95 | 0.3971 | 0.4092 | 0.4213 | 0.4334 | 0.4455 | 0.4576 | 0.4697 |
| | | 0.6150 | 0.6272 | 0.6393 | 0.6490 | 0.6573 | 0.6694 | 0.6815 |
| | 0.90 | 0.4157 | 0.4277 | 0.4398 | 0.4518 | 0.4635 | 0.4747 | 0.4862 |
| | | 0.5964 | 0.6084 | 0.6205 | 0.6325 | 0.6446 | 0.6567 | 0.6688 |
| | 0.80 | 0.4340 | 0.4460 | 0.4580 | 0.4700 | 0.4820 | 0.4940 | 0.5059 |
| | | 0.5780 | 0.5900 | 0.6020 | 0.6140 | 0.6260 | 0.6380 | 0.6500 |
| | 0.99 | 0.3609 | 0.3730 | 0.3852 | 0.3950 | 0.4031 | 0.4152 | 0.4273 |
| | | 0.6391 | 0.6513 | 0.6634 | 0.6736 | 0.6813 | 0.6935 | 0.7057 |
| | 0.98 | 0.3736 | 0.3857 | 0.3977 | 0.4097 | 0.4196 | 0.4274 | 0.4397 |
| | | 0.6264 | 0.6384 | 0.6505 | 0.6626 | 0.6703 | 0.6804 | 0.6925 |
| 84 | 0.95 | 0.3924 | 0.4043 | 0.4163 | 0.4283 | 0.4402 | 0.4522 | 0.4641 |
| | | 0.6076 | 0.6196 | 0.6316 | 0.6436 | 0.6556 | 0.6625 | 0.6733 |
| | 0.90 | 0.4108 | 0.4227 | 0.4343 | 0.4446 | 0.4546 | 0.4642 | 0.4761 |
| | | 0.5892 | 0.6011 | 0.6131 | 0.6250 | 0.6369 | 0.6488 | 0.6607 |
| | 0.80 | 0.4289 | 0.4407 | 0.4526 | 0.4644 | 0.4763 | 0.4881 | 0.5000 |
| | | 0.5711 | 0.5830 | 0.5948 | 0.6067 | 0.6185 | 0.6304 | 0.6423 |

| $n$ | $\beta$ | 0 | 1 | 2 | 3 | 4 | 5 | 6 |
|---|---|---|---|---|---|---|---|---|
| 85 | 0.99 | 0.0000 | 0.0001 | 0.0017 | 0.0051 | 0.0098 | 0.0152 | 0.0214 |
| | | 0.0579 | 0.0819 | 0.1052 | 0.1233 | 0.1412 | 0.1590 | 0.1716 |
| | 0.98 | 0.0000 | 0.0002 | 0.0025 | 0.0067 | 0.0120 | 0.0182 | 0.0250 |
| | | 0.0533 | 0.0771 | 0.0954 | 0.1134 | 0.1311 | 0.1488 | 0.1612 |
| | 0.95 | 0.0000 | 0.0006 | 0.0041 | 0.0096 | 0.0162 | 0.0234 | 0.0311 |
| | | 0.0442 | 0.0628 | 0.0807 | 0.0984 | 0.1159 | 0.1334 | 0.1457 |
| | 0.90 | 0.0000 | 0.0012 | 0.0062 | 0.0130 | 0.0206 | 0.0288 | 0.0350 |
| | | 0.0351 | 0.0532 | 0.0709 | 0.0884 | 0.1058 | 0.1180 | 0.1353 |
| | 0.80 | 0.0000 | 0.0026 | 0.0097 | 0.0181 | 0.0258 | 0.0320 | 0.0434 |
| | | 0.0259 | 0.0435 | 0.0609 | 0.0782 | 0.0903 | 0.1076 | 0.1195 |
| 86 | 0.99 | 0.0000 | 0.0001 | 0.0017 | 0.0051 | 0.0096 | 0.0151 | 0.0211 |
| | | 0.0572 | 0.0809 | 0.1039 | 0.1218 | 0.1396 | 0.1572 | 0.1696 |
| | 0.98 | 0.0000 | 0.0002 | 0.0025 | 0.0066 | 0.0119 | 0.0180 | 0.0247 |
| | | 0.0527 | 0.0762 | 0.0943 | 0.1120 | 0.1296 | 0.1471 | 0.1594 |
| | 0.95 | 0.0000 | 0.0005 | 0.0041 | 0.0095 | 0.0160 | 0.0231 | 0.0308 |
| | | 0.0437 | 0.0621 | 0.0798 | 0.0973 | 0.1146 | 0.1319 | 0.1440 |
| | 0.90 | 0.0000 | 0.0012 | 0.0062 | 0.0128 | 0.0204 | 0.0285 | 0.0346 |
| | | 0.0347 | 0.0526 | 0.0701 | 0.0873 | 0.1045 | 0.1166 | 0.1337 |
| | 0.80 | 0.0000 | 0.0025 | 0.0095 | 0.0178 | 0.0255 | 0.0316 | 0.0429 |
| | | 0.0256 | 0.0430 | 0.0602 | 0.0773 | 0.0893 | 0.1063 | 0.1182 |
| 87 | 0.99 | 0.0000 | 0.0001 | 0.0017 | 0.0050 | 0.0095 | 0.0149 | 0.0209 |
| | | 0.0566 | 0.0800 | 0.1027 | 0.1204 | 0.1379 | 0.1553 | 0.1676 |
| | 0.98 | 0.0000 | 0.0002 | 0.0024 | 0.0065 | 0.0118 | 0.0178 | 0.0244 |
| | | 0.0521 | 0.0754 | 0.0932 | 0.1107 | 0.1281 | 0.1454 | 0.1575 |
| | 0.95 | 0.0000 | 0.0005 | 0.0040 | 0.0094 | 0.0158 | 0.0229 | 0.0304 |
| | | 0.0432 | 0.0614 | 0.0789 | 0.0961 | 0.1133 | 0.1304 | 0.1424 |
| | 0.90 | 0.0000 | 0.0012 | 0.0061 | 0.0127 | 0.0202 | 0.0282 | 0.0342 |
| | | 0.0343 | 0.0520 | 0.0693 | 0.0863 | 0.1033 | 0.1153 | 0.1322 |
| | 0.80 | 0.0000 | 0.0025 | 0.0094 | 0.0176 | 0.0252 | 0.0312 | 0.0424 |
| | | 0.0253 | 0.0425 | 0.0595 | 0.0764 | 0.0883 | 0.1051 | 0.1168 |
| 88 | 0.99 | 0.0000 | 0.0001 | 0.0016 | 0.0049 | 0.0094 | 0.0147 | 0.0206 |
| | | 0.0559 | 0.0791 | 0.1016 | 0.1191 | 0.1364 | 0.1536 | 0.1657 |
| | 0.98 | 0.0000 | 0.0002 | 0.0024 | 0.0064 | 0.0116 | 0.0176 | 0.0241 |
| | | 0.0515 | 0.0745 | 0.0922 | 0.1095 | 0.1266 | 0.1437 | 0.1557 |
| | 0.95 | 0.0000 | 0.0005 | 0.0040 | 0.0093 | 0.0156 | 0.0226 | 0.0301 |
| | | 0.0427 | 0.0607 | 0.0780 | 0.0951 | 0.1120 | 0.1289 | 0.1408 |
| | 0.90 | 0.0000 | 0.0011 | 0.0060 | 0.0125 | 0.0199 | 0.0278 | 0.0338 |
| | | 0.0339 | 0.0514 | 0.0685 | 0.0854 | 0.1022 | 0.1140 | 0.1307 |
| | 0.80 | 0.0000 | 0.0025 | 0.0093 | 0.0174 | 0.0249 | 0.0308 | 0.0420 |
| | | 0.0250 | 0.0421 | 0.0589 | 0.0756 | 0.0873 | 0.1039 | 0.1155 |

| $n$ | $\beta$ | 7 | 8 | 9 | 10 | 11 | 12 | 13 |
|---|---|---|---|---|---|---|---|---|
| | 0.99 | 0.0280 | 0.0350 | 0.0424 | 0.0500 | 0.0578 | 0.0601 | 0.0727 |
| | | 0.1892 | 0.2069 | 0.2192 | 0.2322 | 0.2490 | 0.2612 | 0.2788 |
| | 0.98 | 0.0322 | 0.0397 | 0.0476 | 0.0532 | 0.0618 | 0.0717 | 0.0770 |
| | | 0.1788 | 0.1911 | 0.2086 | 0.2207 | 0.2382 | 0.2503 | 0.2624 |
| 85 | 0.95 | 0.0392 | 0.0441 | 0.0543 | 0.0627 | 0.0698 | 0.0806 | 0.0869 |
| | | 0.1631 | 0.1752 | 0.1926 | 0.2047 | 0.2167 | 0.2341 | 0.2461 |
| | 0.90 | 0.0435 | 0.0531 | 0.0593 | 0.0708 | 0.0779 | 0.0883 | 0.0985 |
| | | 0.1474 | 0.1647 | 0.1767 | 0.1886 | 0.2059 | 0.2178 | 0.2297 |
| | 0.80 | 0.0493 | 0.0608 | 0.0704 | 0.0781 | 0.0902 | 0.0987 | 0.1075 |
| | | 0.1368 | 0.1487 | 0.1606 | 0.1724 | 0.1896 | 0.2015 | 0.2132 |
| | 0.99 | 0.0276 | 0.0346 | 0.0418 | 0.0494 | 0.0571 | 0.0593 | 0.0718 |
| | | 0.1870 | 0.2045 | 0.2166 | 0.2340 | 0.2461 | 0.2581 | 0.2755 |
| | 0.98 | 0.0318 | 0.0393 | 0.0470 | 0.0526 | 0.0610 | 0.0709 | 0.0761 |
| | | 0.1767 | 0.1888 | 0.2061 | 0.2182 | 0.2355 | 0.2474 | 0.2593 |
| 86 | 0.95 | 0.0388 | 0.0436 | 0.0536 | 0.0620 | 0.0689 | 0.0797 | 0.0857 |
| | | 0.1612 | 0.1732 | 0.1904 | 0.2023 | 0.2142 | 0.2314 | 0.2432 |
| | 0.90 | 0.0430 | 0.0525 | 0.0585 | 0.0700 | 0.0769 | 0.0872 | 0.0973 |
| | | 0.1457 | 0.1628 | 0.1746 | 0.1864 | 0.2035 | 0.2153 | 0.2270 |
| | 0.80 | 0.0487 | 0.0601 | 0.0695 | 0.0772 | 0.0892 | 0.0975 | 0.1062 |
| | | 0.1352 | 0.1470 | 0.1587 | 0.1704 | 0.1874 | 0.1991 | 0.2108 |
| | 0.99 | 0.0273 | 0.0342 | 0.0414 | 0.0488 | 0.0565 | 0.0584 | 0.0710 |
| | | 0.1849 | 0.2021 | 0.2141 | 0.2313 | 0.2433 | 0.2552 | 0.2723 |
| | 0.98 | 0.0314 | 0.0388 | 0.0465 | 0.0520 | 0.0603 | 0.0700 | 0.0753 |
| | | 0.1747 | 0.1867 | 0.2038 | 0.2157 | 0.2328 | 0.2446 | 0.2564 |
| 87 | 0.95 | 0.0383 | 0.0431 | 0.0530 | 0.0613 | 0.0680 | 0.0788 | 0.0847 |
| | | 0.1594 | 0.1712 | 0.1882 | 0.2000 | 0.2117 | 0.2287 | 0.2404 |
| | 0.90 | 0.0424 | 0.0519 | 0.0578 | 0.0692 | 0.0760 | 0.0862 | 0.0961 |
| | | 0.1440 | 0.1609 | 0.1726 | 0.1843 | 0.2012 | 0.2128 | 0.2244 |
| | 0.80 | 0.0481 | 0.0594 | 0.0687 | 0.0763 | 0.0882 | 0.0963 | 0.1050 |
| | | 0.1337 | 0.1453 | 0.1569 | 0.1685 | 0.1853 | 0.1969 | 0.2084 |
| | 0.99 | 0.0270 | 0.0338 | 0.0409 | 0.0482 | 0.0558 | 0.0576 | 0.0701 |
| | | 0.1828 | 0.1998 | 0.2117 | 0.2287 | 0.2405 | 0.2522 | 0.2692 |
| | 0.98 | 0.0310 | 0.0384 | 0.0460 | 0.0514 | 0.0596 | 0.0692 | 0.0744 |
| | | 0.1727 | 0.1845 | 0.2014 | 0.2132 | 0.2301 | 0.2418 | 0.2535 |
| 88 | 0.95 | 0.0379 | 0.0426 | 0.0524 | 0.0606 | 0.0672 | 0.0779 | 0.0836 |
| | | 0.1576 | 0.1693 | 0.1861 | 0.1977 | 0.2093 | 0.2261 | 0.2377 |
| | 0.90 | 0.0419 | 0.0513 | 0.0571 | 0.0684 | 0.0751 | 0.0853 | 0.0950 |
| | | 0.1424 | 0.1591 | 0.1707 | 0.1822 | 0.1989 | 0.2104 | 0.2219 |
| | 0.80 | 0.0475 | 0.0588 | 0.0679 | 0.0755 | 0.0872 | 0.0951 | 0.1038 |
| | | 0.1322 | 0.1437 | 0.1551 | 0.1678 | 0.1832 | 0.1946 | 0.2060 |

| $n$ | $\beta$ | 14 | 15 | 16 | 17 | 18 | 19 | 20 |
|---|---|---|---|---|---|---|---|---|
| | 0.99 | 0.0818 | 0.0859 | 0.0986 | 0.1051 | 0.1157 | 0.1232 | 0.1337 |
| | | 0.2909 | 0.3030 | 0.3206 | 0.3327 | 0.3447 | 0.3567 | 0.3687 |
| | 0.98 | 0.0881 | 0.0953 | 0.1053 | 0.1133 | 0.1237 | 0.1310 | 0.1432 |
| | | 0.2799 | 0.2919 | 0.3039 | 0.3159 | 0.3335 | 0.3454 | 0.3574 |
| 85 | 0.95 | 0.0983 | 0.1063 | 0.1158 | 0.1269 | 0.1333 | 0.1456 | 0.1548 |
| | | 0.2580 | 0.2754 | 0.2873 | 0.2992 | 0.3111 | 0.3230 | 0.3404 |
| | 0.90 | 0.1057 | 0.1179 | 0.1262 | 0.1352 | 0.1473 | 0.1568 | 0.1646 |
| | | 0.2470 | 0.2589 | 0.2707 | 0.2826 | 0.2944 | 0.3118 | 0.3236 |
| | 0.80 | 0.1194 | 0.1308 | 0.1367 | 0.1486 | 0.1605 | 0.1723 | 0.1779 |
| | | 0.2250 | 0.2423 | 0.2541 | 0.2659 | 0.2776 | 0.2894 | 0.3011 |
| | 0.99 | 0.0808 | 0.0847 | 0.0974 | 0.1038 | 0.1142 | 0.1217 | 0.1320 |
| | | 0.2875 | 0.2994 | 0.3168 | 0.3288 | 0.3407 | 0.3525 | 0.3644 |
| | 0.98 | 0.0870 | 0.0942 | 0.1040 | 0.1119 | 0.1221 | 0.1295 | 0.1414 |
| | | 0.2766 | 0.2885 | 0.3004 | 0.3130 | 0.3296 | 0.3414 | 0.3532 |
| 86 | 0.95 | 0.0972 | 0.1049 | 0.1145 | 0.1253 | 0.1318 | 0.1439 | 0.1528 |
| | | 0.2550 | 0.2722 | 0.2840 | 0.2957 | 0.3075 | 0.3192 | 0.3365 |
| | 0.90 | 0.1044 | 0.1165 | 0.1246 | 0.1336 | 0.1456 | 0.1548 | 0.1627 |
| | | 0.2442 | 0.2559 | 0.2676 | 0.2793 | 0.2910 | 0.3082 | 0.3198 |
| | 0.80 | 0.1181 | 0.1291 | 0.1351 | 0.1469 | 0.1586 | 0.1703 | 0.1752 |
| | | 0.2224 | 0.2395 | 0.2512 | 0.2628 | 0.2744 | 0.2860 | 0.2976 |
| | 0.99 | 0.0799 | 0.0836 | 0.0962 | 0.1026 | 0.1128 | 0.1203 | 0.1303 |
| | | 0.2842 | 0.2960 | 0.3132 | 0.3250 | 0.3367 | 0.3484 | 0.3632 |
| | 0.98 | 0.0859 | 0.0931 | 0.1027 | 0.1106 | 0.1206 | 0.1280 | 0.1396 |
| | | 0.2735 | 0.2852 | 0.2969 | 0.3141 | 0.3258 | 0.3374 | 0.3491 |
| 87 | 0.95 | 0.0960 | 0.1036 | 0.1132 | 0.1238 | 0.1303 | 0.1423 | 0.1508 |
| | | 0.2521 | 0.2691 | 0.2807 | 0.2923 | 0.3040 | 0.3181 | 0.3326 |
| | 0.90 | 0.1032 | 0.1152 | 0.1230 | 0.1321 | 0.1439 | 0.1528 | 0.1608 |
| | | 0.2414 | 0.2530 | 0.2645 | 0.2761 | 0.2876 | 0.3046 | 0.3162 |
| | 0.80 | 0.1167 | 0.1276 | 0.1336 | 0.1452 | 0.1568 | 0.1684 | 0.1724 |
| | | 0.2199 | 0.2368 | 0.2483 | 0.2598 | 0.2713 | 0.2827 | 0.2942 |
| | 0.99 | 0.0790 | 0.0824 | 0.0950 | 0.1015 | 0.1114 | 0.1190 | 0.1287 |
| | | 0.2809 | 0.2926 | 0.3096 | 0.3212 | 0.3329 | 0.3445 | 0.3615 |
| | 0.98 | 0.0849 | 0.0921 | 0.1014 | 0.1094 | 0.1191 | 0.1265 | 0.1379 |
| | | 0.2703 | 0.2820 | 0.2935 | 0.3105 | 0.3220 | 0.3336 | 0.3451 |
| 88 | 0.95 | 0.0950 | 0.1023 | 0.1119 | 0.1222 | 0.1288 | 0.1407 | 0.1489 |
| | | 0.2492 | 0.2660 | 0.2775 | 0.2890 | 0.3005 | 0.3174 | 0.3288 |
| | 0.90 | 0.1021 | 0.1139 | 0.1214 | 0.1306 | 0.1423 | 0.1509 | 0.1590 |
| | | 0.2386 | 0.2501 | 0.2615 | 0.2730 | 0.2844 | 0.3012 | 0.3126 |
| | 0.80 | 0.1154 | 0.1260 | 0.1321 | 0.1436 | 0.1550 | 0.1663 | 0.1690 |
| | | 0.2174 | 0.2341 | 0.2455 | 0.2569 | 0.2682 | 0.2796 | 0.2909 |

| n | β | 21 | 22 | 23 | 24 | 25 | 26 | 27 |
|---|---|---|---|---|---|---|---|---|
| 85 | 0.99 | 0.1411 | 0.1528 | 0.1589 | 0.1715 | 0.1789 | 0.1891 | 0.2011 |
| | | 0.3831 | 0.3984 | 0.4104 | 0.4224 | 0.4343 | 0.4463 | 0.4582 |
| | 0.98 | 0.1487 | 0.1611 | 0.1703 | 0.1787 | 0.1910 | 0.1991 | 0.2085 |
| | | 0.3693 | 0.3812 | 0.3931 | 0.4108 | 0.4227 | 0.4346 | 0.4465 |
| | 0.95 | 0.1630 | 0.1751 | 0.1852 | 0.1925 | 0.2046 | 0.2166 | 0.2243 |
| | | 0.3523 | 0.3642 | 0.3760 | 0.3878 | 0.3997 | 0.4115 | 0.4233 |
| | 0.90 | 0.1766 | 0.1885 | 0.1964 | 0.2058 | 0.2177 | 0.2296 | 0.2407 |
| | | 0.3354 | 0.3472 | 0.3590 | 0.3707 | 0.3825 | 0.3943 | 0.4060 |
| | 0.80 | 0.1895 | 0.2014 | 0.2131 | 0.2249 | 0.2339 | 0.2422 | 0.2540 |
| | | 0.3139 | 0.3302 | 0.3419 | 0.3537 | 0.3654 | 0.3771 | 0.3888 |
| 86 | 0.99 | 0.1395 | 0.1508 | 0.1571 | 0.1695 | 0.1764 | 0.1869 | 0.1985 |
| | | 0.3819 | 0.3937 | 0.4056 | 0.4174 | 0.4292 | 0.4410 | 0.4528 |
| | 0.98 | 0.1470 | 0.1593 | 0.1680 | 0.1766 | 0.1887 | 0.1963 | 0.2060 |
| | | 0.3650 | 0.3768 | 0.3910 | 0.4060 | 0.4178 | 0.4295 | 0.4413 |
| | 0.95 | 0.1611 | 0.1731 | 0.1827 | 0.1903 | 0.2022 | 0.2141 | 0.2209 |
| | | 0.3482 | 0.3599 | 0.3716 | 0.3833 | 0.3950 | 0.4067 | 0.4184 |
| | 0.90 | 0.1745 | 0.1863 | 0.1936 | 0.2034 | 0.2152 | 0.2269 | 0.2375 |
| | | 0.3315 | 0.3431 | 0.3548 | 0.3664 | 0.3781 | 0.3897 | 0.4020 |
| | 0.80 | 0.1873 | 0.1990 | 0.2107 | 0.2223 | 0.2306 | 0.2394 | 0.2511 |
| | | 0.3148 | 0.3264 | 0.3380 | 0.3496 | 0.3612 | 0.3728 | 0.3843 |
| 87 | 0.99 | 0.1378 | 0.1489 | 0.1552 | 0.1675 | 0.1739 | 0.1848 | 0.1959 |
| | | 0.3774 | 0.3891 | 0.4008 | 0.4125 | 0.4242 | 0.4359 | 0.4476 |
| | 0.98 | 0.1453 | 0.1574 | 0.1658 | 0.1746 | 0.1866 | 0.1935 | 0.2037 |
| | | 0.3608 | 0.3727 | 0.3896 | 0.4013 | 0.4129 | 0.4246 | 0.4362 |
| | 0.95 | 0.1593 | 0.1711 | 0.1803 | 0.1881 | 0.1999 | 0.2116 | 0.2176 |
| | | 0.3442 | 0.3558 | 0.3674 | 0.3789 | 0.3905 | 0.4020 | 0.4163 |
| | 0.90 | 0.1725 | 0.1842 | 0.1909 | 0.2011 | 0.2127 | 0.2243 | 0.2343 |
| | | 0.3277 | 0.3392 | 0.3507 | 0.3622 | 0.3737 | 0.3852 | 0.4023 |
| | 0.80 | 0.1852 | 0.1968 | 0.2083 | 0.2198 | 0.2273 | 0.2367 | 0.2482 |
| | | 0.3112 | 0.3227 | 0.3341 | 0.3456 | 0.3570 | 0.3685 | 0.3799 |
| 88 | 0.99 | 0.1363 | 0.1471 | 0.1535 | 0.1656 | 0.1715 | 0.1827 | 0.1933 |
| | | 0.3731 | 0.3847 | 0.3962 | 0.4078 | 0.4194 | 0.4309 | 0.4427 |
| | 0.98 | 0.1436 | 0.1556 | 0.1636 | 0.1726 | 0.1844 | 0.1908 | 0.2013 |
| | | 0.3567 | 0.3737 | 0.3852 | 0.3967 | 0.4082 | 0.4197 | 0.4312 |
| | 0.95 | 0.1575 | 0.1692 | 0.1780 | 0.1860 | 0.1976 | 0.2092 | 0.2141 |
| | | 0.3403 | 0.3517 | 0.3632 | 0.3746 | 0.3860 | 0.3996 | 0.4144 |
| | 0.90 | 0.1706 | 0.1821 | 0.1881 | 0.1988 | 0.2103 | 0.2218 | 0.2311 |
| | | 0.3240 | 0.3354 | 0.3468 | 0.3581 | 0.3695 | 0.3837 | 0.3978 |
| | 0.80 | 0.1831 | 0.1945 | 0.2059 | 0.2173 | 0.2240 | 0.2340 | 0.2454 |
| | | 0.3077 | 0.3190 | 0.3303 | 0.3417 | 0.3530 | 0.3643 | 0.3757 |

| $n$ | $\beta$ | 28 | 29 | 30 | 31 | 32 | 33 | 34 |
|---|---|---|---|---|---|---|---|---|
| | 0.99 | 0.2068 | 0.2191 | 0.2311 | 0.2361 | 0.2489 | 0.2611 | 0.2708 |
| | | 0.4702 | 0.4821 | 0.4941 | 0.5060 | 0.5180 | 0.5299 | 0.5419 |
| | 0.98 | 0.2206 | 0.2306 | 0.2381 | 0.2502 | 0.2623 | 0.2716 | 0.2798 |
| | | 0.4584 | 0.4703 | 0.4822 | 0.4941 | 0.5060 | 0.5179 | 0.5298 |
| 85 | 0.95 | 0.2340 | 0.2460 | 0.2579 | 0.2681 | 0.2753 | 0.2872 | 0.2991 |
| | | 0.4351 | 0.4477 | 0.4612 | 0.4756 | 0.4882 | 0.5000 | 0.5119 |
| | 0.90 | 0.2469 | 0.2588 | 0.2706 | 0.2825 | 0.2943 | 0.3052 | 0.3117 |
| | | 0.4187 | 0.4353 | 0.4471 | 0.4589 | 0.4706 | 0.4824 | 0.4942 |
| | 0.80 | 0.2658 | 0.2775 | 0.2893 | 0.3010 | 0.3123 | 0.3190 | 0.3301 |
| | | 0.4005 | 0.4123 | 0.4240 | 0.4357 | 0.4474 | 0.4591 | 0.4708 |
| | 0.99 | 0.2044 | 0.2165 | 0.2280 | 0.2339 | 0.2460 | 0.2580 | 0.2668 |
| | | 0.4646 | 0.4764 | 0.4883 | 0.5000 | 0.5118 | 0.5237 | 0.5355 |
| | 0.98 | 0.2181 | 0.2274 | 0.2354 | 0.2473 | 0.2592 | 0.2675 | 0.2765 |
| | | 0.4530 | 0.4648 | 0.4766 | 0.4883 | 0.5000 | 0.5118 | 0.5235 |
| 86 | 0.95 | 0.2313 | 0.2431 | 0.2549 | 0.2643 | 0.2721 | 0.2839 | 0.2956 |
| | | 0.4321 | 0.4475 | 0.4592 | 0.4708 | 0.4825 | 0.4942 | 0.5059 |
| | 0.90 | 0.2441 | 0.2558 | 0.2675 | 0.2792 | 0.2909 | 0.3008 | 0.3081 |
| | | 0.4186 | 0.4303 | 0.4419 | 0.4535 | 0.4652 | 0.4768 | 0.4884 |
| | 0.80 | 0.2627 | 0.2743 | 0.2859 | 0.2975 | 0.3078 | 0.3147 | 0.3263 |
| | | 0.3959 | 0.4075 | 0.4191 | 0.4306 | 0.4422 | 0.4538 | 0.4653 |
| | 0.99 | 0.2020 | 0.2140 | 0.2250 | 0.2312 | 0.2432 | 0.2551 | 0.2628 |
| | | 0.4592 | 0.4709 | 0.4830 | 0.4949 | 0.5066 | 0.5179 | 0.5292 |
| | 0.98 | 0.2156 | 0.2243 | 0.2327 | 0.2445 | 0.2563 | 0.2634 | 0.2734 |
| | | 0.4478 | 0.4594 | 0.4710 | 0.4826 | 0.4942 | 0.5059 | 0.5175 |
| 87 | 0.95 | 0.2286 | 0.2403 | 0.2520 | 0.2605 | 0.2690 | 0.2806 | 0.2922 |
| | | 0.4308 | 0.4423 | 0.4539 | 0.4654 | 0.4770 | 0.4885 | 0.5000 |
| | 0.90 | 0.2413 | 0.2529 | 0.2644 | 0.2760 | 0.2875 | 0.2963 | 0.3045 |
| | | 0.4138 | 0.4253 | 0.4368 | 0.4483 | 0.4598 | 0.4713 | 0.4828 |
| | 0.80 | 0.2597 | 0.2712 | 0.2826 | 0.2941 | 0.3033 | 0.3111 | 0.3226 |
| | | 0.3914 | 0.4028 | 0.4143 | 0.4257 | 0.4372 | 0.4486 | 0.4600 |
| | 0.99 | 0.1997 | 0.2116 | 0.2220 | 0.2286 | 0.2404 | 0.2521 | 0.2587 |
| | | 0.4559 | 0.4695 | 0.4827 | 0.4943 | 0.5058 | 0.5174 | 0.5278 |
| | 0.98 | 0.2131 | 0.2212 | 0.2300 | 0.2417 | 0.2533 | 0.2591 | 0.2702 |
| | | 0.4427 | 0.4541 | 0.4656 | 0.4771 | 0.4886 | 0.5000 | 0.5115 |
| 88 | 0.95 | 0.2260 | 0.2376 | 0.2491 | 0.2567 | 0.2659 | 0.2774 | 0.2889 |
| | | 0.4259 | 0.4373 | 0.4487 | 0.4601 | 0.4715 | 0.4829 | 0.4943 |
| | 0.90 | 0.2385 | 0.2500 | 0.2614 | 0.2729 | 0.2843 | 0.2918 | 0.3011 |
| | | 0.4091 | 0.4205 | 0.4319 | 0.4432 | 0.4546 | 0.4660 | 0.4773 |
| | 0.80 | 0.2568 | 0.2681 | 0.2795 | 0.2908 | 0.2987 | 0.3076 | 0.3189 |
| | | 0.3870 | 0.3983 | 0.4096 | 0.4209 | 0.4322 | 0.4438 | 0.4567 |

| n | β | 35 | 36 | 37 | 38 | 39 | 40 | 41 |
|---|---|---|---|---|---|---|---|---|
| | 0.99 | 0.2787 | 0.2908 | 0.3029 | 0.3140 | 0.3205 | 0.3326 | 0.3446 |
| | | 0.5538 | 0.5658 | 0.5777 | 0.5897 | 0.6017 | 0.6127 | 0.6206 |
| | 0.98 | 0.2918 | 0.3038 | 0.3158 | 0.3244 | 0.3334 | 0.3453 | 0.3573 |
| | | 0.5417 | 0.5536 | 0.5655 | 0.5774 | 0.5893 | 0.5971 | 0.6070 |
| 85 | 0.95 | 0.3110 | 0.3229 | 0.3322 | 0.3403 | 0.3522 | 0.3641 | 0.3759 |
| | | 0.5233 | 0.5329 | 0.5433 | 0.5538 | 0.5650 | 0.5768 | 0.5886 |
| | 0.90 | 0.3235 | 0.3353 | 0.3471 | 0.3589 | 0.3706 | 0.3824 | 0.3942 |
| | | 0.5059 | 0.5177 | 0.5295 | 0.5412 | 0.5530 | 0.5648 | 0.5733 |
| | 0.80 | 0.3418 | 0.3536 | 0.3653 | 0.3770 | 0.3887 | 0.4004 | 0.4122 |
| | | 0.4825 | 0.4942 | 0.5059 | 0.5176 | 0.5293 | 0.5410 | 0.5527 |
| | 0.99 | 0.2754 | 0.2874 | 0.2993 | 0.3093 | 0.3167 | 0.3287 | 0.3406 |
| | | 0.5473 | 0.5591 | 0.5709 | 0.5827 | 0.5945 | 0.6064 | 0.6182 |
| | 0.98 | 0.2884 | 0.3003 | 0.3118 | 0.3188 | 0.3295 | 0.3413 | 0.3531 |
| | | 0.5353 | 0.5471 | 0.5588 | 0.5706 | 0.5823 | 0.5941 | 0.6048 |
| 86 | 0.95 | 0.3074 | 0.3191 | 0.3268 | 0.3364 | 0.3481 | 0.3598 | 0.3715 |
| | | 0.5176 | 0.5293 | 0.5409 | 0.5526 | 0.5620 | 0.5714 | 0.5817 |
| | 0.90 | 0.3197 | 0.3314 | 0.3430 | 0.3547 | 0.3663 | 0.3780 | 0.3896 |
| | | 0.5000 | 0.5117 | 0.5233 | 0.5349 | 0.5466 | 0.5582 | 0.5698 |
| | 0.80 | 0.3379 | 0.3495 | 0.3611 | 0.3727 | 0.3842 | 0.3958 | 0.4074 |
| | | 0.4769 | 0.4885 | 0.5000 | 0.5116 | 0.5232 | 0.5348 | 0.5463 |
| | 0.99 | 0.2722 | 0.2841 | 0.2959 | 0.3046 | 0.3131 | 0.3249 | 0.3366 |
| | | 0.5409 | 0.5525 | 0.5642 | 0.5759 | 0.5876 | 0.5993 | 0.6110 |
| | 0.98 | 0.2851 | 0.2968 | 0.3073 | 0.3140 | 0.3257 | 0.3373 | 0.3490 |
| | | 0.5291 | 0.5407 | 0.5523 | 0.5639 | 0.5755 | 0.5872 | 0.5988 |
| 87 | 0.95 | 0.3039 | 0.3146 | 0.3202 | 0.3325 | 0.3441 | 0.3557 | 0.3673 |
| | | 0.5116 | 0.5231 | 0.5347 | 0.5462 | 0.5578 | 0.5693 | 0.5796 |
| | 0.90 | 0.3161 | 0.3276 | 0.3391 | 0.3506 | 0.3621 | 0.3736 | 0.3851 |
| | | 0.4943 | 0.5058 | 0.5173 | 0.5288 | 0.5403 | 0.5518 | 0.5633 |
| | 0.80 | 0.3340 | 0.3455 | 0.3569 | 0.3684 | 0.3798 | 0.3913 | 0.4027 |
| | | 0.4715 | 0.4832 | 0.4950 | 0.5064 | 0.5175 | 0.5286 | 0.5401 |
| | 0.99 | 0.2691 | 0.2808 | 0.2925 | 0.2998 | 0.3095 | 0.3211 | 0.3328 |
| | | 0.5375 | 0.5476 | 0.5578 | 0.5692 | 0.5807 | 0.5923 | 0.6039 |
| | 0.98 | 0.2819 | 0.2934 | 0.3029 | 0.3104 | 0.3219 | 0.3335 | 0.3450 |
| | | 0.5230 | 0.5345 | 0.5460 | 0.5574 | 0.5689 | 0.5804 | 0.5919 |
| 88 | 0.95 | 0.3004 | 0.3101 | 0.3173 | 0.3287 | 0.3402 | 0.3516 | 0.3631 |
| | | 0.5058 | 0.5172 | 0.5286 | 0.5400 | 0.5514 | 0.5628 | 0.5742 |
| | 0.90 | 0.3125 | 0.3239 | 0.3353 | 0.3467 | 0.3580 | 0.3694 | 0.3795 |
| | | 0.4887 | 0.5000 | 0.5114 | 0.5228 | 0.5341 | 0.5455 | 0.5569 |
| | 0.80 | 0.3302 | 0.3416 | 0.3529 | 0.3642 | 0.3756 | 0.3869 | 0.3982 |
| | | 0.4700 | 0.4831 | 0.4944 | 0.5057 | 0.5170 | 0.5271 | 0.5367 |

| $n$ | $\beta$ | 42 | 43 | 44 | 45 | 46 | 47 | 48 |
|---|---|---|---|---|---|---|---|---|
| | 0.99 | 0.3566 | 0.3686 | 0.3794 | 0.3873 | 0.3983 | 0.4103 | 0.4223 |
| | | 0.6314 | 0.6434 | 0.6554 | 0.6674 | 0.6795 | 0.6860 | 0.6971 |
| | 0.98 | 0.3692 | 0.3811 | 0.3930 | 0.4029 | 0.4107 | 0.4226 | 0.4345 |
| | | 0.6189 | 0.6308 | 0.6427 | 0.6547 | 0.6666 | 0.6756 | 0.6842 |
| 85 | 0.95 | 0.3877 | 0.3996 | 0.4114 | 0.4232 | 0.4350 | 0.4462 | 0.4567 |
| | | 0.6004 | 0.6123 | 0.6241 | 0.6359 | 0.6478 | 0.6597 | 0.6678 |
| | 0.90 | 0.4059 | 0.4171 | 0.4267 | 0.4352 | 0.4470 | 0.4588 | 0.4705 |
| | | 0.5829 | 0.5941 | 0.6058 | 0.6176 | 0.6294 | 0.6411 | 0.6529 |
| | 0.80 | 0.4239 | 0.4356 | 0.4473 | 0.4590 | 0.4707 | 0.4824 | 0.4941 |
| | | 0.5644 | 0.5761 | 0.5878 | 0.5996 | 0.6113 | 0.6230 | 0.6347 |
| | 0.99 | 0.3524 | 0.3643 | 0.3731 | 0.3818 | 0.3936 | 0.4055 | 0.4173 |
| | | 0.6269 | 0.6357 | 0.6476 | 0.6594 | 0.6713 | 0.6833 | 0.6907 |
| | 0.98 | 0.3649 | 0.3767 | 0.3872 | 0.3952 | 0.4059 | 0.4177 | 0.4294 |
| | | 0.6128 | 0.6233 | 0.6351 | 0.6469 | 0.6587 | 0.6705 | 0.6812 |
| 86 | 0.95 | 0.3832 | 0.3949 | 0.4066 | 0.4183 | 0.4286 | 0.4380 | 0.4474 |
| | | 0.5934 | 0.6051 | 0.6168 | 0.6285 | 0.6402 | 0.6519 | 0.6636 |
| | 0.90 | 0.4008 | 0.4100 | 0.4185 | 0.4302 | 0.4418 | 0.4534 | 0.4651 |
| | | 0.5815 | 0.5900 | 0.5992 | 0.6104 | 0.6220 | 0.6337 | 0.6453 |
| | 0.80 | 0.4190 | 0.4305 | 0.4421 | 0.4537 | 0.4652 | 0.4768 | 0.4884 |
| | | 0.5579 | 0.5695 | 0.5810 | 0.5926 | 0.6042 | 0.6158 | 0.6273 |
| | 0.99 | 0.3483 | 0.3589 | 0.3655 | 0.3773 | 0.3890 | 0.4007 | 0.4124 |
| | | 0.6227 | 0.6345 | 0.6411 | 0.6517 | 0.6634 | 0.6751 | 0.6869 |
| | 0.98 | 0.3607 | 0.3722 | 0.3807 | 0.3895 | 0.4012 | 0.4128 | 0.4245 |
| | | 0.6105 | 0.6193 | 0.6278 | 0.6393 | 0.6510 | 0.6627 | 0.6743 |
| 87 | 0.95 | 0.3788 | 0.3904 | 0.4019 | 0.4119 | 0.4204 | 0.4307 | 0.4422 |
| | | 0.5881 | 0.5981 | 0.6096 | 0.6212 | 0.6327 | 0.6443 | 0.6559 |
| | 0.90 | 0.3943 | 0.4022 | 0.4137 | 0.4252 | 0.4367 | 0.4482 | 0.4597 |
| | | 0.5748 | 0.5863 | 0.5978 | 0.6057 | 0.6149 | 0.6264 | 0.6379 |
| | 0.80 | 0.4142 | 0.4256 | 0.4371 | 0.4485 | 0.4599 | 0.4714 | 0.4825 |
| | | 0.5515 | 0.5629 | 0.5744 | 0.5858 | 0.5973 | 0.6087 | 0.6202 |
| | 0.99 | 0.3444 | 0.3533 | 0.3614 | 0.3730 | 0.3846 | 0.3961 | 0.4077 |
| | | 0.6154 | 0.6270 | 0.6386 | 0.6467 | 0.6556 | 0.6672 | 0.6789 |
| | 0.98 | 0.3566 | 0.3665 | 0.3736 | 0.3851 | 0.3966 | 0.4081 | 0.4196 |
| | | 0.6034 | 0.6149 | 0.6264 | 0.6335 | 0.6434 | 0.6550 | 0.6665 |
| 88 | 0.95 | 0.3745 | 0.3859 | 0.3962 | 0.4044 | 0.4143 | 0.4258 | 0.4372 |
| | | 0.5857 | 0.5956 | 0.6038 | 0.6141 | 0.6255 | 0.6369 | 0.6484 |
| | 0.90 | 0.3866 | 0.3977 | 0.4090 | 0.4204 | 0.4318 | 0.4431 | 0.4545 |
| | | 0.5682 | 0.5796 | 0.5910 | 0.6023 | 0.6134 | 0.6205 | 0.6306 |
| | 0.80 | 0.4095 | 0.4208 | 0.4321 | 0.4432 | 0.4533 | 0.4633 | 0.4729 |
| | | 0.5467 | 0.5568 | 0.5679 | 0.5792 | 0.5905 | 0.6018 | 0.6131 |

| $n$ | $\beta$ | 0 | 1 | 2 | 3 | 4 | 5 | 6 |
|---|---|---|---|---|---|---|---|---|
| | 0.99 | 0.0000 | 0.0001 | 0.0016 | 0.0049 | 0.0093 | 0.0145 | 0.0204 |
| | | 0.0553 | 0.0782 | 0.1004 | 0.1177 | 0.1348 | 0.1518 | 0.1638 |
| | 0.98 | 0.0000 | 0.0002 | 0.0024 | 0.0064 | 0.0115 | 0.0174 | 0.0238 |
| | | 0.0509 | 0.0737 | 0.0911 | 0.1083 | 0.1252 | 0.1421 | 0.1540 |
| 89 | 0.95 | 0.0000 | 0.0005 | 0.0040 | 0.0092 | 0.0154 | 0.0223 | 0.0297 |
| | | 0.0422 | 0.0600 | 0.0771 | 0.0940 | 0.1107 | 0.1274 | 0.1392 |
| | 0.90 | 0.0000 | 0.0011 | 0.0059 | 0.0124 | 0.0197 | 0.0275 | 0.0334 |
| | | 0.0335 | 0.0509 | 0.0677 | 0.0844 | 0.1010 | 0.1127 | 0.1292 |
| | 0.80 | 0.0000 | 0.0025 | 0.0092 | 0.0172 | 0.0246 | 0.0305 | 0.0415 |
| | | 0.0247 | 0.0416 | 0.0582 | 0.0747 | 0.0863 | 0.1028 | 0.1142 |
| | 0.99 | 0.0000 | 0.0001 | 0.0016 | 0.0048 | 0.0092 | 0.0144 | 0.0202 |
| | | 0.0547 | 0.0773 | 0.0993 | 0.1164 | 0.1333 | 0.1501 | 0.1620 |
| | 0.98 | 0.0000 | 0.0002 | 0.0023 | 0.0063 | 0.0114 | 0.0172 | 0.0235 |
| | | 0.0504 | 0.0728 | 0.0901 | 0.1070 | 0.1238 | 0.1405 | 0.1523 |
| 90 | 0.95 | 0.0000 | 0.0005 | 0.0039 | 0.0091 | 0.0153 | 0.0221 | 0.0294 |
| | | 0.0418 | 0.0594 | 0.0763 | 0.0929 | 0.1095 | 0.1260 | 0.1376 |
| | 0.90 | 0.0000 | 0.0011 | 0.0059 | 0.0123 | 0.0195 | 0.0272 | 0.0330 |
| | | 0.0331 | 0.0503 | 0.0670 | 0.0835 | 0.0999 | 0.1115 | 0.1278 |
| | 0.80 | 0.0000 | 0.0024 | 0.0091 | 0.0171 | 0.0244 | 0.0301 | 0.0410 |
| | | 0.0245 | 0.0411 | 0.0576 | 0.0739 | 0.0854 | 0.1016 | 0.1130 |
| | 0.99 | 0.0000 | 0.0001 | 0.0016 | 0.0048 | 0.0091 | 0.0142 | 0.0199 |
| | | 0.0583 | 0.0765 | 0.0982 | 0.1151 | 0.1319 | 0.1485 | 0.1602 |
| | 0.98 | 0.0000 | 0.0002 | 0.0023 | 0.0062 | 0.0112 | 0.0170 | 0.0233 |
| | | 0.0498 | 0.0720 | 0.0891 | 0.1059 | 0.1225 | 0.1390 | 0.1506 |
| 91 | 0.95 | 0.0000 | 0.0005 | 0.0039 | 0.0090 | 0.0151 | 0.0218 | 0.0291 |
| | | 0.0413 | 0.0587 | 0.0754 | 0.0919 | 0.1083 | 0.1247 | 0.1361 |
| | 0.90 | 0.0000 | 0.0011 | 0.0058 | 0.0121 | 0.0193 | 0.0269 | 0.0327 |
| | | 0.0328 | 0.0498 | 0.0662 | 0.0826 | 0.0988 | 0.1102 | 0.1264 |
| | 0.80 | 0.0000 | 0.0024 | 0.0090 | 0.0169 | 0.0241 | 0.0298 | 0.0406 |
| | | 0.0242 | 0.0407 | 0.0569 | 0.0731 | 0.0844 | 0.1005 | 0.1117 |
| | 0.99 | 0.0000 | 0.0001 | 0.0016 | 0.0047 | 0.0090 | 0.0141 | 0.0197 |
| | | 0.0577 | 0.0756 | 0.0971 | 0.1139 | 0.1304 | 0.1469 | 0.1592 |
| | 0.98 | 0.0000 | 0.0002 | 0.0023 | 0.0062 | 0.0111 | 0.0168 | 0.0230 |
| | | 0.0493 | 0.0713 | 0.0882 | 0.1047 | 0.1211 | 0.1375 | 0.1490 |
| 92 | 0.95 | 0.0000 | 0.0005 | 0.0038 | 0.0089 | 0.0149 | 0.0216 | 0.0287 |
| | | 0.0409 | 0.0581 | 0.0746 | 0.0909 | 0.1071 | 0.1233 | 0.1347 |
| | 0.90 | 0.0000 | 0.0011 | 0.0057 | 0.0120 | 0.0190 | 0.0266 | 0.0323 |
| | | 0.0324 | 0.0492 | 0.0655 | 0.0817 | 0.0978 | 0.1090 | 0.1251 |
| | 0.80 | 0.0000 | 0.0024 | 0.0089 | 0.0167 | 0.0238 | 0.0295 | 0.0402 |
| | | 0.0239 | 0.0403 | 0.0563 | 0.0723 | 0.0835 | 0.0995 | 0.1105 |

| $n$ | $\beta$ | 7 | 8 | 9 | 10 | 11 | 12 | 13 |
|---|---|---|---|---|---|---|---|---|
| | 0.99 | 0.0267 | 0.0334 | 0.0404 | 0.0477 | 0.0552 | 0.0567 | 0.0693 |
| | | 0.1807 | 0.1975 | 0.2093 | 0.2261 | 0.2378 | 0.2500 | 0.2662 |
| | 0.98 | 0.0307 | 0.0379 | 0.0454 | 0.0508 | 0.0589 | 0.0684 | 0.0736 |
| | | 0.1707 | 0.1825 | 0.1992 | 0.2108 | 0.2275 | 0.2391 | 0.2558 |
| 89 | 0.95 | 0.0374 | 0.0421 | 0.0518 | 0.0599 | 0.0664 | 0.0770 | 0.0826 |
| | | 0.1558 | 0.1674 | 0.1840 | 0.1955 | 0.2089 | 0.2236 | 0.2350 |
| | 0.90 | 0.0415 | 0.0508 | 0.0564 | 0.0676 | 0.0742 | 0.0843 | 0.0939 |
| | | 0.1408 | 0.1573 | 0.1688 | 0.1802 | 0.1967 | 0.2081 | 0.2194 |
| | 0.80 | 0.0470 | 0.0581 | 0.0671 | 0.0746 | 0.0862 | 0.0939 | 0.1027 |
| | | 0.1307 | 0.1421 | 0.1534 | 0.1699 | 0.1812 | 0.1925 | 0.2037 |
| | 0.99 | 0.0264 | 0.0330 | 0.0399 | 0.0471 | 0.0546 | 0.0558 | 0.0685 |
| | | 0.1787 | 0.1953 | 0.2070 | 0.2236 | 0.2351 | 0.2517 | 0.2632 |
| | 0.98 | 0.0303 | 0.0375 | 0.0449 | 0.0503 | 0.0582 | 0.0676 | 0.0727 |
| | | 0.1688 | 0.1804 | 0.1970 | 0.2085 | 0.2250 | 0.2364 | 0.2529 |
| 90 | 0.95 | 0.0370 | 0.0417 | 0.0512 | 0.0593 | 0.0656 | 0.0762 | 0.0816 |
| | | 0.1541 | 0.1655 | 0.1820 | 0.1933 | 0.2098 | 0.2211 | 0.2324 |
| | 0.90 | 0.0410 | 0.0502 | 0.0558 | 0.0669 | 0.0733 | 0.0834 | 0.0928 |
| | | 0.1392 | 0.1556 | 0.1669 | 0.1782 | 0.1945 | 0.2058 | 0.2170 |
| | 0.80 | 0.0464 | 0.0575 | 0.0663 | 0.0738 | 0.0853 | 0.0928 | 0.1015 |
| | | 0.1292 | 0.1405 | 0.1517 | 0.1680 | 0.1792 | 0.1903 | 0.2015 |
| | 0.99 | 0.0261 | 0.0326 | 0.0395 | 0.0466 | 0.0540 | 0.0582 | 0.0677 |
| | | 0.1767 | 0.1932 | 0.2047 | 0.2211 | 0.2325 | 0.2489 | 0.2603 |
| | 0.98 | 0.0300 | 0.0371 | 0.0444 | 0.0497 | 0.0575 | 0.0668 | 0.0719 |
| | | 0.1670 | 0.1785 | 0.1948 | 0.2062 | 0.2225 | 0.2338 | 0.2501 |
| 91 | 0.95 | 0.0366 | 0.0412 | 0.0506 | 0.0586 | 0.0649 | 0.0753 | 0.0806 |
| | | 0.1524 | 0.1637 | 0.1800 | 0.1912 | 0.2075 | 0.2187 | 0.2299 |
| | 0.90 | 0.0405 | 0.0497 | 0.0551 | 0.0661 | 0.0724 | 0.0825 | 0.0917 |
| | | 0.1377 | 0.1539 | 0.1651 | 0.1765 | 0.1924 | 0.2035 | 0.2146 |
| | 0.80 | 0.0459 | 0.0568 | 0.0656 | 0.0730 | 0.0843 | 0.0917 | 0.1004 |
| | | 0.1278 | 0.1390 | 0.1500 | 0.1662 | 0.1772 | 0.1883 | 0.1993 |
| | 0.99 | 0.0258 | 0.0323 | 0.0390 | 0.0461 | 0.0534 | 0.0576 | 0.0669 |
| | | 0.1748 | 0.1910 | 0.2024 | 0.2187 | 0.2300 | 0.2462 | 0.2574 |
| | 0.98 | 0.0297 | 0.0366 | 0.0439 | 0.0492 | 0.0568 | 0.0661 | 0.0712 |
| | | 0.1652 | 0.1779 | 0.1927 | 0.2039 | 0.2201 | 0.2313 | 0.2474 |
| 92 | 0.95 | 0.0362 | 0.0408 | 0.0500 | 0.0580 | 0.0641 | 0.0745 | 0.0796 |
| | | 0.1507 | 0.1619 | 0.1780 | 0.1891 | 0.2052 | 0.2163 | 0.2274 |
| | 0.90 | 0.0401 | 0.0491 | 0.0545 | 0.0654 | 0.0716 | 0.0816 | 0.0906 |
| | | 0.1362 | 0.1522 | 0.1633 | 0.1793 | 0.1903 | 0.2013 | 0.2123 |
| | 0.80 | 0.0454 | 0.0562 | 0.0648 | 0.0722 | 0.0834 | 0.0907 | 0.0994 |
| | | 0.1264 | 0.1375 | 0.1484 | 0.1644 | 0.1753 | 0.1862 | 0.1971 |

| $n$ | $\beta$ | 14 | 15 | 16 | 17 | 18 | 19 | 20 |
|---|---|---|---|---|---|---|---|---|
| | 0.99 | 0.0781 | 0.0813 | 0.0939 | 0.1003 | 0.1101 | 0.1176 | 0.1271 |
| | | 0.2777 | 0.2906 | 0.3061 | 0.3176 | 0.3291 | 0.3416 | 0.3574 |
| | 0.98 | 0.0839 | 0.0910 | 0.1002 | 0.1082 | 0.1177 | 0.1251 | 0.1362 |
| | | 0.2673 | 0.2788 | 0.2902 | 0.3070 | 0.3184 | 0.3298 | 0.3412 |
| 89 | 0.95 | 0.0939 | 0.1011 | 0.1106 | 0.1208 | 0.1273 | 0.1391 | 0.1470 |
| | | 0.2464 | 0.2630 | 0.2744 | 0.2858 | 0.2971 | 0.3138 | 0.3251 |
| | 0.90 | 0.1009 | 0.1126 | 0.1199 | 0.1291 | 0.1407 | 0.1490 | 0.1572 |
| | | 0.2360 | 0.2473 | 0.2586 | 0.2699 | 0.2817 | 0.2978 | 0.3091 |
| | 0.80 | 0.1141 | 0.1245 | 0.1306 | 0.1420 | 0.1533 | 0.1642 | 0.1698 |
| | | 0.2150 | 0.2315 | 0.2428 | 0.2540 | 0.2652 | 0.2764 | 0.2878 |
| | 0.99 | 0.0772 | 0.0802 | 0.0928 | 0.0992 | 0.1087 | 0.1163 | 0.1256 |
| | | 0.2746 | 0.2912 | 0.3027 | 0.3140 | 0.3254 | 0.3420 | 0.3534 |
| | 0.98 | 0.0829 | 0.0900 | 0.0990 | 0.1069 | 0.1162 | 0.1237 | 0.1345 |
| | | 0.2643 | 0.2757 | 0.2870 | 0.3035 | 0.3149 | 0.3262 | 0.3374 |
| 90 | 0.95 | 0.0928 | 0.0998 | 0.1094 | 0.1193 | 0.1259 | 0.1375 | 0.1452 |
| | | 0.2441 | 0.2601 | 0.2714 | 0.2826 | 0.2938 | 0.3103 | 0.3215 |
| | 0.90 | 0.0998 | 0.1114 | 0.1184 | 0.1277 | 0.1391 | 0.1471 | 0.1555 |
| | | 0.2334 | 0.2446 | 0.2557 | 0.2669 | 0.2833 | 0.2945 | 0.3057 |
| | 0.80 | 0.1129 | 0.1230 | 0.1291 | 0.1404 | 0.1516 | 0.1622 | 0.1679 |
| | | 0.2128 | 0.2290 | 0.2401 | 0.2512 | 0.2623 | 0.2734 | 0.2898 |
| | 0.99 | 0.0764 | 0.0792 | 0.0917 | 0.0981 | 0.1074 | 0.1150 | 0.1241 |
| | | 0.2716 | 0.2880 | 0.2993 | 0.3106 | 0.3218 | 0.3383 | 0.3495 |
| | 0.98 | 0.0820 | 0.0890 | 0.0978 | 0.1058 | 0.1149 | 0.1224 | 0.1329 |
| | | 0.2614 | 0.2726 | 0.2853 | 0.3002 | 0.3114 | 0.3226 | 0.3337 |
| 91 | 0.95 | 0.0918 | 0.0986 | 0.1082 | 0.1179 | 0.1246 | 0.1360 | 0.1434 |
| | | 0.2461 | 0.2573 | 0.2684 | 0.2795 | 0.2919 | 0.3069 | 0.3180 |
| | 0.90 | 0.0987 | 0.1101 | 0.1170 | 0.1263 | 0.1376 | 0.1453 | 0.1538 |
| | | 0.2308 | 0.2419 | 0.2529 | 0.2640 | 0.2802 | 0.2913 | 0.3023 |
| | 0.80 | 0.1116 | 0.1216 | 0.1277 | 0.1389 | 0.1499 | 0.1603 | 0.1661 |
| | | 0.2154 | 0.2265 | 0.2375 | 0.2484 | 0.2594 | 0.2704 | 0.2866 |
| | 0.99 | 0.0755 | 0.0781 | 0.0907 | 0.0970 | 0.1062 | 0.1138 | 0.1226 |
| | | 0.2686 | 0.2849 | 0.2960 | 0.3072 | 0.3188 | 0.3345 | 0.3456 |
| | 0.98 | 0.0810 | 0.0881 | 0.0967 | 0.1046 | 0.1135 | 0.1210 | 0.1314 |
| | | 0.2585 | 0.2697 | 0.2858 | 0.2969 | 0.3080 | 0.3190 | 0.3304 |
| 92 | 0.95 | 0.0908 | 0.0975 | 0.1070 | 0.1165 | 0.1232 | 0.1346 | 0.1416 |
| | | 0.2434 | 0.2545 | 0.2655 | 0.2765 | 0.2926 | 0.3036 | 0.3145 |
| | 0.90 | 0.0977 | 0.1089 | 0.1156 | 0.1250 | 0.1361 | 0.1435 | 0.1521 |
| | | 0.2283 | 0.2393 | 0.2502 | 0.2611 | 0.2772 | 0.2881 | 0.2990 |
| | 0.80 | 0.1104 | 0.1201 | 0.1263 | 0.1374 | 0.1483 | 0.1584 | 0.1643 |
| | | 0.2131 | 0.2240 | 0.2349 | 0.2458 | 0.2566 | 0.2675 | 0.2835 |

| $n$ | $\beta$ | 21 | 22 | 23 | 24 | 25 | 26 | 27 |
|---|---|---|---|---|---|---|---|---|
| 89 | 0.99 | 0.1347 | 0.1453 | 0.1517 | 0.1637 | 0.1690 | 0.1806 | 0.1909 |
| | | 0.3689 | 0.3803 | 0.3917 | 0.4032 | 0.4146 | 0.4271 | 0.4430 |
| | 0.98 | 0.1420 | 0.1539 | 0.1615 | 0.1706 | 0.1824 | 0.1881 | 0.1991 |
| | | 0.3526 | 0.3694 | 0.3808 | 0.3922 | 0.4036 | 0.4149 | 0.4263 |
| | 0.95 | 0.1557 | 0.1673 | 0.1757 | 0.1839 | 0.1954 | 0.2066 | 0.2097 |
| | | 0.3365 | 0.3478 | 0.3591 | 0.3704 | 0.3823 | 0.3985 | 0.4098 |
| | 0.90 | 0.1687 | 0.1801 | 0.1854 | 0.1966 | 0.2080 | 0.2193 | 0.2280 |
| | | 0.3204 | 0.3316 | 0.3429 | 0.3541 | 0.3656 | 0.3821 | 0.3933 |
| | 0.80 | 0.1811 | 0.1924 | 0.2036 | 0.2149 | 0.2206 | 0.2314 | 0.2427 |
| | | 0.3042 | 0.3155 | 0.3267 | 0.3379 | 0.3491 | 0.3603 | 0.3715 |
| 90 | 0.99 | 0.1332 | 0.1435 | 0.1500 | 0.1619 | 0.1665 | 0.1786 | 0.1884 |
| | | 0.3647 | 0.3760 | 0.3874 | 0.3987 | 0.4111 | 0.4267 | 0.4380 |
| | 0.98 | 0.1404 | 0.1522 | 0.1594 | 0.1687 | 0.1803 | 0.1853 | 0.1969 |
| | | 0.3521 | 0.3653 | 0.3766 | 0.3878 | 0.3991 | 0.4103 | 0.4215 |
| | 0.95 | 0.1540 | 0.1654 | 0.1734 | 0.1819 | 0.1932 | 0.2040 | 0.2097 |
| | | 0.3327 | 0.3439 | 0.3551 | 0.3663 | 0.3828 | 0.3940 | 0.4052 |
| | 0.90 | 0.1668 | 0.1781 | 0.1827 | 0.1944 | 0.2057 | 0.2169 | 0.2249 |
| | | 0.3168 | 0.3279 | 0.3391 | 0.3502 | 0.3667 | 0.3778 | 0.3889 |
| | 0.80 | 0.1791 | 0.1902 | 0.2014 | 0.2125 | 0.2171 | 0.2289 | 0.2400 |
| | | 0.3009 | 0.3120 | 0.3231 | 0.3341 | 0.3452 | 0.3563 | 0.3674 |
| 91 | 0.99 | 0.1318 | 0.1417 | 0.1484 | 0.1601 | 0.1640 | 0.1766 | 0.1861 |
| | | 0.3607 | 0.3719 | 0.3831 | 0.3948 | 0.4108 | 0.4220 | 0.4331 |
| | 0.98 | 0.1389 | 0.1505 | 0.1574 | 0.1669 | 0.1784 | 0.1824 | 0.1947 |
| | | 0.3501 | 0.3613 | 0.3724 | 0.3835 | 0.3947 | 0.4058 | 0.4169 |
| | 0.95 | 0.1523 | 0.1636 | 0.1712 | 0.1799 | 0.1911 | 0.2014 | 0.2074 |
| | | 0.3291 | 0.3401 | 0.3512 | 0.3632 | 0.3786 | 0.3897 | 0.4007 |
| | 0.90 | 0.1650 | 0.1761 | 0.1797 | 0.1923 | 0.2034 | 0.2145 | 0.2219 |
| | | 0.3133 | 0.3244 | 0.3354 | 0.3464 | 0.3627 | 0.3737 | 0.3847 |
| | 0.80 | 0.1771 | 0.1882 | 0.1992 | 0.2098 | 0.2153 | 0.2264 | 0.2374 |
| | | 0.2976 | 0.3086 | 0.3195 | 0.3305 | 0.3414 | 0.3524 | 0.3633 |
| 92 | 0.99 | 0.1303 | 0.1400 | 0.1468 | 0.1583 | 0.1612 | 0.1747 | 0.1837 |
| | | 0.3567 | 0.3678 | 0.3789 | 0.3952 | 0.4063 | 0.4173 | 0.4284 |
| | 0.98 | 0.1374 | 0.1489 | 0.1554 | 0.1651 | 0.1762 | 0.1789 | 0.1926 |
| | | 0.3463 | 0.3573 | 0.3683 | 0.3793 | 0.3903 | 0.4013 | 0.4123 |
| | 0.95 | 0.1506 | 0.1618 | 0.1690 | 0.1779 | 0.1890 | 0.1989 | 0.2051 |
| | | 0.3255 | 0.3365 | 0.3474 | 0.3636 | 0.3745 | 0.3854 | 0.3964 |
| | 0.90 | 0.1632 | 0.1739 | 0.1792 | 0.1902 | 0.2012 | 0.2122 | 0.2188 |
| | | 0.3099 | 0.3208 | 0.3317 | 0.3446 | 0.3587 | 0.3696 | 0.3805 |
| | 0.80 | 0.1752 | 0.1861 | 0.1970 | 0.2073 | 0.2130 | 0.2239 | 0.2348 |
| | | 0.2944 | 0.3052 | 0.3161 | 0.3269 | 0.3378 | 0.3486 | 0.3594 |

| n | β | 28 | 29 | 30 | 31 | 32 | 33 | 34 |
|---|---|---|---|---|---|---|---|---|
| | 0.99 | 0.1974 | 0.2092 | 0.2191 | 0.2260 | 0.2377 | 0.2491 | 0.2543 |
| | | 0.4544 | 0.4658 | 0.4772 | 0.4886 | 0.5000 | 0.5115 | 0.5229 |
| | 0.98 | 0.2107 | 0.2181 | 0.2274 | 0.2390 | 0.2500 | 0.2557 | 0.2672 |
| | | 0.4377 | 0.4490 | 0.4604 | 0.4717 | 0.4830 | 0.4944 | 0.5057 |
| 89 | 0.95 | 0.2235 | 0.2349 | 0.2463 | 0.2528 | 0.2629 | 0.2743 | 0.2857 |
| | | 0.4211 | 0.4323 | 0.4436 | 0.4549 | 0.4662 | 0.4775 | 0.4888 |
| | 0.90 | 0.2359 | 0.2472 | 0.2585 | 0.2698 | 0.2809 | 0.2870 | 0.2977 |
| | | 0.4046 | 0.4158 | 0.4270 | 0.4383 | 0.4495 | 0.4607 | 0.4720 |
| | 0.80 | 0.2539 | 0.2651 | 0.2763 | 0.2875 | 0.2939 | 0.3041 | 0.3154 |
| | | 0.3826 | 0.3938 | 0.4050 | 0.4162 | 0.4286 | 0.4441 | 0.4553 |
| | 0.99 | 0.1952 | 0.2069 | 0.2162 | 0.2235 | 0.2350 | 0.2458 | 0.2516 |
| | | 0.4493 | 0.4606 | 0.4719 | 0.4831 | 0.4944 | 0.5057 | 0.5170 |
| | 0.98 | 0.2084 | 0.2150 | 0.2249 | 0.2363 | 0.2467 | 0.2528 | 0.2642 |
| | | 0.4328 | 0.4440 | 0.4555 | 0.4679 | 0.4801 | 0.4920 | 0.5033 |
| 90 | 0.95 | 0.2210 | 0.2323 | 0.2435 | 0.2487 | 0.2600 | 0.2713 | 0.2825 |
| | | 0.4164 | 0.4275 | 0.4387 | 0.4499 | 0.4610 | 0.4722 | 0.4833 |
| | 0.90 | 0.2333 | 0.2445 | 0.2556 | 0.2668 | 0.2772 | 0.2832 | 0.2944 |
| | | 0.4001 | 0.4112 | 0.4223 | 0.4334 | 0.4445 | 0.4556 | 0.4667 |
| | 0.80 | 0.2511 | 0.2622 | 0.2733 | 0.2837 | 0.2897 | 0.3008 | 0.3119 |
| | | 0.3784 | 0.3895 | 0.4005 | 0.4131 | 0.4281 | 0.4392 | 0.4503 |
| | 0.99 | 0.1931 | 0.2046 | 0.2134 | 0.2210 | 0.2324 | 0.2426 | 0.2488 |
| | | 0.4443 | 0.4554 | 0.4666 | 0.4778 | 0.4889 | 0.5000 | 0.5112 |
| | 0.98 | 0.2061 | 0.2119 | 0.2224 | 0.2337 | 0.2434 | 0.2500 | 0.2613 |
| | | 0.4280 | 0.4409 | 0.4556 | 0.4667 | 0.4778 | 0.4889 | 0.5000 |
| 91 | 0.95 | 0.2186 | 0.2298 | 0.2404 | 0.2460 | 0.2572 | 0.2683 | 0.2794 |
| | | 0.4118 | 0.4228 | 0.4339 | 0.4449 | 0.4559 | 0.4670 | 0.4780 |
| | 0.90 | 0.2307 | 0.2418 | 0.2528 | 0.2639 | 0.2735 | 0.2801 | 0.2912 |
| | | 0.3957 | 0.4067 | 0.4177 | 0.4287 | 0.4396 | 0.4506 | 0.4616 |
| | 0.80 | 0.2483 | 0.2593 | 0.2703 | 0.2799 | 0.2865 | 0.2975 | 0.3085 |
| | | 0.3743 | 0.3852 | 0.3972 | 0.4125 | 0.4234 | 0.4344 | 0.4453 |
| | 0.99 | 0.1909 | 0.2023 | 0.2106 | 0.2186 | 0.2299 | 0.2395 | 0.2461 |
| | | 0.4394 | 0.4504 | 0.4615 | 0.4725 | 0.4835 | 0.4945 | 0.5056 |
| | 0.98 | 0.2038 | 0.2087 | 0.2200 | 0.2312 | 0.2403 | 0.2473 | 0.2584 |
| | | 0.4265 | 0.4396 | 0.4506 | 0.4616 | 0.4726 | 0.4836 | 0.4946 |
| 92 | 0.95 | 0.2162 | 0.2273 | 0.2373 | 0.2433 | 0.2544 | 0.2654 | 0.2764 |
| | | 0.4073 | 0.4182 | 0.4291 | 0.4401 | 0.4510 | 0.4619 | 0.4728 |
| | 0.90 | 0.2282 | 0.2392 | 0.2501 | 0.2610 | 0.2698 | 0.2771 | 0.2880 |
| | | 0.3914 | 0.4023 | 0.4131 | 0.4240 | 0.4349 | 0.4457 | 0.4566 |
| | 0.80 | 0.2457 | 0.2565 | 0.2674 | 0.2761 | 0.2834 | 0.2943 | 0.3051 |
| | | 0.3702 | 0.3811 | 0.3972 | 0.4080 | 0.4189 | 0.4297 | 0.4405 |

| $n$ | $\beta$ | 35 | 36 | 37 | 38 | 39 | 40 | 41 |
|---|---|---|---|---|---|---|---|---|
| | 0.99 | 0.2661 | 0.2776 | 0.2887 | 0.2945 | 0.3060 | 0.3175 | 0.3290 |
| | | 0.5343 | 0.5457 | 0.5571 | 0.5654 | 0.5748 | 0.5855 | 0.5969 |
| | 0.98 | 0.2787 | 0.2901 | 0.2984 | 0.3069 | 0.3183 | 0.3297 | 0.3411 |
| | | 0.5171 | 0.5284 | 0.5397 | 0.5511 | 0.5624 | 0.5738 | 0.5852 |
| 89 | 0.95 | 0.2970 | 0.3056 | 0.3137 | 0.3250 | 0.3364 | 0.3477 | 0.3590 |
| | | 0.5000 | 0.5113 | 0.5226 | 0.5339 | 0.5452 | 0.5565 | 0.5678 |
| | 0.90 | 0.3090 | 0.3203 | 0.3315 | 0.3428 | 0.3540 | 0.3652 | 0.3734 |
| | | 0.4832 | 0.4944 | 0.5057 | 0.5169 | 0.5281 | 0.5394 | 0.5506 |
| | 0.80 | 0.3266 | 0.3378 | 0.3490 | 0.3602 | 0.3714 | 0.3825 | 0.3937 |
| | | 0.4665 | 0.4777 | 0.4889 | 0.5000 | 0.5112 | 0.5224 | 0.5336 |
| | 0.99 | 0.2631 | 0.2745 | 0.2848 | 0.2911 | 0.3026 | 0.3139 | 0.3253 |
| | | 0.5282 | 0.5395 | 0.5508 | 0.5621 | 0.5734 | 0.5821 | 0.5909 |
| | 0.98 | 0.2756 | 0.2869 | 0.2938 | 0.3034 | 0.3148 | 0.3261 | 0.3373 |
| | | 0.5141 | 0.5246 | 0.5349 | 0.5452 | 0.5561 | 0.5673 | 0.5786 |
| 90 | 0.95 | 0.2937 | 0.3009 | 0.3102 | 0.3214 | 0.3326 | 0.3438 | 0.3550 |
| | | 0.4945 | 0.5056 | 0.5168 | 0.5279 | 0.5391 | 0.5502 | 0.5614 |
| | 0.90 | 0.3056 | 0.3167 | 0.3278 | 0.3390 | 0.3501 | 0.3598 | 0.3666 |
| | | 0.4778 | 0.4889 | 0.5000 | 0.5112 | 0.5223 | 0.5334 | 0.5445 |
| | 0.80 | 0.3230 | 0.3340 | 0.3451 | 0.3562 | 0.3673 | 0.3783 | 0.3894 |
| | | 0.4613 | 0.4724 | 0.4835 | 0.4945 | 0.5056 | 0.5166 | 0.5277 |
| | 0.99 | 0.2602 | 0.2715 | 0.2809 | 0.2879 | 0.2992 | 0.3105 | 0.3217 |
| | | 0.5223 | 0.5335 | 0.5447 | 0.5558 | 0.5670 | 0.5781 | 0.5893 |
| | 0.98 | 0.2725 | 0.2833 | 0.2886 | 0.3001 | 0.3113 | 0.3225 | 0.3336 |
| | | 0.5112 | 0.5223 | 0.5334 | 0.5445 | 0.5530 | 0.5623 | 0.5721 |
| 91 | 0.95 | 0.2901 | 0.2957 | 0.3068 | 0.3179 | 0.3290 | 0.3400 | 0.3511 |
| | | 0.4890 | 0.5000 | 0.5111 | 0.5221 | 0.5331 | 0.5442 | 0.5552 |
| | 0.90 | 0.3022 | 0.3132 | 0.3243 | 0.3353 | 0.3463 | 0.3543 | 0.3626 |
| | | 0.4726 | 0.4836 | 0.4946 | 0.5055 | 0.5165 | 0.5275 | 0.5385 |
| | 0.80 | 0.3194 | 0.3304 | 0.3413 | 0.3523 | 0.3632 | 0.3742 | 0.3851 |
| | | 0.4563 | 0.4672 | 0.4782 | 0.4891 | 0.5000 | 0.5110 | 0.5219 |
| | 0.99 | 0.2573 | 0.2685 | 0.2770 | 0.2848 | 0.2959 | 0.3071 | 0.3180 |
| | | 0.5166 | 0.5276 | 0.5386 | 0.5497 | 0.5607 | 0.5717 | 0.5828 |
| | 0.98 | 0.2696 | 0.2795 | 0.2857 | 0.2968 | 0.3079 | 0.3189 | 0.3298 |
| | | 0.5055 | 0.5165 | 0.5275 | 0.5385 | 0.5495 | 0.5605 | 0.5706 |
| 92 | 0.95 | 0.2862 | 0.2925 | 0.3035 | 0.3144 | 0.3254 | 0.3364 | 0.3473 |
| | | 0.4837 | 0.4946 | 0.5055 | 0.5164 | 0.5273 | 0.5382 | 0.5491 |
| | 0.90 | 0.2989 | 0.3098 | 0.3207 | 0.3316 | 0.3417 | 0.3481 | 0.3586 |
| | | 0.4675 | 0.4783 | 0.4892 | 0.5000 | 0.5109 | 0.5218 | 0.5326 |
| | 0.80 | 0.3160 | 0.3268 | 0.3377 | 0.3485 | 0.3593 | 0.3701 | 0.3810 |
| | | 0.4513 | 0.4622 | 0.4730 | 0.4838 | 0.4946 | 0.5055 | 0.5163 |

| $n$ | $\beta$ | 42 | 43 | 44 | 45 | 46 | 47 | 48 |
|---|---|---|---|---|---|---|---|---|
| 89 | 0.99 | 0.3400 | 0.3472 | 0.3573 | 0.3688 | 0.3802 | 0.3916 | 0.4031 |
| | | 0.6084 | 0.6198 | 0.6312 | 0.6427 | 0.6528 | 0.6600 | 0.6710 |
| | 0.98 | 0.3525 | 0.3606 | 0.3693 | 0.3807 | 0.3921 | 0.4035 | 0.4148 |
| | | 0.5965 | 0.6079 | 0.6193 | 0.6307 | 0.6394 | 0.6475 | 0.6589 |
| | 0.95 | 0.3703 | 0.3812 | 0.3896 | 0.3984 | 0.4097 | 0.4210 | 0.4322 |
| | | 0.5790 | 0.5903 | 0.6016 | 0.6104 | 0.6188 | 0.6297 | 0.6410 |
| | 0.90 | 0.3820 | 0.3932 | 0.4045 | 0.4157 | 0.4269 | 0.4382 | 0.4494 |
| | | 0.5618 | 0.5731 | 0.5843 | 0.5955 | 0.6068 | 0.6180 | 0.6266 |
| | 0.80 | 0.4049 | 0.4161 | 0.4265 | 0.4357 | 0.4440 | 0.4552 | 0.4664 |
| | | 0.5448 | 0.5560 | 0.5643 | 0.5735 | 0.5839 | 0.5951 | 0.6063 |
| 90 | 0.99 | 0.3350 | 0.3419 | 0.3533 | 0.3646 | 0.3759 | 0.3873 | 0.3986 |
| | | 0.6014 | 0.6127 | 0.6241 | 0.6354 | 0.6467 | 0.6581 | 0.6650 |
| | 0.98 | 0.3475 | 0.3531 | 0.3652 | 0.3765 | 0.3877 | 0.3990 | 0.4102 |
| | | 0.5898 | 0.6010 | 0.6123 | 0.6235 | 0.6348 | 0.6469 | 0.6525 |
| | 0.95 | 0.3662 | 0.3755 | 0.3827 | 0.3939 | 0.4051 | 0.4163 | 0.4274 |
| | | 0.5726 | 0.5837 | 0.5949 | 0.6061 | 0.6173 | 0.6245 | 0.6338 |
| | 0.90 | 0.3777 | 0.3888 | 0.4000 | 0.4111 | 0.4222 | 0.4333 | 0.4444 |
| | | 0.5556 | 0.5667 | 0.5778 | 0.5889 | 0.6000 | 0.6112 | 0.6223 |
| | 0.80 | 0.4004 | 0.4106 | 0.4192 | 0.4280 | 0.4391 | 0.4502 | 0.4612 |
| | | 0.5388 | 0.5498 | 0.5609 | 0.5720 | 0.5808 | 0.5894 | 0.5996 |
| 91 | 0.99 | 0.3300 | 0.3382 | 0.3494 | 0.3606 | 0.3718 | 0.3830 | 0.3938 |
| | | 0.5976 | 0.6062 | 0.6170 | 0.6282 | 0.6394 | 0.6506 | 0.6618 |
| | 0.98 | 0.3423 | 0.3500 | 0.3612 | 0.3723 | 0.3834 | 0.3946 | 0.4057 |
| | | 0.5832 | 0.5943 | 0.6054 | 0.6166 | 0.6277 | 0.6388 | 0.6500 |
| | 0.95 | 0.3617 | 0.3692 | 0.3785 | 0.3896 | 0.4006 | 0.4117 | 0.4227 |
| | | 0.5662 | 0.5773 | 0.5883 | 0.5994 | 0.6104 | 0.6215 | 0.6308 |
| | 0.90 | 0.3736 | 0.3846 | 0.3956 | 0.4066 | 0.4176 | 0.4286 | 0.4395 |
| | | 0.5495 | 0.5605 | 0.5714 | 0.5824 | 0.5934 | 0.6044 | 0.6154 |
| | 0.80 | 0.3955 | 0.4038 | 0.4124 | 0.4233 | 0.4343 | 0.4452 | 0.4562 |
| | | 0.5329 | 0.5438 | 0.5548 | 0.5657 | 0.5767 | 0.5876 | 0.5962 |
| 92 | 0.99 | 0.3247 | 0.3344 | 0.3455 | 0.3566 | 0.3677 | 0.3788 | 0.3878 |
| | | 0.5938 | 0.6049 | 0.6122 | 0.6212 | 0.6323 | 0.6434 | 0.6545 |
| | 0.98 | 0.3369 | 0.3462 | 0.3572 | 0.3682 | 0.3792 | 0.3902 | 0.4012 |
| | | 0.5786 | 0.5878 | 0.5988 | 0.6098 | 0.6208 | 0.6318 | 0.6428 |
| | 0.95 | 0.3564 | 0.3635 | 0.3744 | 0.3853 | 0.3963 | 0.4072 | 0.4181 |
| | | 0.5600 | 0.5710 | 0.5819 | 0.5928 | 0.6037 | 0.6147 | 0.6256 |
| | 0.90 | 0.3695 | 0.3804 | 0.3913 | 0.4022 | 0.4130 | 0.4239 | 0.4348 |
| | | 0.5435 | 0.5544 | 0.5652 | 0.5761 | 0.5870 | 0.5978 | 0.6087 |
| | 0.80 | 0.3894 | 0.3971 | 0.4079 | 0.4188 | 0.4296 | 0.4404 | 0.4512 |
| | | 0.5271 | 0.5379 | 0.5488 | 0.5596 | 0.5704 | 0.5812 | 0.5921 |

| $n$ | $\beta$ | 0 | 1 | 2 | 3 | 4 | 5 | 6 |
|---|---|---|---|---|---|---|---|---|
| 93 | 0.99 | 0.0000 | 0.0001 | 0.0016 | 0.0047 | 0.0089 | 0.0139 | 0.0195 |
| | | 0.0571 | 0.0748 | 0.0961 | 0.1126 | 0.1290 | 0.1453 | 0.1615 |
| | 0.98 | 0.0000 | 0.0002 | 0.0023 | 0.0061 | 0.0110 | 0.0166 | 0.0228 |
| | | 0.0488 | 0.0705 | 0.0872 | 0.1036 | 0.1198 | 0.1360 | 0.1474 |
| | 0.95 | 0.0000 | 0.0005 | 0.0038 | 0.0088 | 0.0148 | 0.0214 | 0.0284 |
| | | 0.0404 | 0.0574 | 0.0738 | 0.0900 | 0.1060 | 0.1220 | 0.1332 |
| | 0.90 | 0.0000 | 0.0011 | 0.0057 | 0.0119 | 0.0188 | 0.0263 | 0.0320 |
| | | 0.0321 | 0.0487 | 0.0648 | 0.0808 | 0.0967 | 0.1079 | 0.1237 |
| | 0.80 | 0.0000 | 0.0023 | 0.0088 | 0.0165 | 0.0236 | 0.0291 | 0.0397 |
| | | 0.0237 | 0.0398 | 0.0557 | 0.0716 | 0.0826 | 0.0984 | 0.1093 |
| 94 | 0.99 | 0.0000 | 0.0001 | 0.0015 | 0.0046 | 0.0088 | 0.0138 | 0.0193 |
| | | 0.0565 | 0.0740 | 0.0951 | 0.1114 | 0.1276 | 0.1437 | 0.1597 |
| | 0.98 | 0.0000 | 0.0002 | 0.0022 | 0.0060 | 0.0109 | 0.0164 | 0.0225 |
| | | 0.0482 | 0.0697 | 0.0863 | 0.1025 | 0.1186 | 0.1345 | 0.1458 |
| | 0.95 | 0.0000 | 0.0005 | 0.0037 | 0.0087 | 0.0146 | 0.0211 | 0.0281 |
| | | 0.0400 | 0.0568 | 0.0730 | 0.0890 | 0.1049 | 0.1207 | 0.1318 |
| | 0.90 | 0.0000 | 0.0011 | 0.0056 | 0.0117 | 0.0186 | 0.0260 | 0.0316 |
| | | 0.0317 | 0.0482 | 0.0641 | 0.0799 | 0.0957 | 0.1067 | 0.1224 |
| | 0.80 | 0.0000 | 0.0023 | 0.0087 | 0.0163 | 0.0233 | 0.0288 | 0.0393 |
| | | 0.0234 | 0.0394 | 0.0551 | 0.0708 | 0.0818 | 0.0974 | 0.1082 |
| 95 | 0.99 | 0.0000 | 0.0001 | 0.0015 | 0.0046 | 0.0087 | 0.0136 | 0.0191 |
| | | 0.0559 | 0.0732 | 0.0941 | 0.1103 | 0.1263 | 0.1422 | 0.1581 |
| | 0.98 | 0.0000 | 0.0002 | 0.0022 | 0.0060 | 0.0108 | 0.0163 | 0.0223 |
| | | 0.0477 | 0.0690 | 0.0854 | 0.1014 | 0.1173 | 0.1331 | 0.1443 |
| | 0.95 | 0.0000 | 0.0005 | 0.0037 | 0.0086 | 0.0145 | 0.0209 | 0.0278 |
| | | 0.0396 | 0.0562 | 0.0723 | 0.0881 | 0.1038 | 0.1194 | 0.1304 |
| | 0.90 | 0.0000 | 0.0011 | 0.0056 | 0.0116 | 0.0184 | 0.0258 | 0.0313 |
| | | 0.0314 | 0.0477 | 0.0635 | 0.0791 | 0.0947 | 0.1056 | 0.1211 |
| | 0.80 | 0.0000 | 0.0023 | 0.0086 | 0.0161 | 0.0231 | 0.0285 | 0.0389 |
| | | 0.0232 | 0.0390 | 0.0546 | 0.0701 | 0.0809 | 0.0963 | 0.1071 |
| 96 | 0.99 | 0.0000 | 0.0001 | 0.0015 | 0.0045 | 0.0086 | 0.0135 | 0.0189 |
| | | 0.0553 | 0.0727 | 0.0931 | 0.1091 | 0.1250 | 0.1407 | 0.1564 |
| | 0.98 | 0.0000 | 0.0002 | 0.0022 | 0.0059 | 0.0106 | 0.0161 | 0.0221 |
| | | 0.0472 | 0.0683 | 0.0845 | 0.1004 | 0.1161 | 0.1317 | 0.1428 |
| | 0.95 | 0.0000 | 0.0005 | 0.0037 | 0.0085 | 0.0143 | 0.0207 | 0.0275 |
| | | 0.0392 | 0.0557 | 0.0715 | 0.0872 | 0.1027 | 0.1182 | 0.1291 |
| | 0.90 | 0.0000 | 0.0010 | 0.0055 | 0.0115 | 0.0182 | 0.0255 | 0.0310 |
| | | 0.0311 | 0.0472 | 0.0628 | 0.0783 | 0.0937 | 0.1045 | 0.1199 |
| | 0.80 | 0.0000 | 0.0023 | 0.0085 | 0.0160 | 0.0228 | 0.0282 | 0.0385 |
| | | 0.0229 | 0.0386 | 0.0540 | 0.0693 | 0.0801 | 0.0953 | 0.1060 |

| $n$ | $\beta$ | 7 | 8 | 9 | 10 | 11 | 12 | 13 |
|---|---|---|---|---|---|---|---|---|
| | 0.99 | 0.0255 | 0.0319 | 0.0386 | 0.0456 | 0.0528 | 0.0570 | 0.0661 |
| | | 0.1729 | 0.1890 | 0.2003 | 0.2163 | 0.2275 | 0.2435 | 0.2547 |
| | 0.98 | 0.0293 | 0.0362 | 0.0434 | 0.0487 | 0.0562 | 0.0653 | 0.0704 |
| | | 0.1634 | 0.1794 | 0.1906 | 0.2025 | 0.2177 | 0.2288 | 0.2447 |
| 93 | 0.95 | 0.0358 | 0.0403 | 0.0495 | 0.0573 | 0.0634 | 0.0737 | 0.0787 |
| | | 0.1491 | 0.1602 | 0.1761 | 0.1871 | 0.2030 | 0.2140 | 0.2249 |
| | 0.90 | 0.0396 | 0.0486 | 0.0539 | 0.0647 | 0.0707 | 0.0807 | 0.0896 |
| | | 0.1348 | 0.1506 | 0.1615 | 0.1773 | 0.1883 | 0.1992 | 0.2100 |
| | 0.80 | 0.0449 | 0.0556 | 0.0641 | 0.0715 | 0.0825 | 0.0896 | 0.0983 |
| | | 0.1251 | 0.1360 | 0.1468 | 0.1626 | 0.1734 | 0.1842 | 0.1950 |
| | 0.99 | 0.0252 | 0.0316 | 0.0382 | 0.0451 | 0.0522 | 0.0564 | 0.0654 |
| | | 0.1710 | 0.1870 | 0.1981 | 0.2140 | 0.2251 | 0.2409 | 0.2519 |
| | 0.98 | 0.0290 | 0.0358 | 0.0429 | 0.0481 | 0.0556 | 0.0646 | 0.0696 |
| | | 0.1617 | 0.1775 | 0.1886 | 0.2044 | 0.2154 | 0.2263 | 0.2421 |
| 94 | 0.95 | 0.0354 | 0.0399 | 0.0489 | 0.0567 | 0.0627 | 0.0729 | 0.0777 |
| | | 0.1475 | 0.1585 | 0.1742 | 0.1851 | 0.2008 | 0.2117 | 0.2225 |
| | 0.90 | 0.0392 | 0.0481 | 0.0533 | 0.0640 | 0.0699 | 0.0798 | 0.0886 |
| | | 0.1333 | 0.1490 | 0.1598 | 0.1755 | 0.1863 | 0.1970 | 0.2078 |
| | 0.80 | 0.0444 | 0.0550 | 0.0634 | 0.0707 | 0.0817 | 0.0886 | 0.0973 |
| | | 0.1238 | 0.1346 | 0.1453 | 0.1609 | 0.1716 | 0.1823 | 0.1930 |
| | 0.99 | 0.0250 | 0.0312 | 0.0378 | 0.0446 | 0.0516 | 0.0558 | 0.0647 |
| | | 0.1692 | 0.1850 | 0.1960 | 0.2117 | 0.2227 | 0.2384 | 0.2493 |
| | 0.98 | 0.0287 | 0.0355 | 0.0425 | 0.0476 | 0.0550 | 0.0639 | 0.0689 |
| | | 0.1599 | 0.1756 | 0.1866 | 0.2022 | 0.2131 | 0.2239 | 0.2396 |
| 95 | 0.95 | 0.0350 | 0.0395 | 0.0484 | 0.0561 | 0.0620 | 0.0722 | 0.0768 |
| | | 0.1460 | 0.1568 | 0.1724 | 0.1832 | 0.1987 | 0.2095 | 0.2202 |
| | 0.90 | 0.0388 | 0.0476 | 0.0527 | 0.0634 | 0.0692 | 0.0790 | 0.0876 |
| | | 0.1319 | 0.1474 | 0.1581 | 0.1736 | 0.1843 | 0.1950 | 0.2062 |
| | 0.80 | 0.0439 | 0.0545 | 0.0627 | 0.0700 | 0.0808 | 0.0876 | 0.0962 |
| | | 0.1225 | 0.1331 | 0.1438 | 0.1592 | 0.1698 | 0.1804 | 0.1910 |
| | 0.99 | 0.0247 | 0.0309 | 0.0374 | 0.0441 | 0.0511 | 0.0552 | 0.0640 |
| | | 0.1675 | 0.1831 | 0.1940 | 0.2095 | 0.2203 | 0.2359 | 0.2467 |
| | 0.98 | 0.0284 | 0.0351 | 0.0420 | 0.0471 | 0.0544 | 0.0632 | 0.0682 |
| | | 0.1583 | 0.1738 | 0.1846 | 0.2001 | 0.2109 | 0.2216 | 0.2371 |
| 96 | 0.95 | 0.0347 | 0.0391 | 0.0479 | 0.0556 | 0.0613 | 0.0714 | 0.0760 |
| | | 0.1445 | 0.1552 | 0.1706 | 0.1813 | 0.1967 | 0.2073 | 0.2179 |
| | 0.90 | 0.0384 | 0.0471 | 0.0521 | 0.0627 | 0.0684 | 0.0782 | 0.0866 |
| | | 0.1306 | 0.1459 | 0.1565 | 0.1718 | 0.1824 | 0.1930 | 0.2083 |
| | 0.80 | 0.0434 | 0.0539 | 0.0620 | 0.0692 | 0.0800 | 0.0866 | 0.0952 |
| | | 0.1212 | 0.1318 | 0.1423 | 0.1576 | 0.1681 | 0.1785 | 0.1890 |

| n | β | 14 | 15 | 16 | 17 | 18 | 19 | 20 |
|---|---|---|---|---|---|---|---|---|
| 93 | 0.99 | 0.0747 | 0.0770 | 0.0896 | 0.0960 | 0.1049 | 0.1125 | 0.1211 |
| | | 0.2657 | 0.2818 | 0.2928 | 0.3038 | 0.3199 | 0.3309 | 0.3419 |
| | 0.98 | 0.0801 | 0.0871 | 0.0956 | 0.1035 | 0.1122 | 0.1197 | 0.1298 |
| | | 0.2558 | 0.2668 | 0.2827 | 0.2937 | 0.3047 | 0.3156 | 0.3316 |
| | 0.95 | 0.0899 | 0.0963 | 0.1059 | 0.1152 | 0.1219 | 0.1331 | 0.1399 |
| | | 0.2408 | 0.2517 | 0.2626 | 0.2735 | 0.2894 | 0.3003 | 0.3112 |
| | 0.90 | 0.0966 | 0.1078 | 0.1142 | 0.1236 | 0.1347 | 0.1418 | 0.1505 |
| | | 0.2259 | 0.2367 | 0.2475 | 0.2583 | 0.2742 | 0.2850 | 0.2958 |
| | 0.80 | 0.1092 | 0.1188 | 0.1250 | 0.1359 | 0.1467 | 0.1565 | 0.1625 |
| | | 0.2108 | 0.2216 | 0.2324 | 0.2431 | 0.2539 | 0.2646 | 0.2805 |
| 94 | 0.99 | 0.0739 | 0.0760 | 0.0886 | 0.0950 | 0.1037 | 0.1113 | 0.1197 |
| | | 0.2629 | 0.2788 | 0.2897 | 0.3006 | 0.3165 | 0.3274 | 0.3382 |
| | 0.98 | 0.0792 | 0.0862 | 0.0945 | 0.1024 | 0.1109 | 0.1185 | 0.1283 |
| | | 0.2530 | 0.2639 | 0.2797 | 0.2906 | 0.3014 | 0.3122 | 0.3281 |
| | 0.95 | 0.0889 | 0.0952 | 0.1048 | 0.1139 | 0.1206 | 0.1317 | 0.1381 |
| | | 0.2383 | 0.2491 | 0.2598 | 0.2706 | 0.2864 | 0.2971 | 0.3079 |
| | 0.90 | 0.0956 | 0.1066 | 0.1128 | 0.1223 | 0.1332 | 0.1401 | 0.1489 |
| | | 0.2235 | 0.2342 | 0.2449 | 0.2556 | 0.2713 | 0.2820 | 0.2927 |
| | 0.80 | 0.1081 | 0.1174 | 0.1237 | 0.1345 | 0.1452 | 0.1547 | 0.1608 |
| | | 0.2086 | 0.2193 | 0.2299 | 0.2406 | 0.2512 | 0.2618 | 0.2775 |
| 95 | 0.99 | 0.0731 | 0.0749 | 0.0876 | 0.0940 | 0.1026 | 0.1102 | 0.1183 |
| | | 0.2608 | 0.2758 | 0.2866 | 0.2974 | 0.3131 | 0.3239 | 0.3346 |
| | 0.98 | 0.0783 | 0.0853 | 0.0934 | 0.1013 | 0.1096 | 0.1172 | 0.1269 |
| | | 0.2504 | 0.2611 | 0.2768 | 0.2875 | 0.2982 | 0.3089 | 0.3246 |
| | 0.95 | 0.0880 | 0.0941 | 0.1037 | 0.1126 | 0.1193 | 0.1303 | 0.1365 |
| | | 0.2358 | 0.2464 | 0.2571 | 0.2678 | 0.2833 | 0.2940 | 0.3046 |
| | 0.90 | 0.0946 | 0.1055 | 0.1115 | 0.1210 | 0.1318 | 0.1384 | 0.1473 |
| | | 0.2211 | 0.2317 | 0.2423 | 0.2529 | 0.2685 | 0.2791 | 0.2896 |
| | 0.80 | 0.1070 | 0.1161 | 0.1224 | 0.1330 | 0.1437 | 0.1529 | 0.1591 |
| | | 0.2064 | 0.2170 | 0.2275 | 0.2381 | 0.2486 | 0.2591 | 0.2746 |
| 96 | 0.99 | 0.0724 | 0.0737 | 0.0867 | 0.0930 | 0.1014 | 0.1090 | 0.1170 |
| | | 0.2622 | 0.2729 | 0.2836 | 0.2943 | 0.3099 | 0.3205 | 0.3311 |
| | 0.98 | 0.0775 | 0.0844 | 0.0924 | 0.1003 | 0.1084 | 0.1160 | 0.1254 |
| | | 0.2477 | 0.2584 | 0.2739 | 0.2845 | 0.2951 | 0.3067 | 0.3212 |
| | 0.95 | 0.0871 | 0.0930 | 0.1026 | 0.1113 | 0.1181 | 0.1290 | 0.1348 |
| | | 0.2333 | 0.2439 | 0.2544 | 0.2650 | 0.2804 | 0.2909 | 0.3014 |
| | 0.90 | 0.0936 | 0.1044 | 0.1102 | 0.1198 | 0.1305 | 0.1367 | 0.1458 |
| | | 0.2188 | 0.2293 | 0.2398 | 0.2513 | 0.2657 | 0.2762 | 0.2866 |
| | 0.80 | 0.1059 | 0.1148 | 0.1211 | 0.1317 | 0.1422 | 0.1512 | 0.1575 |
| | | 0.2043 | 0.2147 | 0.2252 | 0.2356 | 0.2460 | 0.2585 | 0.2718 |

| $n$ | $\beta$ | 21 | 22 | 23 | 24 | 25 | 26 | 27 |
|---|---|---|---|---|---|---|---|---|
| | 0.99 | 0.1289 | 0.1384 | 0.1452 | 0.1564 | 0.1614 | 0.1728 | 0.1815 |
| | | 0.3529 | 0.3638 | 0.3765 | 0.3909 | 0.4019 | 0.4128 | 0.4237 |
| | 0.98 | 0.1359 | 0.1473 | 0.1534 | 0.1633 | 0.1741 | 0.1793 | 0.1905 |
| | | 0.3426 | 0.3535 | 0.3644 | 0.3752 | 0.3861 | 0.3970 | 0.4116 |
| 93 | 0.95 | 0.1490 | 0.1601 | 0.1669 | 0.1760 | 0.1870 | 0.1965 | 0.2029 |
| | | 0.3220 | 0.3328 | 0.3438 | 0.3597 | 0.3705 | 0.3813 | 0.3921 |
| | 0.90 | 0.1614 | 0.1719 | 0.1772 | 0.1882 | 0.1991 | 0.2099 | 0.2158 |
| | | 0.3066 | 0.3174 | 0.3282 | 0.3441 | 0.3549 | 0.3657 | 0.3764 |
| | 0.80 | 0.1733 | 0.1841 | 0.1949 | 0.2047 | 0.2107 | 0.2215 | 0.2323 |
| | | 0.2912 | 0.3020 | 0.3127 | 0.3234 | 0.3341 | 0.3449 | 0.3556 |
| | 0.99 | 0.1275 | 0.1368 | 0.1436 | 0.1546 | 0.1596 | 0.1709 | 0.1792 |
| | | 0.3491 | 0.3599 | 0.3759 | 0.3867 | 0.3975 | 0.4083 | 0.4191 |
| | 0.98 | 0.1344 | 0.1457 | 0.1514 | 0.1616 | 0.1721 | 0.1774 | 0.1885 |
| | | 0.3389 | 0.3497 | 0.3605 | 0.3712 | 0.3820 | 0.3949 | 0.4087 |
| 94 | 0.95 | 0.1474 | 0.1584 | 0.1648 | 0.1741 | 0.1850 | 0.1941 | 0.2007 |
| | | 0.3186 | 0.3293 | 0.3451 | 0.3558 | 0.3665 | 0.3772 | 0.3879 |
| | 0.90 | 0.1597 | 0.1699 | 0.1754 | 0.1862 | 0.1969 | 0.2077 | 0.2127 |
| | | 0.3034 | 0.3140 | 0.3247 | 0.3405 | 0.3511 | 0.3618 | 0.3724 |
| | 0.80 | 0.1715 | 0.1822 | 0.1929 | 0.2023 | 0.2085 | 0.2192 | 0.2298 |
| | | 0.2882 | 0.2988 | 0.3094 | 0.3200 | 0.3306 | 0.3412 | 0.3518 |
| | 0.99 | 0.1262 | 0.1352 | 0.1421 | 0.1528 | 0.1580 | 0.1691 | 0.1771 |
| | | 0.3454 | 0.3576 | 0.3719 | 0.3826 | 0.3933 | 0.4040 | 0.4147 |
| | 0.98 | 0.1330 | 0.1442 | 0.1495 | 0.1598 | 0.1700 | 0.1755 | 0.1865 |
| | | 0.3353 | 0.3460 | 0.3566 | 0.3673 | 0.3785 | 0.3937 | 0.4044 |
| 95 | 0.95 | 0.1459 | 0.1567 | 0.1627 | 0.1723 | 0.1831 | 0.1918 | 0.1986 |
| | | 0.3152 | 0.3258 | 0.3415 | 0.3521 | 0.3627 | 0.3733 | 0.3838 |
| | 0.90 | 0.1580 | 0.1679 | 0.1735 | 0.1842 | 0.1949 | 0.2054 | 0.2093 |
| | | 0.3002 | 0.3107 | 0.3231 | 0.3369 | 0.3474 | 0.3580 | 0.3685 |
| | 0.80 | 0.1697 | 0.1803 | 0.1909 | 0.1998 | 0.2063 | 0.2169 | 0.2274 |
| | | 0.2851 | 0.2956 | 0.3062 | 0.3167 | 0.3272 | 0.3376 | 0.3481 |
| | 0.99 | 0.1249 | 0.1336 | 0.1406 | 0.1510 | 0.1563 | 0.1674 | 0.1749 |
| | | 0.3418 | 0.3574 | 0.3680 | 0.3786 | 0.3892 | 0.3998 | 0.4103 |
| | 0.98 | 0.1316 | 0.1427 | 0.1476 | 0.1582 | 0.1681 | 0.1737 | 0.1845 |
| | | 0.3318 | 0.3424 | 0.3529 | 0.3635 | 0.3790 | 0.3896 | 0.4001 |
| 96 | 0.95 | 0.1444 | 0.1551 | 0.1607 | 0.1705 | 0.1812 | 0.1895 | 0.1966 |
| | | 0.3119 | 0.3224 | 0.3379 | 0.3484 | 0.3589 | 0.3694 | 0.3798 |
| | 0.90 | 0.1564 | 0.1660 | 0.1717 | 0.1823 | 0.1929 | 0.2030 | 0.2082 |
| | | 0.2971 | 0.3075 | 0.3229 | 0.3334 | 0.3438 | 0.3543 | 0.3647 |
| | 0.80 | 0.1680 | 0.1784 | 0.1889 | 0.1974 | 0.2042 | 0.2146 | 0.2251 |
| | | 0.2822 | 0.2926 | 0.3030 | 0.3134 | 0.3238 | 0.3342 | 0.3445 |

| $n$ | $\beta$ | 28 | 29 | 30 | 31 | 32 | 33 | 34 |
|---|---|---|---|---|---|---|---|---|
| | 0.99 | 0.1889 | 0.2002 | 0.2079 | 0.2162 | 0.2274 | 0.2364 | 0.2434 |
| | | 0.4346 | 0.4455 | 0.4564 | 0.4673 | 0.4782 | 0.4891 | 0.5000 |
| | 0.98 | 0.2015 | 0.2052 | 0.2176 | 0.2287 | 0.2371 | 0.2446 | 0.2557 |
| | | 0.4240 | 0.4349 | 0.4457 | 0.4566 | 0.4675 | 0.4783 | 0.4892 |
| 93 | 0.95 | 0.2139 | 0.2248 | 0.2343 | 0.2407 | 0.2516 | 0.2625 | 0.2734 |
| | | 0.4029 | 0.4137 | 0.4245 | 0.4353 | 0.4461 | 0.4569 | 0.4677 |
| | 0.90 | 0.2258 | 0.2366 | 0.2474 | 0.2582 | 0.2661 | 0.2741 | 0.2849 |
| | | 0.3872 | 0.3980 | 0.4087 | 0.4195 | 0.4302 | 0.4410 | 0.4517 |
| | 0.80 | 0.2430 | 0.2538 | 0.2645 | 0.2723 | 0.2804 | 0.2911 | 0.3019 |
| | | 0.3663 | 0.3801 | 0.3930 | 0.4037 | 0.4144 | 0.4251 | 0.4358 |
| | 0.99 | 0.1869 | 0.1980 | 0.2052 | 0.2139 | 0.2250 | 0.2333 | 0.2408 |
| | | 0.4299 | 0.4407 | 0.4515 | 0.4623 | 0.4731 | 0.4839 | 0.4947 |
| | 0.98 | 0.1990 | 0.2043 | 0.2153 | 0.2262 | 0.2340 | 0.2420 | 0.2529 |
| | | 0.4195 | 0.4302 | 0.4410 | 0.4517 | 0.4625 | 0.4732 | 0.4839 |
| 94 | 0.95 | 0.2116 | 0.2224 | 0.2313 | 0.2382 | 0.2490 | 0.2597 | 0.2705 |
| | | 0.3986 | 0.4093 | 0.4200 | 0.4307 | 0.4414 | 0.4520 | 0.4628 |
| | 0.90 | 0.2234 | 0.2341 | 0.2448 | 0.2555 | 0.2625 | 0.2712 | 0.2819 |
| | | 0.3831 | 0.3937 | 0.4044 | 0.4150 | 0.4256 | 0.4363 | 0.4469 |
| | 0.80 | 0.2405 | 0.2511 | 0.2617 | 0.2685 | 0.2774 | 0.2881 | 0.2987 |
| | | 0.3626 | 0.3782 | 0.3888 | 0.3994 | 0.4100 | 0.4206 | 0.4312 |
| | 0.99 | 0.1849 | 0.1959 | 0.2025 | 0.2116 | 0.2226 | 0.2303 | 0.2383 |
| | | 0.4254 | 0.4361 | 0.4467 | 0.4579 | 0.4696 | 0.4813 | 0.4926 |
| | 0.98 | 0.1967 | 0.2021 | 0.2130 | 0.2238 | 0.2310 | 0.2395 | 0.2503 |
| | | 0.4150 | 0.4257 | 0.4363 | 0.4469 | 0.4576 | 0.4682 | 0.4788 |
| 95 | 0.95 | 0.2094 | 0.2201 | 0.2284 | 0.2357 | 0.2463 | 0.2570 | 0.2677 |
| | | 0.3944 | 0.4050 | 0.4156 | 0.4261 | 0.4368 | 0.4490 | 0.4630 |
| | 0.90 | 0.2210 | 0.2316 | 0.2422 | 0.2528 | 0.2587 | 0.2684 | 0.2790 |
| | | 0.3791 | 0.3896 | 0.4001 | 0.4106 | 0.4212 | 0.4317 | 0.4422 |
| | 0.80 | 0.2380 | 0.2485 | 0.2590 | 0.2645 | 0.2745 | 0.2850 | 0.2955 |
| | | 0.3637 | 0.3742 | 0.3847 | 0.3952 | 0.4057 | 0.4162 | 0.4267 |
| | 0.99 | 0.1830 | 0.1939 | 0.1998 | 0.2094 | 0.2202 | 0.2273 | 0.2358 |
| | | 0.4209 | 0.4316 | 0.4442 | 0.4578 | 0.4683 | 0.4789 | 0.4895 |
| | 0.98 | 0.1943 | 0.2000 | 0.2108 | 0.2215 | 0.2279 | 0.2370 | 0.2476 |
| | | 0.4107 | 0.4212 | 0.4317 | 0.4422 | 0.4528 | 0.4633 | 0.4738 |
| 96 | 0.95 | 0.2072 | 0.2178 | 0.2256 | 0.2332 | 0.2438 | 0.2543 | 0.2649 |
| | | 0.3903 | 0.4008 | 0.4112 | 0.4223 | 0.4373 | 0.4477 | 0.4582 |
| | 0.90 | 0.2187 | 0.2292 | 0.2397 | 0.2499 | 0.2545 | 0.2656 | 0.2761 |
| | | 0.3751 | 0.3855 | 0.3960 | 0.4064 | 0.4168 | 0.4273 | 0.4400 |
| | 0.80 | 0.2355 | 0.2459 | 0.2559 | 0.2596 | 0.2717 | 0.2821 | 0.2925 |
| | | 0.3600 | 0.3704 | 0.3807 | 0.3911 | 0.4015 | 0.4119 | 0.4223 |

| $n$ | $\beta$ | 35 | 36 | 37 | 38 | 39 | 40 | 41 |
|---|---|---|---|---|---|---|---|---|
| 93 | 0.99 | 0.2546 | 0.2656 | 0.2731 | 0.2817 | 0.2927 | 0.3037 | 0.3137 |
| | | 0.5109 | 0.5219 | 0.5328 | 0.5437 | 0.5546 | 0.5655 | 0.5764 |
| | 0.98 | 0.2667 | 0.2758 | 0.2826 | 0.2936 | 0.3046 | 0.3155 | 0.3253 |
| | | 0.5000 | 0.5109 | 0.5218 | 0.5326 | 0.5435 | 0.5544 | 0.5652 |
| | 0.95 | 0.2824 | 0.2893 | 0.3002 | 0.3111 | 0.3219 | 0.3327 | 0.3435 |
| | | 0.4785 | 0.4893 | 0.5000 | 0.5108 | 0.5216 | 0.5324 | 0.5432 |
| | 0.90 | 0.2957 | 0.3065 | 0.3173 | 0.3281 | 0.3369 | 0.3440 | 0.3548 |
| | | 0.4624 | 0.4732 | 0.4839 | 0.4947 | 0.5054 | 0.5162 | 0.5269 |
| | 0.80 | 0.3126 | 0.3233 | 0.3340 | 0.3448 | 0.3555 | 0.3662 | 0.3756 |
| | | 0.4465 | 0.4572 | 0.4679 | 0.4786 | 0.4893 | 0.5000 | 0.5108 |
| 94 | 0.99 | 0.2518 | 0.2628 | 0.2692 | 0.2787 | 0.2896 | 0.3005 | 0.3094 |
| | | 0.5054 | 0.5162 | 0.5270 | 0.5378 | 0.5486 | 0.5594 | 0.5702 |
| | 0.98 | 0.2638 | 0.2721 | 0.2796 | 0.2905 | 0.3013 | 0.3121 | 0.3208 |
| | | 0.4947 | 0.5054 | 0.5162 | 0.5269 | 0.5376 | 0.5484 | 0.5591 |
| | 0.95 | 0.2786 | 0.2863 | 0.2970 | 0.3078 | 0.3185 | 0.3292 | 0.3388 |
| | | 0.4742 | 0.4855 | 0.4965 | 0.5071 | 0.5174 | 0.5275 | 0.5375 |
| | 0.90 | 0.2926 | 0.3033 | 0.3139 | 0.3246 | 0.3319 | 0.3404 | 0.3510 |
| | | 0.4575 | 0.4682 | 0.4788 | 0.4896 | 0.5003 | 0.5108 | 0.5213 |
| | 0.80 | 0.3093 | 0.3199 | 0.3305 | 0.3411 | 0.3517 | 0.3622 | 0.3699 |
| | | 0.4418 | 0.4524 | 0.4630 | 0.4736 | 0.4842 | 0.4948 | 0.5053 |
| 95 | 0.99 | 0.2492 | 0.2598 | 0.2649 | 0.2757 | 0.2865 | 0.2973 | 0.3050 |
| | | 0.5033 | 0.5136 | 0.5235 | 0.5334 | 0.5431 | 0.5534 | 0.5640 |
| | 0.98 | 0.2610 | 0.2684 | 0.2767 | 0.2874 | 0.2981 | 0.3088 | 0.3161 |
| | | 0.4894 | 0.5000 | 0.5107 | 0.5213 | 0.5319 | 0.5425 | 0.5532 |
| | 0.95 | 0.2747 | 0.2832 | 0.2939 | 0.3045 | 0.3151 | 0.3257 | 0.3341 |
| | | 0.4736 | 0.4842 | 0.4948 | 0.5053 | 0.5159 | 0.5265 | 0.5371 |
| | 0.90 | 0.2895 | 0.3001 | 0.3106 | 0.3205 | 0.3262 | 0.3368 | 0.3473 |
| | | 0.4535 | 0.4653 | 0.4773 | 0.4895 | 0.5000 | 0.5106 | 0.5199 |
| | 0.80 | 0.3061 | 0.3166 | 0.3271 | 0.3375 | 0.3480 | 0.3573 | 0.3636 |
| | | 0.4372 | 0.4476 | 0.4581 | 0.4686 | 0.4791 | 0.4896 | 0.5000 |
| 96 | 0.99 | 0.2466 | 0.2566 | 0.2621 | 0.2728 | 0.2835 | 0.2942 | 0.3005 |
| | | 0.5000 | 0.5106 | 0.5212 | 0.5318 | 0.5423 | 0.5508 | 0.5596 |
| | 0.98 | 0.2583 | 0.2646 | 0.2738 | 0.2844 | 0.2950 | 0.3052 | 0.3111 |
| | | 0.4843 | 0.4948 | 0.5053 | 0.5158 | 0.5263 | 0.5368 | 0.5473 |
| | 0.95 | 0.2707 | 0.2803 | 0.2908 | 0.3013 | 0.3118 | 0.3223 | 0.3292 |
| | | 0.4687 | 0.4791 | 0.4896 | 0.5000 | 0.5105 | 0.5210 | 0.5314 |
| | 0.90 | 0.2865 | 0.2970 | 0.3074 | 0.3163 | 0.3228 | 0.3333 | 0.3437 |
| | | 0.4532 | 0.4636 | 0.4740 | 0.4844 | 0.4948 | 0.5053 | 0.5157 |
| | 0.80 | 0.3029 | 0.3133 | 0.3237 | 0.3341 | 0.3444 | 0.3522 | 0.3599 |
| | | 0.4326 | 0.4430 | 0.4534 | 0.4637 | 0.4741 | 0.4845 | 0.4949 |

| $n$ | $\beta$ | 42 | 43 | 44 | 45 | 46 | 47 | 48 |
|---|---|---|---|---|---|---|---|---|
| | 0.99 | 0.3198 | 0.3308 | 0.3418 | 0.3528 | 0.3637 | 0.3738 | 0.3811 |
| | | 0.5873 | 0.5982 | 0.6092 | 0.6189 | 0.6262 | 0.6363 | 0.6472 |
| | 0.98 | 0.3315 | 0.3425 | 0.3534 | 0.3643 | 0.3751 | 0.3860 | 0.3969 |
| | | 0.5761 | 0.5870 | 0.5940 | 0.6031 | 0.6140 | 0.6249 | 0.6357 |
| 93 | 0.95 | 0.3509 | 0.3596 | 0.3704 | 0.3812 | 0.3920 | 0.4028 | 0.4136 |
| | | 0.5540 | 0.5648 | 0.5756 | 0.5864 | 0.5972 | 0.6080 | 0.6188 |
| | 0.90 | 0.3656 | 0.3763 | 0.3871 | 0.3979 | 0.4086 | 0.4194 | 0.4301 |
| | | 0.5377 | 0.5484 | 0.5591 | 0.5699 | 0.5806 | 0.5914 | 0.6021 |
| | 0.80 | 0.3820 | 0.3929 | 0.4036 | 0.4143 | 0.4250 | 0.4357 | 0.4464 |
| | | 0.5215 | 0.5322 | 0.5429 | 0.5536 | 0.5643 | 0.5750 | 0.5857 |
| | 0.99 | 0.3164 | 0.3273 | 0.3381 | 0.3490 | 0.3598 | 0.3682 | 0.3758 |
| | | 0.5810 | 0.5918 | 0.6026 | 0.6134 | 0.6242 | 0.6318 | 0.6402 |
| | 0.98 | 0.3280 | 0.3388 | 0.3496 | 0.3604 | 0.3711 | 0.3819 | 0.3915 |
| | | 0.5699 | 0.5806 | 0.5914 | 0.6011 | 0.6085 | 0.6181 | 0.6289 |
| 94 | 0.95 | 0.3450 | 0.3557 | 0.3664 | 0.3771 | 0.3878 | 0.3985 | 0.4092 |
| | | 0.5481 | 0.5587 | 0.5694 | 0.5801 | 0.5908 | 0.6015 | 0.6122 |
| | 0.90 | 0.3617 | 0.3723 | 0.3830 | 0.3936 | 0.4043 | 0.4149 | 0.4255 |
| | | 0.5319 | 0.5426 | 0.5532 | 0.5638 | 0.5745 | 0.5851 | 0.5957 |
| | 0.80 | 0.3781 | 0.3887 | 0.3993 | 0.4099 | 0.4205 | 0.4311 | 0.4417 |
| | | 0.5159 | 0.5265 | 0.5371 | 0.5477 | 0.5583 | 0.5689 | 0.5795 |
| | 0.99 | 0.3130 | 0.3238 | 0.3345 | 0.3453 | 0.3553 | 0.3621 | 0.3718 |
| | | 0.5747 | 0.5854 | 0.5961 | 0.6068 | 0.6175 | 0.6282 | 0.6379 |
| | 0.98 | 0.3245 | 0.3352 | 0.3459 | 0.3565 | 0.3672 | 0.3776 | 0.3854 |
| | | 0.5638 | 0.5744 | 0.5851 | 0.5957 | 0.6064 | 0.6146 | 0.6224 |
| 95 | 0.95 | 0.3414 | 0.3520 | 0.3626 | 0.3732 | 0.3837 | 0.3943 | 0.4049 |
| | | 0.5451 | 0.5541 | 0.5634 | 0.5740 | 0.5845 | 0.5951 | 0.6057 |
| | 0.90 | 0.3579 | 0.3684 | 0.3790 | 0.3895 | 0.4000 | 0.4105 | 0.4211 |
| | | 0.5291 | 0.5385 | 0.5480 | 0.5579 | 0.5684 | 0.5789 | 0.5895 |
| | 0.80 | 0.3741 | 0.3846 | 0.3951 | 0.4056 | 0.4161 | 0.4266 | 0.4371 |
| | | 0.5105 | 0.5210 | 0.5315 | 0.5420 | 0.5525 | 0.5629 | 0.5734 |
| | 0.99 | 0.3098 | 0.3204 | 0.3310 | 0.3417 | 0.3503 | 0.3573 | 0.3679 |
| | | 0.5687 | 0.5792 | 0.5898 | 0.6003 | 0.6109 | 0.6215 | 0.6321 |
| | 0.98 | 0.3211 | 0.3317 | 0.3423 | 0.3528 | 0.3634 | 0.3722 | 0.3789 |
| | | 0.5579 | 0.5684 | 0.5789 | 0.5894 | 0.6000 | 0.6105 | 0.6211 |
| 96 | 0.95 | 0.3378 | 0.3483 | 0.3588 | 0.3693 | 0.3797 | 0.3902 | 0.4007 |
| | | 0.5419 | 0.5524 | 0.5628 | 0.5701 | 0.5788 | 0.5889 | 0.5993 |
| | 0.90 | 0.3542 | 0.3646 | 0.3750 | 0.3854 | 0.3959 | 0.4063 | 0.4167 |
| | | 0.5261 | 0.5365 | 0.5469 | 0.5556 | 0.5641 | 0.5730 | 0.5833 |
| | 0.80 | 0.3703 | 0.3806 | 0.3910 | 0.4014 | 0.4118 | 0.4222 | 0.4325 |
| | | 0.5052 | 0.5156 | 0.5260 | 0.5364 | 0.5467 | 0.5571 | 0.5675 |

| $n$ | $\beta$ | 0 | 1 | 2 | 3 | 4 | 5 | 6 |
|---|---|---|---|---|---|---|---|---|
| | 0.99 | 0.0000 | 0.0001 | 0.0015 | 0.0045 | 0.0085 | 0.0133 | 0.0187 |
| | | 0.0547 | 0.0759 | 0.0921 | 0.1080 | 0.1237 | 0.1393 | 0.1548 |
| | 0.98 | 0.0000 | 0.0002 | 0.0022 | 0.0058 | 0.0105 | 0.0159 | 0.0218 |
| | | 0.0467 | 0.0676 | 0.0836 | 0.0993 | 0.1149 | 0.1304 | 0.1413 |
| 97 | 0.95 | 0.0000 | 0.0005 | 0.0036 | 0.0084 | 0.0142 | 0.0205 | 0.0272 |
| | | 0.0388 | 0.0551 | 0.0708 | 0.0863 | 0.1016 | 0.1170 | 0.1277 |
| | 0.90 | 0.0000 | 0.0010 | 0.0054 | 0.0114 | 0.0181 | 0.0252 | 0.0307 |
| | | 0.0308 | 0.0467 | 0.0622 | 0.0775 | 0.0927 | 0.1035 | 0.1186 |
| | 0.80 | 0.0000 | 0.0022 | 0.0085 | 0.0158 | 0.0226 | 0.0279 | 0.0381 |
| | | 0.0227 | 0.0382 | 0.0534 | 0.0686 | 0.0793 | 0.0944 | 0.1049 |
| | 0.99 | 0.0000 | 0.0001 | 0.0015 | 0.0044 | 0.0084 | 0.0132 | 0.0185 |
| | | 0.0542 | 0.0751 | 0.0912 | 0.1069 | 0.1224 | 0.1378 | 0.1532 |
| | 0.98 | 0.0000 | 0.0002 | 0.0021 | 0.0058 | 0.0104 | 0.0158 | 0.0216 |
| | | 0.0463 | 0.0669 | 0.0828 | 0.0983 | 0.1137 | 0.1290 | 0.1398 |
| 98 | 0.95 | 0.0000 | 0.0005 | 0.0036 | 0.0083 | 0.0140 | 0.0203 | 0.0269 |
| | | 0.0384 | 0.0545 | 0.0701 | 0.0854 | 0.1006 | 0.1158 | 0.1264 |
| | 0.90 | 0.0000 | 0.0010 | 0.0054 | 0.0112 | 0.0179 | 0.0250 | 0.0304 |
| | | 0.0305 | 0.0462 | 0.0615 | 0.0767 | 0.0918 | 0.1024 | 0.1174 |
| | 0.80 | 0.0000 | 0.0022 | 0.0084 | 0.0157 | 0.0224 | 0.0276 | 0.0377 |
| | | 0.0225 | 0.0378 | 0.0529 | 0.0679 | 0.0784 | 0.0934 | 0.1038 |
| | 0.99 | 0.0000 | 0.0001 | 0.0015 | 0.0044 | 0.0084 | 0.0131 | 0.0183 |
| | | 0.0536 | 0.0743 | 0.0903 | 0.1058 | 0.1212 | 0.1364 | 0.1516 |
| | 0.98 | 0.0000 | 0.0002 | 0.0021 | 0.0057 | 0.0103 | 0.0156 | 0.0214 |
| | | 0.0458 | 0.0662 | 0.0819 | 0.0973 | 0.1126 | 0.1277 | 0.1384 |
| 99 | 0.95 | 0.0000 | 0.0005 | 0.0036 | 0.0083 | 0.0139 | 0.0201 | 0.0267 |
| | | 0.0380 | 0.0540 | 0.0694 | 0.0845 | 0.0996 | 0.1146 | 0.1252 |
| | 0.90 | 0.0000 | 0.0010 | 0.0053 | 0.0111 | 0.0177 | 0.0247 | 0.0300 |
| | | 0.0301 | 0.0458 | 0.0609 | 0.0759 | 0.0909 | 0.1014 | 0.1163 |
| | 0.80 | 0.0000 | 0.0022 | 0.0083 | 0.0155 | 0.0221 | 0.0273 | 0.0373 |
| | | 0.0222 | 0.0374 | 0.0524 | 0.0672 | 0.0777 | 0.0925 | 0.1028 |
| | 0.99 | 0.0000 | 0.0001 | 0.0014 | 0.0043 | 0.0083 | 0.0129 | 0.0181 |
| | | 0.0531 | 0.0736 | 0.0893 | 0.1047 | 0.1200 | 0.1351 | 0.1501 |
| | 0.98 | 0.0000 | 0.0002 | 0.0021 | 0.0057 | 0.0102 | 0.0154 | 0.0212 |
| | | 0.0453 | 0.0656 | 0.0811 | 0.0963 | 0.1114 | 0.1264 | 0.1370 |
| 100 | 0.95 | 0.0000 | 0.0005 | 0.0035 | 0.0082 | 0.0137 | 0.0199 | 0.0264 |
| | | 0.0376 | 0.0534 | 0.0687 | 0.0837 | 0.0986 | 0.1135 | 0.1239 |
| | 0.90 | 0.0000 | 0.0010 | 0.0053 | 0.0110 | 0.0175 | 0.0245 | 0.0297 |
| | | 0.0298 | 0.0453 | 0.0603 | 0.0752 | 0.0900 | 0.1004 | 0.1151 |
| | 0.80 | 0.0000 | 0.0022 | 0.0082 | 0.0153 | 0.0219 | 0.0270 | 0.0370 |
| | | 0.0220 | 0.0371 | 0.0519 | 0.0666 | 0.0769 | 0.0916 | 0.1017 |

| n | β | 7 | 8 | 9 | 10 | 11 | 12 | 13 |
|---|---|---|---|---|---|---|---|---|
| | 0.99 | 0.0244 | 0.0306 | 0.0370 | 0.0436 | 0.0505 | 0.0546 | 0.0633 |
| | | 0.1657 | 0.1812 | 0.1920 | 0.2074 | 0.2181 | 0.2334 | 0.2441 |
| | 0.98 | 0.0281 | 0.0347 | 0.0416 | 0.0466 | 0.0538 | 0.0625 | 0.0675 |
| | | 0.1566 | 0.1720 | 0.1827 | 0.1981 | 0.2087 | 0.2193 | 0.2346 |
| 97 | 0.95 | 0.0343 | 0.0387 | 0.0474 | 0.0550 | 0.0606 | 0.0707 | 0.0751 |
| | | 0.1430 | 0.1536 | 0.1688 | 0.1794 | 0.1946 | 0.2052 | 0.2157 |
| | 0.90 | 0.0380 | 0.0466 | 0.0515 | 0.0621 | 0.0676 | 0.0774 | 0.0857 |
| | | 0.1292 | 0.1444 | 0.1549 | 0.1701 | 0.1805 | 0.1910 | 0.2062 |
| | 0.80 | 0.0429 | 0.0533 | 0.0614 | 0.0685 | 0.0792 | 0.0857 | 0.0943 |
| | | 0.1200 | 0.1304 | 0.1408 | 0.1559 | 0.1663 | 0.1767 | 0.1870 |
| | 0.99 | 0.0242 | 0.0302 | 0.0366 | 0.0432 | 0.0500 | 0.0541 | 0.0626 |
| | | 0.1640 | 0.1793 | 0.1900 | 0.2052 | 0.2158 | 0.2310 | 0.2416 |
| | 0.98 | 0.0278 | 0.0344 | 0.0412 | 0.0462 | 0.0532 | 0.0619 | 0.0668 |
| | | 0.1550 | 0.1702 | 0.1809 | 0.1960 | 0.2066 | 0.2171 | 0.2322 |
| 98 | 0.95 | 0.0340 | 0.0383 | 0.0469 | 0.0544 | 0.0600 | 0.0700 | 0.0742 |
| | | 0.1415 | 0.1521 | 0.1671 | 0.1776 | 0.1927 | 0.2031 | 0.2135 |
| | 0.90 | 0.0376 | 0.0461 | 0.0510 | 0.0614 | 0.0669 | 0.0766 | 0.0848 |
| | | 0.1279 | 0.1429 | 0.1533 | 0.1683 | 0.1787 | 0.1890 | 0.2041 |
| | 0.80 | 0.0425 | 0.0528 | 0.0607 | 0.0678 | 0.0783 | 0.0847 | 0.0933 |
| | | 0.1188 | 0.1291 | 0.1394 | 0.1544 | 0.1646 | 0.1749 | 0.1852 |
| | 0.99 | 0.0239 | 0.0299 | 0.0362 | 0.0427 | 0.0495 | 0.0535 | 0.0619 |
| | | 0.1624 | 0.1775 | 0.1884 | 0.2031 | 0.2136 | 0.2287 | 0.2391 |
| | 0.98 | 0.0275 | 0.0340 | 0.0407 | 0.0457 | 0.0526 | 0.0612 | 0.0661 |
| | | 0.1535 | 0.1685 | 0.1790 | 0.1940 | 0.2045 | 0.2167 | 0.2299 |
| 99 | 0.95 | 0.0336 | 0.0379 | 0.0464 | 0.0539 | 0.0593 | 0.0693 | 0.0734 |
| | | 0.1401 | 0.1505 | 0.1654 | 0.1758 | 0.1907 | 0.2010 | 0.2113 |
| | 0.90 | 0.0372 | 0.0457 | 0.0504 | 0.0608 | 0.0662 | 0.0758 | 0.0839 |
| | | 0.1266 | 0.1415 | 0.1518 | 0.1666 | 0.1769 | 0.1871 | 0.2020 |
| | 0.80 | 0.0420 | 0.0523 | 0.0601 | 0.0671 | 0.0776 | 0.0838 | 0.0924 |
| | | 0.1176 | 0.1278 | 0.1380 | 0.1528 | 0.1630 | 0.1732 | 0.1833 |
| | 0.99 | 0.0237 | 0.0296 | 0.0358 | 0.0423 | 0.0490 | 0.0530 | 0.0613 |
| | | 0.1607 | 0.1757 | 0.1907 | 0.2011 | 0.2125 | 0.2264 | 0.2367 |
| | 0.98 | 0.0272 | 0.0337 | 0.0403 | 0.0452 | 0.0521 | 0.0606 | 0.0655 |
| | | 0.1519 | 0.1668 | 0.1772 | 0.1921 | 0.2024 | 0.2173 | 0.2276 |
| 100 | 0.95 | 0.0333 | 0.0375 | 0.0459 | 0.0533 | 0.0587 | 0.0686 | 0.0726 |
| | | 0.1387 | 0.1490 | 0.1638 | 0.1741 | 0.1888 | 0.1990 | 0.2092 |
| | 0.90 | 0.0368 | 0.0452 | 0.0499 | 0.0602 | 0.0655 | 0.0751 | 0.0830 |
| | | 0.1254 | 0.1401 | 0.1503 | 0.1650 | 0.1751 | 0.1853 | 0.2000 |
| | 0.80 | 0.0416 | 0.0518 | 0.0595 | 0.0665 | 0.0768 | 0.0829 | 0.0915 |
| | | 0.1164 | 0.1265 | 0.1366 | 0.1513 | 0.1614 | 0.1714 | 0.1815 |

| $n$ | $\beta$ | 14 | 15 | 16 | 17 | 18 | 19 | 20 |
|---|---|---|---|---|---|---|---|---|
| | 0.99 | 0.0716 | 0.0758 | 0.0857 | 0.0920 | 0.1003 | 0.1079 | 0.1157 |
| | | 0.2595 | 0.2701 | 0.2807 | 0.2937 | 0.3066 | 0.3172 | 0.3277 |
| | 0.98 | 0.0766 | 0.0835 | 0.0914 | 0.0992 | 0.1072 | 0.1148 | 0.1240 |
| | | 0.2452 | 0.2561 | 0.2711 | 0.2816 | 0.2921 | 0.3074 | 0.3179 |
| 97 | 0.95 | 0.0862 | 0.0920 | 0.1015 | 0.1101 | 0.1169 | 0.1276 | 0.1332 |
| | | 0.2309 | 0.2414 | 0.2518 | 0.2637 | 0.2775 | 0.2879 | 0.2983 |
| | 0.90 | 0.0926 | 0.1034 | 0.1089 | 0.1185 | 0.1291 | 0.1351 | 0.1443 |
| | | 0.2166 | 0.2270 | 0.2374 | 0.2526 | 0.2630 | 0.2733 | 0.2837 |
| | 0.80 | 0.1048 | 0.1135 | 0.1199 | 0.1303 | 0.1407 | 0.1494 | 0.1558 |
| | | 0.2022 | 0.2125 | 0.2229 | 0.2332 | 0.2435 | 0.2587 | 0.2690 |
| | 0.99 | 0.0708 | 0.0750 | 0.0848 | 0.0911 | 0.0992 | 0.1068 | 0.1144 |
| | | 0.2568 | 0.2673 | 0.2778 | 0.2930 | 0.3035 | 0.3139 | 0.3243 |
| | 0.98 | 0.0758 | 0.0827 | 0.0904 | 0.0982 | 0.1060 | 0.1136 | 0.1227 |
| | | 0.2427 | 0.2578 | 0.2683 | 0.2787 | 0.2891 | 0.3043 | 0.3146 |
| 98 | 0.95 | 0.0853 | 0.0909 | 0.1005 | 0.1089 | 0.1157 | 0.1263 | 0.1316 |
| | | 0.2285 | 0.2389 | 0.2492 | 0.2644 | 0.2747 | 0.2850 | 0.2953 |
| | 0.90 | 0.0917 | 0.1023 | 0.1077 | 0.1173 | 0.1278 | 0.1336 | 0.1428 |
| | | 0.2144 | 0.2247 | 0.2350 | 0.2500 | 0.2603 | 0.2705 | 0.2808 |
| | 0.80 | 0.1037 | 0.1123 | 0.1187 | 0.1290 | 0.1393 | 0.1478 | 0.1543 |
| | | 0.2001 | 0.2104 | 0.2206 | 0.2308 | 0.2410 | 0.2561 | 0.2663 |
| | 0.99 | 0.0701 | 0.0742 | 0.0839 | 0.0902 | 0.0981 | 0.1057 | 0.1131 |
| | | 0.2542 | 0.2646 | 0.2750 | 0.2901 | 0.3004 | 0.3107 | 0.3210 |
| | 0.98 | 0.0750 | 0.0818 | 0.0894 | 0.0972 | 0.1048 | 0.1125 | 0.1213 |
| | | 0.2402 | 0.2552 | 0.2656 | 0.2759 | 0.2861 | 0.3012 | 0.3114 |
| 99 | 0.95 | 0.0844 | 0.0899 | 0.0995 | 0.1077 | 0.1145 | 0.1251 | 0.1301 |
| | | 0.2262 | 0.2365 | 0.2467 | 0.2617 | 0.2719 | 0.2821 | 0.2923 |
| | 0.90 | 0.0908 | 0.1013 | 0.1065 | 0.1162 | 0.1265 | 0.1320 | 0.1414 |
| | | 0.2122 | 0.2224 | 0.2326 | 0.2475 | 0.2577 | 0.2678 | 0.2780 |
| | 0.80 | 0.1027 | 0.1111 | 0.1175 | 0.1277 | 0.1379 | 0.1461 | 0.1527 |
| | | 0.1981 | 0.2083 | 0.2184 | 0.2285 | 0.2386 | 0.2535 | 0.2636 |
| | 0.99 | 0.0694 | 0.0735 | 0.0830 | 0.0892 | 0.0970 | 0.1046 | 0.1119 |
| | | 0.2516 | 0.2619 | 0.2722 | 0.2871 | 0.2974 | 0.3076 | 0.3178 |
| | 0.98 | 0.0742 | 0.0810 | 0.0884 | 0.0962 | 0.1037 | 0.1113 | 0.1200 |
| | | 0.2378 | 0.2527 | 0.2629 | 0.2731 | 0.2833 | 0.2982 | 0.3083 |
| 100 | 0.95 | 0.0836 | 0.0889 | 0.0985 | 0.1065 | 0.1134 | 0.1238 | 0.1286 |
| | | 0.2240 | 0.2341 | 0.2443 | 0.2591 | 0.2692 | 0.2793 | 0.2894 |
| | 0.90 | 0.0899 | 0.1003 | 0.1053 | 0.1150 | 0.1253 | 0.1305 | 0.1400 |
| | | 0.2101 | 0.2202 | 0.2303 | 0.2450 | 0.2551 | 0.2652 | 0.2752 |
| | 0.80 | 0.1016 | 0.1099 | 0.1163 | 0.1264 | 0.1365 | 0.1445 | 0.1512 |
| | | 0.1962 | 0.2062 | 0.2162 | 0.2262 | 0.2362 | 0.2510 | 0.2610 |

| n | β | 21 | 22 | 23 | 24 | 25 | 26 | 27 |
|---|---|---|---|---|---|---|---|---|
| | 0.99 | 0.1236 | 0.1321 | 0.1392 | 0.1493 | 0.1547 | 0.1656 | 0.1728 |
| | | 0.3383 | 0.3537 | 0.3642 | 0.3746 | 0.3851 | 0.3956 | 0.4061 |
| | 0.98 | 0.1303 | 0.1412 | 0.1457 | 0.1565 | 0.1661 | 0.1719 | 0.1826 |
| | | 0.3284 | 0.3388 | 0.3493 | 0.3603 | 0.3751 | 0.3856 | 0.3960 |
| 97 | 0.95 | 0.1429 | 0.1535 | 0.1586 | 0.1687 | 0.1793 | 0.1872 | 0.1945 |
| | | 0.3087 | 0.3220 | 0.3344 | 0.3448 | 0.3552 | 0.3656 | 0.3759 |
| | 0.90 | 0.1548 | 0.1641 | 0.1700 | 0.1804 | 0.1909 | 0.2006 | 0.2061 |
| | | 0.2940 | 0.3044 | 0.3196 | 0.3300 | 0.3403 | 0.3506 | 0.3609 |
| | 0.80 | 0.1662 | 0.1766 | 0.1869 | 0.1951 | 0.2021 | 0.2124 | 0.2228 |
| | | 0.2793 | 0.2896 | 0.2999 | 0.3102 | 0.3205 | 0.3307 | 0.3426 |
| | 0.99 | 0.1223 | 0.1306 | 0.1377 | 0.1477 | 0.1531 | 0.1639 | 0.1707 |
| | | 0.3396 | 0.3500 | 0.3604 | 0.3708 | 0.3812 | 0.3915 | 0.4021 |
| | 0.98 | 0.1289 | 0.1397 | 0.1439 | 0.1549 | 0.1643 | 0.1701 | 0.1808 |
| | | 0.3250 | 0.3353 | 0.3457 | 0.3609 | 0.3713 | 0.3816 | 0.3919 |
| 98 | 0.95 | 0.1414 | 0.1520 | 0.1566 | 0.1670 | 0.1775 | 0.1850 | 0.1926 |
| | | 0.3056 | 0.3207 | 0.3310 | 0.3413 | 0.3516 | 0.3618 | 0.3721 |
| | 0.90 | 0.1532 | 0.1622 | 0.1682 | 0.1786 | 0.1889 | 0.1983 | 0.2040 |
| | | 0.2910 | 0.3013 | 0.3164 | 0.3266 | 0.3368 | 0.3471 | 0.3573 |
| | 0.80 | 0.1645 | 0.1748 | 0.1851 | 0.1928 | 0.2000 | 0.2103 | 0.2205 |
| | | 0.2765 | 0.2867 | 0.2968 | 0.3070 | 0.3172 | 0.3274 | 0.3425 |
| | 0.99 | 0.1211 | 0.1292 | 0.1363 | 0.1460 | 0.1515 | 0.1623 | 0.1687 |
| | | 0.3361 | 0.3464 | 0.3567 | 0.3670 | 0.3773 | 0.3875 | 0.4027 |
| | 0.98 | 0.1276 | 0.1383 | 0.1420 | 0.1534 | 0.1624 | 0.1684 | 0.1789 |
| | | 0.3217 | 0.3319 | 0.3422 | 0.3573 | 0.3675 | 0.3777 | 0.3879 |
| 99 | 0.95 | 0.1400 | 0.1504 | 0.1546 | 0.1653 | 0.1757 | 0.1828 | 0.1906 |
| | | 0.3025 | 0.3175 | 0.3277 | 0.3378 | 0.3480 | 0.3582 | 0.3683 |
| | 0.90 | 0.1517 | 0.1604 | 0.1665 | 0.1768 | 0.1870 | 0.1960 | 0.2019 |
| | | 0.2881 | 0.3002 | 0.3132 | 0.3233 | 0.3334 | 0.3436 | 0.3537 |
| | 0.80 | 0.1629 | 0.1731 | 0.1832 | 0.1905 | 0.1980 | 0.2082 | 0.2183 |
| | | 0.2737 | 0.2838 | 0.2939 | 0.3039 | 0.3140 | 0.3241 | 0.3390 |
| | 0.99 | 0.1199 | 0.1277 | 0.1350 | 0.1444 | 0.1500 | 0.1606 | 0.1667 |
| | | 0.3328 | 0.3430 | 0.3531 | 0.3633 | 0.3735 | 0.3854 | 0.3987 |
| | 0.98 | 0.1263 | 0.1369 | 0.1400 | 0.1518 | 0.1606 | 0.1667 | 0.1771 |
| | | 0.3185 | 0.3286 | 0.3406 | 0.3537 | 0.3638 | 0.3739 | 0.3840 |
| 100 | 0.95 | 0.1386 | 0.1489 | 0.1526 | 0.1637 | 0.1740 | 0.1806 | 0.1887 |
| | | 0.2995 | 0.3143 | 0.3244 | 0.3345 | 0.3445 | 0.3546 | 0.3646 |
| | 0.90 | 0.1502 | 0.1586 | 0.1649 | 0.1750 | 0.1852 | 0.1938 | 0.1999 |
| | | 0.2852 | 0.3000 | 0.3101 | 0.3201 | 0.3301 | 0.3401 | 0.3502 |
| | 0.80 | 0.1613 | 0.1713 | 0.1814 | 0.1883 | 0.1961 | 0.2061 | 0.2161 |
| | | 0.2710 | 0.2810 | 0.2909 | 0.3009 | 0.3109 | 0.3222 | 0.3357 |

| n | β | 28 | 29 | 30 | 31 | 32 | 33 | 34 |
|---|---|---|---|---|---|---|---|---|
| | 0.99 | 0.1811 | 0.1919 | 0.1972 | 0.2073 | 0.2180 | 0.2243 | 0.2333 |
| | | 0.4170 | 0.4321 | 0.4425 | 0.4530 | 0.4635 | 0.4739 | 0.4844 |
| | 0.98 | 0.1921 | 0.1980 | 0.2086 | 0.2192 | 0.2248 | 0.2345 | 0.2451 |
| | | 0.4064 | 0.4168 | 0.4272 | 0.4376 | 0.4481 | 0.4585 | 0.4689 |
| 97 | 0.95 | 0.2051 | 0.2156 | 0.2227 | 0.2308 | 0.2413 | 0.2517 | 0.2618 |
| | | 0.3863 | 0.3966 | 0.4076 | 0.4224 | 0.4328 | 0.4431 | 0.4535 |
| | 0.90 | 0.2165 | 0.2269 | 0.2373 | 0.2469 | 0.2525 | 0.2629 | 0.2732 |
| | | 0.3713 | 0.3816 | 0.3919 | 0.4022 | 0.4128 | 0.4279 | 0.4382 |
| | 0.80 | 0.2331 | 0.2434 | 0.2528 | 0.2586 | 0.2689 | 0.2792 | 0.2895 |
| | | 0.3563 | 0.3666 | 0.3768 | 0.3871 | 0.3974 | 0.4077 | 0.4179 |
| | 0.99 | 0.1792 | 0.1899 | 0.1944 | 0.2051 | 0.2157 | 0.2213 | 0.2309 |
| | | 0.4172 | 0.4276 | 0.4380 | 0.4483 | 0.4587 | 0.4690 | 0.4794 |
| | 0.98 | 0.1898 | 0.1959 | 0.2065 | 0.2170 | 0.2215 | 0.2321 | 0.2426 |
| | | 0.4022 | 0.4125 | 0.4229 | 0.4332 | 0.4434 | 0.4537 | 0.4640 |
| 98 | 0.95 | 0.2030 | 0.2134 | 0.2199 | 0.2284 | 0.2388 | 0.2491 | 0.2586 |
| | | 0.3823 | 0.3926 | 0.4078 | 0.4181 | 0.4283 | 0.4386 | 0.4488 |
| | 0.90 | 0.2143 | 0.2246 | 0.2349 | 0.2440 | 0.2499 | 0.2602 | 0.2704 |
| | | 0.3675 | 0.3777 | 0.3879 | 0.3981 | 0.4133 | 0.4235 | 0.4337 |
| | 0.80 | 0.2307 | 0.2409 | 0.2497 | 0.2560 | 0.2662 | 0.2764 | 0.2866 |
| | | 0.3527 | 0.3628 | 0.3730 | 0.3832 | 0.3933 | 0.4035 | 0.4137 |
| | 0.99 | 0.1774 | 0.1879 | 0.1915 | 0.2030 | 0.2135 | 0.2182 | 0.2286 |
| | | 0.4130 | 0.4232 | 0.4335 | 0.4437 | 0.4540 | 0.4642 | 0.4745 |
| | 0.98 | 0.1876 | 0.1939 | 0.2044 | 0.2145 | 0.2174 | 0.2298 | 0.2401 |
| | | 0.3981 | 0.4084 | 0.4186 | 0.4287 | 0.4390 | 0.4508 | 0.4644 |
| 99 | 0.95 | 0.2009 | 0.2112 | 0.2170 | 0.2261 | 0.2364 | 0.2466 | 0.2555 |
| | | 0.3785 | 0.3923 | 0.4037 | 0.4139 | 0.4240 | 0.4341 | 0.4443 |
| | 0.90 | 0.2121 | 0.2223 | 0.2325 | 0.2410 | 0.2474 | 0.2576 | 0.2677 |
| | | 0.3638 | 0.3739 | 0.3840 | 0.3977 | 0.4091 | 0.4192 | 0.4294 |
| | 0.80 | 0.2284 | 0.2385 | 0.2467 | 0.2534 | 0.2635 | 0.2736 | 0.2837 |
| | | 0.3491 | 0.3592 | 0.3693 | 0.3793 | 0.3894 | 0.3995 | 0.4095 |
| | 0.99 | 0.1756 | 0.1858 | 0.1906 | 0.2010 | 0.2112 | 0.2147 | 0.2263 |
| | | 0.4088 | 0.4190 | 0.4291 | 0.4393 | 0.4494 | 0.4595 | 0.4697 |
| | 0.98 | 0.1855 | 0.1920 | 0.2023 | 0.2120 | 0.2172 | 0.2275 | 0.2377 |
| | | 0.3941 | 0.4042 | 0.4143 | 0.4251 | 0.4395 | 0.4496 | 0.4597 |
| 100 | 0.95 | 0.1989 | 0.2091 | 0.2141 | 0.2239 | 0.2340 | 0.2442 | 0.2524 |
| | | 0.3756 | 0.3896 | 0.3997 | 0.4097 | 0.4198 | 0.4298 | 0.4398 |
| | 0.90 | 0.2100 | 0.2201 | 0.2302 | 0.2381 | 0.2449 | 0.2550 | 0.2651 |
| | | 0.3602 | 0.3702 | 0.3814 | 0.3950 | 0.4051 | 0.4151 | 0.4251 |
| | 0.80 | 0.2261 | 0.2361 | 0.2436 | 0.2509 | 0.2609 | 0.2709 | 0.2809 |
| | | 0.3456 | 0.3556 | 0.3656 | 0.3756 | 0.3855 | 0.3955 | 0.4054 |

| $n$ | $\beta$ | 35 | 36 | 37 | 38 | 39 | 40 | 41 |
|---|---|---|---|---|---|---|---|---|
| 97 | 0.99 | 0.2440 | 0.2534 | 0.2594 | 0.2700 | 0.2806 | 0.2905 | 0.2948 |
| | | 0.4948 | 0.5053 | 0.5157 | 0.5262 | 0.5366 | 0.5471 | 0.5576 |
| | 0.98 | 0.2555 | 0.2606 | 0.2710 | 0.2815 | 0.2920 | 0.3013 | 0.3073 |
| | | 0.4793 | 0.4897 | 0.5000 | 0.5104 | 0.5208 | 0.5312 | 0.5416 |
| | 0.95 | 0.2662 | 0.2774 | 0.2878 | 0.2982 | 0.3086 | 0.3182 | 0.3228 |
| | | 0.4638 | 0.4742 | 0.4845 | 0.4949 | 0.5052 | 0.5156 | 0.5259 |
| | 0.90 | 0.2836 | 0.2939 | 0.3043 | 0.3120 | 0.3195 | 0.3299 | 0.3402 |
| | | 0.4485 | 0.4588 | 0.4691 | 0.4794 | 0.4897 | 0.5000 | 0.5104 |
| | 0.80 | 0.2998 | 0.3101 | 0.3204 | 0.3306 | 0.3402 | 0.3464 | 0.3562 |
| | | 0.4282 | 0.4385 | 0.4487 | 0.4590 | 0.4693 | 0.4795 | 0.4898 |
| 98 | 0.99 | 0.2415 | 0.2503 | 0.2567 | 0.2672 | 0.2777 | 0.2868 | 0.2929 |
| | | 0.4897 | 0.5000 | 0.5104 | 0.5207 | 0.5311 | 0.5414 | 0.5518 |
| | 0.98 | 0.2524 | 0.2577 | 0.2682 | 0.2786 | 0.2890 | 0.2974 | 0.3042 |
| | | 0.4751 | 0.4859 | 0.4965 | 0.5067 | 0.5167 | 0.5264 | 0.5361 |
| | 0.95 | 0.2643 | 0.2746 | 0.2849 | 0.2952 | 0.3055 | 0.3141 | 0.3206 |
| | | 0.4591 | 0.4693 | 0.4796 | 0.4898 | 0.5000 | 0.5103 | 0.5205 |
| | 0.90 | 0.2807 | 0.2909 | 0.3012 | 0.3075 | 0.3163 | 0.3265 | 0.3367 |
| | | 0.4439 | 0.4541 | 0.4643 | 0.4745 | 0.4847 | 0.4949 | 0.5052 |
| | 0.80 | 0.2967 | 0.3069 | 0.3171 | 0.3273 | 0.3357 | 0.3424 | 0.3526 |
| | | 0.4238 | 0.4340 | 0.4442 | 0.4543 | 0.4645 | 0.4747 | 0.4848 |
| 99 | 0.99 | 0.2390 | 0.2472 | 0.2541 | 0.2645 | 0.2749 | 0.2832 | 0.2900 |
| | | 0.4847 | 0.4949 | 0.5052 | 0.5154 | 0.5256 | 0.5359 | 0.5461 |
| | 0.98 | 0.2494 | 0.2551 | 0.2655 | 0.2758 | 0.2860 | 0.2935 | 0.3011 |
| | | 0.4746 | 0.4848 | 0.4950 | 0.5051 | 0.5153 | 0.5255 | 0.5357 |
| | 0.95 | 0.2616 | 0.2718 | 0.2820 | 0.2922 | 0.3024 | 0.3099 | 0.3174 |
| | | 0.4544 | 0.4646 | 0.4747 | 0.4848 | 0.4950 | 0.5051 | 0.5153 |
| | 0.90 | 0.2779 | 0.2880 | 0.2975 | 0.3022 | 0.3131 | 0.3232 | 0.3333 |
| | | 0.4395 | 0.4496 | 0.4597 | 0.4698 | 0.4799 | 0.4900 | 0.5000 |
| | 0.80 | 0.2938 | 0.3038 | 0.3139 | 0.3240 | 0.3311 | 0.3389 | 0.3490 |
| | | 0.4196 | 0.4296 | 0.4397 | 0.4498 | 0.4598 | 0.4699 | 0.4799 |
| 100 | 0.99 | 0.2366 | 0.2441 | 0.2515 | 0.2618 | 0.2721 | 0.2795 | 0.2870 |
| | | 0.4798 | 0.4899 | 0.5000 | 0.5102 | 0.5203 | 0.5304 | 0.5406 |
| | 0.98 | 0.2464 | 0.2526 | 0.2628 | 0.2730 | 0.2832 | 0.2894 | 0.2981 |
| | | 0.4698 | 0.4799 | 0.4900 | 0.5000 | 0.5101 | 0.5202 | 0.5303 |
| | 0.95 | 0.2590 | 0.2691 | 0.2792 | 0.2893 | 0.2994 | 0.3055 | 0.3142 |
| | | 0.4499 | 0.4599 | 0.4699 | 0.4800 | 0.4900 | 0.5000 | 0.5101 |
| | 0.90 | 0.2751 | 0.2851 | 0.2939 | 0.2999 | 0.3100 | 0.3200 | 0.3300 |
| | | 0.4351 | 0.4451 | 0.4551 | 0.4651 | 0.4751 | 0.4851 | 0.4951 |
| | 0.80 | 0.2908 | 0.3008 | 0.3108 | 0.3202 | 0.3258 | 0.3356 | 0.3455 |
| | | 0.4154 | 0.4254 | 0.4353 | 0.4453 | 0.4552 | 0.4652 | 0.4752 |

| $n$ | $\beta$ | 42 | 43 | 44 | 45 | 46 | 47 | 48 |
|---|---|---|---|---|---|---|---|---|
| 97 | 0.99 | 0.3065 | 0.3171 | 0.3276 | 0.3380 | 0.3451 | 0.3536 | 0.3641 |
| | | 0.5680 | 0.5753 | 0.5839 | 0.5940 | 0.6045 | 0.6150 | 0.6255 |
| | 0.98 | 0.3178 | 0.3283 | 0.3387 | 0.3492 | 0.3593 | 0.3666 | 0.3750 |
| | | 0.5520 | 0.5625 | 0.5729 | 0.5833 | 0.5937 | 0.6041 | 0.6145 |
| | 0.95 | 0.3343 | 0.3447 | 0.3551 | 0.3655 | 0.3758 | 0.3862 | 0.3965 |
| | | 0.5363 | 0.5466 | 0.5570 | 0.5673 | 0.5777 | 0.5852 | 0.5935 |
| | 0.90 | 0.3505 | 0.3608 | 0.3712 | 0.3815 | 0.3918 | 0.4021 | 0.4122 |
| | | 0.5207 | 0.5310 | 0.5413 | 0.5516 | 0.5619 | 0.5722 | 0.5793 |
| | 0.80 | 0.3665 | 0.3767 | 0.3870 | 0.3973 | 0.4076 | 0.4178 | 0.4281 |
| | | 0.5000 | 0.5103 | 0.5206 | 0.5308 | 0.5411 | 0.5514 | 0.5616 |
| 98 | 0.99 | 0.3034 | 0.3138 | 0.3242 | 0.3336 | 0.3395 | 0.3499 | 0.3603 |
| | | 0.5621 | 0.5725 | 0.5829 | 0.5901 | 0.5984 | 0.6086 | 0.6189 |
| | 0.98 | 0.3145 | 0.3249 | 0.3352 | 0.3456 | 0.3544 | 0.3608 | 0.3712 |
| | | 0.5464 | 0.5567 | 0.5669 | 0.5772 | 0.5876 | 0.5979 | 0.6082 |
| | 0.95 | 0.3309 | 0.3412 | 0.3515 | 0.3617 | 0.3720 | 0.3822 | 0.3925 |
| | | 0.5308 | 0.5410 | 0.5513 | 0.5615 | 0.5718 | 0.5820 | 0.5923 |
| | 0.90 | 0.3470 | 0.3572 | 0.3674 | 0.3776 | 0.3878 | 0.3980 | 0.4062 |
| | | 0.5154 | 0.5256 | 0.5358 | 0.5460 | 0.5562 | 0.5664 | 0.5766 |
| | 0.80 | 0.3627 | 0.3729 | 0.3831 | 0.3932 | 0.4034 | 0.4136 | 0.4237 |
| | | 0.4950 | 0.5051 | 0.5153 | 0.5254 | 0.5356 | 0.5458 | 0.5559 |
| 99 | 0.99 | 0.3003 | 0.3106 | 0.3209 | 0.3292 | 0.3360 | 0.3463 | 0.3566 |
| | | 0.5564 | 0.5666 | 0.5769 | 0.5871 | 0.5974 | 0.6040 | 0.6126 |
| | 0.98 | 0.3113 | 0.3216 | 0.3318 | 0.3421 | 0.3494 | 0.3572 | 0.3674 |
| | | 0.5434 | 0.5522 | 0.5612 | 0.5714 | 0.5815 | 0.5917 | 0.6020 |
| | 0.95 | 0.3276 | 0.3377 | 0.3479 | 0.3581 | 0.3682 | 0.3784 | 0.3871 |
| | | 0.5254 | 0.5355 | 0.5457 | 0.5558 | 0.5660 | 0.5761 | 0.5862 |
| | 0.90 | 0.3435 | 0.3536 | 0.3637 | 0.3738 | 0.3839 | 0.3925 | 0.3984 |
| | | 0.5101 | 0.5202 | 0.5303 | 0.5404 | 0.5505 | 0.5606 | 0.5707 |
| | 0.80 | 0.3591 | 0.3692 | 0.3792 | 0.3893 | 0.3994 | 0.4094 | 0.4195 |
| | | 0.4900 | 0.5000 | 0.5101 | 0.5202 | 0.5302 | 0.5403 | 0.5503 |
| 100 | 0.99 | 0.2973 | 0.3075 | 0.3177 | 0.3246 | 0.3327 | 0.3429 | 0.3530 |
| | | 0.5507 | 0.5608 | 0.5710 | 0.5811 | 0.5913 | 0.6014 | 0.6104 |
| | 0.98 | 0.3082 | 0.3184 | 0.3285 | 0.3379 | 0.3436 | 0.3536 | 0.3637 |
| | | 0.5404 | 0.5505 | 0.5606 | 0.5677 | 0.5762 | 0.5858 | 0.5959 |
| | 0.95 | 0.3243 | 0.3344 | 0.3444 | 0.3545 | 0.3645 | 0.3741 | 0.3813 |
| | | 0.5201 | 0.5302 | 0.5402 | 0.5502 | 0.5603 | 0.5703 | 0.5803 |
| | 0.90 | 0.3400 | 0.3501 | 0.3601 | 0.3701 | 0.3794 | 0.3865 | 0.3949 |
| | | 0.5050 | 0.5150 | 0.5250 | 0.5350 | 0.5450 | 0.5550 | 0.5650 |
| | 0.80 | 0.3555 | 0.3655 | 0.3755 | 0.3854 | 0.3954 | 0.4053 | 0.4153 |
| | | 0.4851 | 0.4951 | 0.5050 | 0.5150 | 0.5249 | 0.5349 | 0.5449 |

| $n$ | $\beta$ | 49 | 50 | 51 | 52 | 53 | 54 | 55 |
|---|---|---|---|---|---|---|---|---|
| 97 | 0.99 | 0.3745 | 0.3850 | 0.3955 | 0.4060 | 0.4161 | 0.4247 | 0.4320 |
| | | 0.6359 | 0.6464 | 0.6549 | 0.6620 | 0.6724 | 0.6829 | 0.6935 |
| | 0.98 | 0.3855 | 0.3959 | 0.4063 | 0.4167 | 0.4271 | 0.4375 | 0.4480 |
| | | 0.6250 | 0.6334 | 0.6407 | 0.6508 | 0.6613 | 0.6717 | 0.6822 |
| | 0.95 | 0.4065 | 0.4148 | 0.4223 | 0.4327 | 0.4430 | 0.4534 | 0.4637 |
| | | 0.6035 | 0.6138 | 0.6242 | 0.6345 | 0.6449 | 0.6553 | 0.6657 |
| | 0.90 | 0.4207 | 0.4278 | 0.4381 | 0.4484 | 0.4587 | 0.4690 | 0.4793 |
| | | 0.5878 | 0.5979 | 0.6082 | 0.6185 | 0.6288 | 0.6392 | 0.6495 |
| | 0.80 | 0.4384 | 0.4486 | 0.4589 | 0.4692 | 0.4794 | 0.4897 | 0.5000 |
| | | 0.5719 | 0.5822 | 0.5924 | 0.6027 | 0.6130 | 0.6233 | 0.6335 |
| 98 | 0.99 | 0.3707 | 0.3811 | 0.3914 | 0.4016 | 0.4099 | 0.4171 | 0.4275 |
| | | 0.6293 | 0.6397 | 0.6501 | 0.6605 | 0.6664 | 0.6758 | 0.6862 |
| | 0.98 | 0.3815 | 0.3918 | 0.4021 | 0.4124 | 0.4228 | 0.4331 | 0.4433 |
| | | 0.6185 | 0.6288 | 0.6392 | 0.6456 | 0.6544 | 0.6648 | 0.6751 |
| | 0.95 | 0.4006 | 0.4077 | 0.4180 | 0.4282 | 0.4385 | 0.4487 | 0.4590 |
| | | 0.5994 | 0.6075 | 0.6178 | 0.6280 | 0.6383 | 0.6485 | 0.6588 |
| | 0.90 | 0.4132 | 0.4234 | 0.4336 | 0.4438 | 0.4540 | 0.4642 | 0.4744 |
| | | 0.5868 | 0.5938 | 0.6020 | 0.6122 | 0.6224 | 0.6326 | 0.6428 |
| | 0.80 | 0.4339 | 0.4441 | 0.4542 | 0.4644 | 0.4746 | 0.4847 | 0.4949 |
| | | 0.5661 | 0.5763 | 0.5864 | 0.5966 | 0.6068 | 0.6169 | 0.6271 |
| 99 | 0.99 | 0.3669 | 0.3772 | 0.3874 | 0.3960 | 0.4026 | 0.4129 | 0.4231 |
| | | 0.6228 | 0.6331 | 0.6434 | 0.6537 | 0.6640 | 0.6708 | 0.6791 |
| | 0.98 | 0.3776 | 0.3878 | 0.3980 | 0.4083 | 0.4185 | 0.4286 | 0.4388 |
| | | 0.6122 | 0.6224 | 0.6326 | 0.6428 | 0.6506 | 0.6579 | 0.6682 |
| | 0.95 | 0.3928 | 0.4036 | 0.4138 | 0.4239 | 0.4340 | 0.4442 | 0.4543 |
| | | 0.5964 | 0.6072 | 0.6129 | 0.6216 | 0.6318 | 0.6419 | 0.6521 |
| | 0.90 | 0.4090 | 0.4191 | 0.4293 | 0.4394 | 0.4495 | 0.4596 | 0.4697 |
| | | 0.5809 | 0.5910 | 0.6016 | 0.6075 | 0.6161 | 0.6262 | 0.6363 |
| | 0.80 | 0.4295 | 0.4396 | 0.4497 | 0.4597 | 0.4698 | 0.4798 | 0.4899 |
| | | 0.5604 | 0.5705 | 0.5805 | 0.5906 | 0.6006 | 0.6107 | 0.6208 |
| 100 | 0.99 | 0.3632 | 0.3734 | 0.3826 | 0.3896 | 0.3986 | 0.4087 | 0.4189 |
| | | 0.6174 | 0.6266 | 0.6368 | 0.6470 | 0.6571 | 0.6673 | 0.6754 |
| | 0.98 | 0.3738 | 0.3839 | 0.3940 | 0.4041 | 0.4142 | 0.4238 | 0.4323 |
| | | 0.6060 | 0.6161 | 0.6262 | 0.6363 | 0.6464 | 0.6564 | 0.6621 |
| | 0.95 | 0.3895 | 0.3996 | 0.4096 | 0.4197 | 0.4297 | 0.4397 | 0.4498 |
| | | 0.5904 | 0.6004 | 0.6105 | 0.6187 | 0.6259 | 0.6355 | 0.6455 |
| | 0.90 | 0.4050 | 0.4150 | 0.4250 | 0.4350 | 0.4450 | 0.4550 | 0.4650 |
| | | 0.5750 | 0.5850 | 0.5950 | 0.6051 | 0.6135 | 0.6206 | 0.6299 |
| | 0.80 | 0.4253 | 0.4352 | 0.4452 | 0.4551 | 0.4651 | 0.4751 | 0.4850 |
| | | 0.5548 | 0.5648 | 0.5747 | 0.5847 | 0.5947 | 0.6046 | 0.6146 |

# Glossary

**Symbols**

| | |
|---|---|
| $X$ | random variable , p. xvi |
| $\vec{X} = (X_1, \ldots, X_n)$ | random sample, p. xvi |
| $\breve{\theta}$ | distribution parameter, p. xvi |
| $\theta$ | distribution parameter, actual value, p. xvi |
| $F^{(\theta)}_{X_1,\ldots,X_n}(x_1, \ldots, x_n)$ | distribution function, p. 16 |
| $\left(\Omega, \mathcal{S}, \mathbf{P}\right)$ | probability space, p. 17 |
| $B(p)$ | Bernoulli distribution, p. 19 |
| $Bi(n,p)$ | binomial distribution, p. 20 |
| $H(N,M,n)$ | hypergeometric distribution, p. 21 |
| $f_{\nu_1,\nu_2}(x)$ | density function of the $F$-distribution with $(\nu_1, \nu_2)$ degrees of freedom, p. 22 |
| $Po(a)$ | Poisson distribution, p. 24 |
| $\chi^2_{1-\alpha,2(m+1)}$ | $(1-\alpha)$-quantile of the $\chi^2$-distribution with $2(m+1)$ degrees of freedom, p. 24 |
| $H(N,Np,n)$ | hypergeometric distribution, p. 27 |
| $[\underline{p}, \overline{p}] \subset [0,1]$ | measurement range, p. 29 |
| $\mathbf{P}^{(\theta)}_{\vec{X}}$ | probability distribution, p. 35 |
| $\left(L^{(\beta,n)}_{\breve{p}}(\vec{X}), U^{(\beta,n)}_{\breve{p}}(\vec{X})\right)$ | measurement procedure, p. 43 |
| $TF(\vec{X};\breve{p})$ | relative frequency, p. 46 |
| $X_S$ | sum of sample elements, p. 47 |

| | |
|---|---|
| $[\underline{p},\overline{p}]$ | measurement space, p. 60 |
| $\mathcal{X}^n$ | prediction space, p. 60 |
| $\left[\underline{p},\overline{p}\right] \times \mathcal{X}^n$ | measurement & prediction space, p. 61 |
| $\left([\underline{p},\overline{p}] \times \mathcal{X}^n, \mathcal{B}_{[\underline{p},\overline{p}]} \otimes \mathcal{P}_{\mathcal{X}^n}\right)$ | measurable space, p. 61 |
| $\beta$ | confidence level, p. 62 |
| $\Lambda_{[\underline{p},\overline{p}]}$ | measure function on $[\underline{p},\overline{p}]$, p. 63 |
| $\lambda_{[\underline{p},\overline{p}]}(p)$ | marginal density of $\breve{p}$, p. 63 |
| $\mathbf{P}_{\vec{X}}^{(p)}$ | conditional distribution of $\vec{X}$, p. 64 |
| $\Lambda_{[\underline{p},\overline{p}]} \otimes \mathbf{P}_{\vec{X}}^{(p)}$ | joint distribution of $\left(\breve{p}, \vec{X}\right)$, p. 64 |
| $Q_{\{0,1\}^n}$ | marginal probability measure of $\vec{X}$, p. 65 |
| $w\left(\vec{x}\right)$ | weight function, p. 65 |
| $g_{\breve{\theta}}^{(\vec{x})}$ | conditional density of $\breve{p}$, p. 65 |
| $G_{\breve{p}}^{(\vec{x})}$ | conditional measure of $\breve{p}$, p. 66 |
| $Q_{\{0,1\}^n} \otimes G_{\breve{p}}^{(\vec{x})}$ | joint measure of $(\breve{p}, \vec{X})$, p. 66 |
| $\left([\underline{p},\overline{p}] \times \{0,1,\ldots,n\}\right)$ | measurement & prediction space, p. 67 |
| $\mathbf{P}_{X_S}^{(p)}$ | conditional measure of $X_S$, p. 68 |
| $\Lambda_{[\underline{p},\overline{p}]} \otimes \mathbf{P}_{X_S}^{(p)}$ | joint measure on $\mathcal{B}_{[\underline{p},\overline{p}]} \otimes \mathcal{P}_{\{0,1,\ldots,n\}}$, p. 68 |
| $Q_{\{0,1,\ldots,n\}}$ | marginal measure on $\mathcal{P}_{\{0,1,\ldots,n\}}$, p. 68 |
| $G_{\breve{p}}^{(x_S)}$ | conditional measure on $\mathcal{B}_{[\underline{p},\overline{p}]}$, p. 68 |
| $w_S(x_S)$ | weight function, p. 68 |
| $Q_{\{0,1,\ldots,n\}} \otimes G_{\breve{p}}^{(x_S)}$ | joint measure on $\mathcal{P}_{\{0,1,\ldots,n\}} \otimes \mathcal{B}_{[\underline{p},\overline{p}]}$, p. 69 |
| $M_{\vec{X},\breve{p}}^{(\beta,n)}$ | $\beta$-measurement and prediction space, p. 71 |

| | |
|---|---|
| $A_{\vec{X}}^{(\beta,n)}(p)$ | $\beta$-prediction region for $\vec{X}$, p. 71 |
| $A_{\vec{X}}^{(\beta,n)}$ | $\beta$-prediction procedure, p. 71 |
| $\overline{C}_{\breve{p}}^{(\beta,n)}(\vec{x})$ | $\beta$-measurement interval, p. 71 |
| $\overline{C}_{\breve{p}}^{(\beta,n)}$ | $\beta$-measurement procedure, p. 72 |
| $M_{X_S,\breve{p}}^{(\beta,n)}$ | $\beta$-measurement and prediction space for $X_S$, p. 72 |
| $A_{X_S}^{(\beta,n)}(p)$ | $\beta$-prediction region for $X_S$, p. 73 |
| $\left\{A_{X_S}^{(\beta,n)}(p)\right\}_{p\in[\underline{p},\overline{p}]}$ | $\beta$-prediction procedure for $X_S$, p. 73 |
| $\overline{C}_{\breve{p}}^{(\beta,n)}(x_S)$ | $\beta$-measurement interval based on $X_S$, p. 73 |
| $\left\{\overline{C}_{\breve{p}}^{(\beta,n)}(x_S)\right\}_{x_S\in\{0,1,\ldots,n\}}$ | $\beta$-measurement procedure based on $X_S$, p. 73 |
| $V_w(M_{\vec{X},\breve{p}}^{(\beta,n)})$ | weighted volume, p. 76 |
| $V_{prob}\left(M_{\vec{X},\breve{p}}^{(\beta,n)}\right)$ | probability volume, p. 77 |
| $V_{prob}(M_{\vec{X},\breve{p}}^{(\beta,n)}) - \beta$ | reliability excess, p. 78 |
| $\left\{{}^*C_{\breve{p}}^{(\beta,n)}(\vec{x})\right\}_{\vec{x}\in\{0,1\}^n}$ | Neyman $\beta$-measurement procedure for $\vec{X}$, p. 79 |
| $\left\{{}^*C_{\breve{p}}^{(\beta,n)}(x_S)\right\}_{x_S\in\{0,1,\ldots,n\}}$ | Neyman $\beta$-measurement procedure for $X_S$, p. 80 |
| $\dfrac{\mathbf{P}_{\vec{X}}^{(p)}\left(\{\vec{x}\}\right)}{Q_{\{0,1\}^n}\left(\{\vec{x}\}\right)}$ | likelihood ratio, p. 83 |
| $TF_{\breve{p}}^{(n)}(X_S)$ | relative frequency estimator, p. 99 |

| | |
|---|---|
| $T_{\breve{p}}^{(\beta,n)}(\vec{X})$ | $\beta$-estimator, p. 105 |
| $TMS_{\breve{p}}^{(\beta,n)}(\vec{X})$ | minimum MSE-$\beta$-estimator, p. 107 |
| $TMP_{\breve{p}}^{(\beta,n)}(\vec{X})$ | midpoint-$\beta$-estimator, p. 111 |
| $H_0$ | null hypothesis, p. 120 |
| $\phi_{H_0}$ | decision function, p. 123 |
| $A_{\vec{X}}^{(1-\alpha,n)}([\underline{p}^{(0)},\overline{p}^{(0)}])$ | local $(1-\alpha)$-prediction region, p. 126 |
| $E_{\breve{p},\vec{X}}^{(1-\alpha,n)}$ | $\alpha$-exclusion space, p. 126 |
| $Q_{\{0,1\}^n}\left(A_{\vec{X}}^{(1-\alpha,n)}([\underline{p}^{(0)},\overline{p}^{(0)}])\right)$ | failure probability, p. 128 |
| $H_i$ | $i$th alternative, p. 146 |
| $\alpha_i$ | significance level of $i$th alternative, p. 147 |
| $\phi_{H_1H_2\ldots H_m}$ | decision function, p. 147 |
| $Q_{\{0,1\}^n}\left(O_{\vec{X}}^{(\alpha_1,\ldots,\alpha_m,n)}(H_1,\ldots,H_m)\right)$ | failure probability, p. 151 |

# References


[1] Bernoulli, J. (1713). *Ars Conjectandi.* Thurnisorium, Basel.

[2] Blyth, C.R. and Still, H.A. (1983). Binomial Confidence Intervals. *J. Am. Statist. Assoc.* 78, 108-116.

[3] Blyth, C.R. and Hutchinson, D.W. (1960). Tables of Neyman-Shortest Confidence Intervals for the Binomial Parameter. *Biometrika* 47, 381-391.

[4] Bunke, O. (1959/1960). Neue Konfidenzintervalle für den Parameter der Binomialverteilung. *Wissenschaft. Zeitschr. Humboldt-Universität Berlin, Mathematisch-Naturwissenschaft. Reihe*, 9, 335-363.

[5] Chapman, D.G. (1952). On Tests and Estimates for the Ratio of Poisson mMeans. *Ann. Inst. Statist. Math.* 4, 45-49.

[6] Clopper, C., J., and Pearson, E. S. (1934). The Use of Confidence or Fiducial Limits Illustrated in the Case of the Binomial. *Biometrika*, 26, 404-413.

[7] Clunies-Ross, C.W. (1958). Interval Estimation for the Parameter of a Binomial Distribution. *Biometrika* 45, 275-279.

[8] v. Collani, E., Dräger, K., and Hottendorf, J. (1996). *Tables for Optimal Two-Sided Confidence Intervals and Tests for an Unknown Probability.* Monograph Series in Stochastics 1, Würzburg Research Group on Quality Control, Würzburg.

[9] v. Collani, E., Dräger, K., Kiesmüller, G., and Göb, R. (1999). A Four-Parameter Family of Estimators for the Probability of an Event. Research Report No. 81, Würzburg Research Group on Quality Control.

[10] v. Collani, E., Dumitrescu, M., and Lepenis, R. (1999). An Optimal Measurement and Prediction Procedure. Research Report No. 84, Würzburg Research Group on Quality Control.

[11] v. Collani, E., and Dumitrescu, M. (2001). Complete Neyman Measurement Procedure. Appears in METRIKA.

[12] v. Collani, E., and Dumitrescu, M. (2000). Neyman Exclusion Procedure. *Econ. Qual. Control* 15, 15-34.

[13] v. Collani, E., and Dumitrescu, M. (2000). Neyman Comparison Procedure. *Econ. Qual. Control* 15, 35-53.

[14] Crow, E. L. (1956). Confidence Intervals for a Proportion. *Biometrika*, 43, 423-435.

[15] de Finetti, B. (1974). *Theory of Probability.* Vol. I. John Wiley & Sons, London.

[16] Fisher, R. A. (1930). Inverse Probability. *Proc. Cambridge Phil. Soc.*, 26, 528-535.

[17] Gnedenko, B.W. (1991). *Einführung in die Wahrscheinlichkeitstheorie.* Akademie Verlag, Berlin.

[18] Hald, A. (1952). *Statistical Tables and Formulas.* John Wiley & Sons, London.

[19] Kotz, S., and Johnson, N.L. (1986). *Encyclopedia of Statistical Sciences.* John Wiley & Sons, New York.

[20] Laplace, P.-S. (1814); *Essai philosophique sur les probabilités*, Christian Burgois, Paris.

[21] Laplace, P.-S. (1994). *Philosophical Essay on Probabilities.* Translated from the fifth French edition of 1825 by Andrew I. Dale. Springer-Verlag, New York.

[22] Müller, P.H., Neumann, P., and Storm, R. (1977). *Tafeln der mathematischen Statistik.* Carl Hanser, München.

[23] Nelder, J. (1986). Statistics, Science and Technology. *J. Roy. Statist. Soc. A* 149, 109-121.

[24] Neumann, P. (1986). Optimale zweiseitige Konfidenzintervalle für eine unbekannte Wahrscheinlichkeit - ein axiomatischer Zugang. *Wissenschaft. Zeitschr. Technischen Universität Dresden* 35, 29-32.

[25] Neyman, J. (1934). On the Two Different Aspects of the Representation Method: The Method of Stratified Sampling and the Method of Purposive Selection. *J. Roy. Statist. Soc.*, 97, 558-625.

[26] Neyman, J. (1935). On the Problem of Confidence Intervals. *Ann. Math. Statist.*, 6, 111-116.

[27] Neyman, J. (1937). Outline of a Theory of Statistical Estimation Based on the Classical Theory of Probability. *Phil. Trans. Roy. Soc., Ser. A*, 236, 333-380.

[28] Neyman, J. (1941). Fiducial Argument and the Theory of Confidence Intervals. *Biometrika*, 32, 129-150.

[29] Odeh, R.E., and Owen, D.B. (1983). *Attribute Sampling Plans, Tables of Tests and Confidence Limits for Proportions.* Marcel Dekker, New York.

[30] Owen, D.B. (1962). *Handbook of Statistical Tables.* Addison-Wesley, Reading, MA.

[31] Rao, C.R. (1995). Comments on Sir David's Random Thoughts on Survival. *Student*, 1, 127-128.

[32] Staff of Reseach and Education Association (1978). *The Statistics Problem Solver.* Research and Education Association, New York.

[33] v. Seckendorff, V. (2000). Detection Limits of Selected Rare-Earth Elements in Electron-Probe Microanalysis. *Eur. J. Mineral.* 12, 73-93.

[34] Sterne, T.E. (1954). Some Remarks on Confidence or Fiducial Limits. *Biometrika*, 41, 275 -278.

# Index

MIX
Papier aus verantwortungsvollen Quellen
Paper from responsible sources
FSC® C105338

If you have any concerns about our products,
you can contact us on
**ProductSafety@springernature.com**

In case Publisher is established outside the EU,
the EU authorized representative is:
**Springer Nature Customer Service Center GmbH**
**Europaplatz 3, 69115 Heidelberg, Germany**

Printed by Libri Plureos GmbH
in Hamburg, Germany